T0255718

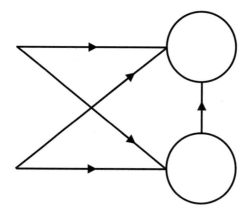

Stephen Lynch

Dynamical Systems with Applications using Mathematica®

Second Edition

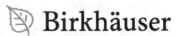 Birkhäuser

Stephen Lynch
School of Computing, Mathematics and Digital Technology
Manchester Metropolitan University
Manchester
UK

ISBN 978-3-319-87089-2   ISBN 978-3-319-61485-4 (eBook)
DOI 10.1007/978-3-319-61485-4

Mathematics Subject Classification (2010): 37-01, 49K15, 78A60, 28A80, 80A30, 34H10, 34K18, 70K05, 34C07, 34D06, 92B20, 94C05

1st edition: © Birkhäuser Boston 2007
2nd edition: © Springer International Publishing AG 2017
Softcover reprint of the hardcover 1st edition 2017

Printed on acid-free paper

This book is published under the trade name Birkhauser (www.birkhauser-science.com)
The registered company is Springer International Publishing AG
The registered company address is: Gewerbestrasse 11, 6330 Cham, Switzerland

Preface

Since the first printing of this book in 2007, Mathematica® has evolved from Mathematica version 6.0 to Mathematica version 11.2 in 2017. Accordingly, the second edition has been thoroughly updated and new material has been added. In this edition, there are many more applications, examples and exercises, all with solutions, and new sections on series solutions of ordinary differential equations and Newton fractals, have been added. There are also new chapters on delay differential equations, image processing, binary oscillator computing, and simulation with Wolfram SystemModeler.

This book provides an introduction to the theory of dynamical systems with the aid of Mathematica. It is written for both senior undergraduates and graduate students. Chapter 1 provides a tutorial introduction to Mathematica–new users should go through this chapter carefully whilst those moderately familiar and experienced users will find this chapter a useful source of reference. The first part of the book deals with continuous systems using differential equations, including both ordinary and delay differential equations (Chapters 2–12), the second part is devoted to the study of discrete systems (Chapters 13–17), and Chapters 18–22 deal with both continuous and discrete systems. Chapter 23 gives examples of coursework and also lists three Mathematica-based examinations to be sat in a computer laboratory with access to Mathematica. Chapter 24 lists answers to all of the exercises given in the book. It should be pointed out that dynamical systems theory is not limited to these topics but also encompasses partial differential equations, integral and integro-differential equations, and stochastic systems, for instance. References [1]–[6] given at the end of the Preface provide more information for the interested reader. The author has gone for breadth of coverage rather than fine detail and theorems with proofs are kept at a minimum. The material is not clouded by functional analytic and group theoretical definitions, and so is intelligible to readers with a general mathematical background. Some of the topics covered are scarcely covered elsewhere. Most of the material in Chapters 9–12 and 16–22 is at postgraduate level and has been influenced by the author's own research interests. There is more theory in these chapters than in the rest of the book since it is not easily accessed anywhere else. It has been found that these chapters are especially useful as reference material for senior undergraduate project work. The theory in other chapters of the book is dealt with more comprehensively in other texts, some of which may be found in the references section of the corresponding chapter. The book has a very hands-on approach and takes the reader from the basic theory right through to recently published research material.

Mathematica is extremely popular with a wide range of researchers from all sorts of disciplines, it has a very user-friendly interface and has extensive visualization and numerical computation capabilities. It is an ideal package to adopt for the study of nonlinear dynamical systems; the numerical algorithms work very quickly, and complex pictures can be plotted within seconds. The Wolfram SystemModeler^{TM} package is used for simulating dynamical processes. It is as close as one can get to building apparatus and investigating the output for a given input without the need for an actual physical model. For this reason, SystemModeler is very popular in the field of engineering.

The first chapter provides an efficient tutorial introduction to Mathematica. New users will find the tutorials will enable them to become familiar with Mathematica within a few hours. Both engineering and mathematics students appreciate this method of teaching and I have found that it generally works well with one staff member to about twenty students in a computer laboratory. In most cases, I have chosen to list the Mathematica notebook commands at the end of each chapter, this avoids unnecessary cluttering in the text. The Mathematica programs have been kept as simple as possible and should run under later versions of the package. All files for the book (including updates) can even be downloaded from the Web at

http://library.wolfram.com/infocenter/Books/9563/.

Readers will find that they can reproduce the figures given in the text, and then it is not too difficult to change parameters or equations to investigate other systems.

Chapters 3–12 deal with continuous dynamical systems. Chapters 2 and 3 cover some theory of ordinary differential equations and applications to models in the real world are given. The theory of differential equations applied to chemical kinetics and electric circuits is introduced in some detail. The memristor is introduced and one of the most remarkable stories in the history of mathematics is relayed. Chapter 2 ends with the existence and uniqueness theorem for the solutions of certain types of differential equations. The theory behind the construction of phase plane portraits for two-dimensional systems is dealt with in Chapter 3. Applications are taken from chemical kinetics, economics, electronics, epidemiology, mechanics, and population dynamics. The modeling of the populations of interacting species are discussed in some detail in Chapter 4 and domains of stability are discussed for the first time. Limit cycles, or isolated periodic solutions, are introduced in Chapter 5. Since we live in a periodic world, these are the most common type of solution found when modeling nonlinear dynamical systems. They appear extensively when modeling both the technological and natural sciences. Hamiltonian, or conservative, systems and stability are discussed in Chapter 6 and Chapter 7 is concerned with how planar systems vary depending upon a parameter. Bifurcation, bistability, multistability, and normal forms are discussed.

The reader is first introduced to the concept of chaos in continuous systems in Chapters 8 and 9, where three-dimensional systems and Poincaré maps are investigated. These higher-dimensional systems can exhibit strange attractors and chaotic dynamics. One can rotate the three-dimensional objects in Mathematica and plot time series plots to get a better understanding of the dynamics involved. Once again, the theory can be applied to chemical kinetics (including stiff systems), electric circuits, and epidemiology; a simplified model for the weather is also briefly discussed. Chapter 9 deals with Poincaré first return maps that can be used to untangle complicated interlacing trajectories in higher-dimensional spaces. A periodically driven nonlinear pendulum is also investigated by means of a nonautonomous differential equation. Both local and global bifurcations are investigated in Chapter 10. The main results and statement of the famous second part of David Hilbert's sixteenth problem are listed in Chapter 11. In order to understand these results, Poincaré compactification is introduced. There is some detail on Liénard systems, in particular, in this part of the book, but they do have a ubiquity for systems in the plane. Chapter 12 provides an introduction to delay differential equations with applications in biology, nonlinear optics and other dynamical systems.

Chapters 13–17 deal with discrete dynamical systems. Chapter 13 starts with a general introduction to iteration and linear recurrence (or difference) equations. The bulk of the chapter is concerned with the Leslie model used to investigate the population of a single species split into different age classes. Harvesting and culling policies are then investigated and optimal solutions are sought. Nonlinear discrete dynamical systems are dealt with in Chapter 14. Bifurcation diagrams, chaos, intermittency, Lyapunov exponents, periodicity, quasiperiodicity, and universality are some of the topics introduced. The theory is then applied to real-world problems from a broad range of disciplines including population dynamics, biology, economics, nonlinear optics, and neural networks. Chapter 15 is concerned with complex iterative maps, Julia sets and the now-famous Mandelbrot set are plotted. Basins of attraction are investigated for these complex systems and Newton fractals are introduced. As a simple introduction to optics, electromagnetic waves and Maxwell's equations are studied at the beginning of Chapter 16. Complex iterative equations are used to model the propagation of light waves through nonlinear optical fibers. A brief history of nonlinear bistable optical resonators is discussed and the simple fiber ring resonator is analyzed in particular. Chapter 16 is devoted to the study of these optical resonators and phenomena such as bistability, chaotic attractors, feedback, hysteresis, instability, linear stability analysis, multistability, nonlinearity, and steady-states are discussed. The first and second iterative methods are defined in this chapter. Some simple fractals may be constructed using pencil and paper in Chapter 17, and the concept of fractal dimension is introduced. Fractals may be thought of as identical motifs repeated on ever-reduced scales. Unfortunately, most of the fractals appearing in nature are not homogeneous but are more heterogeneous, hence the need for the multifractal theory given later in the chapter. It has been found that the distribution of stars and galaxies in our universe are multifractal, and there is even evidence of multifractals in rainfall, stock markets, and heartbeat rhythms. Applications in geoscience, materials science, microbiology, and image processing are briefly discussed. Chapter 18 provides a brief introduction to image processing which is being used more and more by a diverse range of scientific disciplines, especially medical imaging. The fast Fourier transform is introduced and has a wide range of applications throughout the realms of science.

Chapter 19 is devoted to the new and exciting theory behind chaos control and synchronization. For most systems, the maxim used by engineers in the past has been "stability good, chaos bad", but more and more nowadays this is being replaced with "stability good, chaos better". There are exciting and novel applications in cardiology, communications, engineering, laser technology, and space research, for example. A brief introduction to the enticing field of neural networks is presented in Chapter 20. Imagine trying to make a computer mimic the human brain. One could ask the question: In the future will it be possible for computers to think and even be conscious? The human brain will always be more powerful than traditional, sequential, logic-based digital computers and scientists are trying to incorporate some features of the brain into modern computing. Neural networks perform through learning and no underlying equations are required. Mathematicians and computer scientists are attempting to mimic the way neurons work together via synapses; indeed, a neural network can be thought of as a crude multidimensional model of the human brain. The expectations are high for future applications in a broad range of disciplines. Neural networks are already being used in machine learning and pattern recognition (computer vision, credit card fraud, prediction and forecasting, disease recognition, facial and speech recognition), the consumer home entertainment market, psychological profiling, predicting wave over-topping events, and control problems, for example. They also provide a parallel architecture allowing for very fast computational and response times. In recent years, the disciplines of neural networks and nonlinear dynamics have increasingly coalesced and a new branch of science called neurodynamics is emerging. Lyapunov functions can be used to determine the stability of certain types of neural network. There is also evidence of chaos, feedback, nonlinearity, periodicity, and chaos synchronization in the brain.

Chapter 21 focuses on binary oscillator computing, the subject of UK, International and Taiwanese patents. The author and his co-inventor, Jon Borresen, came up with the idea when modeling connected biological neurons. Binary oscillator technology can be applied to the design of Arithmetic Logic Units (ALU)s, memory and other basic computing components. It has the potential to provide revolutionary computational speed-up, energy saving and novel applications and may be applicable to a variety of technological paradigms including biological neurons, Complementary Metal-Oxide-Semiconductor (CMOS), memristors, optical oscillators, and superconducting materials. The research has the potential for MMU and industrial partners to develop super fast, low-power computers, and may provide an assay for neuronal degradation for brain malfunctions such as Alzheimer's, epilepsy and Parkinson's disease!

Examples of SystemModeler, referred to in earlier chapters of the book, are presented in Chapter 22. It is possible to change the type of input into the system, or parameter values, and investigate the output very quickly. This is as close as one can get to experimentation without the need for expensive equipment.

Examples of coursework and three examination-type papers are listed in Chapter 23, and a set of outline solutions for the book is listed in Chapter 24.

Both textbooks and research papers are presented in the list of references. The textbooks can be used to gain more background material, and the research papers have been given to encourage further reading and independent study.

This book is informed by the research interests of the author which are currently nonlinear ordinary differential equations, nonlinear optics, multifractals, neural networks, and binary oscillator computing. Some references include recently published research articles by the author along with two patents.

The prerequisites for studying dynamical systems using this book are undergraduate courses in linear algebra, real and complex analysis, calculus, and ordinary differential equations; a knowledge of a computer language such as C or Fortran would be beneficial but not essential.

Recommended Textbooks

[1] K. Adzievski and A.H. Siddiqi, *Introduction to Partial Differential Equations for Scientists and Engineers using Mathematica, Second Edition*, Chapman & Hall/CRC (Applied Mathematics & Nonlinear Science), New York, 2016.

[2] B. Bhattacharya and M. Majumdar, *Random Dynamical Systems in Finance*, Chapman & Hall/CRC, New York, 2016.

[3] L.C. de Barros, R.C. Bassanezi and W.A. Lodwick, *A First Course in Fuzzy Logic, Fuzzy Dynamical Systems, and Biomathematics: Theory and Applications*, Springer, New York, 2016.

[4] V. Volterra, *Theory of Functionals and of Integral and Integro-Differential Equations*, Dover Publications, New York, 2005.

[5] J. Mallet-Paret (Editor), J. Wu (Editor), H. Zhu (Editor), Y. Yi (Editor), *Infinite Dimensional Dynamical Systems (Fields Institute Communications)*, Springer, New York, 2013.

[6] C. Bernido, M.V. Carpio-Bernido, M. Grothaus et al., *Stochastic and Infinite Dimensional Analysis*, Birkhäuser, New York, 2016.

I would like to express my sincere thanks to Wolfram for supplying me with the latest versions of Mathematica and SystemModeler. Thanks also go to the reviewer of this book for his thorough checking and all of the reviewers from the other editions of my books. Special thanks go to Birkhäuser and Springer International Publishers. Finally, thanks to my family and especially my wife Gaynor, and our children, Sebastian and Thalia, for their continuing love, inspiration, and support.

Stephen Lynch FIMA SFHEA

Contents

Chapter 1
A Tutorial Introduction to Mathematica

Aims and Objectives

- To provide a tutorial guide to Mathematica.

- To give practical experience in using the package.

- To promote self-help using the online help facilities.

- To provide a concise source of reference for experienced users.

On completion of this chapter the reader should be able to

- use Mathematica as a tool;

- produce simple Mathematica notebooks;

- access some Mathematica commands and notebooks over the World Wide Web.

It is assumed that the reader is familiar with either the *Windows*, *Mac OS*, or *UNIX* platform. This book was prepared using Mathematica (Version 11.2) but most programs should work under earlier and later versions of the package. Note that the online version of the Mathematica commands for this book will be written using the most up-to-date version of the package.

The command lines and programs listed in this chapter have been chosen to allow the reader to become familiar with Mathematica within a few hours. They provide a concise summary of the type of commands that will be used throughout the text. New users should be able to start on their own problems after completing the chapter, and experienced users should find this chapter an excellent source of reference. Of course, there are many Mathematica textbooks on the market for those who require further applications or more detail.

If you experience any problems, there are several options for you to take. There is an excellent index within Mathematica, and Mathematica commands,

© Springer International Publishing AG 2017
S. Lynch, *Dynamical Systems with Applications using Mathematica®*,
DOI 10.1007/978-3-319-61485-4_1

notebooks, programs, and output can also be viewed in color over the Web at
Mathematica's Information Center

http://library.wolfram.com/infocenter/Books/AppliedMathematics/.

The notebook files can be found at the links Calculus Analysis and Dynamical
Systems. Download the notebook files and save the relevant files from the archive
onto your computer.

1.1 A Quick Tour of Mathematica

To start Mathematica, simply double-click on the Mathematica icon. In the Unix
environment, one types **mathematica** as a shell command. The author has used
the Windows platform in the preparation of this material. When Mathematica
starts up, a blank notebook appears on the computer screen entitled **Untitled-
1** as shown in Figure 1.1. Click on Format tab and **Input Form**, then type in your
Mathematica commands and type **SHIFT-RETURN** to get an output. Mathematica comes with some built-in palettes to help with inputting mathematics. Some
examples of palettes are given in Figure 1.2. The buttons on the palettes serve
essentially as additional keys on the users keyboard. Input to the Mathematica
notebook can either be performed by typing in text commands or pointing and
clicking on the symbols provided by various palettes and subpalettes. As many
of the Mathematica programs in subsequent chapters of this book are necessarily text based, the author has decided to present the material using commands
in their text format. However, the use of palettes can save some time in typing, and the reader may wish to experiment with the "clicking on palettes"

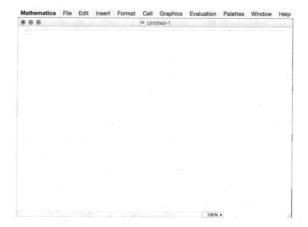

Figure 1.1:   A blank Mathematica notebook Untitled-1

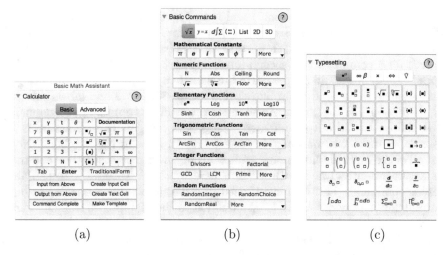

(a)   (b)   (c)

Figure 1.2: Some Basic Math Assistant palettes: (a) Calculator. (b) Basic Commands. (c) Typesetting

approach. Mathematica even allows users to create their own palettes, again readers might find this useful. Mathematica notebooks can be used to generate full publication-quality documents. In fact, all of the Mathematica help pages have been created with interactive notebooks. The **Help** menu also includes an online version of the Mathematica book, simply click on Help and Mathematica Documentation. New users should start with *Hand's-On Start to Wolfram Mathematica* [3] or Wolfram's *Elementary Introduction* [6]. The author recommends a brief tour of some of the help pages to give the reader an idea of how the notebooks can be used. For example, click on the **Help** toolbar at the top of the Mathematica graphical user interface and scroll down to **Help browser...**. Simply type in Solve and ENTER, an interactive Mathematica notebook will be opened showing the syntax, some related commands, and examples of the Solve command.

The interactive notebooks from each chapter of this book can be downloaded from Mathematica's Information center at

http://library.wolfram.com/infocenter/Books/AppliedMathematics/.

The author has provided the reader with a tutorial introduction to Mathematica in Sections 1.2 and 1.3. Each tutorial should take no more than one hour to complete. The author highly recommends that new users go through these tutorials line by line; however, readers already familiar with the package will probably use this chapter as reference material only.

Tutorial One provides a basic introduction to the Mathematica package. The first command line shows the reader how to input comments, which are extremely useful when writing long or complicated programs. The reader will type in (*This is a comment*) and then type SHIFT-ENTER or SHIFT-RETURN, (hold down the SHIFT key and press ENTER). Mathematica will label the first input with In[1]:=(*This is a comment*). Note that no output is given for a comment. The second input line is simple arithmetic. The reader types 2+3-36/2+2^3, and types SHIFT-ENTER to compute the result. Mathematica labels the second input with In[2]:=2+3-36/2+2^3 and labels the corresponding output Out[2]=-5. As the reader continues to input new command lines, the input and output numbers change accordingly. This allows users to easily label input and output that may be useful later in the notebook. Note that all of the built-in Mathematica functions begin with capital letters and that the arguments are always enclosed in square brackets. Tutorial Two contains graphic commands, and commands used to solve simple differential equations. Tutorial Three provides a simple introduction to the Manipulate command and programming with Mathematica.

The tutorials are intended to give the reader a concise and efficient introduction to the Mathematica package. Many more commands are listed in other chapters of the book, where the output has been included. Of course, there are many Mathematica textbooks on the market for those who require further applications or more detail. A list of some textbooks is given in the reference section of this Chapter [1]–[6].

1.2 Tutorial One: The Basics (One Hour)

There is no need to copy the comments, they are there to help you. Click on the Mathematica icon and copy the commands. Hold down the SHIFT key and press ENTER at the end of a line to see the answer or use a semicolon to suppress the output. You can interrupt a calculation at any time by typing ALT-COMMA or clicking on the Evaluation toolbar and Quit Kernel. A working Mathematica notebook of Tutorial One can be downloaded from the Mathematica Information Center, as indicated at the start of the chapter.

Mathematica Command Lines	Comments
In[1]:=(*This is a comment*)	(*Helps when writing programs.*)
In[2]:=2+3-36/2+2^3	(*Simple arithmetic.*)
In[3]:=2 3*7	(*Use space or * to multiply.*)
In[4]:=2/3+4/5	(*Fraction arithmetic.*)
In[5]:=2/3+4/5 //N	(*Approximate decimal.*)
In[6]:=Sqrt[16]	(*Square root.*)
In[7]:=Sin[Pi]	(*Trigonometric function.*)
In[8]:=z1=1+2 I;z2=3-4*I;z3=z1-z2/z1	(*Complex arithmetic.*)

```
In[9]:=ComplexExpand[Exp[z1]]              (*Express in form x+iy.*)

In[10]:=Factor[x^3-y^3]                    (*Factorize.*)

In[11]:=Expand[%]                          (*Expand the last result
                                           generated.*)

In[12]:=f=mu x (1-x) /. {mu->4,x->0.2}     (*Evaluate f when mu=4 and
                                           x=0.2.*)

In[13]:=Clear[mu,x]                        (*Clear values.*)

In[14]:=Simplify[(x^3-y^3)/(x-y)]          (*Simplify an expression.*)

In[15]:=Dt[x^2-3 x+6,x]                    (*Total differentiation.*)

In[16]:=D[x^3 y^5,{x,2},{y,3}]             (*Partial differentiation.*)

In[17]:=Integrate[Sin[x] Cos[x],x]         (*Indefinite integration.*)

In[18]:=Integrate[Exp[-x^2],{x,0,Infinity}]  (*Definite integration.*)

In[19]:=Sum[1/n^2,{n,1,Infinity}]          (*An infinite sum.*)

In[20]:=Solve[x^2-5 x+6==0,x]              (*Solving equations. Roots
                                           are in a list.*)

In[21]:=Solve[x^2-5 x+8==0,x]              (*A quadratic with complex
                                           roots.*)

In[22]:=Abs[x] /. Out[21]                  (*Find the modulus of the
                                           roots (see Out[21]).*)

In[23]:=Solve[{x^2+y^2==1,x+3 y==0}]       (*Solving simultaneous
                                           equations.*)

In[24]:=Series[Exp[x],{x,0,5}]             (*Taylor series expansion.*)

In[25]:=Limit[x/Sin[x],x->0]               (*Limits.*)

In[26]:=f=Function[x,4 x (1-x)]            (*Define a function.*)

In[27]:=f[0.2]                             (*Evaluate f(0.2).*)

In[28]:=fofof=Nest[f,x,3]                  (*Calculates f(f(f(x))).*)

In[29]:=fnest=NestList[f,x,4]              (*Generates a list of
                                           composite functions.*)

In[30]:=fofofof=fnest[[4]]                 (*Extract the 4th element
                                           of the list.*)

In[31]:=u=Table[2 i-1,{i,5}]               (*List the first 5 odd
                                           natural numbers.*)

In[32]:=%^2                                (*Square the elements of
                                           the last result generated.*)

In[33]:=a={2,3,4};b={5,6,7};               (*Two vectors.*)

In[34]:=3 a                                (*Scalar multiplication.*)

In[35]:=a.b                                (*Dot product.*)

In[36]:=Cross[a,b]                         (*Cross product.*)

In[37]:=Norm[a]                            (*Norm of a vector.*)

In[38]:=A={{1,2},{3,4}};B={{5,6},{7,8}};   (*Two matrices.*)

In[39]:=A.B-B.A                            (*Matrix arithmetic.*)
```

```
In[40]:=MatrixPower[A,3]                (*Powers of a matrix.*)

In[41]:=Inverse[A]                      (*The inverse of a matrix.*)

In[42]:=Det[B]                          (*The determinant of a
                                        matrix.*)

In[43]:=Tr[B]                           (*The trace of a matrix.*)

In[44]:=M={{1,2,3},{4,5,6},{7,8,9}}     (*A matrix.*)

In[45]:=Eigenvalues[M]                  (*The eigenvalues of a
                                        matrix.*)

In[46]:=Eigenvectors[M]                 (*The eigenvectors of a
                                        matrix.*)

In[47]:=LaplaceTransform[t^3,t,s]       (*Laplace transform.*)

In[48]:=InverseLaplaceTransform[6/s^4,s,t]  (*Inverse Laplace
                                        transform.*)

In[49]:=FourierTransform[t^4 Exp[-t^2],t,w]  (*Fourier transform.*)

In[50]:=InverseFourierTransform[%,w,t]  (*Inverse Fourier
                                        transform.*)

In[51]:=Quit[ ]                         (*Terminates Mathematica
                                        kernel session.*)
```

1.3 Tutorial Two: Plots and Differential Equations (One Hour)

Mathematica has excellent graphical capabilities and many solutions of nonlinear systems are best portrayed graphically. The graphs produced from the input text commands listed below may be found in the Tutorial Two Notebook, which can be downloaded from the Mathematica Information Center. Plots in other chapters of the book are referred to in many of the Mathematica programs at the end of each chapter.

```
(*Plotting graphs.*)

(*Set up the domain and plot a simple function.*)
In[1]:=Plot[Sin[x],{x,-Pi,Pi}]

(*Plot two curves on one graph.*)
In[2]:=Plot[{Cos[x],Exp[-.1 x] Cos[x]},{x,0,60}]

(*Plotting with labels.*)
In[3]:=Plot[Exp[-.1 t] Sin[t],{t,0,60},AxesLabel->{"t","Current"},
       PlotRange->{-1,1}]

(*Contour plot with shading.*)
In[4]:=ContourPlot[y^2/2-x^2/2+x^4/4,{x,-2,2},{y,-2,2}]
```

```
(*Contour plot with no shading.*)
In[5]:=ContourPlot[y^2/2-x^2/2+x^4/4,{x,-2,2},{y,-2,2},
        ContourShading->False]
```

```
(*Contour plot with 20 contours.*)
In[6]:=ContourPlot[y^2/2-x^2/2+x^4/4,{x,-2,2},{y,-2,2},Contours->20]
```

```
(*Surface plot.*)
In[7]:=Plot3D[y^2/2-x^2/2+x^4/4,{x,-2,2},{y,-2,2}]
```

```
(*A 2-D parametric plot.*)
In[8]:=ParametricPlot[{t^3-4 t,t^2},{t,-3,3}]
```

```
(*A 3-D parametric curve.*)
In[9]:=ParametricPlot3D[{Sin[t],Cos[t],t/3},{t,-10,10}]
```

```
(*Plot an implicit curve.*)
In[10]:=ContourPlot[2 x^2+3 y^2==12,{x,-3,3},{y,-3,3}]
```

```
(*Solve a simple separable differential equation.*)
In[11]:=DSolve[x'(t)==-x[t]/t,x[t],t]
```

```
(*Solve an initial value problem (IVP).*)
In[12]:=DSolve[{x'(t)==-t/x[t],x[0]==1},x[t],t]
```

```
(*Solve a second order ordinary differential equation (ODE).*)
In[13]:=DSolve[{x''[t]+5 x'[t]+6 x[t]==10 Sin[t],x[0]==0,x'[0]==0},
        x[t],t]
```

```
(*Solve a system of two ODEs.*)
In[14]:=DSolve[{x'[t]==3 x[t]+4 y[t],y'[t]==-4 x[t]+3 y[t]},
        {x[t],y[t]},t]
```

```
(*Solve a system of three ODEs.*)
In[15]:=DSolve[{x'[t]==x[t],y'[t]==y[t],z'[t]==-z[t]},
        {x[t],y[t],z[t]},t]
```

```
(*Solve an IVP using numerical methods and plot a solution curve.*)
In[16]:=u=NDSolve[{x'[t]==x[t] (.1-.01 x[t]),x[0]==50},x,{t,0,100}]
In[17]:=Plot[Evaluate[x[t]/.u],{t,0,100},PlotRange->All]
```

```
(*Plot a phase plane portrait.*)
In[18]:=v=NDSolve[{x'[t]==.1 x[t]+y[t],y'[t]==-x[t]+.1 y[t],
        x[0]==.01,y[0]==0},{x[t],y[t]},{t,0,50}]
In[19]:=ParametricPlot[{x[t],y[t]}/.v,{t,0,50},PlotRange->All,
        PlotPoints->1000]
```

```
(*Plot a 3-dimensional phase portrait.*)
In[20]:=w=NDSolve[{x'[t]==z[t]-x[t],y'[t]==-y[t],z'[t]==z[t]-17 x[t]+16,
        x[0]==.8,y[0]==.8,z[0]==.8},{x[t],y[t],z[t]},{t,0,20}]
In[21]:=ParametricPlot3D[Evaluate[{x[t],y[t],z[t]}/.w],{t,0,20},
        PlotPoints->1000,PlotRange->All]


(*A stiff van der Pol system of ODEs.*)
In[22]:=Needs["DifferentialEquations'InterpolatingFunctionAnatomy'"];
In[23]:=vanderpol=NDSolve[{Derivative[1][x][t]==y[t],Derivative[1][y][t]
        ==1000*(1-x[t]^2)*y[t]-x[t],x[0]==2,y[0]==0},{x,y},{t,5000}];
In[24]:=T=First[InterpolatingFunctionCoordinates[First[x /. vanderpol]]];
In[25]:=ListPlot[Transpose[{x[T],y[T]} /. First[vanderpol]],
        PlotRange->All]
```

1.4 The Manipulate Command and Simple Mathematica Programs

Sections 1.1, 1.2, and 1.3 illustrate the interactive nature of Mathematica. More involved tasks will require more code. Note that Mathematica is very different from procedural languages such as C, Pascal, or Fortran. An afternoon spent browsing through the pages listed at the Mathematica Information Center will convince readers of this fact.

The Manipulate command was first introduced in Version 6.0 and is very useful in the field of dynamical systems. Manipulate[$expr, \{u, u_{min}, u_{max}\}$] generates a version of $expr$ with controls added to allow interactive manipulation of the value of u.

The Manipulate Command

On execution of the Manipulate command, a parameter slider appears in the notebook. Solutions change as the slider is moved left and right.

```
(*Example 1: Solving simultaneous equations as the parameter a varies.
See Exercise 6 in Chapter 4.*)
In[1]:=Manipulate[Solve[{x (1-y-a x)==0,y (-1+x-a y)==0}],{a,0,1}]
(*Showing the solution lying wholly in the first quadrant.*)
In[2]:=Manipulate[Plot[{1-a x,(x-1)/a},{x,0,2}],{a,0.01,1}]
```

```
(*Example 2: Evaluating functions of functions to the fifth iteration.*)
In[3]:=f=Function[x,4 x (1-x)];
In[4]:=Manipulate[Expand[Nest[f,x,d]],{d,1,5,1}] (*Steps of 1.*)
```

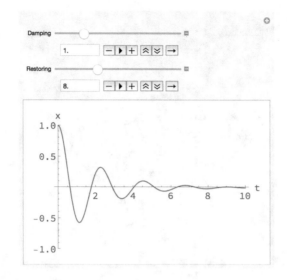

Figure 1.3: Solutions to the pendulum problem, $\ddot{x} + k\dot{x} + rx = 0$, for varying parameter values of the damping coefficient, k, and the restoring coefficient, r, say. In this case, there are two sliders

```
(* Example 3: Solving a differential equation as two parameters vary. *)
(* See Figure 1.3. *)
Manipulate[
Module[{Tmax, xmin, xmax, ode1, sol1, p1},
Tmax = 10; xmin = -1; xmax = 1; ode1[x0_, y0_] :=
NDSolve[{x'[t] == y[t], y'[t] == -k*y[t] - r*x[t], x[0] == x0, y[0] == y0},
{x[t], y[t]}, {t, 0, Tmax}]; sol1[1] = ode1[1, 0];
p1 = Plot[Evaluate[Table[{x[t]} /. sol1[i], {i, 1}]], {t, 0, Tmax},
PlotRange -> {xmin, xmax}, PlotPoints -> 100];
Show[{p1}, PlotRange -> {xmin, xmax}, AxesLabel -> {"t", "x"},
Axes -> True, TextStyle -> {FontSize -> 15}] ], {{k, 0, "Damping"}, 0, 5}
, {{r, 0, "Restoring"}, 0, 25}]
```

```
(* Example 4: Animation of a surface as a parameter varies. *)
(* See Figure 1.4. *)
In[10]:=Manipulate[Plot3D[a*y^2/2-x^2/2+x^4/4,{x,-2,2},{y,-2,2},
AxesLabel->{"x","y","z"},Axes->True,BaseStyle->{FontSize->18}],{a,-5,5}]
In[11]:=Quit[ ]
```

Simple Mathematica Programs

Each Mathematica program is displayed between horizontal lines and kept short to aid in understanding; the output is also included.

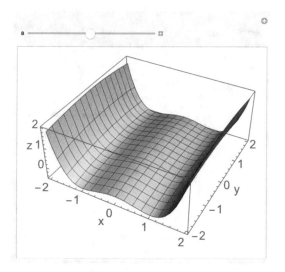

Figure 1.4: Animation of the surface, $z(x,y) = a\frac{y^2}{2} - \frac{x^2}{2} + \frac{x^4}{4}$. The slider is used to vary the parameter a

Modules and local variables. A global variable t is unaffected by the local variable t used in a module.

`In[1]:=Norm3d[a_,b_,c_]=Module[{t},t=Sqrt[a^2+b^2+c^2]]`

$Out[1]=\sqrt{a^2+b^2+c^2}$

`In[2]:=Norm3d[3,4,5]`

$Out[2]=5\sqrt{2}$

The Do command. The first ten terms of the Fibonacci sequence.

```
In[3]:=F[1]=1;F[2]=1;Nmax=10;
In[4]:=Do[F[i]=F[i-1]+F[i-2],{i,3,Nmax}]
In[5]:=Table[F[i],{i,1,Nmax}]
```

$Out[5]=\{1, 1, 2, 3, 5, 8, 13, 21, 34, 55\}$

If, then, else construct.

```
In[6]:=Num=1000;If[Num>0,Print["Num is positive"],If[Num==0,
     Print["Num is zero"],Print["Num is negative"]]]
```

$Out[6]=$Num is positive

Conditional statements with more than two alternatives.

```
In[7]:=r[x_]=Switch[Mod[x, 3],0,a,1,b,2,c]
```

Out[7]=Switch[Mod[x, 3],
0, a,
1, b,
2, c]

```
In[8]:=r[7]
```

Out[8]=b

Which. Defining the tent function.

```
In[9]:=T[x_]=Which[0<=x<1/2,mu x,1/2<=x<=1,mu (1-x)]
```

Out[9]=Which$[0 \leq x < 1/2, mux, 1/2 \leq x \leq 1, mu(1 - x)]$

```
In[10]:=T[4/5]
```

Out[10]=$\frac{mu}{5}$

For loops. Compute f(x), f(f(x)), and f(f(f(x))).

```
In[11]:=For[i=1;t=x,i<4,i++,t=mu t (1-t);Print[Factor[t]]]
```

Out[11]=$-mu(-1 + x)x$

$$-mu^2(-1 + x)x(1 - mux + mux^2)$$

$$-mu^3(-1 + x)x(1 - mux + mux^2)(1 - mu^2x + mu^2x^2 + mu^3x^2 - 2mu^3x^3 + mu^3x^4)$$

1.5 Hints for Programming

The Mathematica language contains very powerful commands, which means that some complex programs may contain only a few lines of code. Of course, the only way to learn programming is to sit down and try it yourself. This section has been included to point out common errors and give advice on how to troubleshoot. Remember to check the help and index pages in Mathematica and the Web if the following does not help you with your particular problem. Throughout the book, I have attempted to make the programs as simple as possible, experienced users may find better ways to program and Mathematica is full of shortcuts to further reduce the numbers of lines of code. I have tried to avoid these shortcuts where possible.

Common typing errors. The author strongly advises new users to type Tutorials One, Two, and Three into their own notebooks, this should reduce typing errors.

- Type SHIFT-ENTER at the end of every command line.

- If a command line is ended with a semicolon, the output will not be displayed.

- Make sure brackets, parentheses, etc. match up in correct pairs.

- Remember that all of the built-in Mathematica functions begin with capital letters and that the arguments are always enclosed in square brackets.

- Remember Mathematica is case sensitive.

- Check the syntax, type ?Solve to list syntax for the Solve command, for example.

Programming tips. The reader should use the Mathematica programs listed in Section 1.3 to practice simple programming techniques.

- It is best to clear values at the start of a large program.

- Use comments throughout the program. You will find them extremely useful in the future.

- Use Modules or Blocks to localize variables. This is especially useful for very large programs.

- If a program involves a large number of iterations, for example, 50,000, then run it for three iterations first and list all output.

- If the computer is not responding hold ALT-COMMA and try reducing the size of the problem.

- Read the error message printed by Mathematica and click on **More information** if necessary.

- Find a similar Mathematica program in a book or on the Web, and edit it to meet your needs.

- Check which version of Mathematica you are using. The syntax of some commands may have altered.

1.6 Mathematica Exercises

1. Evaluate the following:

    (a) $4 + 5 - 6$;

    (b) 3^{12};

(c) $\sin(0.1\pi)$;

(d) $(2 - (3 - 4(3 + 7(1 - (2(3 - 5)))))))$;

(e) $\frac{2}{5} - \frac{3}{4} \times \frac{2}{3}$.

2. Given that

$$A = \begin{pmatrix} 1 & 2 & -1 \\ 0 & 1 & 0 \\ 3 & -1 & 2 \end{pmatrix}, \qquad B = \begin{pmatrix} 1 & 2 & 3 \\ 1 & 1 & 2 \\ 0 & 1 & 2 \end{pmatrix}, \qquad C = \begin{pmatrix} 2 & 1 & 1 \\ 0 & 1 & -1 \\ 4 & 2 & 2 \end{pmatrix},$$

determine the following:

(a) $A + 4BC$;

(b) the inverse of each matrix if it exists;

(c) A^3;

(d) the determinant of C;

(e) the eigenvalues and eigenvectors of B.

3. Given that $z_1 = 1 + i$, $z_2 = -2 + i$ and $z_3 = -i$, evaluate the following:

(a) $z_1 + z_2 - z_3$;

(b) $\frac{z_1 z_2}{z_3}$;

(c) e^{z_1};

(d) $\ln(z_1)$;

(e) $\sin(z_3)$.

4. Evaluate the following limits if they exist:

(a) $\lim_{x \to 0} \frac{\sin x}{x}$;

(b) $\lim_{x \to \infty} \frac{x^3 + 3x^2 - 5}{2x^3 - 7x}$;

(c) $\lim_{x \to \pi} \frac{\cos x + 1}{x - \pi}$;

(d) $\lim_{x \to 0^+} \frac{1}{x}$;

(e) $\lim_{x \to 0} \frac{2\sinh x - 2\sin x}{\cosh x - 1}$.

5. Find the derivatives of the following functions:

(a) $y = 3x^3 + 2x^2 - 5$;

(b) $y = \sqrt{1 + x^4}$;

(c) $y = e^x \sin x \cos x$;

    (d) $y = \tanh x$;

    (e) $y = x^{\ln x}$.

6. Evaluate the following definite integrals:

    (a) $\int_{x=0}^{1} 3x^3 + 2x^2 - 5\,dx$;

    (b) $\int_{x=1}^{\infty} \frac{1}{x^2}\,dx$;

    (c) $\int_{-\infty}^{\infty} e^{-x^2}\,dx$;

    (d) $\int_{0}^{1} \frac{1}{\sqrt{x}}\,dx$;

    (e) $\int_{0}^{\frac{2}{\pi}} \frac{\sin(1/t)}{t^2}\,dt$.

7. Graph the following:

    (a) $y = 3x^3 + 2x^2 - 5$;

    (b) $y = e^{-x^2}$, for $-5 \leq x \leq 5$;

    (c) $x^2 - 2xy - y^2 = 1$;

    (d) $z = 4x^2 e^y - 2x^4 - e^{4y}$, for $-3 \leq x \leq 3$ and $-1 \leq y \leq 1$;

    (e) $x = t^2 - 3t$, $y = t^3 - 9t$, for $-4 \leq t \leq 4$.

8. Solve the following differential equations:

    (a) $\frac{dy}{dx} = \frac{x}{2y}$, given that $y(1) = 1$;

    (b) $\frac{dy}{dx} = \frac{-y}{x}$, given that $y(2) = 3$;

    (c) $\frac{dy}{dx} = \frac{x^2}{y^3}$, given that $y(0) = 1$;

    (d) $\frac{d^2x}{dt^2} + 5\frac{dx}{dt} + 6x = 0$, given that $x(0) = 1$ and $\dot{x}(0) = 0$;

    (e) $\frac{d^2x}{dt^2} + 5\frac{dx}{dt} + 6x = sin(t)$, given that $x(0) = 1$ and $\dot{x}(0) = 0$.

9. Carry out one hundred iterations on the recurrence relation

$$x_{n+1} = 4x_n(1 - x_n),$$

given that (a) $x_0 = 0.2$ and (b) $x_0 = 0.2001$. List the final ten iterates in each case.

10. Type ?While to read the help page on the While command. Use a while loop to program Euclid's algorithm for finding the greatest common divisor of two integers. Use your program to find the greatest common divisor of 12348 and 14238.

Bibliography

[1] J. Borwein and M.P. Skerritt, *An Introduction to Modern Mathematical Computing: With Mathematica*, Springer, New York, 2016.

[2] J.G.S. Leon, *Mathematica Beyond Mathematics: The Wolfram Language in the Real World*, Chapman and Hall, London, 2017.

[3] C. Hastings and K. Mischo, *Hands-On Start to Wolfram Mathematica*, Wolfram Media Inc., Illinois, 2015.

[4] Scientific Books, *MATHEMATICA: Graphical Programming*, CreateSpace Independent Publishing Platform, Amazon, 2016.

[5] P. Wellin, *Essentials of Programming in Mathematica*, Cambridge University Press, Cambridge, 2016.

[6] S. Wolfram, *An Elementary Introduction to the Wolfram Language*, Wolfram Media Inc., Illinois, 2015.

Chapter 2
Differential Equations

Aims and Objectives

- To review basic methods for solving some differential equations.

- To apply the theory to simple mathematical models.

- To introduce an existence and uniqueness theorem.

On completion of this chapter, the reader should be able to

- solve certain first- and second-order differential equations;

- apply the theory to chemical kinetics and electric circuits;

- interpret the solutions in physical terms;

- understand the existence and uniqueness theorem and its implications.

The basic theory of ordinary differential equations (ODEs) and analytical methods for solving some types of ODEs are reviewed. This chapter is not intended to be a comprehensive study on differential equations, but more an introduction to the theory that will be used in later chapters. Most of the material will be covered in first- and second-year undergraduate mathematics courses. The differential equations are applied to all kinds of models, but this chapter concentrates on chemical kinetics and electric circuits in particular.

This chapter ends with the existence and uniqueness theorem and some analysis.

© Springer International Publishing AG 2017
S. Lynch, *Dynamical Systems with Applications using Mathematica*®,
DOI 10.1007/978-3-319-61485-4_2

2.1 Simple Differential Equations and Applications

Definition 1. A differential equation that involves only one independent variable is called an *ordinary differential equation* (ODE). Those involving two or more independent variables are called *partial differential equations* (PDEs). This chapter will be concerned with ODEs only.

The subject of ODEs encompasses analytical, computational, and applicable fields of interest. There are many textbooks written from the elementary to the most advanced, with some focusing on applications and others concentrating on existence theorems and rigorous methods of solution. This chapter is intended to introduce the reader to all three branches of the subject. For more information, the reader should consult the ODE textbooks in the bibliography [2, 7, 10, 15]. To solve ODEs using Mathematica, the reader is directed to [1, 6].

Separable Differential Equations
Consider the differential equation

$$\frac{dx}{dt} = f(t, x) \tag{2.1}$$

and suppose that the function $f(t, x)$ can be factored into a product $f(t, x) = g(t)h(x)$, where $g(t)$ is a function of t and $h(x)$ is a function of x. If f can be factored in this way, then equation (2.1) can be solved by the method of *separation of variables*.

To solve the equation, divide both sides by $h(x)$ to obtain

$$\frac{1}{h(x)} \frac{dx}{dt} = g(t);$$

and integration with respect to t gives

$$\int \frac{1}{h(x)} \frac{dx}{dt} \, dt = \int g(t) \, dt.$$

Changing the variables in the integral gives

$$\int \frac{dx}{h(x)} = \int g(t) \, dt.$$

An analytic solution to (2.1) can be found only if both integrals can be evaluated. The method can be illustrated with some simple examples.

Example 1. Solve the differential equation $\frac{dx}{dt} = -\frac{t}{x}$.

Solution. The differential equation is separable. Separate the variables and integrate both sides with respect to t. Therefore,

$$\int x \frac{dx}{dt}\, dt = -\int t\, dt,$$

and so

$$\int x\, dx = -\int t\, dt.$$

Integration of both sides yields

$$t^2 + x^2 = r^2,$$

where r^2 is a constant. There are an infinite number of solutions. The *solution curves* are concentric circles of radius r centered at the origin. There are an infinite number of solution curves that would fill the plane if they were all plotted. Three such solution curves are plotted in Fig. 2.1.

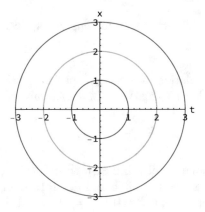

Figure 2.1: Three of an infinite number of solution curves for Example 1.

Example 2. Solve the differential equation $\frac{dx}{dt} = \frac{t}{x^2}$.

Solution. The differential equation is separable. Separate the variables and integrate both sides with respect to t to give

$$\int x^2\, dx = \int t\, dt.$$

Integration of both sides yields

$$\frac{x^3}{3} = \frac{t^2}{2} + C,$$

where C is a constant. Six of an infinite number of solution curves are plotted in Fig. 2.2.

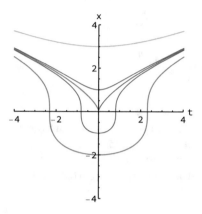

Figure 2.2:   Five solution curves for Example 2.

Example 3. The population of a certain species of fish living in a large lake at time t can be modeled using *Verhulst's equation*, otherwise known as the *logistic equation*,

$$\frac{dP}{dt} = P(\beta - \delta P),$$

where $P(t)$ is the population of fish measured in tens of thousands, and β and δ are constants representing the birth and death rates of the fish living in the lake, respectively. Suppose that $\beta = 0.1$, $\delta = 10^{-3}$, and the initial population is 50×10^4. Solve this *initial value problem* and interpret the results in physical terms.

Solution. Using the methods of separation of variables gives

$$\int \frac{dP}{P(\beta - \delta P)} = \int dt.$$

The solution to the integral on the left may be determined using partial fractions. The general solution is

$$\ln \left| \frac{P}{\beta - \delta P} \right| = \beta t + C_1,$$

or

$$P(t) = \frac{\beta}{\delta + C_2 \beta e^{-\beta t}},$$

computed using Mathematica, where C_1 and C_2 are constants. Substituting the initial conditions, the solution is

$$P(t) = \frac{100}{1 + e^{-0.1t}}.$$

Thus as time increases, the population of fish tends to a value of 100×10^4. The solution curve is plotted in Fig. 2.3.

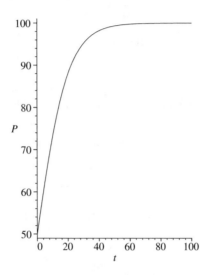

Figure 2.3: Solution curve for the initial value problem in Example 3. Note that the axes would be scaled by 10^4 in applications.

Note the following:

• The quantity $\frac{\beta}{\delta}$ is the ratio of births to deaths and is called the *carrying capacity* of the environment.

• Take care when interpreting the solutions. This and similar continuous models only work for large species populations. The solutions give approximate numbers. Even though time is continuous, the population size is not. For example, you cannot have a fractional living fish, so population sizes have to be rounded out to whole numbers in applications.

• Discrete models can also be applied to population dynamics (see Chap. 13).

Exact Differential Equations
A differential equation of the form

$$M(t, x) + N(t, x)\frac{dx}{dt} = 0 \tag{2.2}$$

is said to be *exact* if there exists a function, say, $F(t, x)$, with continuous second partial derivatives such that

$$\frac{\partial F}{\partial t} = M(t, x), \quad \text{and} \quad \frac{\partial F}{\partial x} = N(t, x).$$

Such a function exists as long as

$$\frac{\partial M}{\partial x} = \frac{\partial N}{\partial t},$$

and then the solution to (2.2) satisfies the equation

$$F(t, x) = C,$$

where C is a constant. Differentiate this equation with respect to t to obtain (2.2).

Example 4. Solve the differential equation

$$\frac{dx}{dt} = \frac{9 - 12t - 5x}{5t + 2x - 4}.$$

Solution In this case, $M(t, x) = -9 + 12t + 5x$ and $N(t, x) = 5t + 2x - 4$. Now

$$\frac{\partial M}{\partial x} = \frac{\partial N}{\partial t} = 5$$

and integration gives the solution $F(t, x) = x^2 + 6t^2 + 5tx - 9t - 4x = C$. There are an infinite number of solution curves, some of which are shown in Fig. 2.4.

Homogeneous Differential Equations
Consider differential equations of the form

$$\frac{dx}{dt} = f\left(\frac{x}{t}\right). \tag{2.3}$$

Substitute $v = \frac{x}{t}$ into (2.3) to obtain

$$\frac{d}{dt}(vt) = f(v).$$

Therefore,

$$v + t\frac{dv}{dt} = f(v),$$

and so

$$\frac{dv}{dt} = \frac{f(v) - v}{t},$$

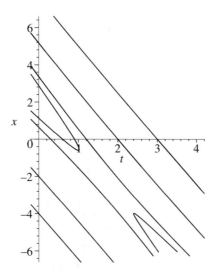

Figure 2.4: Some solution curves for Example 4.

which is separable. A complete solution can be found as long as the equations are integrable, and then, v may be replaced with $\frac{x}{t}$.

Example 5. Solve the differential equation

$$\frac{dx}{dt} = \frac{t-x}{t+x}.$$

Solution. The equation may be rewritten as

$$\frac{dx}{dt} = \frac{1 - \frac{x}{t}}{1 + \frac{x}{t}}. \tag{2.4}$$

Let $v = \frac{x}{t}$. Then, (2.4) becomes

$$\frac{dv}{dt} = \frac{1 - 2v - v^2}{t(1+v)}.$$

This is a separable differential equation. The general solution is given by

$$x^2 + 2tx - t^2 = C,$$

where C is a constant. Some solution curves are plotted in Fig. 2.5.

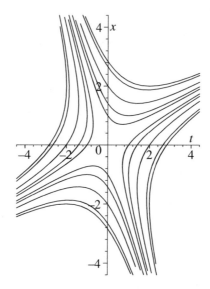

Figure 2.5: Some solution curves for Example 5.

Linear Differential Equations

Consider differential equations of the form

$$\frac{dx}{dt} + P(t)x = Q(t). \tag{2.5}$$

Multiplying through by an integrating factor, say, $J(t)$, (2.5) becomes

$$J\frac{dx}{dt} + JPx = JQ. \tag{2.6}$$

Find J such that (2.6) can be written as

$$\frac{d}{dt}(Jx) = J\frac{dx}{dt} + x\frac{dJ}{dt} = JQ.$$

In order to achieve this, set

$$\frac{dJ}{dt} = JP$$

and integrate to get

$$J(t) = \exp\left(\int P(t)\,dt\right).$$

Thus, the solution to system (2.5) may be found by solving the differential equation

$$\frac{d}{dt}(Jx) = JQ,$$

as long as the right-hand side is integrable.

Example 6. A chemical company pumps v liters of solution containing mass m grams of solute into a large lake of volume V per day. The inflow and outflow of the water is constant. The concentration of solute in the lake, say, σ, satisfies the differential equation

$$\frac{d\sigma}{dt} + \frac{v}{V}\sigma = \frac{m}{V}. \tag{2.7}$$

Determine the concentration of solute in the lake at time t assuming that $\sigma = 0$ when $t = 0$. What happens to the concentration in the long term?

Solution. This is a linear differential equation, and the integrating factor is given by

$$J = \exp\left(\int \frac{v}{V} dt\right) = e^{\frac{vt}{V}}.$$

Multiply (2.7) by the integrating factor to obtain

$$\frac{d}{dt}\left(e^{\frac{vt}{V}}\sigma\right) = e^{\frac{vt}{V}}\frac{m}{V}.$$

Integration gives

$$\sigma(t) = \frac{m}{v} - ke^{-\frac{vt}{V}},$$

where k is a constant. Substituting the initial conditions, the final solution is

$$\sigma(t) = \frac{m}{v}\left(1 - e^{-\frac{vt}{V}}\right).$$

As $t \to \infty$, the concentration settles to $\frac{m}{v}\ gl^{-1}$.

Series Solutions

Another very useful method for determining the solutions to some ODEs is the series solution method. The basic idea is to seek a series solution (assuming that the series converge) of the form

$$x(t) = \sum_{n=0}^{\infty} a_n(t - t_0)^n,$$

about the point t_0. The method holds for infinitely differentiable functions (that is, functions that can be differentiated as often as desired), and is outlined using two simple examples.

Example 7. Determine a series solution to the initial value problem

$$\frac{dx}{dt} + tx = t^3, \tag{2.8}$$

given that $x(0) = 1$.

Solution. Given that $t_0 = 0$, set $x(t) = \sum_{n=0}^{\infty} a_n t^n$. Substituting into (2.8) gives

$$\sum_{n=1}^{\infty} n a_n t^{n-1} + t \left(\sum_{n=0}^{\infty} a_n t^n \right) = t^3.$$

Combining the terms into a single series

$$a_1 + \sum_{n=1}^{\infty} \left((n+1) a_{n+1} + a_{n-1} \right) t^n = t^3.$$

Equating coefficients gives

$$a_1 = 0, 2a_2 + a_0 = 0, 3a_3 + a_1 = 0, 4a_4 + a_2 = 1, 5a_5 + a_3 = 0, \ldots$$

and solving these equations gives $a_{2n+1} = 0$, for $n = 0, 1, 2, \ldots,$

$$a_2 = -\frac{a_0}{2}, a_4 = \frac{1 - a_2}{4},$$

and

$$a_{2n} = -\frac{a_{2n-2}}{2n},$$

where $n = 3, 4, 5, \ldots$. Based on the assumption that $x(t) = \sum_{n=0}^{\infty} a_n t^n$, substituting $x(0) = 1$ gives $a_0 = 1$. Hence, the series solution to the ODE (2.8) is

$$x(t) = 1 - \frac{1}{2} t^2 + \frac{3}{8} t^4 + \sum_{n=3}^{\infty} (-1)^n \left(\frac{1}{(2n)} \frac{1}{(2n-2)} \cdots \frac{1}{6} \frac{3}{8} \right) t^{2n}.$$

Note that the analytic solution can be found in this case and is equal to

$$x(t) = -2 + t^2 + 3e^{-\frac{t^2}{2}},$$

which is equivalent to the series solution above.

Example 8. Consider the van der Pol equation given by

$$\frac{d^2x}{dt^2} + 2\left(x^2 - 1\right)\frac{dx}{dt} + x = 0, \tag{2.9}$$

where $x(0) = 5$ and $\dot{x}(0) = 0$. Use Mathematica to plot a numerical solution against a series solution up to order 6 near to the point $x(0) = 5$.

Solution. Using Mathematica, the series solution is computed to be

$$x(t) = 5 - \frac{5}{2}t^2 + 40t^3 - \frac{11515}{24}t^4 + \frac{9183}{2}t^5 + O(t^6).$$

Figure 2.6 shows the truncated series and numerical solutions for the ODE (2.9) near to $x(0) = 5$. The upper curve is the truncated series approximation that diverges quite quickly away from the numerical solution. Of course, one must also take care that the numerical solution is correct.

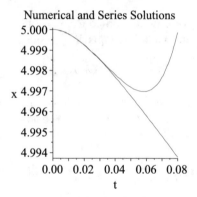

Figure 2.6: [Mathematica] Numerical and truncated series solutions for the van der Pol equation (2.9) near $x(0) = 5$.

2.2 Applications to Chemical Kinetics

Even the simplest chemical reactions can be highly complex and difficult to model. Physical parameters such as temperature, pressure, and mixing are ignored in this text, and differential equations are constructed that are dependent only on the concentrations of the chemicals involved in the reaction. This is potentially a very difficult subject, and some assumptions have to be made to make progress.

The Chemical Law of Mass Action. The rates at which the concentrations of the various chemical species change with time are proportional to their concentrations.

Consider the simple chemical reaction

$$A + \beta B \rightleftharpoons \gamma C,$$

where β and γ are the stoichiometric coefficients, A and B are the reactants, C is the product, and k_1 is the rate constant of the equation. The rate of reaction, say, r, is given by

$$r = \frac{\text{change in concentration}}{\text{change in time}}.$$

For this simple example,

$$r = k_1[A][B] = -\frac{d[A]}{dt} = -\frac{1}{\beta}\frac{d[B]}{dt} = \frac{1}{\gamma}\frac{d[C]}{dt},$$

where $[A], [B]$, and $[C]$ represents the concentrations of A, B, and C, respectively. By adding a second chemical equation, a slightly more complex system is produced,

$$\alpha A \rightleftharpoons \delta D,$$

where k_2 is the rate constant of the second equation and α and δ are the stoichiometric coefficients. Two of the possible reaction rate equations for this system now become

$$\frac{d[A]}{dt} = -k_1\beta[A][B]^{\beta} - k_2\alpha[A]^{\alpha}, \quad \frac{d[D]}{dt} = k_2\delta[A]^{\alpha}.$$

Consider the following example, where one molecule of hydrogen reacts with one molecule of oxygen to produce two molecules of hydroxyl (OH):

$$H_2 + O_2 \rightarrow 2OH.$$

Suppose that the concentration of hydrogen is $[H_2]$ and the concentration of oxygen is $[O_2]$. Then from the chemical law of mass action, the rate equation is given by

$$\text{Rate} = k[H_2][O_2],$$

where k is called the *rate constant*, and the reaction rate equation is

$$\frac{d[OH]}{dt} = 2k[H_2][O_2].$$

Table 2.1: One of the possible reaction rate equations for each chemical reaction.

Chemical reaction	The reaction rate equation for one species may be expressed as follows:
A+B → C	$\frac{dc}{dt} = k_f ab = k_f(a_0 - c)(b_0 - c)$
2A ⇌ B	$\frac{db}{dt} = k_f(a_0 - 2b)^2 - k_r b$
A ⇌ 2B	$\frac{db}{dt} = k_f\left(a_0 - \frac{b}{2}\right) - k_r b^2$
A ⇌ B+C	$\frac{dc}{dt} = k_f(a_0 - c) - k_r(b_0 + c)(c_0 + c)$
A+B ⇌ C	$\frac{dc}{dt} = k_f(a_0 - c)(b_0 - c) - k_r c$
A+B ⇌ C+D	$\frac{dc}{dt} = k_f(a_0 - c)(b_0 - c) - k_r(c_0 + c)(d_0 + c)$

Unfortunately, it is not possible to write down the reaction rate equations based on the stoichiometric (balanced) chemical equations alone. There may be many mechanisms involved in producing OH from hydrogen and oxygen in the above example. Even simple chemical equations can involve a large number of steps and different rate constants. Suppose in this text that the chemical equations give the *rate-determining steps*.

Suppose that species A, B, C, and D have concentrations $a(t), b(t), c(t)$, and $d(t)$ at time t and initial concentrations a_0, b_0, c_0, and d_0, respectively. Table 2.1 lists some reversible chemical reactions and one of the corresponding reaction rate equations, where k_f and k_r are the forward and reverse rate constants, respectively.

Example 9. A reaction equation for sulfate and hydrogen ions to form bisulfite ions is given by

$$SO_3^{2-} + H^+ \rightleftharpoons HSO_3^-,$$

where k_f and k_r are the forward and reverse rate constants, respectively. Denote the concentrations by $a = [SO_3^{2-}]$, $b = [H^+]$, and $c = [HSO_3^-]$, and let the initial concentrations be a_0, b_0, and c_0. Assume that there is much more of species H^+ than the other two species, so that its concentration b can be regarded as constant. The reaction rate equation for $c(t)$ is given by

$$\frac{dc}{dt} = k_f(a_0 - c)b - k_r(c_0 + c).$$

Find a general solution for $c(t)$.

Solution. The differential equation is separable and

$$\int \frac{dc}{k_f(a_0 - c)b - k_r(c_0 + c)} = \int dt.$$

Integration yields

$$c(t) = \frac{k_f a_0 b - k_r c_0}{k_f b + k_r} - \frac{k_r c_0}{k_f b + k_r} + A e^{(-k_f a_0 - k_r)t},$$

where A is a constant.

Example 10. The chemical equation for the reaction between nitrous oxide and oxygen to form nitrogen dioxide at $25^\circ C$,

$$2NO + O_2 \rightarrow 2NO_2$$

obeys the law of mass action. The rate equation is given by

$$\frac{dc}{dt} = k(a_0 - c)^2 \left(b_0 - \frac{c}{2} \right),$$

where $c = [NO_2]$ is the concentration of nitrogen dioxide, k is the rate constant, a_0 is the initial concentration of NO, and b_0 is the initial concentration of O_2. Find the concentration of nitrogen dioxide after time t given that $k = 0.00713l^2 M^{-2} s^{-1}$, $a_0 = 4Ml^{-1}$, $b_0 = 1Ml^{-1}$, and $c(0) = 0Ml^{-1}$.

Solution. The differential equation is separable and

$$\int \frac{dc}{(4 - c)^2(1 - c/2)} = \int k\, dt.$$

Integrating using partial fractions gives

$$kt = \frac{1}{c - 4} + \frac{1}{2} \ln |c - 4| - \frac{1}{2} \ln |c - 2| + \frac{1}{4} - \frac{1}{2} \ln 2.$$

It is not possible to obtain $c(t)$ explicitly, so numerical methods are employed using Mathematica. The concentration of nitrogen dioxide levels off at two moles per liter as time increases, as depicted in Fig. 2.7.

Chemical reactions displaying periodic behavior will be dealt with in Chapter 8. There may be a wide range of timescales involved in chemical reactions, and this can lead to *stiff* systems. Loosely speaking, a stiff system of differential equations is one in which the velocity or magnitude of the vector field changes rapidly in phase space.

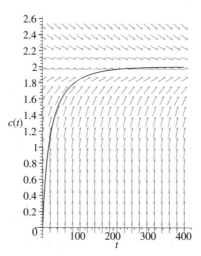

Figure 2.7: [Mathematica] The Concentration of NO_2 in moles per liter against time in seconds.

2.3 Applications to Electric Circuits

For many years, differential equations have been applied to model simple electrical and electronic circuits. If an oscilloscope is connected to the circuit, then the results from the analysis can be seen to match very well with what happens physically. As a simple introduction to electric circuits, linear systems will be considered and the basic definitions and theory will be introduced. The section ends with an introduction to the nonlinear circuit element known as the memristor.

Current and Voltage
The *current* I flowing through a conductor is proportional to the number of positive charge carriers that pass a given point per second. The unit of current is the *ampere* A. A *coulomb* is defined to be the amount of charge that flows through a cross section of wire in 1 second when a current of $1A$ is flowing, so 1 amp is 1 coulomb per second. As the current passes through a circuit element, the charge carriers exchange energy with the circuit elements, and there is a *voltage drop* or *potential difference* measured in joules per coulomb, or *volts* V.

Consider simple electric circuits consisting of voltage sources, resistors, inductors, and capacitors, or RLC circuits. A series RLC circuit is shown schematically in Fig. 2.8. The voltage drop across a resistor and the current flowing through it are related by Ohm's Law.

Ohm's Law. The voltage drop V across a resistor is proportional to the current I flowing through it:

$$V = IR,$$

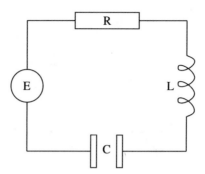

Figure 2.8:   Schematic of a simple RLC series circuit.

where R is the *resistance* of the resistor measured in ohms (Ω).

A changing electric current can create a changing magnetic field that induces a voltage drop across a circuit element, such as a coil.

Faraday's Law. The voltage drop across an inductor is proportional to the rate of change of the current:

$$V = L\frac{dI}{dt},$$

where L is the *inductance* of the inductor measured in henries (H).

A capacitor consists of two plates insulated by some medium. When connected to a voltage source, charges of opposite sign build up on the two plates, and the total charge on the capacitor is given by

$$q(t) = q_0 + \int_{t_0}^{t} I(s)\,ds,$$

where q_0 is the initial charge.

Coulomb's Law. The voltage drop across a capacitor is proportional to the charge on the capacitor:

$$V(t) = \frac{1}{C}q(t) = \frac{1}{C}\left(q_0 + \int_{t_0}^{t} I(s)\,ds\right),$$

where C is the *capacitance* of the capacitor measured in farads (F).

The physical laws governing electric circuits were derived by G.R. Kirchhoff in 1859.

Kirchhoff's Current Law. The algebraic sum of the currents flowing into any junction of an electric circuit must be zero.

Kirchhoff's Voltage Law. The algebraic sum of the voltage drops around any closed loop in an electric circuit must be zero.

Applying Kirchhoff's voltage law to the RLC circuit gives

$$V_L + V_R + V_C = E(t),$$

where V_R, V_L, and V_C are the voltage drops across R, L, and C, respectively, and $E(t)$ is the voltage source, or applied electromotive force (EMF). Substituting for the voltages across the circuit components gives

$$L\frac{dI}{dt} + RI + \frac{1}{C}q = E(t).$$

Since the current is the instantaneous rate of change in charge, $I = \frac{dq}{dt}$, this equation becomes

$$L\frac{d^2q}{dt^2} + R\frac{dq}{dt} + \frac{1}{C}q = E(t). \tag{2.10}$$

This differential equation is called a *linear second-order differential equation*. It is linear because there are no powers of the derivatives, and second order since the order of the highest occurring derivative is two. This equation can be solved by the method of *Laplace transforms* [12]; there are other methods available, and readers should use whichever method they feel most comfortable with. The method of Laplace transforms can be broken down into four distinct steps when finding the solution of a differential equation:

- rewrite equation (2.10) in terms of Laplace transforms;

- insert any given initial conditions;

- rearrange the equation to give the transform of the solution;

- find the inverse transform.

The method is illustrated in the following examples.

Example 11. Consider a series resistor-inductor electrical circuit. Kirchhoff's voltage law gives

$$L\frac{dI}{dt} + RI = E.$$

Given that $L = 10H$; $R = 2\Omega$, and $E = 50\sin(t)V$, find an expression for the current in the circuit if $I(0) = 0$.

Solution. Take Laplace transforms of both sides. Then,

$$10(s\bar{I} - I(0)) + 2\bar{I} = \frac{50}{s^2 + 1}.$$

Inserting the initial condition and rearranging,

$$\bar{I}(5s + 1) = \frac{25}{s^2 + 1},$$

and splitting into partial fractions,

$$\bar{I} = \frac{25}{26}\frac{1}{s^2 + 1} - \frac{125}{26}\frac{s}{s^2 + 1} - \frac{125}{126}\frac{1}{(s - 1/5)}.$$

Take inverse Laplace transforms to give

$$I(t) = \frac{25}{26}\sin(t) - \frac{125}{26}\cos(t) - \frac{125}{126}e^{-\frac{t}{5}}.$$

The periodic expression $\frac{25}{26}\sin(t) - \frac{125}{26}\cos(t)$ is called the *steady state*, and the term $\frac{125}{126}e^{-\frac{t}{5}}$ is called the *transient*. Note that the transient decays to zero as $t \to \infty$.

Example 12. Differentiate equation (2.10) with respect to time and substitute for $\frac{dq}{dt}$ to obtain

$$L\frac{d^2I}{dt^2} + R\frac{dI}{dt} + \frac{1}{C}I = \frac{dE}{dt}.$$

The second-order differential equation for a certain RLC circuit is given by

$$\frac{d^2I}{dt^2} + 5\frac{dI}{dt} + 6I = 10\sin(t).$$

Solve this differential equation given that $I(0) = \dot{I}(0) = 0$ (a *passive circuit*).

Solution. Take Laplace transforms of both sides:

$$(s^2\bar{I} - sI(0) - \dot{I}(0)) + 5(s\bar{I} - I(0)) + 6\bar{I} = \frac{10}{s^2 + 1}.$$

Substitute the initial conditions to obtain

$$\bar{I}(s^2 + 5s + 6) = \frac{10}{s^2 + 1}.$$

Splitting into partial fractions gives

$$\bar{I} = \frac{2}{s + 2} - \frac{1}{s + 3} + \frac{1}{s^2 + 1} - \frac{s}{s^2 + 1}.$$

Take inverse transforms to get

$$I(t) = 2e^{-2t} - e^{-3t} + \sin(t) - \cos(t).$$

The Memristor. The examples discussed thus far have concerned electric circuits with linear elements; however, nonlinear electric circuits are now coming to the fore. It is now widely acknowledged that Professor Leon Chua is the father of nonlinear circuit theory. Chua's famous nonlinear electric circuit is discussed in Chapter 8, and the circuit is easy to construct even in school physics laboratories. It has long been believed that there are only three fundamental passive circuit elements such as the capacitor, the inductor, and the resistor. In 1971, Chua [3] used mathematics to prove the existence of a fourth fundamental nonlinear element which acts like a resistor with memory, and he called the new device the memristor. The three well-known circuit elements are described by the equations

$$\frac{1}{C} = \frac{dv}{dq}, \ L = \frac{d\phi}{di}, \ R = \frac{dv}{di},$$

where $\frac{1}{C}$ is the inverse capacitance, L is inductance, R is incremental resistance, v is voltage, i is current, q is charge, and ϕ is flux. In addition, the current and voltage are described by the following physical laws

$$i = \frac{dq}{dt}, \ v = \frac{d\phi}{dt}.$$

This gives five relationships on three elements and leaves a gap in the harmonic symmetry of Chua's aesthetics. Chua discovered the missing functional relationship between charge and flux which is given by

$$M = \frac{d\phi}{dq},$$

where M is the memristance. Fig. 2.9 displays the relationships between the four fundamental elements.

In 1976, Chua and Kang [4] discovered that a memristor displays a pinched hysteresis and suggested that this effect could be used as a test to determine if a device could be truly categorized as a memristor. A pinched hysteresis loop is demonstrated in Chapter 21, and the Mathematica program for plotting the loop is listed within the Mathematica commands section of that chapter.

In 2008, a team at HP Laboratories [13] announced that they had evidence that many nanoscale electronic devices which involve the motion of charged atomic or molecular species act as memristors. Their analysis was based on results from a thin film of titanium dioxide, and they are currently building devices for computer logic, nanoelectronic memories, and neuromorphic computer architectures. A long-term project of HP Labs Research has been the development of *The Machine*, which was supposed to reinvent the fundamental architecture of computing. Among the principal components to be used were the memristor and silicon photonics using optical communications; unfortunately, in June 2015, HP Labs announced that memristors were to be removed from The

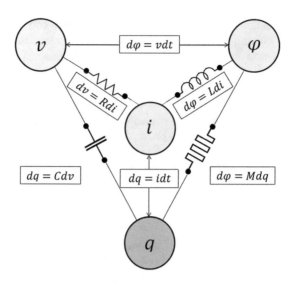

Figure 2.9: The memristor: the missing link discovered.

Machine's road map. Some researchers believe that Strukov's memristor modeling equations [13] do not simulate the devices physics very well but believe that Chang's and Yakopcic's models [8] provide a good compromise.

It is now understood that man-made memristive devices have been around for over two hundred years. In 2012, Prodromakis et al. [9] published a paper entitled "Two centuries of memristors". Indeed, it is now known that the first demonstration of a memristor device took place at the Royal Institution in 1808. Sir Humphrey Davy produced a 1000 V carbon arc discharge, and modern technology has demonstrated a pinched hysteresis effect in this system.

Incredibly, natural memristors have been around for hundreds of millions of years, and there are memristors in plants and early life forms. Chua [5] shows that sodium and potassium ion channel memristors are the key to generating action potentials in the Hodgkin-Huxley equations (see Chapter 21), and he explains some unresolved anomalies with the original equations. In terms of neurobiology, the tutorial shows that synapses are locally passive memristors and that neurons act as locally active memristors. Chua also shows that the circuits used to model the Josephson junction effect should include memristor elements to explain the workings of these devices accurately. The author and Borresen believe it is possible to make superfast low-power computers using Josephson junctions acting as neurons connected together with memristors acting as axons and synapses. More detail is provided in Chapter 21.

2.4 Existence and Uniqueness Theorem

Definition 2. A function $\mathbf{f}(\mathbf{x})$ with $\mathbf{f} : \Re^n \rightarrow \Re^n$ is said to satisfy a *Lipschitz condition* in a domain $D \subset \Re^n$ if there exists a constant, say, L, such that

$$\| \mathbf{f}(\mathbf{x_1}) - \mathbf{f}(\mathbf{x_2}) \| \leq L \| \mathbf{x_1} - \mathbf{x_2} \|,$$

where $\mathbf{x_1}, \mathbf{x_2} \in D$.

If the function $\mathbf{f}$ satisfies the Lipschitz condition, then it is said to be *Lipschitz continuous*. Note that Lipschitz continuity in $\mathbf{x}$ implies continuity in $\mathbf{x}$, but the converse is not always true.

Existence and Uniqueness Theorem. *Suppose that $\mathbf{f}$ is continuously Lipschitz; then for an initial point $\mathbf{x_0} \in D$, the* autonomous *differential equation*

$$\frac{d\mathbf{x}}{dt} = \dot{\mathbf{x}} = \mathbf{f}(\mathbf{x}) \tag{2.11}$$

has a unique solution, say, $\phi_t(\mathbf{x_0})$, that is defined on the maximal interval of existence.

Note that (2.11) is called autonomous as long as $\mathbf{f}$ is independent of t. The proof of this theorem can be found in most textbooks that specialize in the theory of ODEs. As far as the reader is concerned, this theorem implies that as long as $\mathbf{f}$ is continuously differentiable, i.e., $\mathbf{f} \in C^1(D)$, then two distinct solutions cannot intersect in finite time.

The following simple examples involving first-order ODEs illustrate the theorem quite well.

Example 13. Solve the following linear differential equations, and state the maximal interval of existence for each solution:

(a) $\dot{x} = x$, $x(0) = 1$;

(b) $\dot{x} = x^2$, $x(0) = 1$;

(c) $\dot{x} = 3x^{\frac{2}{3}}$, $x(0) = 0$.

Solutions.

(a) The solution to this elementary differential equation is $x(t) = e^t$, which is unique and defined for all t. The maximal interval of existence in this case is $-\infty < t < \infty$. Note that $f(x) = x$ is continuously differentiable.

(b) The solution is given by

$$x(t) = \frac{1}{1-t},$$

which is not defined for $t = 1$. Therefore, there is a unique solution on the maximal interval of existence given by $-\infty < t < 1$.

(c) The function $f(x) = 3x^{\frac{2}{3}}$ is not continuously differentiable and does not satisfy the Lipschitz condition at $x = 0$; $\frac{\partial f}{\partial x} = 2x^{-\frac{1}{3}}$ is not continuous at $x = 0$. Integration gives

$$\int \frac{1}{3} x^{-\frac{2}{3}}\, dx = \int dt,$$

with general solution $x(t) = t^3 + C$. The solution to the initial value problem is therefore $x(t) = t^3$. The point $x = 0$ is zero when $\dot{x} = 0$. This means that a solution starting at this point should stay there for all t. Thus, there are two solutions starting at $x_0 = 0$, namely $\phi_1(t) = t^3$ and $\phi_2(t) = 0$. In fact, there are infinitely many solutions starting at $x_0 = 0$. In this case, there exist solutions but they are not unique.

Note that the solution would be unique on the maximal interval of existence $0 < t < \infty$ if the initial condition were $x(1) = 1$.

Consider autonomous differential equations of the form

$$\dot{\mathbf{x}} = \mathbf{f}(\mathbf{x}), \qquad\qquad (2.12)$$

where $\mathbf{x} \in \Re^n$.

Definition 3. A *critical point* (*equilibrium point, fixed point, stationary point*) is a point that satisfies the equation $\dot{\mathbf{x}} = \mathbf{f}(\mathbf{x}) = 0$. If a solution starts at this point, it remains there forever.

Definition 4. A critical point, say, $\mathbf{x_0}$, of the differential equation (2.12) is called *stable* if given $\epsilon > 0$, there is a $\delta > 0$, such that for all $t \geq t_0$, $\| \mathbf{x}(t) - \mathbf{x_0}(t) \| < \epsilon$, whenever $\| \mathbf{x}(t_0) - \mathbf{x_0}(t_0) \| < \delta$, where $\mathbf{x}(t)$ is a solution of (2.12).

A critical point that is not stable is called an *unstable* critical point.

Example 14. Find and classify the critical points for the following one-dimensional differential equations.

(a) $\dot{x} = x$;

(b) $\dot{x} = -x$;

(c) $\dot{x} = x^2 - 1$.

Solutions.

(a) There is one critical point at $x_0 = 0$. If $x < 0$, then $\dot{x} < 0$, and if $x > 0$, then $\dot{x} > 0$. Therefore, x_0 is an unstable critical point. Solutions starting either side of x_0 are repelled away from it.

(b) There is one critical point at $x_0 = 0$. If $x < 0$, then $\dot{x} > 0$, and if $x > 0$, then $\dot{x} < 0$. Solutions starting either side of x_0 are attracted toward it. The critical point is stable.

(c) There are two critical points, one at $x_1 = -1$ and the other at $x_2 = 1$. If $x > 1$, then $\dot{x} > 0$; if $-1 < x < 1$, then $\dot{x} < 0$; and if $x < -1$, then $\dot{x} > 0$. Therefore, solutions starting near to x_1 but not on it are attracted toward this point, and x_1 is a stable critical point. Solutions starting near x_2 but not on it move away from this point, and x_2 is an unstable critical point.

By linearizing near a critical point, one can obtain a quantitative measure of stability as demonstrated below. Consider one-dimensional systems here; higher-dimensional systems will be investigated in other chapters.

Linear Stability Analysis
Let x^* be a critical point of $\dot{x} = f(x)$, $x \in \Re$. Consider a *small perturbation*, say, $\xi(t)$, away from the critical point at x^* to give $x(t) = x^* + \xi(t)$. A simple analysis is now applied to determine whether the perturbation grows or decays as time evolves. Now

$$\dot{\xi} = \dot{x} = f(x) = f(x^* + \xi),$$

and after a Taylor series expansion,

$$\dot{\xi} = f(x^*) + \xi f'(x^*) + \frac{\xi^2}{2} f''(x^*) + \cdots .$$

In order to apply a linear stability analysis, the nonlinear terms are ignored. Hence,

$$\dot{\xi} = \xi f'(x^*),$$

since $f(x^*) = 0$. Therefore, the perturbation $\xi(t)$ grows exponentially if $f'(x^*) > 0$ and decays exponentially if $f'(x^*) < 0$. If $f'(x^*) = 0$, then higher-order derivatives must be considered to determine the stability of the critical point.

A linear stability analysis is used extensively throughout the realms of non-linear dynamics and will appear in other chapters of this book.

Example 15. Use a linear stability analysis to determine the stability of the critical points for the following differential equations:

(a) $\dot{x} = \sin(x)$;

(b) $\dot{x} = x^2$;

(c) $\dot{x} = e^{-x} - 1$.

Solutions.

(a) There are critical points at $x_n = n\pi$, where n is an integer. When n is even, $f'(x_n) = 1 > 0$, and these critical points are unstable. When n is odd, $f'(x_n) = -1 < 0$, and these critical points are stable.

(b) There is one critical point at $x_0 = 0$ and $f'(x) = 2x$ in this case. Now $f'(0) = 0$ and $f''(0) = 2 > 0$. Therefore, x_0 is attracting when $x < 0$ and repelling when $x > 0$. The critical point is called *semistable*.

(c) There is one critical point at $x_0 = 0$. Now $f'(0) = -1 < 0$, and therefore, the critical point at the origin is stable.

The theory of autonomous systems of ODEs in two dimensions will be discussed in the next chapter.

2.5 Mathematica Commands in Text Format

For more information on solving differential equations using Mathematica, the reader should look for **DSolve** and **NDSolve** under the Help tab.

```
(* See Example 1: Solving simple ODEs. *)
In[1]:=DSolve[x'[t]==-t/x[t],x[t],t]
```

```
(* See Example 8 and Fig. 2.6: Series solutions for van der Pol. *)
In[2]:=DEqn=x''[t]+2*(x[t]^2-1)*x'[t]+x[t];
In[3]:=Ser=Series[DEqn,{t,0,6}];
In[4]:=Serone=Ser/.{x[0]->5,x'[0]->0};
In[5]:=Eqns=LogicalExpand[Serone==0];
In[6]:=Coeffs=Solve[Eqns];
In[7]:=Serx=Series[x[t],{t,0,5}];
In[8]:=ApproxSol=Normal[Serx]/.{x[0]->5,x'[0]->0}/.Coeffs[[1]]
In[9]:=Plot[ApproxSol,{t,0,0.08}]
```

```
(* See Example 10 and Fig. 2.7: Chemical kinetics. *)
In[10]:=k=.00713;a0=4;b0=1;sol=NDSolve[{c'[t]==k (a0-c[t]^2)(b0-c[t]/2),
c[0]==0},c,{t,0,700}]
In[11]:=Plot[Evaluate[c[t]/.sol],{t,0,700}]
```

```
(* See Example 12: A second-order ODE. *)
In[12]:=DSolve[{i''[t]+5i'[t]+6i[t]==10Sin[t],i[0]==0,i'[0]==0},i[t],t]
```

```
(* Solving an IVP. *)
In[13]:=DSolve[{x'[t]==2 Sqrt[x[t]],x[t0]==x0},x[t],t]
In[14]:=Simplify[%]
```

```
(* See Exercise 7: Solving a system of 3 ODEs *)
In[15]:=DSolve[{x'[t]==-a x[t],y'[t]==a x[t]-b y[t],z'[t]==b y[t],x[0]==M,
y[0]==0,z[0]==0},{x[t],y[t],z[t]},t];
In[16]:=Simplify[%]
```

2.6 Exercises

1. Sketch some solution curves for the following differential equations

   (a) $\frac{dy}{dx} = -\frac{y}{x}$;

   (b) $\frac{dy}{dx} = \frac{2y}{x}$;

   (c) $\frac{dy}{dx} = \frac{y}{2x}$;

   (d) $\frac{dy}{dx} = \frac{y^2}{x}$;

   (e) $\frac{dy}{dx} = -\frac{xy}{x^2+y^2}$;

   (f) $\frac{dy}{dx} = \frac{y}{x^2}$.

2. Fossils are often dated using the differential equation

$$\frac{dA}{dt} = -\alpha A,$$

   where A is the amount of radioactive substance remaining, α is a constant, and t is measured in years. Assuming that $\alpha = 1.5 \times 10^{-7}$, determine the age of a fossil containing radioactive substance A if only 30% of the substance remains.

3. Write down the chemical reaction rate equations for the reversible reaction equations

   (a) $A + B + C \rightleftharpoons D$,

   (b) $A + A + A \rightleftharpoons A_3$,

   given that the forward rate constant is k_f and the reverse rate constant is k_r, in each case. Assume that the chemical equations are the rate-determining steps.

4. (a) Consider a series resistor-inductor circuit with $L = 2\,H$, $R = 10\,\Omega$ and an applied EMF of $E = 100\sin(t)$. Use an integrating factor to solve the differential equation, and find the current in the circuit after 0.2 seconds given that $I(0) = 0$.

   (b) The differential equation used to model a series resistor-capacitor circuit is given by

$$R\frac{dQ}{dt} + \frac{Q}{C} = E,$$

   where Q is the charge across the capacitor. If a variable resistance $R = 1/(5+t)\,\Omega$ and a capacitance $C = 0.5\,F$ are connected in series with an applied EMF, $E = 100\,V$, find the charge on the capacitor given that $Q(0) = 0$.

5. (a) A forensic scientist is called to the scene of a murder. The temperature of the corpse is found to be $75°F$, and one hour later, the temperature has dropped to $70°F$. If the temperature of the room in which the body was discovered is a constant $68°F$, how long before the first temperature reading was taken did the murder occur? Assume that the body obeys Newton's Law of Cooling,

$$\frac{dT}{dt} = \beta(T - T_R),$$

where T is the temperature of the corpse, β is a constant, and T_R is room temperature.

(b) The differential equation used to model the concentration of glucose in the blood, say, $g(t)$, when it is being fed intravenously into the body, is given by

$$\frac{dg}{dt} + kg = \frac{G}{100V},$$

where k is a constant, G is the rate at which glucose is admitted, and V is the volume of blood in the body. Solve the differential equation and discuss the results.

(c) Single fiber muscle can be modeled using simple differential equations [11]. Download our preprint paper on "Hysteresis in muscle" from ResearchGate, and use Mathematica to reproduce the results of the Hill model given in that paper.

6. Show that the series solution of the Airy equation

$$\frac{d^2x}{dt^2} - tx = 0,$$

where $x(0) = a_0$ and $\dot{x}(0) = a_1$, used in physics to model the diffraction of light, is given by

$$x(t) = a_0 \left(1 + \sum_{1}^{\infty} \left(\frac{t^{3k}}{(2.3)(5.6)\cdots((3k-1)(3k))}\right)\right)$$

$$+ a_1 \left(t + \sum_{1}^{\infty} \left(\frac{t^{3k+1}}{(3.4)(6.7)\cdots((3k)(3k+1))}\right)\right).$$

7. A chemical substance A changes into substance B at a rate α times the amount of A present. Substance B changes into C at a rate β times the amount of B present. If initially only substance A is present and its amount is M, show that the amount of C present at time t is

$$M + M \left(\frac{\beta e^{-\alpha t} - \alpha e^{-\beta t}}{\alpha - \beta} \right).$$

8. Two tanks A and B, each of volume V, are filled with water at time $t = 0$. For $t > 0$, volume v of solution containing mass m of solute flows into tank A per second; mixture flows from tank A to tank B at the same rate; and mixture flows away from tank B at the same rate. The differential equations used to model this system are given by

$$\frac{d\sigma_A}{dt} + \frac{v}{V}\sigma_A = \frac{m}{V}, \quad \frac{d\sigma_B}{dt} + \frac{v}{V}\sigma_B = \frac{v}{V}\sigma_A,$$

where $\sigma_{A,B}$ are the concentrations of solute in tanks A and B, respectively. Show that the mass of solute in tank B is given by

$$\frac{mV}{v}\left(1 - e^{-vt/V}\right) - mte^{-vt/V}.$$

9. In an epidemic the rate at which healthy people become infected is a times their number, the rates of recovery and death are, respectively, b and c times the number of infected people. If initially there are N healthy people and no sick people, find the number of deaths up to time t. Is this a realistic model? What other factors should be taken into account?

10. (a) Determine the maximal interval of existence for each of the following initial value problems:

   (i) $\dot{x} = x^4, x(0) = 1$;

   (ii) $\dot{x} = \frac{x^2-1}{2}, x(0) = 2$;

   (iii) $\dot{x} = x(x-2), x(0) = 3$.

   (b) For what values of t_0 and x_0 does the initial value problem

$$\dot{x} = 2\sqrt{x}, \ x(t_0) = x_0,$$

   have a unique solution?

Bibliography

[1] M.L. Abell and J.P. Braselton, *Differential Equations with Mathematica*, 4th ed., Academic Press, Cambridge MA, 2016.

[2] R. Bronson and G. Costa, *Schaum's Outline of Differential Equations*, 3rd ed., McGraw-Hill, New York, 2006.

[3] L.O. Chua, Memristor-missing circuit element, *IEEE Transactions on Circuit Theory* **CT18**, (1971), 507–519.

[4] L.O. Chua and S.M. Kang, Memristive devices and systems, *Proceedings of the IEEE* **64**, (1976), 209–223.

[5] L.O. Chua, Memristor, Hodgkin-Huxley and the edge of chaos, *Nanotechnology* **24**, (2013).

[6] P. Dechaumphai, *Calculus and Differential Equations with Mathematica*, Alpha Science International Limited, Oxford, UK, 2016.

[7] S.G. Krantz, *Differential Equations Demystified*, McGraw-Hill, New York, 2004.

[8] Linn E, Siemon A, Waser R, Menzel S. Applicability of well-established memristive models for simulations of resistive switching devices. Circuits and Systems I: Regular Papers. IEEE Transactions on 2014; 61: 2402–2410.

[9] T. Prodromakis, C. Toumazou, and L.O. Chua, Two centuries of memristors, *Nature Materials* **11**, (2012), 478–481.

[10] B. Rai and D.P. Choudhury, *Elementary Ordinary Differential Equations*, Alpha Science Intl. Ltd., Oxford, UK, 2005.

[11] J. Ramos, S. Lynch, D.A. Jones and H. Degens, Hysteresis in muscle, *Int. J. of Bifurcation and Chaos* **27**, (2017), 1730003.

[12] M.R. Spiegel, *Schaum's Outline of Laplace Transforms*, McGraw-Hill, New York, 1965.

[13] D.B. Strukov, G.S. Snider, D.R. Stewart, et al., The missing memristor found, *Nature* **453**, (2008), 80–83.

[14] G. Turrell, *Mathematics for Chemistry and Physics*, Academic Press, New York, 2001.

[15] G. Zill, *A First Course in Differential Equations with Modeling Applications*, 10th ed., Cengage Learning, Belmont, CA, 2012.

Chapter 3
Planar Systems

Aims and Objectives

- To introduce the theory of planar autonomous linear differential equations.

- To extend the theory of linear systems to that of nonlinear systems.

On completion of this chapter, the reader should be able to

- find and classify critical points in the plane;

- carry out simple linear transformations;

- construct phase plane diagrams using isoclines, vector fields, and eigenvectors;

- apply the theory to simple modeling problems.

Basic analytical methods for solving two-dimensional linear autonomous differential equations are reviewed, and simple phase portraits are constructed in the plane.

The method of linearization is introduced, and both hyperbolic and non-hyperbolic critical points are defined. Phase portraits are constructed using *Hartman's theorem*. The linearization technique used here is based on a linear stability analysis. There are many textbooks on planar systems; for example, see [1]–[9].

© Springer International Publishing AG 2017
S. Lynch, *Dynamical Systems with Applications using Mathematica®*,
DOI 10.1007/978-3-319-61485-4_3

3.1 Canonical Forms

Consider linear two-dimensional autonomous systems of the form

$$\frac{dx}{dt} = \dot{x} = a_{11}x + a_{12}y, \quad \frac{dy}{dt} = \dot{y} = a_{21}x + a_{22}y, \quad (3.1)$$

where the a_{ij} are constants. The system is linear as the terms in $x, y, \dot{x}$, and $\dot{y}$ are all linear. System (3.1) can be written in the equivalent matrix form as

$$\dot{\mathbf{x}} = A\mathbf{x}, \quad (3.2)$$

where $\mathbf{x} \in \Re^2$ and

$$A = \begin{pmatrix} a_{11} & a_{12} \\ a_{21} & a_{22} \end{pmatrix}.$$

Definition 1. Every solution of (3.1) and (3.2), say, $\phi(t) = (x(t), y(t))$, can be represented as a curve in the plane. The solution curves are called *trajectories* or *orbits*.

The existence and uniqueness theorem guarantees that trajectories do not cross. Note that there is an infinite number of trajectories that would fill the plane if they were all plotted. However, the *qualitative behavior* can be determined by plotting just a few of the trajectories given the appropriate number of initial conditions.

Definition 2. The *phase portrait* is a two-dimensional figure showing how the qualitative behavior of system (3.1) is determined as x and y vary with t.

With the appropriate number of trajectories plotted, it should be possible to determine where any trajectory will end up from any given initial condition.

Definition 3. The *direction field* or *vector field* gives the *gradients* $\frac{dy}{dx}$ and *direction vectors* of the trajectories in the phase plane.

The slope of the trajectories can be determined using the chain rule,

$$\frac{dy}{dx} = \frac{\dot{y}}{\dot{x}},$$

and the direction of the vector field is given by $\dot{x}$ and $\dot{y}$ at each point in the xy plane.

Definition 4. The contour lines for which $\frac{dy}{dx}$ is a constant are called *isoclines*.

Isoclines may be used to help with the construction of the phase portrait. For example, the isoclines for which $\dot{x} = 0$ and $\dot{y} = 0$ are used to determine where the trajectories have vertical and horizontal tangent lines, respectively. If $\dot{x} = 0$, then there is no motion horizontally, and trajectories are either stationary or move vertically. A similar argument is used when $\dot{y} = 0$.

Using linear algebra, the phase portrait of any linear system of the form (3.2) can be transformed to a so-called *canonical form* $\dot{\mathbf{y}} = J\mathbf{y}$ by applying a transformation $\mathbf{x} = P\mathbf{y}$, where P is to be determined, and $J = P^{-1}AP$ is of one of the following forms:

$$J_1 = \begin{pmatrix} \lambda_1 & 0 \\ 0 & \lambda_2 \end{pmatrix}, \quad J_2 = \begin{pmatrix} \alpha & \beta \\ -\beta & \alpha \end{pmatrix},$$
$$J_3 = \begin{pmatrix} \lambda_1 & 0 \\ 0 & \lambda_1 \end{pmatrix}, \quad J_4 = \begin{pmatrix} \lambda_1 & \mu \\ 0 & \lambda_1 \end{pmatrix},$$

where $\lambda_{1,2}$, α, β, and μ are real constants. Matrix J_1 has two real distinct eigenvalues, matrix J_2 has complex eigenvalues, and matrices J_3 and J_4 have repeated eigenvalues. The qualitative type of phase portrait is determined from each of these canonical forms.

Nonsimple Canonical Systems

The linear system (3.2) is *nonsimple* if the matrix A is singular (i.e., $\det(A) = 0$, and at least one of the eigenvalues is zero). The system then has critical points other than the origin.

Example 1. Sketch a phase portrait of the system $\dot{x} = x$, $\dot{y} = 0$.

Solution. The critical points are found by solving the equations $\dot{x} = \dot{y} = 0$, which has the solution $x = 0$. Thus, there is an infinite number of critical points lying along the y-axis. The direction field has gradient given by

$$\frac{dy}{dx} = \frac{\dot{y}}{\dot{x}} = \frac{0}{x} = 0$$

for $x \neq 0$. This implies that the direction field is horizontal for points not on the y-axis. The direction vectors may be determined from the equation $\dot{x} = x$ since if $x > 0$, then $\dot{x} > 0$, and the trajectories move from left to right; if $x < 0$, then $\dot{x} < 0$, and trajectories move from right to left. A phase portrait is plotted in Figure 3.1.

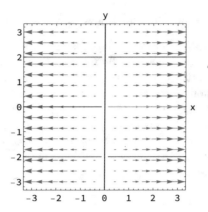

Figure 3.1: Six trajectories and a vector field plot for Example 1. Note that there is an infinite number of critical points lying on the y-axis. The vector field is plotted using the VectorPlot command (see Section 3.6).

Simple Canonical Systems

System (3.2) is *simple* if $\det(A) \neq 0$, and the the origin is then the only critical point. The critical points may be classified depending upon the type of eigenvalues.

3.1.I Real Distinct Eigenvalues. Suppose that system (3.2) can be diagonalized to obtain

$$\dot{x} = \lambda_1 x, \quad \dot{y} = \lambda_2 y.$$

The solutions to this system are $x(t) = C_1 e^{\lambda_1 t}$ and $y(t) = C_2 e^{\lambda_2 t}$, where C_1 and C_2 are constants. The solution curves may be found by solving the differential equation given by

$$\frac{dy}{dx} = \frac{\dot{y}}{\dot{x}} = \frac{\lambda_2 y}{\lambda_1 x},$$

which is integrable. The solution curves are given by $|y|^{\lambda_1} = K|x|^{\lambda_2}$. The type of phase portrait depends on the values of λ_1 and λ_2, as summarized below:

- If the eigenvalues are distinct, real, and positive, then the critical point is called an *unstable node*.

- If the eigenvalues are distinct, real, and negative, then the critical point is called a *stable node*.

- If one eigenvalue is positive and the other negative, then the critical point is called a *saddle point* or *col*.

Possible phase portraits for these canonical systems along with vector fields superimposed are shown in Figure 3.2.

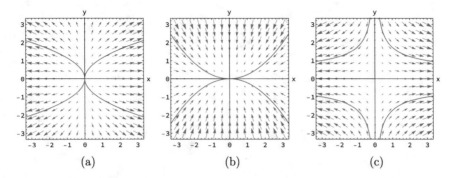

Figure 3.2:   Possible phase portraits for canonical systems with two real distinct eigenvalues: (a) unstable node; (b) stable node; and (c) saddle point or col.

3.1.II Complex Eigenvalues ($\lambda = \alpha \pm i\beta$). Consider a canonical system of the form

$$\dot{x} = \alpha x + \beta y, \quad \dot{y} = -\beta x + \alpha y. \tag{3.3}$$

Convert to *polar coordinates* by making the transformations $x = r\cos\theta$ and $y = r\sin\theta$. Then, elementary calculus gives

$$r\dot{r} = x\dot{x} + y\dot{y}, \quad r^2\dot{\theta} = x\dot{y} - y\dot{x}.$$

System (3.3) becomes
$$\dot{r} = \alpha r, \quad \dot{\theta} = -\beta.$$

The type of phase portrait depends on the values of α and β:

- If $\alpha > 0$, then the critical point is called an unstable focus.

- If $\alpha = 0$, then the critical point is called a center.

- If $\alpha < 0$, then the critical point is called a stable focus.

- If $\dot{\theta} > 0$, then the trajectories spiral counterclockwise around the origin.

- If $\dot{\theta} < 0$, then the trajectories spiral clockwise around the origin.

Phase portraits of the canonical systems with the vector fields superimposed are shown in Figure 3.3.

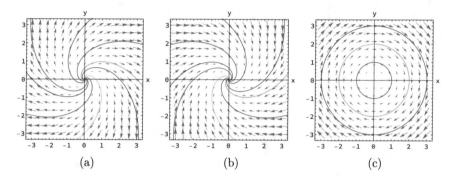

Figure 3.3: Possible phase portraits for canonical systems with complex eigenvalues: (a) unstable focus; (b) stable focus; and (c) center.

3.1.III Repeated Real Eigenvalues. Suppose that the canonical matrices are of the form J_3 or J_4. The type of phase portrait is determined by the following:

- If there are two linearly independent eigenvectors, then the critical point is called a *singular node*.

- If there is one linearly independent eigenvector, then the critical point is called a *degenerate node*.

Possible phase portraits with vector fields superimposed are shown in Figure 3.4.

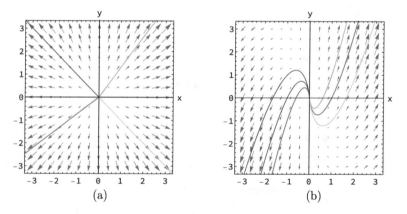

Figure 3.4: Possible phase portraits for canonical systems with repeated eigenvalues: (a) an unstable singular node and (b) an unstable degenerate node.

The classifications given in this section may be summarized using the trace and determinant of the matrix A as defined in system (3.2). If the eigenvalues are $\lambda_{1,2}$, then the characteristic equation is given by $(\lambda - \lambda_1)(\lambda - \lambda_2) = \lambda^2 -$

$(\lambda_1 + \lambda_2)\lambda + \lambda_1\lambda_2 = \lambda^2 - \text{trace}(A)\lambda + \det(A) = 0$. Therefore,

$$\lambda_{1,2} = \frac{\text{trace}(A) \pm \sqrt{(\text{trace}(A))^2 - 4\det(A)}}{2}.$$

The summary is depicted in Figure 3.5.

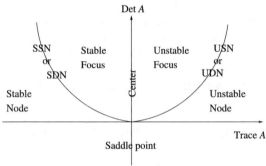

SSN - Stable Singular Node
SDN - Stable Degenerate Node
USN - Unstable Singular Node
UDN - Unstable Degenerate Node

Figure 3.5: Classification of critical points for system (3.2). The parabola has equation $T^2 - 4D = 0$, where $D = \det(A)$ and $T = \text{trace}(A)$.

3.2 Eigenvectors Defining Stable and Unstable Manifolds

Consider Figure 3.5. Apart from the region $T^2 - 4D > 0$, where the trajectories spiral, the phase portraits of the canonical forms of (3.2) all contain straight line trajectories that remain on the coordinate axes forever and exhibit exponential growth or decay along it. These special trajectories are determined by the eigenvectors of the matrix A and are called the *manifolds*. If the trajectories move toward the critical point at the origin as $t \to \infty$ along the axis, then there is exponential decay, and the axis is called a *stable manifold*. If trajectories move away from the critical point as $t \to \infty$, then the axis is called an *unstable manifold*.

In the general case, the manifolds do not lie along the axes. Suppose that a trajectory is of the form

$$\mathbf{x}(t) = \exp(\lambda t)\mathbf{e},$$

where $\mathbf{e} \neq 0$ is a vector and λ is a constant. This trajectory satisfies equation (3.2) since it is a solution curve. Therefore, substituting into (3.2),

$$\lambda \exp(\lambda t)\mathbf{e} = \exp(\lambda t)A\mathbf{e}$$

or

$$\lambda \mathbf{e} = A\mathbf{e}.$$

From elementary linear algebra, if there exists a nonzero column vector $\mathbf{e}$ satisfying this equation, then λ is called an eigenvalue of A, and $\mathbf{e}$ is called an eigenvector of A corresponding to the eigenvalue λ. If λ is negative, then the corresponding eigenvector gives the direction of the stable manifold, and if λ is positive, then the eigenvector gives the direction of the unstable manifold.

When $\lambda_1 \neq \lambda_2$, it is known from elementary linear algebra that the eigenvectors $\mathbf{e_1}$ and $\mathbf{e_2}$, corresponding to the eigenvalues λ_1 and λ_2, are linearly independent. Therefore, the general solution to the differential equations given by (3.1) is given by

$$\mathbf{x}(t) = C_1 \exp(\lambda_1 t)\mathbf{e_1} + C_2 \exp(\lambda_2 t)\mathbf{e_2},$$

where $C_1, and\, C_2$ are constants. In fact, for any given initial condition, this solution is unique by the existence and uniqueness theorem.

Definition 5. Suppose that $\mathbf{0} \in \Re^2$ is a critical point of the linear system (3.2). Then, the stable and unstable manifolds of the critical point $\mathbf{0}$ are denoted by $E_S(\mathbf{0})$ and $E_U(\mathbf{0})$, respectively and are determined by the eigenvectors of the critical point at $\mathbf{0}$.

Consider the following two simple illustrations.

Example 2. Determine the stable and unstable manifolds for the linear system

$$\dot{x} = 2x + y, \quad \dot{y} = x + 2y.$$

Solution. The system can be written as $\dot{\mathbf{x}} = A\mathbf{x}$, where

$$A = \begin{pmatrix} 2 & 1 \\ 1 & 2 \end{pmatrix}.$$

The characteristic equation for matrix A is given by $\det(A - \lambda I) = 0$, or in this case,

$$\begin{vmatrix} 2 - \lambda & 1 \\ 1 & 2 - \lambda \end{vmatrix} = 0.$$

Therefore, the characteristic equation is $\lambda^2 - 4\lambda + 3 = 0$, which has roots $\lambda_1 = 1$ and $\lambda_2 = 3$. Since both eigenvalues are real and positive, the critical point at the origin is an unstable node. The manifolds are determined from the eigenvectors ' corresponding to these eigenvalues. The eigenvector for λ_1 is $\mathbf{e_1} = (1, -1)^T$, and the eigenvector for λ_2 is $\mathbf{e_2} = (1, 1)^T$, where T represents the transpose matrix. The manifolds are shown in Figure 3.6.

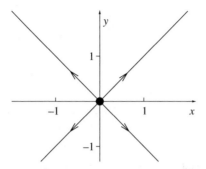

Figure 3.6: The two unstable manifolds, defined by the eigenvectors $\mathbf{e_1}$ and $\mathbf{e_2}$, for Example 2.

For the sake of completeness, the general solution in this case is given by

$$\mathbf{x}(t) = C_1 \exp(t)(1, -1)^T + C_2 \exp(3t)(1, 1)^T.$$

Example 3. Determine the stable and unstable manifolds for the linear system

$$\dot{\mathbf{x}} = \begin{pmatrix} -3 & 4 \\ -2 & 3 \end{pmatrix} \mathbf{x}.$$

Solution. The characteristic equation for matrix A is given by

$$\begin{vmatrix} -3 - \lambda & 4 \\ -2 & 3 - \lambda \end{vmatrix} = 0.$$

Therefore, the characteristic equation is $\lambda^2 - 1 = 0$, which has roots $\lambda_1 = 1$ and $\lambda_2 = -1$. Since one eigenvalue is real and positive and the other is real and negative, the critical point at the origin is a saddle point. The manifolds are derived from the eigenvectors corresponding to these eigenvalues. The eigenvector for λ_1 is $(1, 1)^T$, and the eigenvector for λ_2 is $(2, 1)^T$. The manifolds are shown in Figure 3.7.

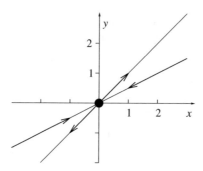

Figure 3.7:   The stable and unstable manifolds for Example 3. The trajectories lying on the stable manifold tend to the origin as $t \to \infty$ but never reach it.

For the sake of completeness, the general solution in this case is given by

$$\mathbf{x}(t) = C_1 \exp(t)(1,1)^T + C_2 \exp(-t)(2,1)^T.$$

Notation. The stable and unstable manifolds of linear systems will be denoted by E_S and E_U, respectively. Center manifolds (where the eigenvalues have zero real part) will be discussed in Chapter 8.

3.3 Phase Portraits of Linear Systems in the Plane

Definition 6. Two systems of first-order autonomous differential equations are said to be *qualitatively* (or *topologically*) *equivalent* if there exists an invertible mapping that maps one phase portrait onto the other while preserving the orientation of the trajectories.

Phase portraits can be constructed using isoclines, vector fields, and eigenvectors (for real eigenvalues).

Example 4. Consider the system

$$\dot{\mathbf{x}} = \begin{pmatrix} 2 & 1 \\ 1 & 2 \end{pmatrix} \mathbf{x}.$$

Find (a) the eigenvalues and corresponding eigenvectors of A; (b) a nonsingular matrix P such that $J = P^{-1}AP$ is diagonal; (c) new coordinates (u,v) such that substituting $x = x(u,v), y = y(u,v)$, converts the linear dynamical system

$$\dot{x} = 2x + y, \quad \dot{y} = x + 2y, \quad \text{into} \quad \dot{u} = \lambda_1 u, \quad \dot{v} = \lambda_2 v$$

for suitable λ_1, λ_2; and (d) sketch phase portraits for these qualitatively equivalent systems.

Solutions The origin is a unique critical point.

(a) From Example 2, the eigenvalues and corresponding eigenvectors are given by $\lambda_1 = 1$, $(1, -1)^T$ and $\lambda_2 = 3$, $(1, 1)^T$; the critical point is an unstable node.

(b) Using elementary linear algebra, the columns of matrix P are these eigenvectors and so

$$P = \begin{pmatrix} 1 & 1 \\ -1 & 1 \end{pmatrix},$$

and

$$J = P^{-1}AP = \begin{pmatrix} 1 & 0 \\ 0 & 3 \end{pmatrix}.$$

(c) Take the linear transformation $\mathbf{x} = P\mathbf{u}$ to obtain the system $\dot{u} = u$, $\dot{v} = 3v$.

(d) Consider the isoclines. In the xy plane, the flow is horizontal on the line where $\dot{y} = 0$ and hence on the line $y = -x/2$. On this line, $\dot{x} = 3x/2$; thus $\dot{x} > 0$ if $x > 0$ and $\dot{x} < 0$ if $x < 0$. The flow is vertical on the line $y = -2x$. On this line, $\dot{y} < 0$ if $x > 0$ and $\dot{y} > 0$ if $x < 0$.

Vector fields: The directions of the vector fields can be determined from $\dot{x}$ and $\dot{y}$ at points (x, y) in the plane.

Consider the slope of the trajectories. If $x + 2y > 0$ and $2x + y > 0$, then $\frac{dy}{dx} > 0$; if $x + 2y < 0$ and $2x + y > 0$, then $\frac{dy}{dx} < 0$; if $x + 2y > 0$ and $2x + y < 0$, then $\frac{dy}{dx} < 0$; and if $x + 2y < 0$ and $2x + y < 0$, then $\frac{dy}{dx} > 0$.

Manifolds: From the eigenvectors, both manifolds are unstable. One passes through $(0, 0)$ and $(1, 1)$ and the other through $(0, 0)$ and $(1, -1)$.

Putting all of this information together gives the phase portrait in Figure 3.8(a). The canonical phase portrait is shown in Figure 3.8(b).

Example 5. Sketch a phase portrait for the system

$$\dot{x} = -x - y, \quad \dot{y} = x - y.$$

Solution. The origin is the only critical point. The characteristic equation is given by

$$|A - \lambda I| = \lambda^2 + 2\lambda + 2 = 0,$$

which has complex solutions $\lambda_{1,2} = -1 \pm i$. The critical point at the origin is a stable focus.

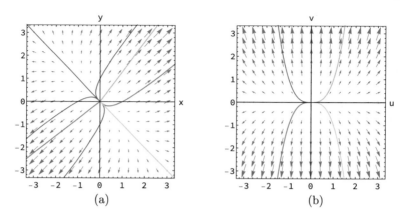

Figure 3.8: [Mathematica] Qualitatively equivalent phase portraits for Example 4.

Consider the isoclines. In the xy plane, the flow is horizontal on the line where $\dot{y} = 0$ and hence on the line $y = x$. On this line, $\dot{x} = -2x$; thus $\dot{x} < 0$ if $x > 0$ and $\dot{x} > 0$ if $x < 0$. The flow is vertical on the line where $\dot{x} = 0$ and hence on the line $y = -x$. On this line, $\dot{y} < 0$ if $x > 0$, and $\dot{y} > 0$ if $x < 0$.

Vector fields: The directions of the vector fields can be determined from $\dot{x}$ and $\dot{y}$ at points (x, y) in the plane.

Consider the slope of the trajectories. If $y > x$ and $y > -x$, then $\frac{dy}{dx} > 0$; if $y > x$ and $y < -x$, then $\frac{dy}{dx} < 0$; if $y < x$ and $y > -x$, then $\frac{dy}{dx} < 0$; and if $y < x$ and $y < -x$, then $\frac{dy}{dx} > 0$.

Manifolds: The eigenvectors are complex, and there are no real manifolds.

Converting to polar coordinates gives $\dot{r} = -r, \dot{\theta} = 1$. Putting all of this information together gives the phase portrait in Figure 3.9.

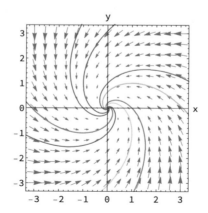

Figure 3.9: Some trajectories for Example 5. The critical point is a stable focus.

Example 6. Sketch a phase portrait for the system

$$\dot{x} = -2x, \quad \dot{y} = -4x - 2y.$$

Solution. The origin is the only critical point. The characteristic equation is given by

$$|A - \lambda I| = \lambda^2 - 4\lambda + 4 = 0,$$

which has repeated roots $\lambda_{1,2} = -2$.

Consider the isoclines. In the xy plane, the flow is horizontal on the line where $\dot{y} = 0$ and hence on the line $y = -2x$. Trajectories which start on the y-axis remain there forever.

Vector fields: The directions of the vector fields can be determined from $\dot{x}$ and $\dot{y}$ at points (x, y) in the plane.

Consider the slope of the trajectories. The slopes are given by $\frac{dy}{dx}$ at each point (x, y) in the plane.

Manifolds: There is one linearly independent eigenvector, $(0, 1)^T$. Therefore, the critical point is a stable degenerate node. The stable manifold E_S is the y-axis.

Putting all of this together gives the phase portrait in Figure 3.10.

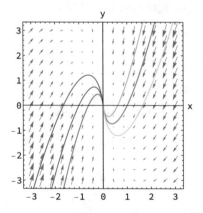

Figure 3.10: Some trajectories for Example 6. The critical point is a stable degenerate node.

Phase portraits of nonlinear planar autonomous systems will be considered in the following sections, where stable and unstable manifolds do not necessarily lie on straight lines. However, all is not lost as the manifolds for certain critical points are tangent to the eigenvectors of the linearized system at that point.

Manifolds in three-dimensional systems will be discussed in Chapter 14.

3.4 Linearization and Hartman's Theorem

Suppose that the nonlinear autonomous system

$$\dot{x} = P(x,y), \quad \dot{y} = Q(x,y) \tag{3.4}$$

has a critical point at (u,v), where P and Q are at least quadratic in x and y. Take a linear transformation which moves the critical point to the origin. Let $X = x - u$ and $Y = y - v$. Then system (3.4) becomes

$$\dot{X} = P(X+u, Y+v) = P(u,v) + X\frac{\partial P}{\partial x}\Big|_{x=u,y=v} + Y\frac{\partial P}{\partial y}\Big|_{x=u,y=v} + R(X,Y)$$

$$\dot{Y} = Q(X+u, Y+v) = Q(u,v) + X\frac{\partial Q}{\partial x}\Big|_{x=u,y=v} + Y\frac{\partial Q}{\partial y}\Big|_{x=u,y=v} + S(X,Y)$$

after a Taylor series expansion. The nonlinear terms R and S satisfy the conditions $\frac{R}{r} \to 0$ and $\frac{S}{r} \to 0$ as $r = \sqrt{X^2 + Y^2} \to 0$. The functions R and S are said to be "big Oh of r^2," or in mathematical notation, $R = \mathrm{O}(r^2)$ and $S = \mathrm{O}(r^2)$. Discard the nonlinear terms in the system, and note that $P(u,v) = Q(u,v) = 0$ since (u,v) is a critical point of system (3.4). The *linearized* system is then of the form

$$\dot{X} = X\frac{\partial P}{\partial x}\Big|_{x=u,y=v} + Y\frac{\partial P}{\partial y}\Big|_{x=u,y=v}$$

$$\dot{Y} = X\frac{\partial Q}{\partial x}\Big|_{x=u,y=v} + Y\frac{\partial Q}{\partial y}\Big|_{x=u,y=v} \tag{3.5}$$

and the Jacobian matrix is given by

$$J(u,v) = \begin{pmatrix} \frac{\partial P}{\partial x} & \frac{\partial P}{\partial y} \\ \frac{\partial Q}{\partial x} & \frac{\partial Q}{\partial y} \end{pmatrix}\Bigg|_{x=u,y=v}.$$

Definition 7. A critical point is called *hyperbolic* if the real part of the eigenvalues of the Jacobian matrix $J(u,v)$ is nonzero. If the real part of either of the eigenvalues of the Jacobian are equal to zero, then the critical point is called *nonhyperbolic*.

Hartman's Theorem. *Suppose that (u,v) is a hyperbolic critical point of system (3.4). Then there is a neighborhood of this critical point on which the phase portrait for the nonlinear system resembles that of the linearized system (3.5). In other words, there is a curvilinear continuous change of coordinates* *taking one phase portrait to the other, and in a small region around the critical point, the portraits are qualitatively equivalent.*

A proof of this theorem may be found in Hartman's book [2]. Note that the stable and unstable manifolds of the nonlinear system will be tangent to the manifolds of the linearized system near the relevant critical point. These trajectories diverge as one moves away from the critical point; this is illustrated in Examples 7 and 8.

Notation. Stable and unstable manifolds of a nonlinear system are labeled W_S and W_U, respectively.

Hartman's theorem implies that W_S and W_U are tangent to E_S and E_U at the relevant critical point. If any of the critical points are nonhyperbolic, then other methods must be used to sketch a phase portrait, and numerical solvers may be required.

3.5 Constructing Phase Plane Diagrams

The method for plotting phase portraits for nonlinear planar systems having hyperbolic critical points may be broken down into three distinct steps:

- Locate all of the critical points.

- Linearize and classify each critical point according to Hartman's theorem.

- Determine the isoclines, and use $\frac{dy}{dx}$ to obtain slopes of trajectories.

The method can be illustrated with some simple examples. Examples 10–12 illustrate possible approaches when a critical point is not hyperbolic.

Example 7. Sketch a phase portrait for the nonlinear system

$$\dot{x} = x, \quad \dot{y} = x^2 + y^2 - 1.$$

Solution. Locate the critical points by solving the equations $\dot{x} = \dot{y} = 0$. Hence, $\dot{x} = 0$ if $x = 0$ and $\dot{y} = 0$ if $x^2 + y^2 = 1$. If $x = 0$, then $\dot{y} = 0$ if $y^2 = 1$, which has solutions $y = 1$ and $y = -1$. Therefore, there are two critical points: $(0, 1)$ and $(0, -1)$.

Linearize by finding the Jacobian matrix; hence

$$J = \begin{pmatrix} \frac{\partial P}{\partial x} & \frac{\partial P}{\partial y} \\ \frac{\partial Q}{\partial x} & \frac{\partial Q}{\partial y} \end{pmatrix} = \begin{pmatrix} 1 & 0 \\ 2x & 2y \end{pmatrix}.$$

Linearize at each critical point; hence

$$J_{(0,1)} = \begin{pmatrix} 1 & 0 \\ 0 & 2 \end{pmatrix}.$$

The matrix is in diagonal form. There are two distinct positive eigenvalues, and hence, the critical point is an unstable node.

For the other critical point,

$$J_{(0,-1)} = \begin{pmatrix} 1 & 0 \\ 0 & -2 \end{pmatrix}.$$

There is one positive and one negative eigenvalue, and so this critical point is a saddle point or col.

Note that the matrices $J_{(0,1)}$ and $J_{(0,-1)}$ are in diagonal form. The eigenvectors for both critical points are $(1,0)^T$ and $(0,1)^T$. Thus, in a small neighborhood around each critical point, the stable and unstable manifolds are tangent to the lines generated by the eigenvectors through each critical point. Therefore, near each critical point, the manifolds are horizontal and vertical. Note that the manifolds of the nonlinear system W_S and W_U need not be straight lines but are tangent to E_S and E_U at the relevant critical point.

Consider the isoclines. Now $\dot{x} = 0$ on $x = 0$, and on this line, $\dot{y} = y^2 - 1$. Thus, if $|y| < 1$, then $\dot{y} < 0$, and if $|y| > 1$, then $\dot{y} > 0$. Also, $\dot{y} = 0$ on the circle $x^2 + y^2 = 1$, and on this curve, $\dot{x} = x$. Thus, if $x > 0$, then $\dot{x} > 0$, and if $x < 0$, then $\dot{x} < 0$. The slope of the trajectories is given by

$$\frac{dy}{dx} = \frac{x^2 + y^2 - 1}{x}.$$

Putting all of this information together gives a phase portrait as depicted in Figure 3.11.

Example 8. Sketch a phase portrait for the nonlinear system

$$\dot{x} = y, \quad \dot{y} = x(1 - x^2) + y.$$

Solution. Locate the critical points by solving the equations $\dot{x} = \dot{y} = 0$. Hence, $\dot{x} = 0$ if $y = 0$, and $\dot{y} = 0$ if $x(1 - x^2) + y = 0$. If $y = 0$, then $\dot{y} = 0$ if $x(1 - x^2) = 0$, which has solutions $x = 0$, $x = 1$, and $x = -1$. Therefore, there are three critical points, $(0,0)$, $(1,0)$, and $(-1,0)$.

Linearize by finding the Jacobian matrix; hence

$$J = \begin{pmatrix} \frac{\partial P}{\partial x} & \frac{\partial P}{\partial y} \\ \frac{\partial Q}{\partial x} & \frac{\partial Q}{\partial y} \end{pmatrix} = \begin{pmatrix} 0 & 1 \\ 1 - 3x^2 & 1 \end{pmatrix}.$$

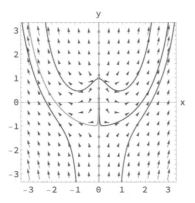

Figure 3.11: A phase portrait for Example 7. The stable and unstable manifolds (W_S, W_U) are tangent to horizontal or vertical lines (E_S, E_U) in a small neighborhood of each critical point.

Linearize at each critical point; hence

$$J_{(0,0)} = \begin{pmatrix} 0 & 1 \\ 1 & 1 \end{pmatrix}.$$

The eigenvalues are

$$\lambda_1 = \frac{1 + \sqrt{5}}{2} \quad \text{and} \quad \lambda_2 = \frac{1 - \sqrt{5}}{2}.$$

The corresponding eigenvectors are $(1 \ \lambda_1)^T$ and $(1 \ \lambda_2)^T$. Thus, the critical point at the origin is a saddle point or col.

For the other critical points,

$$J_{(1,0)} = J_{(-1,0)} = \begin{pmatrix} 0 & 1 \\ -2 & 1 \end{pmatrix}.$$

The eigenvalues are

$$\lambda = \frac{1 \pm i\sqrt{7}}{2},$$

and so both critical points are unstable foci.

Consider the isoclines. Now $\dot{x} = 0$ on $y = 0$, and on this line, $\dot{y} = x(1 - x^2)$. Thus, if $0 < x < 1$, then $\dot{y} > 0$; if $x > 1$, then $\dot{y} < 0$; if $-1 < x < 0$, then $\dot{y} < 0$, and if $x < -1$, then $\dot{y} > 0$. Also, $\dot{y} = 0$ on the curve $y = x - x^3$, and on this curve, $\dot{x} = y$. Thus, if $y > 0$, then $\dot{x} > 0$, and if $y < 0$, then $\dot{x} < 0$. The slope of the trajectories is given by

$$\frac{dy}{dx} = \frac{x - x^3 + y}{y}.$$

Note that on $x = 0$ and $x = \pm 1$, $\frac{dy}{dx} = 1$. Putting all of this information together gives a phase portrait as depicted in Figure 3.12.

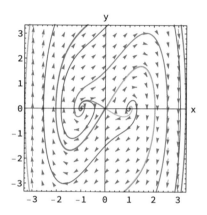

Figure 3.12: [Mathematica] A phase portrait for Example 8. Note that, in a small neighborhood of the origin, the unstable manifold (W_U) is tangent to the line E_U given by $y = \lambda_1 x$, and the stable manifold (W_S) is tangent to the line E_S given by $y = \lambda_2 x$.

Example 9. Plot a phase portrait for the system

$$\dot{x} = x\left(1 - \frac{x}{2} - y\right), \quad \dot{y} = y\left(x - 1 - \frac{y}{2}\right).$$

Solution. Locate the critical points by solving the equations $\dot{x} = \dot{y} = 0$. Hence, $\dot{x} = 0$ if either $x = 0$ or $y = 1 - \frac{x}{2}$. Suppose that $x = 0$. Then, $\dot{y} = 0$ if $y\left(-1 - \frac{y}{2}\right) = 0$, which has solutions $y = 0$ or $y = -2$. Suppose that $y = 1 - \frac{x}{2}$. Then, $\dot{y} = 0$ if either $1 - \frac{x}{2} = 0$ or $1 - \frac{x}{2} = 2x - 2$, which has solutions $x = 2$ or $x = \frac{6}{5}$. Thus, there are four critical points at $(0,0)$, $(2,0)$, $(0,-2)$, and $\left(\frac{6}{5}, \frac{2}{5}\right)$. Notice that $\dot{x} = 0$ when $x = 0$, which means that the flow is vertical on the y-axis. Similarly, $\dot{y} = 0$ when $y = 0$, and the flow is horizontal along the x-axis. In this case, the axes are *invariant*.

Linearize by finding the Jacobian matrix; hence

$$J = \begin{pmatrix} \frac{\partial P}{\partial x} & \frac{\partial P}{\partial y} \\ \frac{\partial Q}{\partial x} & \frac{\partial Q}{\partial y} \end{pmatrix} = \begin{pmatrix} 1 - x - y & -x \\ y & x - 1 - y \end{pmatrix}.$$

Linearize around each of the critical points, and apply Hartman's theorem. Consider the critical point at (0,0). The eigenvalues are $\lambda = \pm 1$, and the critical point is a saddle point or col. Next, consider the critical point at (2,0); now the eigenvalues are $\lambda_1 = 1$ and $\lambda_2 = -1$. The corresponding eigenvectors are $(-1,1)^T$ and $(1,0)^T$, respectively. This critical point is also a saddle point or col. Consider the critical point at $(0,-2)$. Now the eigenvalues are $\lambda_1 = 3$ and $\lambda_2 = 1$; the corresponding eigenvectors are $(1,-1)^T$ and $(0,1)^T$, respectively. The critical point at $(0,-2)$ is therefore an unstable node. Finally, consider the critical point at $\left(\frac{6}{5},\frac{2}{5}\right)$. The eigenvalues in this case are

$$\lambda = \frac{-2 \pm i\sqrt{11}}{5}$$

and the critical point is a stable focus. There is no need to find the eigenvectors; they are complex in this case.

Consider the isoclines. Now $\dot{x} = 0$ on $x = 0$ or on $y = 1 - \frac{x}{2}$, and $\dot{y} = 0$ on $y = 0$ or on $y = 2x - 2$. The directions of the flow can be found by considering $\dot{y}$ and $\dot{x}$ on these curves.

The slope of the trajectories is given by

$$\frac{dy}{dx} = \frac{y\left(x - 1 - \frac{y}{2}\right)}{x\left(1 - \frac{x}{2} - y\right)}.$$

A phase portrait, indicating the stable and unstable manifolds of the critical points, is shown in Figure 3.13.

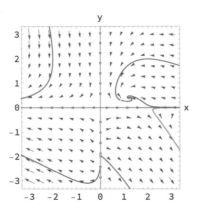

Figure 3.13: A phase portrait for Example 9. The axes are invariant.

Example 10. Sketch a phase portrait for the nonlinear system

$$\dot{x} = y^2, \quad \dot{y} = x.$$

Solution. Locate the critical points by solving the equations $\dot{x} = \dot{y} = 0$. Therefore, $\dot{x} = 0$ if $y = 0$, and $\dot{y} = 0$ if $x = 0$. Thus, the origin is the only critical point.

Attempt to linearize by finding the Jacobian matrix; hence

$$J = \begin{pmatrix} \frac{\partial P}{\partial x} & \frac{\partial P}{\partial y} \\ \frac{\partial Q}{\partial x} & \frac{\partial Q}{\partial y} \end{pmatrix} = \begin{pmatrix} 0 & 2y \\ 1 & 0 \end{pmatrix}.$$

Linearize at the origin to obtain

$$J_{(0,0)} = \begin{pmatrix} 0 & 0 \\ 1 & 0 \end{pmatrix}.$$

The origin is a nonhyperbolic critical point. To sketch a phase portrait, solve the differential equation

$$\frac{dy}{dx} = \frac{\dot{y}}{\dot{x}} = \frac{x}{y^2},$$

using the method of separation of variables highlighted in the previous chapter.

Consider the isoclines. Now $\dot{x} = 0$ on $y = 0$, and on this line $\dot{y} = x$. Thus if $x > 0$, then $\dot{y} > 0$, and if $x < 0$, then $\dot{y} < 0$. Also, $\dot{y} = 0$ on $x = 0$, and on this line $\dot{x} = y^2$. Thus $\dot{x} > 0$ for all y. The slope of the trajectories is given by $\frac{dy}{dx} = \frac{x}{y^2}$. Putting all of this information together gives a phase portrait as depicted in Figure 3.14.

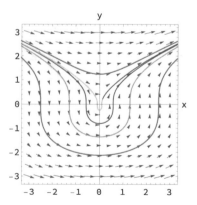

Figure 3.14:   A phase portrait for Example 10 that has a nonhyperbolic critical point at the origin. There is a *cusp* at the origin.

Example 11. A simple model for the spread of an *epidemic* in a city is given by

$$\dot{S} = -\tau SI, \quad \dot{I} = \tau SI - rI,$$

where $S(t)$ and $I(t)$ represent the numbers of susceptible and infected individuals scaled by 1000, respectively; τ is a constant measuring how quickly the disease is transmitted; r measures the rate of recovery (assume that those who recover become immune); and t is measured in days. Determine a value for S at which the infected population is maximum.

Given that $\tau = 0.003$ and $r = 0.5$, sketch a phase portrait showing three trajectories whose initial points are at $(1000, 1)$, $(700, 1)$, and $(500, 1)$. Give a physical interpretation in each case.

Solution. The maximum number of infected individuals occurs when $\frac{dI}{dS} = 0$. Now

$$\frac{dI}{dS} = \frac{\dot{I}}{\dot{S}} = \frac{\tau S - r}{-\tau S}.$$

Therefore, $\frac{dI}{dS} = 0$ when $S = \frac{r}{\tau}$. The number $\frac{r}{\tau}$ is called the *threshold value*.

The critical points for this system are found by solving the equations $\dot{S} = \dot{I} = 0$. Therefore, there is an infinite number of critical points lying along the horizontal axis. A phase portrait is plotted in Figure 3.15.

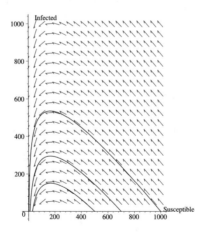

Figure 3.15: A phase portrait showing three trajectories for Example 11. The axes are scaled by 10^3 in each case. Trajectories are only plotted in the first quadrant since populations cannot be negative.

In each case, the population of susceptibles decreases to a constant value, and the population of infected individuals increases and then decreases to zero.

Note that in each case, the maximum number of infected individuals occurs at $S = \frac{r}{\tau} \approx 167,000$.

Example 12. Chemical kinetics involving the derivation of one differential equation were introduced in Chapter 8. This example will consider a system of two differential equations. Consider the isothermal chemical reaction

$$A + B \rightleftharpoons C,$$

in which one molecule of A combines with one molecule of B to form one molecule of C. In the reverse reaction, one molecule of C returns to $A + B$. Suppose that the rate of the forward reaction is k_f and the rate of the backward reaction is k_r. Let the concentrations of A, B, and C be a, b, and c, respectively. Assume that the concentration of A is much larger than the concentrations of B and C and can therefore be thought of as constant. From the law of mass action, the equations for the kinetics of b and c are

$$\dot{b} = k_r c - k_f ab, \quad \dot{c} = k_f ab - k_r c.$$

Find the critical points, and sketch a typical trajectory for this system. Interpret the results in physical terms.

Solution. The critical points are found by determining where $\dot{b} = \dot{c} = 0$. Clearly, there is infinite number of critical points along the line $c = \frac{k_f a}{k_r} b$. The slope of the trajectories is given by

$$\frac{dc}{db} = \frac{\dot{c}}{\dot{b}} = -1.$$

If $c < \frac{k_f a}{k_r} b$, then $\dot{b} < 0$ and $\dot{c} > 0$. Similarly, if $c > \frac{k_f a}{k_r} b$, then $\dot{b} > 0$ and $\dot{c} < 0$. Two typical solution curves are plotted in Figure 3.16.

Thus the final concentrations of B and C depend upon the initial concentrations of these chemicals. Two trajectories starting from the initial points at $(b_0, 0)$ and (b_0, c_0) are plotted in Figure 3.16. Note that the chemical reaction obeys the law of *conservation of mass*; this explains why the trajectories lie along the lines $b + c = $ constant.

Example 13. Suppose that H is a population of healthy rabbits and I is the subpopulation of infected rabbits that never recover once infected, both

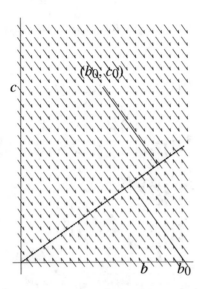

Figure 3.16:   Two solution curves for the chemical kinetic equation in Example 12, where a is assumed to be constant. The dotted line represents the critical points lying on the line $c = \frac{k_f a}{k_r} b$.

measured in millions. The following differential equations can be used to model the dynamics of the system:

$$\dot{H} = (b - d)H - \delta I, \quad \dot{I} = \tau I(H - I) - (\delta + d)I,$$

where b is the birth rate, d is the natural death rate, δ is the rate of death of the diseased rabbits, and τ is the rate at which the disease is transmitted.

Given that $b = 4$, $d = 1$, $\delta = 6$, and $\tau = 1$, and given an initial population of $(H_0, I_0) = (2, 2)$, plot a phase portrait, and explain what happens to the rabbits in real-world terms.

Solution. There are two critical points in the first quadrant at $0 = (0, 0)$ and $P = (14, 7)$. The Jacobian matrix is given by

$$J = \begin{pmatrix} (b - d) & -\delta \\ \tau I & \tau H - 2\tau I - (\delta + b) \end{pmatrix}.$$

The critical point at the origin is a col with eigenvalues and corresponding eigenvectors given by $\lambda_1 = 3$, $(1, 0)^T$ and $\lambda_2 = -7$, $(3, 5)^T$. The critical point at $P = (14, 7)$ has eigenvalues $\lambda = -2 \pm i\sqrt{17}$ and is therefore a stable focus. A phase portrait is plotted in Figure 3.17. Either the population of rabbits stabilizes to the values at P, or they become extinct, depending on the initial

populations. For example, plot a solution curve for the trajectory starting at
$(H_0, I_0) = (2, 2)$.

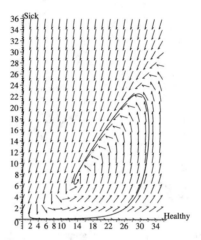

Figure 3.17:   A trajectory starting from the initial point $(2, 2)$. The population
stabilizes to 14 million healthy rabbits and 7 million infected rabbits.

Models of interacting species will be considered in Chapter 4.

3.6 Mathematica Commands

```
(* Linear system, phase portrait with vector plot. See Figure 3.8(a). *)
a=2;b=1;c=1;d=2;p1=VectorPlot[{a*x+b*y,c*x+d*y},{x,-3,3},{y,-3,3}] ;
ode1[x0_,y0_]:=NDSolve[{x'[t]==a*x[t]+b*y[t],y'[t]==c*x[t]+d*y[t],
x[0]==x0,y[0]==y0},{x[t],y[t]},{t,-50,10}];
sol[1]=ode1[-1,0];sol[2]=ode1[1,1];sol[3]=ode1[0,-1];sol[4]=ode1[1,0];
sol[5]=ode1[0,1];sol[6]=ode1[-1,-1];sol[7]=ode1[-1,1];sol[8]=ode1[1,-1];
p2=ParametricPlot[Evaluate[Table[{x[t],y[t]}/.sol[i],{i,8}]],{t,-50,10},
PlotStyle->{Thick}, PlotRange->{{-3,3},{-3,3}},PlotPoints->100,
AxesLabel->{"x","y"}];
Show[{p1,p2},PlotRange->{{-3,3},{-3,3}},AxesLabel->{"x","y"},Axes->True,
BaseStyle->{FontSize->18}]
```

```
(* Nonlinear system, phase portrait with vector plot. See Figure 3.12. *)
p1=VectorPlot[{y,x (1-x^2)+y},{x,-3,3},{y,-3,3},VectorScale->{Small,0.8}] ;
ode1[x0_,y0_]:=NDSolve[{x'[t]==y[t],y'[t]==x[t] (1-(x[t])^2)+y[t],
```

```
x[0]==x0,y[0]==y0},{x[t],y[t]},{t,-10,10},MaxSteps->Infinity];
sol[1]=ode1[0.1,0.16] ;sol[2]=ode1[0.1,-0.06];sol[3]=ode1[-0.1,-0.16];
sol[4]=ode1[-0.1,0.06];sol[5]=ode1[1.3,0] ;sol[6]=ode1[-1.3,0];
sol[7]=ode1[-1,-1];
p2=ParametricPlot[Evaluate[Table[{x[t],y[t]}/.sol[i],{i,7}]],{t,-10,10},
PlotStyle->{Thick},PlotRange->{{-3,3},{-3,3}},PlotPoints->100,
AxesLabel->{"x","y"}];
Show[{p1,p2},PlotRange->{{-3,3},{-3,3}},AxesLabel->{"x","y"},
Axes->True,BaseStyle->{FontSize->18}]
```

3.7 Exercises

1. (a) Find the eigenvalues and eigenvectors of the matrix

$$B = \begin{pmatrix} -7 & 6 \\ 2 & -6 \end{pmatrix}.$$

   Sketch a phase portrait for the system $\dot{\mathbf{x}} = B\mathbf{x}$ and its corresponding canonical form.

   (b) Carry out the same procedures as in part (a) for the system

$$\dot{x} = -4x - 8y, \quad \dot{y} = -2y.$$

2. Sketch phase portraits for the following linear systems:

   (a) $\dot{x} = 0, \quad \dot{y} = x + 2y$;

   (b) $\dot{x} = x + 2y, \quad \dot{y} = 0$;

   (c) $\dot{x} = 3x + 4y, \quad \dot{y} = 4x - 3y$;

   (d) $\dot{x} = 3x + y, \quad \dot{y} = -x + 3y$;

   (e) $\dot{x} = y, \quad \dot{y} = -x - 2y$;

   (f) $\dot{x} = x - y, \quad \dot{y} = y - x$.

3. A very simple mechanical oscillator can be modeled using the second-order differential equation

$$\frac{d^2x}{dt^2} + \mu \frac{dx}{dt} + 25x = 0,$$

   where x measures displacement from equilibrium.

   (a) Rewrite this equation as a linear first-order system by setting $\dot{x} = y$.

(b) Sketch phase portraits when (i) $\mu = -8$, (ii) $\mu = 0$, (iii) $\mu = 8$, and (iii) $\mu = 26$.

(c) Describe the dynamical behavior in each case given that $x(0) = 1$ and $\dot{x}(0) = 0$.

Plot the corresponding solutions in the tx plane.

4. Plot phase portraits for the following systems:

(a) $\dot{x} = y$, $\dot{y} = x - y + x^3$;

(b) $\dot{x} = -2x - y + 2$, $\dot{y} = xy$;

(c) $\dot{x} = x^2 - y^2$, $\dot{y} = xy - 1$;

(d) $\dot{x} = 2 - x - y^2$, $\dot{y} = -y(x^2 + y^2 - 3x + 1)$;

(e) $\dot{x} = y^2$, $\dot{y} = x^2$;

(f) $\dot{x} = x^2$, $\dot{y} = y^2$;

(g) $\dot{x} = y$, $\dot{y} = x^3$;

(h) $\dot{x} = x$, $\dot{y} = \mu - y^2$, for $\mu < 0$, $\mu = 0$, and $\mu > 0$.

5. Construct a nonlinear system that has four critical points: two saddle points, one stable focus, and one unstable focus.

6. A nonlinear capacitor-resistor electrical circuit can be modeled using the differential equations

$$\dot{x} = y, \quad \dot{y} = -x + x^3 - (a_0 + x)y,$$

where a_0 is a nonzero constant and $x(t)$ represents the current in the circuit at time t. Sketch phase portraits when $a_0 > 0$ and $a_0 < 0$, and give a physical interpretation of the results.

7. An age-dependent population can be modeled by the differential equations

$$\dot{p} = \beta + p(a - bp), \quad \dot{\beta} = \beta(c + (a - bp)),$$

where p is the population, β is the birth rate, and $a, b,$ and c are all positive constants. Find the critical points of this system, and determine the long-term solution.

8. The power, say, P, generated by a water wheel of velocity V can be modeled by the system

$$\dot{P} = -\alpha P + PV, \quad \dot{V} = 1 - \beta V - P^2,$$

where α and β are both positive. Describe the qualitative behavior of this system as α and β vary, and give physical interpretations of the results.

9. A very simple model for the economy is given by

$$\dot{I} = I - KS, \quad \dot{S} = I - CS - G_0,$$

where I represents income, S is the rate of spending, G_0 denotes constant government spending, and C and K are positive constants.

(a) Plot possible solution curves when $C = 1$, and interpret the solutions in economic terms. What happens when $C \neq 1$?

(b) Plot the solution curve when $K = 4, C = 2, G_0 = 4, I(0) = 15$, and $S(0) = 5$. What happens for other initial conditions?

10. Given that

$$\frac{d^3\eta}{d\tau^3} = -\eta\frac{d^2\eta}{d\tau^2}$$

and

$$x = \frac{\eta\frac{d\eta}{d\tau}}{\frac{d^2\eta}{d\tau^2}}, \quad y = \frac{\left(\frac{d\eta}{d\tau}\right)^2}{\eta\frac{d^2\eta}{d\tau^2}} \quad \text{and } t = log\left|\frac{d\eta}{d\tau}\right|,$$

prove that

$$\dot{x} = x(1 + x + y), \quad \dot{y} = y(2 + x - y).$$

Plot a phase portrait in the xy plane.

Bibliography

[1] F. Dumortier, J. Llibre, and J.C. Arts, *Qualitative Theory of Planar Differential Systems*, Springer-Verlag, New York, 2006.

[2] P. Hartman, *Ordinary Differential Equations*, John Wiley, New York, 1964.

[3] D.W. Jordan and P. Smith, *Nonlinear Ordinary Differential Equations* 4th ed., Oxford University Press, 2007.

[4] W. Kelley and A. Peterson, *The Theory of Differential Equations: Classical and Qualitative*, Prentice Hall, 2003.

[5] A.C. King, J. Billingham and S.R. Otto, *Differential Equations: Linear, Nonlinear, Ordinary, Partial*, Cambridge University Press, 2003.

[6] J.H. Liu, *A First Course in the Qualitative Theory of Differential Equations*, Prentice Hall, Upper Saddle River, NJ, 2002.

[7] J.C. Robinson, *An Introduction to Ordinary Differential Equations*, Cambridge University Press, 2004.

[8] D. Schaeffer and J.W. Cain, *Ordinary Differential Equations: Basics and Beyond (Texts in Applied Mathematics)*, Springer, New York, 2016.

[9] B. West, S. Strogatz, J.M. McDill, J. Cantwell, and H. Holn, *Interactive Differential Equations*, Version 2.0, Addison-Wesley, Reading, MA, 1997.

Chapter 4
Interacting Species

Aims and Objectives

- To apply the theory of planar systems to modeling interacting species.

On completion of this chapter the reader should be able to

- plot solution curves to modeling problems for planar systems;

- interpret the results in terms of species behavior.

The theory of planar ODEs is applied to the study of interacting species. The models are restricted in that only two species are considered and external factors such as pollution, environment, refuge, age classes, and other species interactions, for example, are ignored. However, even these restricted systems give useful results. These simple models can be applied to species living in our oceans and to both animal and insect populations on land. Note that the continuous differential equations used in this chapter are only relevant if the species populations under consideration are large, typically scaled by 10^4, 10^5, or 10^6 in applications.

A host-parasite system is presented subject to different types of predation by a predator species.

4.1 Competing Species

Suppose that there are two species in competition with one another in an environment where the common food supply is limited. For example, sea lions and penguins, red and gray squirrels, and ants and termites are all species which fall into this category. There are two particular types of outcome that are often observed in the real world. In the first case, there is *coexistence*, in which the

© Springer International Publishing AG 2017   73
S. Lynch, *Dynamical Systems with Applications using Mathematica®*,
DOI 10.1007/978-3-319-61485-4_4

two species live in harmony. (In nature, this is the most likely outcome; other-wise, one of the species would be extinct.) In the second case, there is *mutual exclusion*, in which one of the species becomes extinct. (For example, American gray squirrels imported into the UK are causing the extinction of the smaller native red squirrels.)

Both coexistence and mutual exclusion can be observed when plotting solution curves on a phase plane diagram. Consider the following general model for two competing species.

Example 1. Sketch possible phase plane diagrams for the following system:

$$\dot{x} = x(\beta - \delta x - \gamma y), \quad \dot{y} = y(b - dy - cx), \tag{4.1}$$

where $\beta, \delta, \gamma, a, b,$ and c are all positive constants with $x(t)$ and $y(t)$—both positive—representing the two species populations measured in tens or hundreds of thousands.

Solution. The terms appearing in the right-hand sides of equation (4.1) have a physical meaning as follows:

- The terms $\beta x - \delta x^2$ and $by - dy^2$ represent the usual logistic growth of one species (Verhulst's equation).

- Both species suffer as a result of competition over a limited food supply, hence the terms $-\gamma xy$ and $-cxy$ in $\dot{x}$ and $\dot{y}$.

Construct a phase plane diagram in the usual way. Find the critical points, linearize around each one, determine the isoclines, and plot the phase plane portrait.

Locate the critical points by solving the equations $\dot{x} = \dot{y} = 0$. There are four critical points at

$$O = (0,0), \quad P = \left(0, \frac{b}{d}\right), \quad Q = \left(\frac{\beta}{\delta}, 0\right), \quad \text{and} \quad R = \left(\frac{\gamma b - \beta d}{\gamma c - \delta d}, \frac{\beta c - \delta b}{\gamma c - \delta d}\right).$$

Suppose that $C_1 = \gamma c - \delta d$, $C_2 = \gamma b - \beta d$, and $C_3 = \beta c - \delta b$. For the critical point to lie in the first quadrant, one of the following conditions must hold: Either

(i) $C_1, C_2,$ and C_3 are all negative, or

(ii) $C_1, C_2,$ and C_3 are all positive.

Linearize by finding the Jacobian matrix. Therefore,

$$J = \begin{pmatrix} \beta - 2\delta x - \gamma y & -\gamma x \\ -cy & b - 2dy - cx \end{pmatrix}.$$

Linearize at each critical point. Thus

$$J_O = \begin{pmatrix} \beta & 0 \\ 0 & b \end{pmatrix}.$$

For the critical point at P,

$$J_P = \begin{pmatrix} \beta - \gamma b/d & 0 \\ -bc/d & -b \end{pmatrix}.$$

For the critical point at Q,

$$J_Q = \begin{pmatrix} -\beta & -\gamma\beta/\delta \\ 0 & b - \beta c/\delta \end{pmatrix}.$$

Finally, for the critical point at R,

$$J_R = \frac{1}{C_1} \begin{pmatrix} \delta C_2 & \gamma C_2 \\ cC_3 & dC_3 \end{pmatrix}.$$

Consider case (i) first. The fixed points are all simple, and it is not difficult to show that O is an unstable node, P and Q are cols, and for certain parameter values R is a stable fixed point. A phase portrait is plotted in Figure 4.1(a), where eight of an infinite number of solution curves are plotted. Each trajectory is plotted numerically for both positive and negative time steps; in this way, critical points are easily identified in the phase plane. For the parameter values chosen here, the two species coexist and the populations stabilize to constant values after long time periods. The arrows in the Figure 4.1(a) show the vector field plot and define the direction of the trajectories for system (4.1). The slope of each arrow is given by $\frac{dy}{dx}$ at the point, and the direction of the arrows is determined from $\dot{x}$ and $\dot{y}$. There is a stable node lying wholly in the first quadrant at R, and the nonzero populations $x(t)$ and $y(t)$ tend to this critical point with increasing time no matter what the initial populations are. The domain of stability for the critical point at R is therefore $S_R = \{(x, y) \in \Re^2 : x > 0, y > 0\}$. Now consider case (ii). The fixed points are all simple, and it is not difficult to show that O is an unstable node, P and Q are stable or improper nodes, and R is a col. A phase portrait is shown in Figure 4.1(b), where nine of an infinite number of solution curves are plotted. Once more the trajectories are plotted for both positive and negative time iterations. In this case, one of the species becomes extinct.

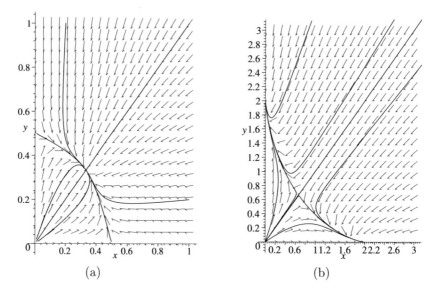

Figure 4.1:   (a) A possible phase portrait showing coexistence and (b) a possible phase portrait depicting mutual exclusion. Note that the axes are invariant in both cases.

In Figure 4.1(b), the critical point lying wholly in the first quadrant is a saddle point or col, which is unstable. The long-term behavior of the system is divided along the diagonal in the first quadrant. Trajectories starting to the right of the diagonal will tend to the critical point at $Q = (2,0)$, which implies that species y becomes extinct. Trajectories starting to the left of the diagonal will tend to the critical point at $P = (0,2)$, which means that species x will become extinct. Numerically, the trajectories lying on the stable manifold of the saddle point in the first quadrant will tend toward the critical point at R. However, in the real world, populations cannot remain exactly on the stable manifold, and trajectories will be diverted from this critical point leading to extinction of one of the species. The domain of stability for the critical point at $P = (0,2)$ is given by, $S_P = \{(x,y) \in \Re^2 : x > 0, y > 0, y > x\}$. The domain of stability for the critical point at $Q = (2,0)$ is given by, $S_Q = \{(x,y) \in \Re^2 : x > 0, y > 0, y < x\}$.

4.2 Predator-Prey Models

Consider a two-species predator-prey model in which one species preys on another. Examples in the natural world include sharks and fish, lynx and snow-shoe hares, and ladybirds and aphids. A very simple differential equation—first used by Volterra in 1926 [7], [10] and known as the *Lotka–Volterra model*—is given in Example 2.

Example 2. Sketch a phase portrait for the system

$$\dot{x} = x(\alpha - cy), \quad \dot{y} = y(\gamma x - \delta), \tag{4.2}$$

where α, c, γ, and δ are all positive constants, with $x(t)$ and $y(t)$ representing the scaled population of prey and predator, respectively, and t is measured in years.

Solution. The terms appearing in the right-hand sides of equation (4.2) have a physical meaning as follows:

- The term αx represents the growth of the population of prey in the absence of any predators. This is obviously a crude model; the population of a species cannot increase forever.

- The terms $-cxy$ and $+\gamma xy$ represent species interaction. The population of prey suffers and predators gain from the interaction.

- The term $-\delta y$ represents the extinction of predators in the absence of prey.

Attempt to construct a phase plane diagram in the usual way. Find the critical points, linearize around each one, determine the isoclines, and plot the phase plane portrait.

The critical points are found by solving the equations $\dot{x} = \dot{y} = 0$. There are two critical points, one at $O = (0,0)$ and the other at $P = \left(\frac{\delta}{\gamma}, \frac{\alpha}{c}\right)$.

Linearize to obtain

$$J = \begin{pmatrix} \alpha - cy & -cx \\ \gamma y & -\delta + \gamma x \end{pmatrix}.$$

The critical point at the origin is a saddle point, and the stable and unstable manifolds lie along the axes. The stable manifold lies on the positive y-axis, and the unstable manifold lies on the x-axis. The critical point at P is not hyperbolic, and so Hartman's Theorem cannot be applied. System (4.2) has solution curves (the differential equation is separable) given by $x^{\delta} y^{\alpha} e^{-\gamma x} e^{-cy} = K$, where K is a constant. These solution curves may be plotted in the phase plane. The isoclines are given by $x = 0$, $y = \frac{\alpha}{c}$, where the flow is vertical, and $y = 0$, $x = \frac{\delta}{\gamma}$, where the flow is horizontal. The vector fields are found by considering $\dot{x}$, $\dot{y}$, and $\frac{dy}{dx}$. A phase portrait is shown in Figure 4.2.

The population fluctuations can also be represented in the tx and ty planes. The graphs shown in Figure 4.3 show how the populations of predator and prey typically oscillate. Note that the oscillations are dependent on the initial conditions. In Figure 4.3, the period of both cycles is about 5 years. Different sets of initial conditions can give solutions with different amplitudes. For example, plot the solution curves in the tx and ty planes for the initial conditions $x(0) = 3$ and $y(0) = 5$.

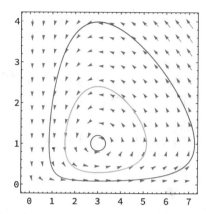

Figure 4.2: [Mathematica] A phase portrait for the Lotka–Volterra model.

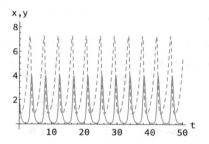

Figure 4.3: [Mathematica] Time series plots, periodic behavior of the prey and predators for one set of initial conditions, namely $x(0) = 1$, $y(0) = 2$. The population of prey is shown as the dashed curve and the population of predator is a solid curve.

How can this system be interpreted in terms of species behavior? Consider the trajectory passing through the point $(1, 1)$ in Figure 4.2. At this point, the ratio of predators to prey is relatively high; as a result, the population of predators drops. The ratio of predators to prey drops, and so the population of prey increases. Once there are lots of prey, the predator numbers will again start to increase. The resulting cyclic behavior is repeated over and over and is shown as the largest closed trajectory in Figure 4.2.

If small perturbations are introduced into system (4.2)—to model other factors, for example—then the qualitative behavior changes. The periodic cycles can be destroyed by adding small terms into the right-hand sides of system (4.2). The system is said to be *structurally unstable* (or *not robust*).

Many predator-prey interactions have been modeled in the natural world. For example, there are data dating back over 150 years for the populations of lynx and snowshoe hares from the Hudson Bay Company in Canada. The data clearly show that the populations periodically rise and fall (with a period of about 10 years) and that the maximum and minimum values (amplitudes) are relatively constant. This is not true for the Lotka–Volterra model (see Figure 4.2). Different initial conditions can give solutions with different amplitudes. In 1975, Holling and Tanner constructed a system of differential equations whose solutions have the same amplitudes in the long term, no matter what the initial populations. Two particular examples of the Holling-Tanner model for predator-prey interactions are given in Example 3.

The reader is encouraged to compare the terms (and their physical meaning) appearing in the right-hand sides of the differential equations in Examples 1–3.

Example 3. Consider the specific Holling-Tanner model

$$\dot{x} = x\left(1 - \frac{x}{7}\right) - \frac{6xy}{(7+7x)}, \quad \dot{y} = 0.2y\left(1 - \frac{Ny}{x}\right), \tag{4.3}$$

where N is a constant with $x(t) \neq 0$ and $y(t)$ representing the populations of prey and predators, respectively. Sketch phase portraits when (i) $N = 2.5$ and (ii) $N = 0.5$.

Solution. The terms appearing in the right-hand sides of equation (4.3) have a physical meaning as follows:

- The term $x\left(1 - \frac{x}{7}\right)$ represents the usual logistic growth in the absence of predators.

- The term $-\frac{6xy}{(7+7x)}$ represents the effect of predators subject to a maximum predation rate.

- The term $0.2y\left(1 - \frac{Ny}{x}\right)$ denotes the predator growth rate when a maximum of x/N predators is supported by x prey.

Construct a phase plane diagram in the usual way. Find the critical points, linearize around each one, determine the isoclines, and plot a phase plane portrait.

Consider case (i). The critical points are found by solving the equations $\dot{x} = \dot{y} = 0$. There are two critical points in the first quadrant, $A = (5,2)$ and $B = (7,0)$. The Jacobian matrices are given by

$$J_A = \begin{pmatrix} -1 & -3/4 \\ 0 & 1/5 \end{pmatrix}$$

and

$$J_B = \begin{pmatrix} -10/21 & -5/7 \\ 2/25 & -1/5 \end{pmatrix}.$$

The eigenvalues and eigenvectors of J_A are given by $\lambda_1 = -1$; $(1,0)^T$ and $\lambda_2 = 1/5$; $\left(-\frac{5}{8}, 1\right)^T$. Therefore, this critical point is a saddle point or col with the stable manifold lying along the x-axis and the unstable manifold tangent to the line with slope $-\frac{8}{5}$ in a small neighborhood around the critical point. The eigenvalues of J_B are given by $\lambda \approx -0.338 \pm 0.195i$. Therefore, the critical point at B is a stable focus.

A phase portrait showing four trajectories and the vector field is shown in Figure 4.4(a).

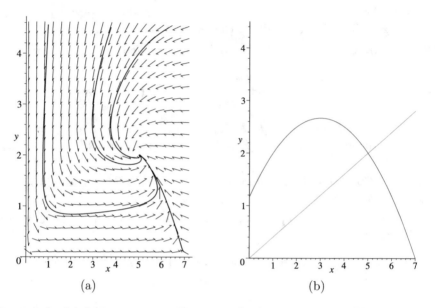

(a)   (b)

Figure 4.4:   (a) A phase portrait for system (4.3) when $N = 2.5$. (b) Intersection of the isoclines.

The populations eventually settle down to constant values. If there are any natural disasters or diseases, for example, the populations would both decrease but eventually return to the stable values. This is, of course, assuming that neither species becomes extinct. There is no periodic behavior in this model.

Consider case (ii). The critical points are found by solving the equations $\dot{x} = \dot{y} = 0$. There are two critical points in the first quadrant, $A = (1, 2)$ and $B = (7, 0)$. The Jacobian matrices are given by

$$J_A = \begin{pmatrix} -1 & -3/4 \\ 0 & 1/5 \end{pmatrix}$$

and

$$J_B = \begin{pmatrix} 2/7 & -3/7 \\ 2/5 & -1/5 \end{pmatrix}.$$

The eigenvalues and eigenvectors of J_A are given by $\lambda_1 = -1$; $(1,0)^T$ and $\lambda_2 = 1/5$; $(-\frac{5}{8},1)^T$. Therefore, this critical point is a saddle point or col with the stable manifold lying along the x-axis and the unstable manifold tangent to the line with slope $-\frac{8}{5}$ near to the critical point. The eigenvalues of J_B are given by $\lambda \approx 0.043 \pm 0.335i$. Therefore, the critical point at B is an unstable focus.

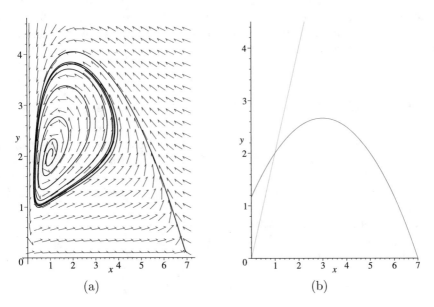

(a) $\qquad\qquad\qquad\qquad$ (b)

Figure 4.5: (a) A phase portrait for system (4.3) when $N = 0.5$. (b) Intersection of the isoclines.

All trajectories lying in the first quadrant are drawn to the closed periodic cycle shown in Figure 4.5(a). Therefore, no matter what the initial values of $x(t)$ and $y(t)$, the populations eventually rise and fall periodically. This isolated periodic trajectory is known as a *stable limit cycle*. In the long term, all trajectories in the first quadrant are drawn to this periodic cycle, and once there, remain there forever. Definitions and the theory of limit cycles will be introduced in Chapter 5. The isoclines are plotted in Figure 4.5(b), these curves show where the flow is horizontal or vertical, in this case.

The limit cycle persists if small terms are added to the right-hand sides of the differential equations in system (4.3). The system is *structurally stable* (or *robust*) since small perturbations do not affect the qualitative behavior. Again the populations of both predator and prey oscillate in a similar manner to the Lotka–Volterra model with another major exception. The final steady-state solution for the Holling-Tanner model is independent of the initial conditions. Use Mathematica to plot time series plots for the solutions plotted in Figure 4.5(a) as in the Mathematica program listed in Section 4.4. The period of the limit cycle can be easily established from the time series plot. This model appears to match very well with what happens for many predator-prey species in the natural world—for example, house sparrows and sparrow hawks in Europe, muskrat, and mink in Central North America, and white-tailed deer and wolf in Ontario.

From the time series plot, the period, say, T, of the limit cycle is approximately 19 units of time. Thus if t is measured in six-month intervals, then this would be a good model for the lynx and snowshoe hare populations, which have a natural period of about 10 years. Periodicity of limit cycles will be discussed in the next chapter.

4.3 Other Characteristics Affecting Interacting Species

A simple model of one species infected with a disease was considered in Chapter 3. The models considered thus far for interacting species have been limited to only two populations, and external factors have been ignored. Hall et al. [3] consider a stable host-parasite system subject to selective predation by a predator species. They consider a microparasite—zooplankton—fish system where the host is Daphnia dentifera and the predator fish species is bluegill sunfish. They investigate how predator selectivity on parasitized and nonparasitized hosts affects the populations. The differential equations are given by

$$\dot{S} = bS[1 - c(S + I)] - dS - \beta SI - f_S(S, I, P),$$
$$\dot{I} = \beta SI - (d + \alpha)I - f_I(S, I, P), \tag{4.4}$$

where S is the susceptible population, I is the infected population, b is the birth rate, c is the density dependence of birth rates, d is the mortality rate, β represents contact with infected hosts, and α is the parasite induced mortality rate. The functions f_S and f_I represent predator interaction with a saturating functional response, given by

$$f_S(S, I, P) = \frac{PS}{h_S + S + \theta \gamma I}, \quad f_I(S, I, P) = \frac{P\theta I}{h_S + S + \theta \gamma I},$$

where P is a predation intensity term, θ represents the selectivity of the predator, h_S represents a half-saturation constant of predators for susceptible hosts, and γ is a handling time for susceptible and infected hosts. More details and bifurcation diagrams are plotted in the research paper [3], and it is shown how predation selectivity can affect the host-parasite system. For example, for the parameter values $b = 0.4$, $c = \frac{1}{20}$, $\theta = 5$, $\alpha = \beta = d = 0.05$, $P = 1$, and $\gamma = h_S = 1$, it is shown that the host-parasite system coexists in a periodic manner as depicted in Figure 4.6. Mathematica command lines for producing time series data are listed in Section 4.4.

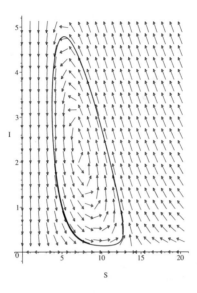

Figure 4.6: Coexistence of the host-parasite species when $P = 1$ and the productivity term, $\frac{1}{c} = 20$. There is a limit cycle in the SI plane.

Note that for other parameter values, predation can catalyze extinction of both hosts and parasites.

There are a great many research papers published every year on interacting species, and the author hopes that this chapter will inspire the reader to investigate further. To conclude Chapter 4, some other characteristics ignored here will be listed. Of course, the differential equations will become more complicated and are beyond the scope of this chapter.

- Age classes—for example, young, mature, and old; time lags need to be introduced into the differential equations (see Chapters 12 and 13).

- Diseases—epidemics affecting one or more species (see Chapter 3).

- Environmental effects.

- Enrichment of prey—this can lead to extinction of predators.

- Harvesting and culling policies (see Chapter 12).

- Pollution—persistence and extinction.

- Refuge—for example, animals in Africa find refuge in the bush.

- Seasonal effects—for example, some animals hibernate in winter.

- Three or more species interactions (see the exercises in Section 4.5).

One interesting example is discussed by Lenbury et al. [6], where predator-prey interaction is coupled to parasitic infection. One or both of the species can become infected, and this can lead to mathematical problems involving four systems of differential equations. The dynamics become far more complicated, and more interesting behavior is possible. Higher-dimensional systems will be discussed later in the book. Different types of species interaction are investigated in [8].

4.4 Mathematica Commands

See Section 3.6 for help with plotting phase portraits.

```
(* Lotka--Volterra model phase portrait. See Figure 4.2. *).
p1=VectorPlot[{x (1-y),y (x-3)},{x,.1,7},{y,.1,5},VectorScale->{Small,0.8}];
ode1[x0_,y0_]:=NDSolve[{x'[t]==x[t] (1-y[t]),y'[t]==y[t] (x[t]-3),
x[0]==x0,y[0]==y0},
{x[t],y[t]},{t,0,200}];sol[3]=ode1[3,1.2];sol[1]=ode1[1,2];sol[2]=ode1[2,2];
p2=ParametricPlot[Evaluate[Table[{x[t],y[t]}/.sol[i],{i,3}]],{t,0,200},
PlotStyle->{Red,Thick},PlotRange->{{0,7},{0,4}}, PlotPoints->100,
AxesLabel->{"x","y"}];
Show[{p1,p2},PlotRange->{{0,7},{0,4}},AxesLabel->{"x","y"},
Axes->True,BaseStyle->{FontSize->18}]

(* Plot time series. See Figure 4.3. *)
p3=Plot[Evaluate[Table[{x[t]}/.sol[i],{i,1}]],{t,0,50},PlotRange->All,
PlotPoints->100,
PlotStyle->Dashing[{.02}]];
p4=Plot[Evaluate[Table[{y[t]}/.sol[i],{i,1}]],{t,0,50},PlotRange->All,
PlotPoints->100];
Show[{p3,p4},PlotRange->{{0,50},{0,8}},AxesLabel->{"t","x,y"},Axes->True,
AxesOrigin->Automatic,BaseStyle->{FontSize->18}]
```

4.5 Exercises

1. Plot a phase portrait for the following competing species model

$$\dot{x} = 2x - x^2 - xy, \quad \dot{y} = 3y - y^2 - 2xy$$

and describe what happens in terms of species behavior.

2. Plot a phase plane diagram for the following predator-prey system and interpret the solutions in terms of species behavior:

$$\dot{x} = 2x - xy, \quad \dot{y} = -3y + xy.$$

3. Plot a phase portrait for the following system and describe what happens to the population for different initial conditions:

$$\dot{x} = 2x - x^2 - xy, \quad \dot{y} = -y - y^2 + xy.$$

4. The differential equations used to model a competing species are given by

$$\dot{x} = x(2 - x - y), \quad \dot{y} = y\left(\mu - y - \mu^2 x\right),$$

where μ is a constant. Describe the qualitative behavior of this system as the parameter μ varies.

5. (a) Sketch a phase portrait for the system

$$\dot{x} = x(4 - y - x), \quad \dot{y} = y(3x - 1 - y), \quad x \geq 0, \ y \geq 0,$$

given that the critical points occur at $O = (0,0)$, $A = (4,0)$, and $B = (5/4, 11/4)$.

(b) Sketch a phase portrait for the system

$$\dot{x} = x(2 - y - x), \quad \dot{y} = y(3 - 2x - y), \quad x \geq 0, \ y \geq 0,$$

given that the critical points occur at $O = (0,0)$, $C = (0,3)$, $D = (2,0)$, and $E = (1,1)$.

One of the systems can be used to model predator-prey interactions and the other competing species. Describe which system applies to which model and interpret the results in terms of species behavior.

6. A predator-prey system may be modeled using the differential equations

$$\dot{x} = x(1 - y - \epsilon x), \quad \dot{y} = y(-1 + x - \epsilon y),$$

where $x(t)$ is the population of prey and $y(t)$ is the predator population size at time t, respectively. Classify the critical points for $\epsilon \geq 0$ and plot phase portraits for the different types of qualitative behavior. Interpret the results in physical terms.

7. A predator-prey model is given by

$$\dot{x} = x(x - x^2 - y), \quad \dot{y} = y(x - 0.6).$$

Sketch a phase portrait and interpret the results in physical terms.

8. Use Mathematica to plot a trajectory for the predator-prey system

$$\dot{x} = x(x - x^2 - y), \quad \dot{y} = y(x - 0.48)$$

using the initial condition $(0.6, 0.1)$. What can you deduce about the long-term populations?

9. Suppose that there are three species of insect X, Y, and Z, say. Give rough sketches to illustrate the possible ways in which these species can interact with one another. You should include the possibility of a species being cannibalistic. Three-dimensional systems will be discussed later.

10. The following three differential equations are used to model a combined predator-prey and competing species system:

$$\dot{x} = x(a_{10} - a_{11}x + a_{12}y - a_{13}z),$$
$$\dot{y} = y(a_{20} - a_{21}x - a_{22}y - a_{23}z),$$
$$\dot{z} = z(a_{30} + a_{31}x - a_{32}y - a_{33}z),$$

where a_{ij} are positive constants. Give a physical interpretation for the terms appearing in the right-hand sides of these differential equations.

Bibliography

[1] F. Brauer and C. Castillo-Chavez, *Systems for Biological Modeling: An Introduction (Advances in Applied Mathematics)*, CRC Press, Florida, 2015.

[2] L. Edelstein-Keshet, *Mathematical Models in Biology (Classics in Applied Mathematics)*, SIAM, Philadelphia, 2005.

[3] S.R. Hall, M.A. Duffy and C.E. Cáceres, Selective predation and productivity jointly drive complex behavior in host-parasite systems, *The American Naturalist*, **165**(1) (2005), 70–81.

[4] A. Hastings, *Population Biology: Concepts and Models*, Springer-Verlag, New York, 2005.

[5] S.B. Hsu, T.W. Hwang, Hopf bifurcation analysis for a predator-prey system of Holling and Leslie type, *Taiwan J. Math.* **3** (1999), 35–53.

[6] Y. Lenbury, S. Rattanamongkonkul, N. Tumravsin, and S. Amornsamakul, Predator-prey interaction coupled by parasitic infection: limit cycles and chaotic behavior, *Math. Comput. Model.*, **30**-9/10 (1999), 131–146.

[7] A.J. Lotka, *Elements of Physical Biology*, William and Wilkins, Baltimore, 1925.

[8] F. Lutscher, T. Iljon, Competition, facilitation and the Allee effect, *OIKOS*, **122**(4) (2013), 621–631.

[9] H.R. Thieme, *Mathematics in Population Biology (Princeton Series in Theoretical and Computational Biology)*, Princeton University Press, Princeton, NJ, 2003.

[10] V. Volterra, Variazioni e fluttuazioni del numero d'individui in specie animalicanniventi, *Mem. R. Com. Tolassogr. Ital.*, **431** (1927), 1–142.

Chapter 5
Limit Cycles

Aims and Objectives

- To give a brief historical background.

- To define features of phase plane portraits.

- To introduce the theory of planar limit cycles.

- To introduce perturbation methods.

On completion of this chapter, the reader should be able to

- prove existence and uniqueness of a limit cycle;

- prove that certain systems have no limit cycles;

- interpret limit cycle behavior in physical terms;

- find approximate solutions for perturbed systems.

Limit cycles, or isolated periodic solutions, are the most common form of solution observed when modeling physical systems in the plane. Early investigations were concerned with mechanical and electronic systems, but periodic behavior is evident in all branches of science. Two limit cycles were plotted in Chapter 4 when considering the modeling of interacting species.

The chapter begins with a historical introduction, and then, the theory of planar limit cycles is introduced.

© Springer International Publishing AG 2017
S. Lynch, *Dynamical Systems with Applications using Mathematica®*,
DOI 10.1007/978-3-319-61485-4_5

5.1 Historical Background

Definition 1. A *limit cycle* is an isolated periodic solution.

Limit cycles in planar differential systems commonly occur when modeling both the technological and natural sciences. Most of the early history in the theory of limit cycles in the plane was stimulated by practical problems. For example, the differential equation derived by Rayleigh in 1877 [14], related to the oscillation of a violin string, is given by

$$\ddot{x} + \epsilon \left(\frac{1}{3} \dot{x}^2 - 1 \right) \dot{x} + x = 0,$$

where $\ddot{x} = \frac{d^2 x}{dt^2}$ and $\dot{x} = \frac{dx}{dt}$. Let $\dot{x} = y$. Then, this differential equation can be written as a system of first-order autonomous differential equations in the plane

$$\dot{x} = y, \quad \dot{y} = -x - \epsilon \left(\frac{y^2}{3} - 1 \right) y. \tag{5.1}$$

A phase portrait is shown in Figure 5.1.

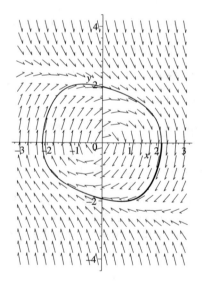

Figure 5.1:   Periodic behavior in the Rayleigh system (5.1) when $\epsilon = 1.0$.

Following the invention of the triode vacuum tube, which was able to produce stable self-excited oscillations of constant amplitude, van der Pol [17] obtained the following differential equation to describe this phenomenon

$$\ddot{x} + \epsilon \left(x^2 - 1\right) \dot{x} + x = 0,$$

which can be written as a planar system of the form

$$\dot{x} = y, \quad \dot{y} = -x - \epsilon \left(x^2 - 1\right) y. \tag{5.2}$$

A phase portrait is shown in Figure 5.2.

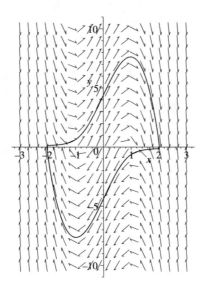

Figure 5.2: Periodic behavior for system (5.2) when $\epsilon = 5.0$.

The basic model of a cell membrane is that of a resistor and capacitor in parallel. The equations used to model the membrane are a variation of the van der Pol equation. The famous Fitzhugh–Nagumo oscillator [3], [8], [15], used to model the action potential of a neuron is a two-variable simplification of the Hodgkin–Huxley equations [5] (see Chapter 21). The Fitzhugh–Nagumo model creates quite accurate action potentials and models the qualitative behavior of the neurons. The differential equations are given by

$$\dot{u} = -u(u - \theta)(u - 1) - v + w, \quad \dot{v} = \epsilon(u - \gamma v),$$

where u is a voltage, v is the recovery of voltage, θ is a threshold, γ is a shunting variable, and ω is a constant voltage. For certain parameter values, the solution demonstrates a slow collection and fast release of voltage; this kind of behavior has been labeled integrate and fire. Note that, for biological systems, neurons cannot collect voltage immediately after firing and need to rest. Oscillatory

behavior for the Fitzhugh–Nagumo system is shown in Figure 5.3. Mathematica
command lines for producing Figure 5.3 are listed in Section 5.4.

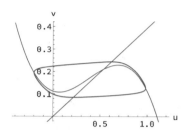

Figure 5.3: [Mathematica] A limit cycle for the Fitzhugh–Nagumo oscillator.
In this case, $\gamma = 2.54$, $\theta = 0.14$, $\omega = 0.112$ and $\epsilon = 0.01$. The blue curves are
the isoclines, where the trajectories cross horizontally and vertically.

Note that when $\omega = \omega(t)$ is a periodic external input, the system becomes
nonautonomous and can display chaotic behavior [15]. The reader can investigate these systems via the exercises in Chapter 9.

Perhaps, the most famous class of differential equations that generalize (5.2)
are those first investigated by Liénard in 1928 [6],

$$\ddot{x} + f(x)\dot{x} + g(x) = 0,$$

or in the phase plane

$$\dot{x} = y, \quad \dot{y} = -g(x) - f(x)y. \tag{5.3}$$

This system can be used to model mechanical systems, where $f(x)$ is known
as the *damping* term and $g(x)$ is called the *restoring force* or *stiffness*. Equation
(5.3) is also used to model resistor-inductor-capacitor circuits (see Chapter 2)
with nonlinear circuit elements. Limit cycles of Liénard systems will be discussed
in some detail in Chapters 10 and 11.

Possible physical interpretations for limit cycle behavior of certain dynamical
systems are listed below:

- For an economic model, Bella [2] considers a Goodwin model of a class
  struggle and demonstrates emerging multiple limit cycles of different orientation.

- For predator-prey and epidemic models, the populations oscillate in phase
  with one another and the systems are robust (see Examples in Chapter 4,
  and Exercise 8 in Chapter 8).

- Periodic behavior is present in integrate and fire neurons (see Figure 5.3). Indeed, the human body is full of oscillatory behavior as described in Chapter 12.

- For mechanical systems, examples include the motion of simple nonlinear pendula (see Section 9.3), wing rock oscillations in aircraft flight dynamics [11], and surge oscillations in axial flow compressors [1], for example.

- For periodic chemical reactions, examples include the Landolt clock reaction and the Belousov-Zhabotinski reaction (see Chapter 8).

- For electrical or electronic circuits, it is possible to construct simple electronic oscillators (Chua's circuit, for example) using a nonlinear circuit element; a limit cycle can be observed if the circuit is connected to an oscilloscope.

Limit cycles are common solutions for all types of dynamical systems. Sometimes it becomes necessary to prove the existence and uniqueness of a limit cycle, as described in the next section.

5.2 Existence and Uniqueness of Limit Cycles in the Plane

To understand the existence and uniqueness theorem, it is necessary to define some features of phase plane portraits. Assume that the existence and uniqueness theorem from Chapter 2 holds for all solutions considered here.

The definitions listed in Chapter 2 can be extended to nonlinear planar systems of the form $\dot{x} = P(x,y), \dot{y} = Q(x,y)$; thus, every solution, say, $\phi(t) = (x(t), y(t))$, can be represented as a curve in the plane and is called a trajectory. The phase portrait shows how the qualitative behavior is determined as x and y vary with t. The trajectory can also be defined in terms of the spatial coordinates $\mathbf{x}$, as in Definition 3 below. A brief look at Example 1 will help the reader to understand Definitions 1–7 in this section.

Definition 2. A *flow* on $\Re^2$ is a mapping $\pi : \Re^2 \to \Re^2$ such that

1. π is continuous;
2. $\pi(\mathbf{x}, 0) = \mathbf{x}$ for all $\mathbf{x} \in \Re^2$;
3. $\pi(\pi(\mathbf{x}, t_1), t_2) = \pi(\mathbf{x}, t_1 + t_2)$.

Definition 3. Suppose that $I_\mathbf{x}$ is the maximal interval of existence. The *trajectory* (or *orbit*) through $\mathbf{x}$ is defined as $\gamma(\mathbf{x}) = \{\pi(\mathbf{x}, t) : t \in I_\mathbf{x}\}$.
The *positive semiorbit* is defined as $\gamma^+(\mathbf{x}) = \{\pi(\mathbf{x}, t) : t > 0\}$.
The *negative semiorbit* is defined as $\gamma^-(\mathbf{x}) = \{\pi(\mathbf{x}, t) : t < 0\}$.

Definition 4. The *positive limit set* of a point $\mathbf{x}$ is defined as

$$\Lambda^+(\mathbf{x}) = \{\mathbf{y} : \text{there exists a sequence } t_n \to \infty \text{ such that } \pi(\mathbf{x}, t) \to \mathbf{y}\}.$$

The *negative limit set* of a point $\mathbf{x}$ is defined as

$$\Lambda^-(\mathbf{x}) = \{\mathbf{y} : \text{there exists a sequence } t_n \to -\infty \text{ such that } \pi(\mathbf{x}, t) \to \mathbf{y}\}.$$

In the phase plane, trajectories tend to a critical point, a closed orbit, or infinity.

Definition 5. A set S is *invariant* with respect to a flow if $\mathbf{x} \in S$ implies that $\gamma(\mathbf{x}) \subset S$.

A set S is *positively invariant* with respect to a flow if $\mathbf{x} \in S$ implies that $\gamma^+(\mathbf{x}) \subset S$.

A set S is *negatively invariant* with respect to a flow if $\mathbf{x} \in S$ implies that $\gamma^-(\mathbf{x}) \subset S$.

A general trajectory can be labeled γ for simplicity.

Definition 6. A limit cycle, say, Γ, is

- a *stable limit cycle* if $\Lambda^+(\mathbf{x}) = \Gamma$ for all $\mathbf{x}$ in some neighborhood; this implies that nearby trajectories are attracted to the limit cycle;

- an *unstable limit cycle* if $\Lambda^-(\mathbf{x}) = \Gamma$ for all $\mathbf{x}$ in some neighborhood; this implies that nearby trajectories are repelled away from the limit cycle;

- a *semistable limit cycle* if it is attracting on one side and repelling on the other.

The stability of limit cycles can also be deduced analytically using the Poincaré map (see Chapter 9). The following example will be used to illustrate each of the Definitions 1–6 above and 7 below.

Definition 7. The period, say, T, of a limit cycle is given by $\mathbf{x}(t) = \mathbf{x}(t+T)$, where T is the minimum period. The period can be found by plotting a time series plot of the limit cycle (see the Mathematica command lines in Chapter 4).

Example 1. Describe some of the features for the following set of polar differential equations in terms of Definitions 1–7:

$$\dot{r} = r(1-r)(2-r)(3-r), \quad \dot{\theta} = -1. \tag{5.4}$$

Solution. A phase portrait is shown in Figure 5.4. There is a unique critical point at the origin since $\dot{\theta}$ is nonzero. There are three limit cycles that may be determined from the equation $\dot{r} = 0$. They are the circles of radii one, two, and three, all centered at the origin. Let Γ_i denotes the limit cycle of radius $r = i$.

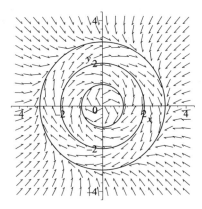

Figure 5.4: Three limit cycles for system (5.4).

There is one critical point at the origin. If a trajectory starts at this point, it remains there forever. A trajectory starting at $(1, 0)$ will reach the point $(-1, 0)$ when $t_1 = \pi$ and the motion is clockwise. Continuing on this path for another time interval $t_2 = \pi$, the orbit returns to $(1, 0)$. Using part 3 of Definition 2, one can write $\pi\left(\pi((1, 0), t_1), t_2\right) = \pi\left((1, 0), 2\pi\right)$ since the limit cycle is of period 2π (see below). On the limit cycle Γ_1, both the positive and negative semiorbits lie on Γ_1.

Suppose that $P = \left(\frac{1}{2}, 0\right)$ and $Q = (4, 0)$ are two points in the plane. The limit sets are given by $\Lambda^+(P) = \Gamma_1$, $\Lambda^-(P) = (0, 0)$, $\Lambda^+(Q) = \Gamma_3$, and $\Lambda^-(Q) = \infty$.

The annulus $A_1 = \{r \in \Re^2 : 0 < r < 1\}$ is positively invariant, and the annulus $A_2 = \{r \in \Re^2 : 1 < r < 2\}$ is negatively invariant.

If $0 < r < 1$, then $\dot{r} > 0$ and the critical point at the origin is unstable. If $1 < r < 2$, then $\dot{r} < 0$ and Γ_1 is a stable limit cycle. If $2 < r < 3$, then $\dot{r} > 0$ and Γ_2 is an unstable limit cycle. Finally, if $r > 3$, then $\dot{r} < 0$ and Γ_3 is a stable limit cycle.

Integrate both sides of $\dot{\theta} = -1$ with respect to time to show that the period of all of the limit cycles is 2π.

The Poincaré-Bendixson Theorem. *Suppose that γ^+ is contained in a bounded region in which there are finitely many critical points. Then $\Lambda^+(\gamma)$ is either*

- *a single critical point;*

- *a single closed orbit;*

- *a graphic—critical points joined by heteroclinic orbits.*

A heteroclinic orbit connects two separate critical points and takes an infinite amount of time to make the connection; more detail is provided in Chapter 6.

Corollary. *Let D be a bounded closed set containing no critical points and suppose that D is positively invariant. Then there exists a limit cycle contained in D.*

A proof to this theorem involves topological arguments and can be found in [13], for example.

Example 2. By considering the flow across the rectangle with corners at $(-1, 2)$, $(1, 2)$, $(1, -2)$, and $(-1, -2)$, prove that the following system has at least one limit cycle:

$$\dot{x} = y - 8x^3, \quad \dot{y} = 2y - 4x - 2y^3. \tag{5.5}$$

Solution. The critical points are found by solving the equations $\dot{x} = \dot{y} = 0$. Set $y = 8x^3$. Then $\dot{y} = 0$ if $x(1 - 4x^2 + 256x^8) = 0$. The graph of the function $y = 1 - 4x^2 + 256x^8$ is given in Figure 5.5(a). The graph has no roots, and the origin is the only critical point.

Linearize at the origin in the usual way. It is not difficult to show that the origin is an unstable focus.

Consider the flow on the sides of the given rectangle:

- On $y = 2, |x| \leq 1, \dot{y} = -4x - 12 < 0$.

- On $y = -2, |x| \leq 1, \dot{y} = -4x + 12 > 0$.

- On $x = 1, |y| \leq 2, \dot{x} = y - 8 < 0$.

- On $x = -1, |y| \leq 2, \dot{y} = y + 8 > 0$.

The flow is depicted in Figure 5.5(b). The rectangle is positively invariant, and there are no critical points other than the origin, which is unstable. Consider a small deleted neighborhood, say, N_ϵ, around this critical point. For example, the boundary of N_ϵ could be a small ellipse. On this ellipse, all trajectories will cross outward. Therefore, there exists a stable limit cycle lying inside the rectangular region and outside of N_ϵ by the corollary to the Poincaré-Bendixson theorem.

Definition 8. A planar simple closed curve is called a *Jordan curve*.

Consider the system

$$\dot{x} = P(x, y), \quad \dot{y} = Q(x, y), \tag{5.6}$$

where P and Q have continuous first-order partial derivatives. Let the vector field be denoted by $\mathbf{X}$ and let ψ be a weighting factor that is continuously differentiable. Recall Green's Theorem, which will be required to prove the following two theorems.

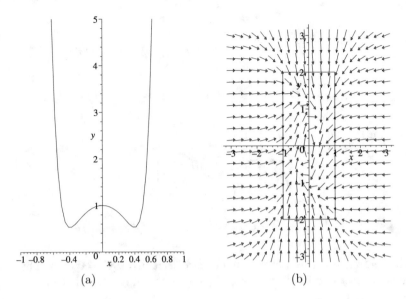

Figure 5.5:   (a) Polynomial of degree 8. (b) Flow across the rectangle for system (5.5).

Green's Theorem. *Let J be a Jordan curve of finite length. Suppose that P and Q are two continuously differentiable functions defined on the interior of J, say, D. Then*

$$\iint_D \left[\frac{\partial P}{\partial x} + \frac{\partial Q}{\partial y} \right] dx\, dy = \oint_J P\, dy - Q\, dx.$$

Dulac's Criteria. *Consider an annular region, say, A, contained in an open set E. If*

$$\nabla.(\psi \mathbf{X}) = \operatorname{div}(\psi \mathbf{X}) = \frac{\partial}{\partial x}(\psi P) + \frac{\partial}{\partial y}(\psi Q)$$

does not change sign in A, then there is at most one limit cycle entirely contained in A.

Proof. Suppose that Γ_1 and Γ_2 are limit cycles encircling K, as depicted in Figure 5.6, of periods T_1 and T_2, respectively. Apply Green's Theorem to the region R shown in Figure 5.6.

$$\iint_R \left[\frac{\partial(\psi P)}{\partial x} + \frac{\partial(\psi Q)}{\partial y} \right] dx\, dy = \oint_{\Gamma_2} \psi P dy - \psi Q dx +$$

$$\int_L \psi P dy - \psi Q dx - \oint_{\Gamma_1} \psi P dy - \psi Q dx - \int_L \psi P dy - \psi Q dx.$$

Now on Γ_1 and Γ_2, $\dot{x} = P$ and $\dot{y} = Q$, so

$$\iint_R \left[\frac{\partial(\psi P)}{\partial x} + \frac{\partial(\psi Q)}{\partial y} \right] dx\, dy$$

$$= \int_0^{T_2} (\psi PQ - \psi QP)\, dt - \int_0^{T_1} (\psi PQ - \psi QP)\, dt,$$

which is zero and contradicts the hypothesis that $\mathrm{div}(\psi \mathbf{X}) \neq 0$ in A. Therefore, there is at most one limit cycle entirely contained in the annulus A. $\qquad\square$

Example 3. Use Dulac's criteria to prove that the system

$$\dot{x} = -y + x(1 - 2x^2 - 3y^2), \quad \dot{y} = x + y(1 - 2x^2 - 3y^2) \qquad (5.7)$$

has a unique limit cycle in an annulus.

Solution. Convert to polar coordinates using the transformations

$$r\dot{r} = x\dot{x} + y\dot{y}, \quad r^2\dot{\theta} = x\dot{y} - y\dot{x}.$$

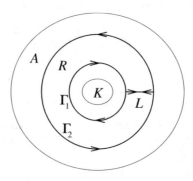

Figure 5.6: Two limit cycles encircling the region K.

Therefore, system (5.7) becomes

$$\dot{r} = r(1 - 2r^2 - r^2 \sin^2\theta), \quad \dot{\theta} = 1.$$

Since $\dot{\theta} = 1$, the origin is the only critical point. On the circle $r = \frac{1}{2}$, $\dot{r} = \frac{1}{2}(\frac{1}{2} - \frac{1}{4}\sin^2\theta)$. Hence, $\dot{r} > 0$ on this circle. On the circle $r = 1$, $\dot{r} = -1 - \sin^2\theta$. Hence, $\dot{r} < 0$ on this circle. If $r \geq 1$, then $\dot{r} < 0$, and if $0 < r \leq \frac{1}{2}$, then $\dot{r} > 0$. Therefore, there exists a limit cycle in the annulus $A = \{r : \frac{1}{2} < r < 1\}$ by the corollary to the Poincaré-Bendixson theorem.

Consider the annulus A. Now $\text{div}(\mathbf{X}) = 2(1 - 4r^2 - 2r^2\sin^2\theta)$. If $\frac{1}{2} < r < 1$, then $\text{div}(\mathbf{X}) < 0$. Since the divergence of the vector field does not change sign in the annulus A, there is at most one limit cycle in A by Dulac's criteria.

A phase portrait is given in Figure 5.7.

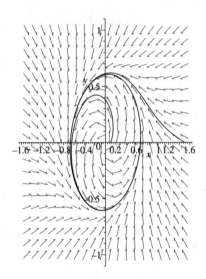

Figure 5.7:   A phase portrait for system (5.7) showing the unique limit cycle.

Example 4. Plot a phase portrait for the Liénard system

$$\dot{x} = y, \quad \dot{y} = -x - y(a_2 x^2 + a_4 x^4 + a_6 x^6 + a_8 x^8 + a_{10} x^{10} + a_{12} x^{12} + a_{14} x^{14}),$$

where $a_2 = 90, a_4 = -882, a_6 = 2598.4, a_8 = -3359.997, a_{10} = 2133.34, a_{12} = -651.638$, and $a_{14} = 76.38$.

Solution. Not all limit cycles are convex closed curves as Figure 5.8 demonstrates.

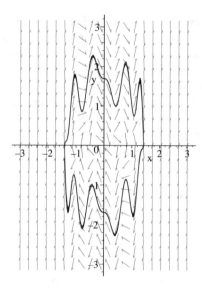

Figure 5.8: A phase portrait for Example 4. The limit cycle is a nonconvex closed curve.

5.3 Nonexistence of Limit Cycles in the Plane

Bendixson's Criteria. *Consider system (5.6) and suppose that D is a simply connected domain (no holes in D) and that*

$$\nabla.(\psi\mathbf{X}) = \operatorname{div}(\psi\mathbf{X}) = \frac{\partial}{\partial x}(\psi P) + \frac{\partial}{\partial y}(\psi Q) \neq 0$$

in D. Then there are no limit cycles entirely contained in D.

Proof. Suppose that D contains a limit cycle Γ of period T. Then from Green's Theorem

$$\iint_D \left[\frac{\partial(\psi P)}{\partial x} + \frac{\partial(\psi Q)}{\partial y}\right] dx\, dy = \oint_\Gamma (\psi P dy - \psi Q dx)$$

$$= \int_0^T \left(\psi P \frac{dy}{dt} - \psi Q \frac{dx}{dt}\right) dt = 0$$

since on Γ, $\dot{x} = P$ and $\dot{y} = Q$. This contradicts the hypothesis that $\operatorname{div}(\psi\mathbf{X}) \neq 0$, and therefore, D contains no limit cycles entirely. $\qquad\square$

Definition 9. Suppose there is a compass on a Jordan curve C and that the needle points in the direction of the vector field. The compass is moved in a counterclockwise direction around the Jordan curve by 2π radians. When it returns to its initial position, the needle will have moved through an angle, say, Θ. The *index*, say, $I_{\mathbf{X}}(C)$, is defined as

$$I_{\mathbf{X}}(C) = \frac{\Delta\Theta}{2\pi},$$

where $\Delta\Theta$ is the overall change in the angle Θ.

The above definition can be applied to isolated critical points. For example, the index of a node, focus, or center is $+1$, and the index of a col is -1. The following result is clear.

Theorem 1. *The sum of the indices of the critical points contained entirely within a limit cycle is $+1$.*

The next theorem then follows.

Theorem 2. *A limit cycle contains at least one critical point.*

When proving that a system has no limit cycles, the following items should be considered:

1. Bendixson's criteria;
2. indices;
3. invariant lines;
4. critical points.

Example 5. Prove that none of the following systems have any limit cycles:

(a) $\dot{x} = 1 + y^2 - e^{xy}, \quad \dot{y} = xy + \cos^2 y.$

(b) $\dot{x} = y^2 - x, \quad \dot{y} = y + x^2 + yx^3.$

(c) $\dot{x} = y + x^3, \quad \dot{y} = x + y + y^3.$

(d) $\dot{x} = 2xy - 2y^4, \quad \dot{y} = x^2 - y^2 - xy^3.$

(e) $\dot{x} = x(2 - y - x), \quad \dot{y} = y(4x - x^2 - 3), \quad \text{given} \quad \psi = \frac{1}{xy}.$

Solutions.

(a) The system has no critical points and hence no limit cycles by Theorem 2.

(b) The origin is the only critical point, and it is a saddle point or col. Since the index of a col is -1, there are no limit cycles from Theorem 1.

(c) Find the divergence, $\text{div}\mathbf{X} = \frac{\partial P}{\partial x} + \frac{\partial Q}{\partial y} = 3x^2 + 3y^2 + 1 \neq 0$. Hence, there are no limit cycles by Bendixson's criteria.

(d) Find the divergence, $\text{div}\mathbf{X} = \frac{\partial P}{\partial x} + \frac{\partial Q}{\partial y} = -3x^2y$. Now $\text{div}\mathbf{X} = 0$ if either $x = 0$ or $y = 0$. However, on the line $x = 0$, $\dot{x} = -2y^4 \leq 0$, and on the line $y = 0$, $\dot{y} = x^2 \geq 0$. Therefore, a limit cycle must lie wholly in one of the four quadrants. This is not possible since $\text{div}\mathbf{X}$ is nonzero here. Hence, there are no limit cycles by Bendixson's criteria. Draw a small diagram to help you understand the solution.

(e) The axes are invariant since $\dot{x} = 0$ if $x = 0$ and $\dot{y} = 0$ if $y = 0$. The weighted divergence is given by $\text{div}(\psi\mathbf{X}) = \frac{\partial}{\partial x}(\psi P) + \frac{\partial}{\partial y}(\psi Q) = -\frac{1}{y}$. Therefore, there are no limit cycles contained entirely in any of the quadrants, and since the axes are invariant, there are no limit cycles in the whole plane.

Example 6. Prove that the system

$$\dot{x} = x(1 - 4x + y), \quad \dot{y} = y(2 + 3x - 2y)$$

has no limit cycles by applying Bendixson's criteria with $\psi = x^m y^n$.

Solution. The axes are invariant since $\dot{x} = 0$ on $x = 0$ and $\dot{y} = 0$ on $y = 0$. Now

$$\text{div}(\psi\mathbf{X}) = \frac{\partial}{\partial x}\left(x^{m+1}y^n - 4x^{m+2}y^n + x^{m+1}y^{n+1}\right) +$$

$$\frac{\partial}{\partial y}\left(2x^m y^{n+1} + 3x^{m+1}y^{n+1} - 2x^m y^{n+2}\right),$$

which simplifies to

$$\text{div}(\psi\mathbf{X}) = (m + 2n + 2)x^m y^n + (-4m + 3n - 5)x^{m+1}y^n + (m - 2n - 3)x^m y^{n+1}.$$

Select $m = \frac{1}{2}$ and $n = -\frac{5}{4}$. Then

$$\text{div}(\psi\mathbf{X}) = -\frac{43}{4}x^{\frac{3}{2}}y^{-\frac{5}{4}}.$$

Therefore, there are no limit cycles contained entirely in any of the four quadrants, and since the axes are invariant, there are no limit cycles at all.

5.4 Perturbation Methods

This section introduces the reader to some basic perturbation methods by means of example. The theory involves mathematical methods for finding series expansion approximations for perturbed systems. Perturbation theory can be applied to algebraic equations, boundary value problems, difference equations, Hamiltonian systems, ODEs, PDEs, and in modern times the theory underlies almost all of quantum field theory and quantum chemistry. There are whole books devoted to the study of perturbation methods, and the reader is directed to the references [4], [9], and [16], for more detailed theory and more in-depth explanations.

The main idea begins with the assumption that the solution to the perturbed system can be expressed as an *asymptotic* or Poincaré expansion of the form

$$x(t, \epsilon) = x_0(t) + \epsilon x_1(t) + \epsilon^2 x_2(t) + \dots. \tag{5.8}$$

Definition 10. The sequence $f(\epsilon) \sim \sum_{n=0}^{\infty} a_n \phi_n(\epsilon)$ is an asymptotic expansion of the continuous function $f(\epsilon)$ if and only if, for all $n \geq 0$,

$$f(\epsilon) = \sum_{n=0}^{N} a_n \phi_n(\epsilon) + O\left(\phi_{N+1}(\epsilon)\right) \quad \text{as } \epsilon \to 0, \tag{5.9}$$

where the sequence constitutes an asymptotic scale such that for every $n \geq 0$,

$$\phi_{n+1}(\epsilon) = o\left(\phi_n(\epsilon)\right) \text{ as } \quad \epsilon \to 0.$$

Definition 11. An asymptotic expansion (5.9) is said to be *uniform* if in addition

$$|R_N(x, \epsilon)| \leq K|\phi_{N+1}(\epsilon)|,$$

for ϵ in a neighborhood of 0, where the Nth remainder $R_N(x, \epsilon) = O(\phi_{N+1}(\epsilon))$ as $\epsilon \to 0$, and K is a constant.

In this particular case, we will be looking for asymptotic expansions of the form

$$x(t, \epsilon) \sim \sum_{k} x_k(t) \delta_k(\epsilon),$$

where $\delta_k(\epsilon) = \epsilon^k$ is an asymptotic scale. It is important to note that the asymptotic expansions often do not converge; however, one-term and two-term approximations provide an analytical expression that is dependent on the parameter, ϵ, and some initial conditions. The major advantage that the perturbation analysis has over numerical analysis is that a general solution is available through perturbation methods where numerical methods only lead to a single solution.

As a simple introduction, consider the following first-order ordinary differential equation:

Example 7. Suppose that for $x \geq 0$,

$$\frac{dx}{dt} + x - \epsilon x^2 = 0, \quad x(0) = 2. \tag{5.10}$$

Solution. This equation can be solved directly, use the DSolve command in Mathematica to solve the system analytically. Set

$$x(t, \epsilon) = x_0(t) + \epsilon x_1(t) + \epsilon^2 x_2(t) + \dots,$$

where in order to satisfy the initial condition $x(0) = 2$, we will have $x_1(0) = 0$, $x_2(0) = 0$, and so on. To compute to $O(\epsilon^2)$, substitute the first three terms into system (5.10) and collect powers of ϵ using the CoefficientList command in Mathematica. The commands are:

```
In[1]:=x=x0[t]+epsilon*x1[t]+epsilon^2*x2[t];
```

```
In[2]:=CoefficientList[Dt[x,t,Constants->{epsilon}]+x-epsilon*x^2,
       epsilon]
```

one obtains:

$$\epsilon^0 : \dot{x}_0(t) + x_0(t) = 0$$
$$\epsilon^1 : \dot{x}_1(t) + x_1(t) - x_0(t)^2 = 0,$$
$$\epsilon^2 : \dot{x}_2(t) + x_2(t) - 2x_0(t)x_1(t) = 0,$$
$$\vdots \quad \vdots$$

and we solve at each order, applying the initial conditions as we proceed.
    For $O(1)$:
$$\dot{x}_0(t) + x_0(t) = 0, \quad x_0(0) = 2,$$

and the solution using Mathematica:

```
In[3]:=DSolve[{x0'[t]+x0[t]==0,x0[0]==2},x0[t],t]
```

is $x_0(t) = 2 * exp(-t)$.
    For $O(\epsilon)$:
$$\dot{x}_1(t) + x_1(t) = 4e^{-2t}, \quad x_1(0) = 0,$$

and the solution using Mathematica is $x_1(t) = 4\left(e^{-t} - e^{-2t}\right)$.
    For $O(\epsilon^2)$:
$$\dot{x}_2(t) + x_2(t) = 4e^{-t}x_1(t), \quad x_2(0) = 0,$$

and the solution using Mathematica is $x_2(t) = 8\left(e^{-t} - 2e^{-2t} + e^{-3t}\right)$.
    Therefore, the solution to second order is:

$$x(t) \approx 2e^{-t} + 4\epsilon\left(e^{-t} - e^{-2t}\right) + 8\epsilon^2\left(e^{-t} - 2e^{-2t} + e^{-3t}\right).$$

To keep the theory simple and in relation to other material in this chapter, the author has decided to focus on perturbed ODEs of the form

$$\ddot{x} + x = \epsilon f(x, \dot{x}), \tag{5.11}$$

where $0 \leq \epsilon \ll 1$ and $f(x, \dot{x})$ is an arbitrary smooth function. The unperturbed system represents a linear oscillator and when $0 < \epsilon \ll 1$, system (5.11) becomes a weakly nonlinear oscillator. Systems of this form include the Duffing equation

$$\ddot{x} + x = \epsilon x^3, \tag{5.12}$$

and the van der Pol equation

$$\ddot{x} + x = \epsilon \left(x^2 - 1 \right) \dot{x}. \tag{5.13}$$

Example 8. Use perturbation theory to find a one-term and two-term asymptotic expansion of Duffing's equation (5.12) with initial conditions $x(0) = 1$ and $\dot{x}(0) = 0$.

Solution. Substitute (5.8) into (5.12) to get

$$\frac{d^2}{dt^2} \left(x_0 + \epsilon x_1 + \ldots \right) + \left(x_0 + \epsilon x_1 + \ldots \right) = \epsilon \left(x_0 + \epsilon x_1 + \ldots \right)^3.$$

Use the Collect command in Mathematica to group terms according to powers of ϵ, thus

$$[\ddot{x}_0 + x_0] + \epsilon \left[\ddot{x}_1 + x_1 - x_0^3 \right] + O(\epsilon^2) = 0.$$

The order equations are

$$O(1): \quad \ddot{x}_0 + x_0 = 0, \qquad\qquad x_0(0) = 1, \quad \dot{x}_0(0) = 0,$$
$$O(\epsilon): \quad \ddot{x}_1 + x_1 = x_0^3, \qquad\qquad x_1(0) = 0, \quad \dot{x}_1(0) = 0.$$

$$\vdots \qquad\qquad\qquad\qquad\qquad \vdots$$

The $O(1)$ solution is $x_0 = \cos(t)$. Let us compare this solution with the numerical solution, say, x_N, when $\epsilon = 0.01$. Figure 5.9 shows the time against the error, $x_N - x_0$, for $0 \leq t \leq 100$.

Using Mathematica, the $O(\epsilon)$ solution is computed to be

$$x_1 = \frac{3}{8} t \sin(t) + \frac{1}{8} \cos(t) - \frac{1}{8} \cos^3(t).$$

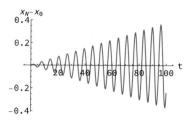

Figure 5.9: [Mathematica] The error between the numerical solution x_N and the one-term expansion x_0 for the Duffing system (5.12) when $\epsilon = 0.01$.

Using the TrigReduce command in Mathematica:

```
In[1]:=x1=3*t*Sin[t]/8+Cos[t]/8-Cos[t]^3/8;
```

```
In[2]:=TrigReduce[x1]
```

we have

$$x \sim x_P = x_0 + \epsilon x_1 = \cos(t) + \epsilon \left(\frac{1}{32} \cos(t) - \frac{1}{32} \cos(3t) + \frac{3}{8} t \sin(t) \right),$$

where x_P represents the Poincaré expansion up to the second term. The term $t \sin(t)$ is called a *secular* term and is an oscillatory term of growing amplitude. Unfortunately, the secular term leads to a nonuniformity for large t. Figure 5.10 shows the error for the two-term Poincaré expansion, $x_N - x_P$, when $\epsilon = 0.01$.

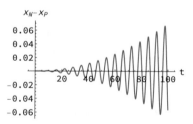

Figure 5.10:   The error between the numerical solution x_N and the two-term expansion x_P for the Duffing system (5.12) when $\epsilon = 0.01$.

By introducing a strained coordinate, the nonuniformity may be overcome and this is the idea behind the Lindstedt–Poincaré technique for periodic systems. The idea is to introduce a straining transformation of the form

$$\frac{\tau}{t} = 1 + \epsilon w_1 + \epsilon^2 w_2 + \ldots, \tag{5.14}$$

and seek values $w_1, w_2, \ldots$, that avoid secular terms appearing in the expansion.

Example 9. Use the Lindstedt–Poincaré technique to determine a two-term uniform asymptotic expansion of Duffing's equation (5.12) with initial conditions $x(0) = 1$ and $\dot{x}(0) = 0$.

Solution. [Mathematica] Using the transformation given in (5.14)

$$\frac{d}{dt} = \frac{d\tau}{dt}\frac{d}{d\tau} = \left(1 + \epsilon w_1 + \epsilon^2 w_2 + \cdots\right)\frac{d}{d\tau},$$

$$\frac{d^2}{dt^2} = \left(1 + \epsilon w_1 + \epsilon^2 w_2 + \cdots\right)^2 \frac{d^2}{d\tau^2}.$$

Applying the transformation to equation (5.12) leads to

$$\left(1 + 2\epsilon w_1 + \epsilon^2 \left(w_1^2 + 2w_2\right) + \cdots\right)\frac{d^2 x}{d\tau^2} + x = \epsilon x^3,$$

where x is now a function of the strained variable τ. Assume that

$$x(\tau, \epsilon) = x_0(\tau) + \epsilon x_1(\tau) + \epsilon^2 x_2(\tau) + \ldots. \tag{5.15}$$

Substituting (5.15) into (5.12) using Mathematica gives the following order equations:

$$O(1): \quad \frac{d^2 x_0}{d\tau^2} + x_0 = 0,$$

$$x_0(\tau = 0) = 1, \quad \frac{dx_0}{d\tau}(\tau = 0) = 0,$$

$$O(\epsilon): \quad \frac{d^2 x_1}{d\tau^2} + x_1 = x_0^3 - 2w_1\frac{d^2 x_0}{d\tau^2},$$

$$x_1(0) = 0, \quad \frac{dx_1}{d\tau}(0) = 0,$$

$$O(\epsilon^2): \quad \frac{d^2 x_2}{d\tau^2} + x_2 = 3x_0^2 x_1 - 2w_1\frac{d^2 x_1}{d\tau^2} - (w_1^2 + 2w_2)\frac{d^2 x_0}{d\tau^2},$$

$$x_2(0) = 0, \quad \frac{dx_2}{d\tau}(0) = 0.$$

The $O(1)$ solution is $x_0(\tau) = \cos(\tau)$. Using Mathematica, the solution to the $O(\epsilon)$ equation is

$$x_1(\tau) = \frac{1}{8}\sin(\tau)\left(3\tau + 8w_1\tau + \cos(\tau)\sin(\tau)\right).$$

To avoid secular terms, select $\omega_1 = -\frac{3}{8}$, then the $O(\epsilon)$ solution is

$$x_1(\tau) = \frac{1}{8}\sin^2(\tau)\cos(\tau).$$

Using Mathematica, the $O(\epsilon^2)$ solution is

$$x_2(\tau) = \frac{1}{512}\sin(\tau)\left(42\tau + 512\omega_2\tau + 23\sin(2\tau) - \sin(4\tau)\right),$$

and selecting $\omega_2 = -\frac{21}{256}$ avoids secular terms.

The two-term uniformly valid expansion of equation (5.12) is

$$x(\tau,\epsilon) \sim x_{LP} = \cos(\tau) + \frac{\epsilon}{8}\sin^2(\tau)\cos(\tau),$$

where

$$\tau = t\left(1 - \frac{3}{8}\epsilon - \frac{21}{256}\epsilon^2 + O(\epsilon^3)\right),$$

as $\epsilon \to 0$. Note that the straining transformation is given to a higher order than the expansion of the solution. The difference between the two-term uniform asymptotic expansion and the numerical solution is depicted in Figure 5.11.

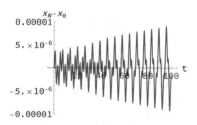

Figure 5.11:   The error between the numerical solution x_N and the two-term Lindstedt–Poincaré expansion x_{LP} for the Duffing system (5.12) when $\epsilon = 0.01$.

Unfortunately, the Lindstedt–Poincaré technique does not always work for oscillatory systems. An example of its failure is provided by the van der Pol equation (5.13).

Example 10. Show that the Lindstedt–Poincaré technique fails for the ODE (5.13) with initial conditions $x(0) = 1$ and $\dot{x}(0) = 0$.

Solution. Substituting (5.15) into (5.13) using Mathematica gives the following order equations:

$$O(1): \quad \frac{d^2 x_0}{d\tau^2} + x_0 = 0,$$

$$x_0(\tau = 0) = 1, \quad \frac{dx_0}{d\tau}(\tau = 0) = 0,$$

$$O(\epsilon): \quad \frac{d^2 x_1}{d\tau^2} + x_1 = \frac{dx_0}{d\tau} - x_0^2 \frac{dx_0}{d\tau} - 2\omega_1 \frac{d^2 x_0}{d\tau^2},$$

$$x_1(0) = 0, \quad \frac{dx_1}{d\tau}(0) = 0,$$

The $O(1)$ solution is $x_0(\tau) = \cos(\tau)$. Using Mathematica, the solution to the $O(\epsilon)$ equation can be simplified to

$$x_1(\tau) = \frac{1}{16}\left(6\tau \cos(\tau) - (5 - 16\tau\omega_1 + \cos(2\tau))\sin(\tau)\right)$$

or

$$x_1(\tau) = \frac{1}{16}\left(\{6\tau \cos(\tau) + 16\tau\omega_1 \sin(\tau)\} - (5 + \cos(2\tau))\sin(\tau)\right).$$

To remove secular terms set $\omega_1 = -\frac{3}{8}\cot(\tau)$, then

$$x(\tau, \epsilon) = \cos(\tau) + O(\epsilon),$$

where

$$\tau = t - \frac{3}{8}\epsilon t \cot(t) + O(\epsilon^2).$$

This is invalid since the cotangent function is singular when $t = n\pi$, where n is an integer. Unfortunately, the Lindstedt–Poincaré technique does not work for all ODEs of the form (5.11); it cannot be used to obtain approximations that evolve aperiodically on a slow time scale.

Consider the van der Pol equation (5.13), Figure 5.12 shows a trajectory starting at $x(0) = 0.1, \dot{x}(0) = 0$ for $\epsilon = 0.05$ and $0 \le t \le 800$. The trajectory spirals around the origin, and it takes many cycles for the amplitude to grow substantially. As $t \to \infty$, the trajectory asymptotes to a limit cycle of approximate radius two. This is an example of a system whose solutions depend simultaneously on widely different scales. In this case, there are two time scales: a fast time scale for the sinusoidal oscillations $\sim O(1)$, and a slow time scale over which the amplitude grows $\sim O(\frac{1}{\epsilon})$. The *method of multiple scales* introduces new slow time variables for each time scale of interest in the problem.

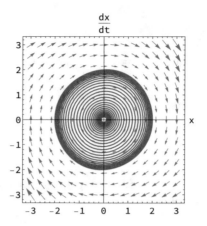

Figure 5.12: A trajectory for the van der Pol equation (5.13) when $\epsilon = 0.05$.

The Method of Multiple Scales.

Introduce new time scales, say, $\tau_0 = t$ and $\tau_1 = \epsilon t$, and seek approximate solutions of the form

$$x(t, \epsilon) \sim x_0\left(\tau_0, \tau_1\right) + \epsilon x_1\left(\tau_0, \tau_1\right) + \cdots . \tag{5.16}$$

Substitute into the ODE and solve the resulting PDEs. An example is given below.

Example 11. Use the method of multiple scales to determine a uniformly valid one-term expansion for the van der Pol equation (5.13) with initial conditions $x(0) = a$ and $\dot{x}(0) = 0$.

Solution. Substitute equation (5.16) into (5.13) using Mathematica gives the following order equations:

$$O(1): \quad \frac{\partial^2 x_0}{\partial \tau_0^2} + x_0 = 0,$$

$$O(\epsilon): \quad \frac{\partial^2 x_1}{\partial \tau_0^2} + x_1 = -2\frac{\partial x_0}{\partial \tau_0 \tau_1} - \left(x_0^2 - 1\right)\frac{\partial x_0}{\partial \tau_0}.$$

The general solution to the $O(1)$ PDE may be found using Mathematica,

$$x_0\left(\tau_0, \tau_1\right) = c_1(\tau_1)\cos(\tau_0) + c_2(\tau_1)\sin(\tau_0)$$

which using trigonometric identities can be expressed as

$$x_0\left(\tau_0, \tau_1\right) = R(\tau_1)\cos(\tau_0 + \theta(\tau_1)), \tag{5.17}$$

where $R(\tau_1)$ and $\theta(\tau_1)$ are the slowly varying amplitude and phase of x_0, respectively. Substituting (5.17), the $O(\epsilon)$ equation becomes

$$\frac{\partial^2 x_1}{\partial \tau_0^2} + x_1 = -2\left(\frac{dR}{d\tau_1}\sin(\tau_0 + \theta(\tau_1)) + R(\tau_1)\frac{d\theta}{d\tau_1}\cos(\tau_0 + \theta(\tau_1))\right)$$
$$- R(\tau_1)\sin(\tau_0 + \theta(\tau_1))\left(R^2(\tau_1)\cos^2(\tau_0 + \theta(\tau_1)) - 1\right). \quad (5.18)$$

In order to avoid resonant terms on the right-hand side which lead to secular terms in the solution, it is necessary to remove the linear terms $\cos(\tau_0 + \theta(\tau_1))$ and $\sin(\tau_0 + \theta(\tau_1))$ from the equation. Use the TrigReduce command in Mathematica to reduce an expression to a form linear in the trigonometric function. Equation (5.18) then becomes

$$\frac{\partial^2 x_1}{\partial \tau_0^2} + x_1 = \left\{-2\frac{dR}{d\tau_1} + R - \frac{R^3}{4}\right\}\sin(\tau_0 + \theta(\tau_1))$$
$$\left\{-2R\frac{d\theta}{d\tau_1}\right\}\cos(\tau_0 + \theta(\tau_1)) - \frac{R^3}{4}\sin(3\tau_0 + 3\theta(\tau_1)).$$

To avoid secular terms set

$$-2\frac{dR}{d\tau_1} + R - \frac{R^3}{4} = 0 \quad \text{and} \quad \frac{d\theta}{d\tau_1} = 0. \quad (5.19)$$

The initial conditions are $x_0(0,0) = a$ and $\frac{\partial x_0}{\partial \tau_0} = 0$ leading to $\theta(0) = 0$ and $R(0) = \frac{a}{2}$. The solutions to system (5.19) with these initial conditions are easily computed with Mathematica, thus

$$R(\tau_1) = \frac{2}{\sqrt{1 + \left(\frac{4}{a^2} - 1\right)e^{-\tau_1}}} \quad \text{and} \quad \theta(\tau_1) = 0.$$

Therefore, the uniformly valid one-term solution is

$$x_0(\tau_0, \tau_1) = \frac{2\cos(\tau_0)}{\sqrt{1 + \left(\frac{4}{a^2} - 1\right)e^{-\tau_1}}} + O(\epsilon)$$

or

$$x(t) = \frac{2\cos(t)}{\sqrt{1 + \left(\frac{4}{a^2} - 1\right)e^{-\epsilon t}}} + O(\epsilon).$$

As $t \to \infty$, the solution tends asymptotically to the limit cycle $x = 2\cos(t) + O(\epsilon)$, for all initial conditions. Notice that only the initial condition $a = 2$ gives a periodic solution.

Figure 5.13 shows the error between the numerical solution and the one-term multiple scale approximation, say, x_{MS}, when $\epsilon = 0.01$, and $x(0) = 1, \dot{x}(0) = 0$.

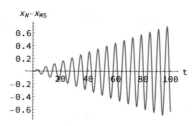

Figure 5.13: The error between the numerical solution x_N and the one-term multiple scale expansion x_{MS} for the van der Pol equation (5.13) when $\epsilon = 0.01$, and $x(0) = 1, \dot{x}(0) = 0$.

5.5 Mathematica Commands

See Chapter 3 for help with plotting phase portraits.

```
(* Limit cycle for Fitzhugh--Nagumo system and isoclines. See Figure 5.3.*)
theta=.14;gamma=2.54;epsilon=.01;omega=.112;
(* Plot isoclines in red. *)
p1=Plot[{-u (u-theta) (u-1)+omega,u/gamma},{u,-.4,1.2},PlotStyle->{Blue}];
sol=NDSolve[{u'[t]==-u[t] (u[t]-theta) (u[t]-1)-v[t]+omega,v'[t]==epsilon
(u[t]-gamma v[t]),u[0]==.5,v[0]==.1},{u[t],v[t]},{t,50,150}];
p2=ParametricPlot[Evaluate[{u[t],v[t]}/.sol],{t,50,150},
PlotStyle->{Red,Thick},
PlotRange->{{-.4,1.2},{0,.4}}, PlotPoints->100,AxesLabel->{"u","v"}];
Show[{p1,p2},PlotRange->{{-.4,1.2},{0,.4}},AxesLabel->{"u","v"},
Axes->True,BaseStyle->{FontSize->18}]
```

```
(* Perturbation Methods. Error between numerical and 2-term solutions. *)
(* See Example 8 and Figure 5.10. *)
Clear[x,y,epsilon]
epsilon=0.01;tmax=0.5;
Duffing=NDSolve[{x'[t]==y[t],y'[t]==epsilon*x[t]^3-x[t],x[0]==1,y[0]==0},
{x,y},{t,0,100},MaxSteps->Infinity];
ParametricPlot[Evaluate[{t,x[t]-Cos[t]-
epsilon*(Cos[t]/8-(Cos[t]^3)/8+3*t*Sin[t]/8)}/.Duffing],
{t,0,100},PlotPoints->1000,PlotRange->
All,AxesLabel->{"t",Subscript[x,N]-Subscript[x,0]},
Axes->True,PlotStyle->{Red,Thick},BaseStyle->{FontSize->18},
ImageSize->{360,222},AspectRatio->Full]
```

```
(* Perturbation Methods. The Lindstedt--Poincare technique. *)
(* See Example 9. *)
Clear[x,x0,x1,x2,w1,w2,epsilon];
SetAttributes[{w1,w2,epsilon},Constant];
x=x0+epsilon*x1+epsilon^2*x2;
Collect[(1+epsilon*w1+epsilon^2*w2)^2*Dt[x,{t,2}]+x-epsilon*x^3,epsilon]

DSolve[{x0''[t]+x0[t]==0,x0[0]==1,x0'[0]==0},x0[t],t]
x0[t]=Cos[t];

DSolve[{x1''[t]+x1[t]==x0[t]^3-2*w1*Dt[x0[t],{t,2}],x1[0]==0,x1'[0]==0},
x1[t],t];
Simplify[%]

x1[t]=Sin[t]^2*Cos[t]/8;w1=-3/8;

DSolve[{x2''[t]+x2[t]==3*x0[t]^2*x1[t]-2*w1*Dt[x1[t],{t,2}]-
(w1^2+2*w2)*Dt[x0[t],{t,2}],x2[0]==0,x2'[0]==0},x2[t],t];
Simplify[%]
```

5.6 Exercises

1. Prove that the system

$$\dot{x} = y + x\left(\frac{1}{2} - x^2 - y^2\right), \quad \dot{y} = -x + y\left(1 - x^2 - y^2\right)$$

   has a stable limit cycle. Plot the limit cycle.

2. By considering the flow across the square with coordinates $(1,1)$, $(1,-1)$, $(-1,-1)$, $(-1,1)$, centered at the origin, prove that the system

$$\dot{x} = -y + x\cos(\pi x), \quad \dot{y} = x - y^3$$

   has a stable limit cycle. Plot the vector field, limit cycle, and square.

3. Prove that the following systems have a unique limit cycle:

   (a) $\dot{x} = x - y - x^3$, $\dot{y} = x + y - y^3$;
   (b) $\frac{dx}{dt} = -y + x\left(1 - \mu x^2 - (\mu + \rho)y^2\right)$, $\frac{dy}{dt} = x + y\left(1 - \mu x^2 - (\mu + \rho)y^2\right)$,
   where $\mu > \rho > 0$.

4. Prove that the system.

$$\dot{x} = y + x(\alpha - x^2 - y^2), \quad \dot{y} = -x + y(1 - x^2 - y^2),$$

where $0 < \alpha < 1$, has a limit cycle and determines its stability.

5. For which parameter values does the Holling-Tanner model

$$\dot{x} = x\beta\left(1 - \frac{x}{k}\right) - \frac{rxy}{(a + ax)}, \quad \dot{y} = by\left(1 - \frac{Ny}{x}\right)$$

have a limit cycle?

6. Plot phase portraits for the Liénard system

$$\dot{x} = y - \mu(-x + x^3), \quad \dot{y} = -x,$$

when (a) $\mu = 0.01$, and (b) $\mu = 10$.

7. Prove that none of the following systems have limit cycles:

   (a) $\dot{x} = y, \quad \dot{y} = -x - (1 + x^2 + x^4)y$;
   (b) $\dot{x} = x - x^2 + 2y^2, \quad \dot{y} = y(x + 1)$;
   (c) $\dot{x} = y^2 - 2x, \quad \dot{y} = 3 - 4y - 2x^2y$;
   (d) $\dot{x} = -x + y^3 - y^4, \quad \dot{y} = 1 - 2y - x^2y + x^4$;
   (e) $\dot{x} = x^2 - y - 1, \quad \dot{y} = y(x - 2)$;
   (f) $\dot{x} = x - y^2(1 + x^3), \quad \dot{y} = x^5 - y$;
   (g) $\dot{x} = 4x - 2x^2 - y^2, \quad \dot{y} = x(1 + xy)$.

8. Prove that neither of the following systems have limit cycles using the given multipliers:

   (a) $\dot{x} = x(4 + 5x + 2y), \quad \dot{y} = y(-2 + 7x + 3y), \quad \psi = \frac{1}{xy^2}$.
   (b) $\dot{x} = x(\beta - \delta x - \gamma y), \quad \dot{y} = y(b - dy - cx), \quad \psi = \frac{1}{xy}$.

   In case (b), prove that there are no limit cycles in the first quadrant only. These differential equations were used as a general model for competing species in Chapter 10.

9. Use the Lindstedt–Poincaré technique to obtain:

   (a) a one-term uniform expansion for the ODE $\frac{d^2x}{dt^2} + x = \epsilon x\left(1 - \left(\frac{dx}{dt}\right)^2\right)$, with initial conditions $x(0) = a$ and $\dot{x}(0) = 0$.

   (b) The $O(\epsilon^2)$ solution to the van der Pol equation: $\frac{d^2x}{dt^2} + \epsilon(x^2 - 1)\frac{dx}{dt} + x = 0$, given that $x(0) = a, \dot{x}(0) = 0$.

   Hint: Show that secular terms are removed by choosing $w_1 = 0, a = 2$ and $w_2 = -\frac{1}{16}$.

   (c) The $O(\epsilon^2)$ solution to the nonlinear spring equation:

$$\frac{d^2x}{dt^2} + \epsilon x^3 + x = 0,$$

given that $x(0) = b, \dot{x}(0) = 0$.

Hint: Show that secular terms are removed by choosing $w_1 = \frac{3b^2}{8}$ and $w_2 = -\frac{21b^4}{256}$.

10 Using the method of multiple scales, show that the one-term uniform valid expansion of the ODE

$$\frac{d^2x}{dt^2} + x = -\epsilon\frac{dx}{dt},$$

with initial conditions $x(0) = b, \dot{x}(0) = 0$ is

$$x(t, \epsilon) \sim x_{MS} = be^{-\frac{\epsilon t}{2}} \cos(t),$$

as $\epsilon \to 0$.

Bibliography

[1] A. Agarwal and N. Ananthkrishnan, Bifurcation analysis for onset and cessation of surge in axial flow compressors, *International Journal of Turbo & Jet Engines*, **17**(3) (2000), 207–217.

[2] G. Bella, Multiple cycles and the Bautin bifurcation in the Goodwin model of a class struggle, *Nonlinear Analysis: Modelling and Control*, **18**(3), (2013), 265–274.

[3] R. Fitzhugh, Impulses and physiological states in theoretical models of nerve membranes, *J. Biophys.*, **1182** (1961), 445–466.

[4] E.J. Hinch, *Perturbation Methods*, Cambridge University Press, 2002.

[5] A.L. Hodgkin and A.F. Huxley, A qualitative description of membrane current and its application to conduction and excitation in nerve, J. Physiol. **117** (1952), 500–544. Reproduced in Bull. Math. Biol. **52** (1990) 25–71.

[6] A. Liénard, Étude des oscillations entrenues, *Revue Génerale de Électricité*, **23** (1928), 946–954.

[7] Han Mao'an, *Bifurcation Theory of Limit Cycles*, Alpha Science International Limited, Oxford, 2016.

[8] J. Nagumo, S. Arimoto and S. Yoshizawa, An active pulse transmission line simulating 1214-nerve axons, *Proc. IRL*, **50** (1970), 2061–2070.

[9] A.H. Nayfeh, *Perturbation Methods*, Wiley-Interscience, 2000.

[10] J.C. Neu, *Singular Perturbation in the Physical Sciences (Graduate Studies in Mathematics)*, American Mathematical Society, Rhode Island, 2015.

[11] D.B. Owens, F.J. Capone, R.M. Hall, J.M. Brandon and J.R. Chambers, Transonic free-to-roll analysis of abrupt wing stall on military aircraft, *Journal of Aircraft*, **41**(3) (2004), 474–484.

[12] M.S. Padin, F.I. Robbio, J.L. Moiola and G.R. Chen, On limit cycle approximations in the van der Pol oscillator, *Chaos, Solitons and Fractals*, **23** (2005), 207–220.

[13] L. Perko, *Differential Equations and Dynamical Systems*, 3rd ed., Springer-Verlag, Berlin, New York, Heidelberg, 2006.

[14] J. Rayleigh, *The Theory of Sound*, Dover, New York, 1945.

[15] C. Rocsoreanu, A. Georgeson and N. Giurgiteanu, *The Fitzhugh-Nagumo Model: Bifurcation and Dynamics*, Kluwer, Dordrecht, Netherlands, 2000.

[16] B. Shivamoggi, *Perturbation Methods for Differential Equations*, Birkhäuser, Boston, 2006.

[17] B. van der Pol, On relaxation oscillations, *Philos. Magazine* **7** (1926), 901–912, 946–954.

[18] Ye Yan Qian, *Theory of Limit Cycles, Translations of Mathematical Monographs*, **66**, American Mathematical Society, Rhode Island, 1986.

[19] J.A. Yorke (Contributor), K. Alligood (Ed.), and T. Sauer (Ed.), *Chaos: An Introduction to Dynamical Systems*, Springer-Verlag, New York, 1996.

Chapter 6
Hamiltonian Systems, Lyapunov Functions, and Stability

Aims and Objectives

- To study Hamiltonian systems in the plane.

- To investigate stability using Lyapunov functions.

On completion of this chapter, the reader should be able to

- prove whether or not a system is Hamiltonian;

- sketch phase portraits of Hamiltonian systems;

- use Lyapunov functions to determine the stability of a critical point; and

- distinguish between stability and asymptotic stability.

The theory of Hamiltonian (or conservative) systems in the plane is introduced. The differential equations are used to model dynamical systems in which there is no energy loss. Hamiltonian systems are also used extensively when bifurcating limit cycles in the plane (see Chapters 16 and 17).

Sometimes it is not possible to apply the linearization techniques to determine the stability of a critical point or invariant set. In certain cases, the flow across level curves, defined by Lyapunov functions, can be used to determine the stability.

6.1 Hamiltonian Systems in the Plane

Definition 1. A system of differential equations on $\Re^2$ is said to be *Hamiltonian* with one degree of freedom if it can be expressed in the form

© Springer International Publishing AG 2017   119
S. Lynch, *Dynamical Systems with Applications using Mathematica®*,
DOI 10.1007/978-3-319-61485-4_6

$$\frac{dx}{dt} = \frac{\partial H}{\partial y}, \quad \frac{dy}{dt} = -\frac{\partial H}{\partial x}, \tag{6.1}$$

where $H(x, y)$ is a twice-continuously differentiable function. The system is said to be *conservative*, and there is no dissipation. In applications, the Hamiltonian is defined by

$$H(x, y) = K(x, y) + V(x, y),$$

where K is the kinetic energy, and V is the potential energy. Hamiltonian systems with two degrees of freedom will be discussed in Chapter 9.

Theorem 1 (Conservation of Energy). *The total energy $H(x, y)$ is a first integral and a constant of the motion.*

Proof. The total derivative along a trajectory is given by

$$\frac{dH}{dt} = \frac{\partial H}{\partial x}\frac{dx}{dt} + \frac{\partial H}{\partial y}\frac{dy}{dt} = 0$$

from the chain rule and (6.1). Therefore, $H(x, y)$ is constant along the solution curves of (6.1), and the trajectories lie on the contours defined by $H(x, y) = C$, where C is a constant. □

Consider a simple mechanical system which is Hamiltonian in the plane.

The Simple Nonlinear Pendulum. The differential equation used to model the motion of a pendulum in the plane (see Figure 6.1) may be derived using Newton's law of motion:

$$\frac{d^2\theta}{dt^2} + \frac{g}{l}\sin\theta = 0, \tag{6.2}$$

where θ is the angular displacement from the vertical, l is the length of the arm of the pendulum, which swings in the plane, and g is the acceleration due to gravity.

This model does not take into account any resistive forces, so once the pendulum is set into motion, it will swing periodically forever, thus obeying the conservation of energy. The system is called conservative since no energy is lost. A periodically forced pendulum will be discussed in Chapter 9.

Let $\dot\theta = \phi$. Then, system (6.2) can be written as a planar system in the form

$$\dot\theta = \phi, \quad \dot\phi = -\frac{g}{l}\sin\theta. \tag{6.3}$$

The critical points occur at $(n\pi, 0)$ in the (θ, ϕ) plane, where n is an integer. It is not difficult to show that the critical points are hyperbolic if n is odd and nonhyperbolic if n is even. Therefore, Hartman's theorem cannot be applied when n is even. However, system (6.3) is a Hamiltonian system with $H(\theta, \phi) =$

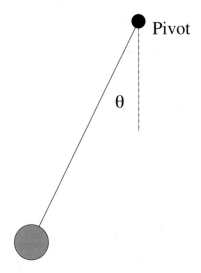

Figure 6.1:   A simple nonlinear pendulum.

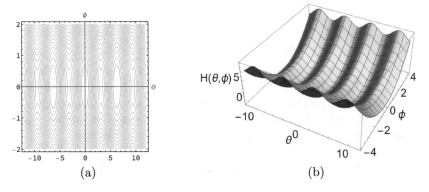

Figure 6.2: [Mathematica] (a) A contour plot for system (6.3) when $-4\pi \leq \theta \leq 4\pi$. (b) The surface $z = H(\theta, \phi)$.

$\frac{\phi^2}{2} - \frac{g}{l} \cos \theta$ (kinetic energy+potential energy), and therefore the solution curves may be plotted. The direction field may be constructed by considering $\frac{d\phi}{d\theta}$, $\dot{\theta}$, and $\dot{\phi}$. Solution curves and direction fields are given in Figure 6.2(a).

The axes of Figure 6.2(a) are the angular displacement (θ) and angular velocity ($\dot{\theta}$). The closed curves surrounding the critical points $(2n\pi, 0)$ represent periodic oscillations, and the wavy lines for large angular velocities correspond to motions in which the pendulum spins around its pivotal point. The closed curves correspond to local minima on the surface $z = H(\theta, \phi)$, and the unstable critical points correspond to local maxima on the same surface.

Definition 2. A critical point of the system

$$\dot{\mathbf{x}} = \mathbf{f}(\mathbf{x}), \quad \mathbf{x} \in \Re^2, \tag{6.4}$$

at which the Jacobian matrix has no zero eigenvalues is called a *nondegenerate critical point*; otherwise, it is called a *degenerate critical point*.

Theorem 2 *Any nondegenerate critical point of an analytic Hamiltonian system is either a saddle point or a center.*

Proof. Assume that the critical point is at the origin. The Jacobian matrix is equal to

$$J_O = \begin{pmatrix} \frac{\partial^2 H}{\partial x \partial y}(0,0) & \frac{\partial^2 H}{\partial y^2}(0,0) \\ -\frac{\partial^2 H}{\partial x^2}(0,0) & -\frac{\partial^2 H}{\partial y \partial x}(0,0) \end{pmatrix}.$$

Now $\mathrm{trace}(J_0) = 0$ and

$$\det(J_0) = \frac{\partial^2 H}{\partial x^2}(0,0)\frac{\partial^2 H}{\partial y^2}(0,0) - \left(\frac{\partial^2 H}{\partial x \partial y}(0,0)\right)^2.$$

The origin is a saddle point if $\det(J_0) < 0$. If $\det(J_0) > 0$, then the origin is either a center or a focus. Note that the critical points of system (6.1) correspond to the stationary points on the surface $z = H(x,y)$. If the origin is a focus, then the origin is not a strict local maximum or minimum of the Hamiltonian function. Suppose that the origin is a stable focus, for instance. Then

$$H(x_0, y_0) = \lim_{t \to \infty} H(x(t, x_0, y_0), y(t, x_0, y_0)) = H(0,0),$$

for all $(x_0, y_0) \in N_\epsilon(0,0)$, where N_ϵ denotes a small deleted neighborhood of the origin. However, $H(x,y) > H(0,0)$ at a local minimum and $H(x,y) < H(0,0)$ at a local maximum, a contradiction. A similar argument can be applied when the origin is an unstable focus.

Therefore, a nondegenerate critical point of a Hamiltonian is either a saddle point or a center. □

Example 1. Find the Hamiltonian for each of the following systems and sketch the phase portraits:
  (a) $\dot{x} = y, \quad \dot{y} = x + x^2$;
  (b) $\dot{x} = y + x^2 - y^2, \quad \dot{y} = -x - 2xy.$

Solution. (a) Integration gives $H(x,y) = \frac{y^2}{2} - \frac{x^2}{2} - \frac{x^3}{3}$; the solution curves are given by $H(x,y) = C$. There are two critical points at $(0,0)$ and $(-1,0)$, which are both nondegenerate. The critical point at the origin is a saddle point or col from linearization, and the eigenvectors are $(1, -1)^T$ and $(1, 1)^T$. The critical point at $(-1, 0)$ is a center from Theorem 1. If $y > 0$, then $\dot{x} > 0$, and if $y < 0$, then $\dot{x} < 0$. A phase portrait is given in Figure 6.3.

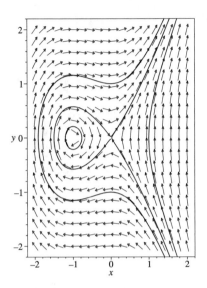

Figure 6.3: A phase portrait for Example 1(a).

(b) Integration gives $H(x, y) = \frac{x^2}{2} + \frac{y^2}{2} + x^2 y - \frac{y^3}{3}$; the solution curves are given by $H(x, y) = C$. There are four critical points at $O = (0, 0)$, $A = (0, 1)$, $B = \left(\frac{\sqrt{3}}{2}, -\frac{1}{2} \right)$, and $C = \left(-\frac{\sqrt{3}}{2}, -\frac{1}{2} \right)$, which are all nondegenerate. The critical point at the origin is a center by Theorem 1, and the critical points at A, B, and C are saddle points or cols from linearization. The eigenvectors determine the stable and unstable manifolds of the cols. The eigenvectors for point A are $(1, \sqrt{3})^T$ and $(1, -\sqrt{3})^T$; the eigenvectors for B are $(1, -\sqrt{3})^T$ and $(1, 0)^T$; and the eigenvectors for C are $(1, 0)^T$ and $(1, \sqrt{3})^T$. The solution curves and direction fields are shown in Figure 6.4.

Definition 3. Suppose that $\mathbf{x_0}$ is a critical point of system (6.4). If $\Lambda^+(\gamma) = \Lambda^-(\gamma) = \mathbf{x_0}$, then γ is a *homoclinic orbit*.

An example of a homoclinic orbit is given in Figure 6.3. The unstable and stable manifolds from the origin form a homoclinic loop around the critical point at $(-1, 0)$. A homoclinic orbit connects a critical point to itself and takes an infinite amount of time to make the connection.

Definition 4. Suppose that $\mathbf{x_0}$ and $\mathbf{y_0}$ are distinct critical points. If $\Lambda^+(\gamma) = \mathbf{x_0}$ and $\Lambda^-(\gamma) = \mathbf{y_0}$, then γ is called a *heteroclinic orbit*.

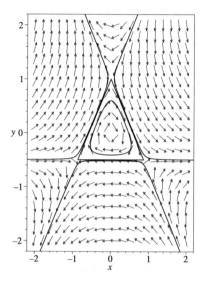

Figure 6.4:   A phase portrait for Example 1.(b). The lines $y = -\frac{1}{2}$, $y = -\sqrt{3}x + 1$, and $y = \sqrt{3}x + 1$ are invariant.

Examples of heteroclinic orbits are given in Figure 6.4. They are the three orbits lying on the line segments $\{y = -\frac{1}{2}, -\frac{\sqrt{3}}{2} < x < \frac{\sqrt{3}}{2}\}$, $\{y = -\sqrt{3}x + 1, -\frac{\sqrt{3}}{2} < x < \frac{\sqrt{3}}{2}\}$, and $\{y = \sqrt{3}x + 1, -\frac{\sqrt{3}}{2} < x < \frac{\sqrt{3}}{2}\}$.

Definition 5. A *separatrix* is an orbit that divides the phase plane into two distinctly different types of qualitative behavior. The homoclinic and hetero-clinic orbits are examples of separatrix cycles.

For example, in Figure 6.3, orbits are bounded inside the homoclinic orbit surrounding the point $(-1, 0)$ and unbounded outside it.

6.2 Lyapunov Functions and Stability

Consider nonlinear systems of the form (6.4). The stability of hyperbolic critical points may be determined from the eigenvalues of the Jacobian matrix. The critical point is stable if the real part of all of the eigenvalues is negative and unstable otherwise. If a critical point is nonhyberbolic, then a method due to Lyapunov may sometimes be used to determine the stability of the critical point.
Imagine a system defined by the potential function $V(x, y)$, where

$$\dot{x} = -\frac{\partial V}{\partial x}, \quad \dot{y} = -\frac{\partial V}{\partial y}.$$

The negative signs arise from the analogies with potential energy from physics. Now

$$\frac{dV}{dt} = \frac{\partial V}{\partial x}\frac{dx}{dt} + \frac{\partial V}{\partial y}\frac{dy}{dt} = -\left(\frac{\partial V}{\partial x}\right)^2 - \left(\frac{\partial V}{\partial y}\right)^2 \le 0.$$

This implies that $V(t)$ decreases along trajectories, and the motion is always toward lower potentials. Now $\dot{x} = \dot{y} = 0$ when $\frac{\partial V}{\partial x} = \frac{\partial V}{\partial y} = 0$, corresponding to local maxima, minima, or saddle points on $V(x,y)$. Local maxima correspond to unstable critical points, and local minima correspond to stable critical points.

Example 2. Plot a phase portrait for the system $\dot{x} = x - x^3$, $\dot{y} = -y$, and plot the potential function for this system.

Solution. There are three critical points at $O = (0,0)$, $A = (-1,0)$, and $B = (1,0)$. The origin is unstable, and the critical points A and B are stable, as seen in Figure 6.5(a). The function $z = V(x,y) = -x^2/2 + x^4/4 + y^2/2$, plotted in Figure 6.5(b) is known as the *double-well potential*. The system is *multistable* since it has two stable critical points.

The local minima in Figure 6.5(b) correspond to the stable critical points at A and B. The local maximum at the origin corresponds to the saddle point in Figure 6.5(a).

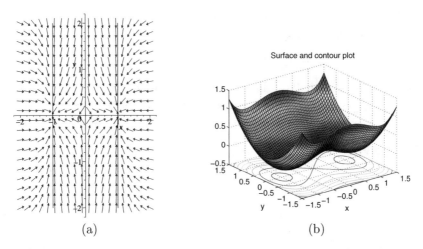

(a)   (b)

Figure 6.5: (a) A phase portrait for Example 2.. (b) The double-well potential.

Definition 6. A critical point, say, $\mathbf{x_0}$, of system (6.4) is called *stable* if given $\epsilon > 0$ there is a $\delta > 0$ such that for all $t \ge t_0$, $\| \mathbf{x}(t) - \mathbf{x_0}(t) \| < \epsilon$ whenever $\| \mathbf{x}(t_0) - \mathbf{x_0}(t_0) \| < \delta$, where $\mathbf{x}(t)$ is a solution of (6.4).

Definition 7. A critical point, say, $\mathbf{x_0}$, of system (6.4) is called *asymptotically stable* if it is stable and there is an $\eta > 0$ such that

$$\lim_{t \to \infty} \| \mathbf{x}(t) - \mathbf{x_0}(t) \| = 0,$$

whenever $\| \mathbf{x}(t_0) - \mathbf{x_0}(t_0) \| < \eta$.

A trajectory near a stable critical point will remain close to that point, whereas a trajectory near an asymptotically stable critical point will move closer and closer to the critical point as $t \to \infty$.

The following theorem holds for system (6.4) when $\mathbf{x} \in \Re^n$. Examples in $\Re^3$ are given in Chapter 8.

The Lyapunov Stability Theorem. *Let E be an open subset of $\Re^n$ containing an isolated critical point $\mathbf{x_0}$. Suppose that $\mathbf{f}$ is continuously differentiable and that there exists a continuously differentiable function, say, $V(\mathbf{x})$, which satisfies the conditions:*

- *$V(\mathbf{x_0}) = 0$;*

- *$V(\mathbf{x}) > 0$, if $\mathbf{x} \neq \mathbf{x_0}$,*

where $\mathbf{x} \in \Re^n$. Then

1. *if $\dot{V}(\mathbf{x}) \leq 0$ for all $\mathbf{x} \in E$, $\mathbf{x_0}$ is stable;*
2. *if $\dot{V}(\mathbf{x}) < 0$ for all $\mathbf{x} \in E$, $\mathbf{x_0}$ is asymptotically stable;*
3. *if $\dot{V}(\mathbf{x}) > 0$ for all $\mathbf{x} \in E$, $\mathbf{x_0}$ is unstable.*

Proof. 1. Choose a small neighborhood N_ϵ surrounding the critical point $\mathbf{x_0}$. In this neighborhood, $\dot{V}(\mathbf{x}) \leq 0$, so a positive semiorbit starting inside N_ϵ remains there forever. The same conclusion is drawn no matter how small ϵ is chosen to be. The critical point is therefore stable.

2. Since $\dot{V}(\mathbf{x}) < 0$, the Lyapunov function must decrease monotonically on every positive semiorbit $\mathbf{x}(t)$. Let ϕ_t be the flow defined by $\mathbf{f}(\mathbf{x})$. Then, either $V(\phi_t) \to \mathbf{x_0}$ as $t \to \infty$ or there is a positive semiorbit $\mathbf{x}(t)$ such that

$$V(\phi_t) \geq n > 0, \quad \text{for all } t \geq t_0, \tag{6.5}$$

for some $n > 0$. Since $\mathbf{x_0}$ is stable, there is an annular region A, defined by $n \leq V(\mathbf{x}) \leq c$, containing this semiorbit. Suppose that $\dot{V}$ attains its upper bound in A, say, $-N$, so

$$\dot{V}(\mathbf{x}) \leq -N < 0, \quad \mathbf{x} \in A, \quad N > 0.$$

Integration gives

$$V(\mathbf{x}(t)) - V(\mathbf{x}(t_0)) \leq -N(t - t_0),$$

where $t > t_0$. This contradicts (6.5), and therefore no path fails to approach the critical point at $\mathbf{x_0}$. The critical point is asymptotically stable.

3. Since $\dot{V}(\mathbf{x}) > 0$, $V(\mathbf{x})$ is strictly increasing along trajectories of (11.4). If ϕ_t is the flow of (11.4), then

$$V(\phi_t) > V(\mathbf{x_0}) > 0$$

for $t > 0$ in a small neighborhood of $\mathbf{x_0}$, N_ϵ. Therefore,

$$V(\phi_t) - V(\mathbf{x_0}) \geq kt$$

for some constant k and $t \geq 0$. Hence for sufficiently large t,

$$V(\phi_t) > kt > K,$$

where K is the maximum of the continuous function $V(\mathbf{x})$ on the compact set $\overline{N_\epsilon}$. Therefore, ϕ_t lies outside the closed set N_ϵ and $\mathbf{x_0}$ is unstable. $\quad\square$

Definition 8. The function $V(\mathbf{x})$ is called a *Lyapunov function*.

Unfortunately, there is no systematic way to construct a Lyapunov function. The Lyapunov functions required for specific examples will be given in this book. Note that if $\dot{V}(\mathbf{x}) = 0$, then all trajectories lie on the curves (surfaces in $\Re^n$) defined by $V(\mathbf{x}) = C$, where C is a constant. The quantity $\dot{V}$ gives the rate of change of V along trajectories, or in other words, $\dot{V}$ gives the direction that trajectories cross the level curves $V(\mathbf{x}) = C$.

Example 3. Determine the stability of the origin for the system

$$\dot{x} = -y^3, \quad \dot{y} = x^3.$$

Solution. The eigenvalues are both zero and the origin is a degenerate critical point. A Lyapunov function for this system is given by $V(x, y) = x^4 + y^4$, and furthermore

$$\frac{dV}{dt} = \frac{\partial V}{\partial x}\frac{dx}{dt} + \frac{\partial V}{\partial y}\frac{dy}{dt} = 4x^3(-y^3) + 4y^3(x^3) = 0.$$

Hence the solution curves lie on the closed curves given by $x^4 + y^4 = C$. The origin is thus stable but not asymptotically stable. The trajectories that start near to the origin remain there but do not approach the origin asymptotically. If $y > 0$, then $\dot{x} < 0$, and if $y < 0$, then $\dot{x} > 0$. The level curves and direction fields are given in Figure 6.6.

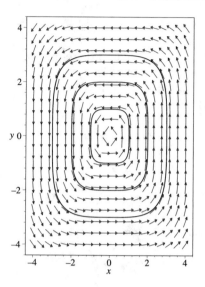

Figure 6.6:   A phase portrait for Example 3.

Example 4. Investigate the stability of the origin for the system

$$\dot{x} = y, \quad \dot{y} = -x - y(1 - x^2)$$

using the Lyapunov function $V(x, y) = x^2 + y^2$.

Solution. Now

$$\frac{dV}{dt} = \frac{\partial V}{\partial x}\frac{dx}{dt} + \frac{\partial V}{\partial y}\frac{dy}{dt} = 2x(y) + 2y(-x - y + yx^2),$$

so

$$\frac{dV}{dt} = 2y^2(x^2 - 1)$$

and $\dot{V} \leq 0$ if $|x| \leq 1$. Therefore, $\dot{V} = 0$, if either $y = 0$ or $x = \pm 1$. When $y = 0$, $\dot{x} = 0$ and $\dot{y} = -x$, which means that a trajectory will move off the line $y = 0$ when $x \neq 0$. Hence if a trajectory starts inside the circle of radius one centered at the origin, then it will approach the origin asymptotically. The origin is asymptotically stable.

Definition 9. Given a Lyapunov function $V(x, y)$, the *Lyapunov domain of stability* is defined by the region for which $\dot{V}(x, y) < 0$.

Example 5. Prove that the origin of the system

$$\dot{x} = -8x - xy^2 - 3y^3, \quad \dot{y} = 2x^2y + 2xy^2$$

is asymptotically stable using the Lyapunov function $V(x, y) = 2x^2 + 3y^2$. Determine the Lyapunov domain of stability based on $V(x, y)$.

Solution. Now

$$\dot{V} = 4x(-8x - xy^2 - 3y^3) + 6y(2x^2y + 2xy^2) = 8x^2(y^2 - 4)$$

and $\dot{V} \leq 0$ if $|y| \leq 2$. Therefore, $\dot{V} = 0$ if either $x = 0$ or $y = \pm 2$. When $x = 0$, $\dot{x} = -3y^3$, and $\dot{y} = 0$, which means that a trajectory will move off the line $x = 0$ when $y \neq 0$. Now $\dot{V} < 0$ if $|y| < 2$. This implies that $\dot{V} < 0$ as long as $V(x, y) = 2x^2 + 3y^2 < 12$. This region defines the domain of Lyapunov stability. Therefore, if a trajectory lies wholly inside the ellipse $2x^2 + 3y^2 = 12$, it will move to the origin asymptotically. Hence the origin is asymptotically stable.

An approximation of the true domain of stability for the origin of the system in Example 5. is indicated in Figure 6.7(a). Notice that it is larger than the Lyapunov domain of stability (Figure 6.7(b)) and that the x-axis is invariant.

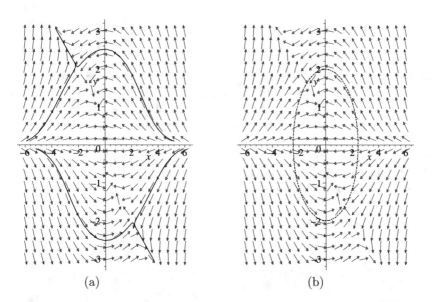

(a)   (b)

Figure 6.7:   (a) A phase portrait for Example 5.. (b) The domain of Lyapunov stability.

Example 6. Consider the system defined by:

$$\frac{dx}{dt} = -x + y^2 + 2x^2, \quad \frac{dy}{dt} = -y + y^2.$$

Prove that the origin is asymptotically stable within a suitable basin of attraction. Determine this basin of attraction given that $V(x, y) = x^2 + y^2$ is a suitable Lyapunov function.

Solution. Now

$$\dot{V} = (2x)(-x + y^2 + 2x^2) + (2y)(-y + y^2) = 2\left(x^2(-1 + 2x) + y^2(y - 1 + x)\right).$$

Therefore, $\frac{dV}{dt} < 0$, as long as $x < \frac{1}{2}$ and $y < 1 - x$. Plotting these lines, it is not difficult to see that the basin of attraction is estimated by $V(x, y) < \frac{1}{4}$.

Example 7. A suitable Lyapunov function for the recurrent Hopfield network modeled using the differential equations

$$\dot{x} = -x + 2\left(\frac{2}{\pi}\tan^{-1}\left(\frac{\gamma\pi x}{2}\right)\right), \quad \dot{y} = -y + 2\left(\frac{2}{\pi}\tan^{-1}\left(\frac{\gamma\pi y}{2}\right)\right),$$

is given by

$$V(a_1, a_2) = -\left(a_1^2 + a_2^2\right) - \frac{4}{\gamma\pi^2}\left(\ln\left(\cos\left(\frac{\pi a_1}{2}\right)\right) + \ln\left(\cos\left(\frac{\pi a_2}{2}\right)\right)\right),$$

where

$$a_1(t) = \frac{2}{\pi}\tan^{-1}\left(\frac{\gamma\pi x}{2}\right) \quad \text{and} \quad a_2(t) = \frac{2}{\pi}\tan^{-1}\left(\frac{\gamma\pi y}{2}\right).$$

Set $\gamma = 0.7$. A vector field plot for the recurrent Hopfield network is given in Chapter 18. There are nine critical points, four are stable, and five are unstable.

Plot the function $V(a_1, a_2)$ and the corresponding contour plot when $|a_i| \leq 1, i = 1, 2$. Continuous Hopfield models are discussed in Chapter 20.

Solution. Figure 6.8(a) shows the surface plot $V(a_1, a_2)$ when $\gamma = 0.7$, there is one local maximum, and there are four local minima. Figure 6.8(b) shows the corresponding contour plot.

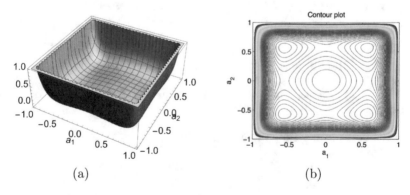

(a)   (b)

Figure 6.8: [Mathematica] The Lyapunov function $V(a_1, a_2)$ when $\gamma = 0.7$. (a) Surface plot; (b) contour plot.

6.3 Mathematica Commands

See Chapter 3 for help with plotting phase portraits.

```
(* Contour plot. See Figure 6.2(a). *)
ContourPlot[y^2/2-Cos[x],{x,-12,12},{y,-2,2},PlotPoints->50,
ContourShading->False,Contours->20,Axes->True,AxesLabel->
{"\[Theta]","\[Phi]"},BaseStyle->{FontSize->18}]

(* Surface plot. See Figure 6.2(b). *)
gamma=.7;
Plot3D[y^2/2-Cos[x],{x,-12,12},{y,-4,4},PlotPoints->100,AxesLabel->
{"\[Theta]","\[Phi]","H(\[Theta],\[Phi])"},BaseStyle->{FontSize->18}]
```

```
(* Three-D plot of surface. See Figure 6.8(a). *)
gamma=0.7;Plot3D[- ((x^2+y^2)+(4/(gamma Pi^2))(Log[Cos[Pi x/2]]+
Log[Cos[Pi y/2]])),{x,-1,1},{y,-1,1},PlotPoints->50,Axes->True,
AxesLabel->{Subscript[a, 1],Subscript[a, 2]}]

(* Contour plot. See Figure 6.8(b). *)
gamma=0.7;ContourPlot[- ((x^2+y^2)+(4/(gamma Pi^2))(Log[Cos[Pi x/2]]+
Log[Cos[Pi y/2]])),{x,-1,1},{y,-1,1},PlotPoints->50,ContourShading->False,
Contours->50,Axes->True,AxesLabel->{Subscript[a, 1],Subscript[a, 2]}]

(* Density plot. Variation of Figure 6.8(b). *)
DensityPlot[-(x^2+y^2)-(4/(gamma Pi^2))(Log[Cos[Pi x/2]]+
Log[Cos[Pi y/2]]),{x,-1,1},{y,-1,1},PlotPoints->50,
ColorFunction->(GrayLevel[#^.4]&),Mesh->False,Axes->True,
AxesLabel->{Subscript[a, 1],Subscript[a, 2]}]
```

6.4 Exercises

1. Find the Hamiltonian of the system

$$\dot{x} = y, \quad \dot{y} = x - x^3$$

   and sketch a phase portrait.

2. Given the Hamiltonian function $H(x,y) = \frac{y^2}{2} + \frac{x^2}{2} - \frac{x^4}{4}$, sketch a phase portrait for the Hamiltonian system.

3. Plot a phase portrait for the damped pendulum equation

$$\ddot{\theta} + 0.15\dot{\theta} + \sin\theta = 0$$

and describe what happens physically.

4. Plot a phase portrait of the system

$$\dot{x} = y(y^2 - 1), \quad \dot{y} = x(1 - x^2).$$

5. Investigate the stability of the critical points at the origin for the systems:

(a) $\dot{x} = -y - x^3$, $\dot{y} = x - y^3$, using the Lyapunov function $V(x, y) = x^2 + y^2$;

(b) $\dot{x} = x(x - \alpha)$, $\dot{y} = y(y - \beta)$, using the Lyapunov function

$$V(x, y) = \left(\frac{x}{\alpha}\right)^2 + \left(\frac{y}{\beta}\right)^2;$$

(c) $\dot{x} = y$ $\dot{y} = y - x^3$, using the Lyapunov function $V(x, y) = ax^4 + bx^2 + cxy + dy^2$.

6. Prove that the origin is a unique critical point of the system

$$\dot{x} = -\frac{1}{2}y(1 + x) + x(1 - 4x^2 - y^2), \quad \dot{y} = 2x(1 + x) + y(1 - 4x^2 - y^2).$$

Determine the stability of the origin using the Lyapunov function $V(x, y) = (1 - 4x^2 - y^2)^2$. Find $\Lambda^+(p)$ for each $p \in \Re^2$. Plot a phase portrait.

7. Determine the values of a for which $V(x, y) = x^2 + ay^2$ is a Lyapunov function for the system

$$\dot{x} = -x + y - x^2 - y^2 + xy^2, \quad \dot{y} = -y + xy - y^2 - x^2y.$$

8. Determine the basin of attraction of the origin for the system

$$\dot{x} = x(x^2 + y^2 - 4) - y, \quad \dot{y} = x + y(x^2 + y^2 - 4)$$

using the Lyapunov function $V(x, y) = x^2 + y^2$.

9. Plot a phase portrait for the system in Exercise 8.

10. Given the parameters $k, \epsilon > 0$, show that the origin of the damped harmonic oscillator modeled with the differential equations:

$$\dot{x} = y, \quad \dot{y} = -kx - \epsilon y^3 \left(1 + x^2\right),$$

is asymptotically stable using the Lyapunov function, $V(x, y) = kx^2 + y^2$.

Bibliography

[1] A. Bacciotti and L. Rosier, *Liapunov Functions and Stability in Control Theory*, Springer-Verlag, New York, 2001.

[2] A.V. Bolsinov and A.T. Fomenko, *Integrable Hamiltonian Systems: Geometry, Topology, Classification*, CRC Press, 2004.

[3] R.L. Devaney, M. Hirsch and S. Smale, *Differential Equations, Dynamical Systems, and an Introduction to Chaos*, 2nd ed., Academic Press, New York, 2003.

[4] P. Giesl, *Construction of Global Lyapunov Functions Using Radial Basis Functions (Lecture Notes in Mathematics, No. 1904)*, Springer-Verlag, New York, 2007.

[5] W.H. Haddad and V. Chellaboina, *Nonlinear Dynamical Systems and Control: A Lyapunov-Based Approach*, Princeton University Press, Princeton, NJ, 2008.

[6] J.P. Lasalle, *Stability by Liapunov's Direct Method: With Applications*, Academic Press, New York, 1961.

[7] J. Llibre, R. Moeckel, C. Simó, *Central Configurations, Periodic Orbits, and Hamiltonian Systems (Advanced Courses in Mathematics - CRM Barcelona)*, Birkhäuser, Boston, 2015.

[8] J.H. Lowenstein, *Essentials of Hamiltonian Dynamics*, Cambridge University Press, 2012.

[9] A.N. Michel, Ling Hou and Derong Liu, *Stability of Dynamical Systems: On the Role of Monotonic and Non-Monotonic Lyapunov Functions (Systems & Control: Foundations & Applications)*, Birkhäuser, Boston, 2015.

[10] J. Moser, Recent developments in the theory of Hamiltonian systems, *Siam Review*, **28**-4 (1986), 459–485.

[11] V.V. Nemitskii and V.V. Stepanov, *Qualitative Theory of Differential Equations*, Princeton University Press, Princeton, NJ, 1960.

[12] G.M. Zaslavsky, *Hamiltonian Chaos and Fractional Dynamics*, Oxford University Press, Oxford, UK, 2008.

[13] G.M. Zaslavsky, *Physics of Chaos in Hamiltonian Systems*, World Scientific, Singapore, 1998.

Chapter 7
Bifurcation Theory

Aims and Objectives

- To introduce bifurcation theory of continuous systems in the plane.

- To introduce the notion of steady-state solution and investigate multistability and bistability.

- To introduce the theory of normal forms.

On completion of this chapter, the reader should be able to

- describe how a phase portrait changes as a parameter changes;

- plot bifurcation diagrams;

- take transformations to obtain simple normal forms; and

- interpret the bifurcation diagrams in terms of physical behavior.

If the behavior of a dynamical system changes suddenly as a parameter is varied, then it is said to have undergone a bifurcation. At a point of bifurcation, stability may be gained or lost.

It may be possible for a nonlinear system to have more than one steady-state solution. For example, different initial conditions can lead to different stable solutions. A system of this form is said to be multistable. Bifurcations of so-called large-amplitude limit cycles are discussed. By introducing a feedback mechanism into the system it is possible to obtain hysteresis or bistable behavior.

© Springer International Publishing AG 2017
S. Lynch, *Dynamical Systems with Applications using Mathematica®*,
DOI 10.1007/978-3-319-61485-4_7

7.1 Bifurcations of Nonlinear Systems in the Plane

Definition 1. A vector field $\mathbf{f} \in \Re^2$, which is continuously differentiable, is called *structurally stable* if small perturbations in the system $\dot{\mathbf{x}} = \mathbf{f}(\mathbf{x})$ leave the qualitative behavior unchanged. If small perturbations cause a change in the qualitative behavior of the system, then $\mathbf{f}$ is called *structurally unstable*.

For example, the Lotka-Volterra model (Example 2, Chapter 4) is structurally unstable, while the Holling-Tanner model (Example 3, Chapter 4) is structurally stable.

Peixoto's Theorem in the Plane. *Let the vector field $\mathbf{f}$ be continuously differentiable on a compact set, say, D. Then $\mathbf{f}$ is structurally stable on D if and only if*

- *the number of critical points and limit cycles is finite and each is hyperbolic;*

- *there are no trajectories connecting saddle points to saddle points.*

Consider systems of the form

$$\dot{\mathbf{x}} = \mathbf{f}(\mathbf{x}, \mu), \tag{7.1}$$

where $\mathbf{x} \in \Re^2$ and $\mu \in \Re$. A value, say, μ_0, for which the vector field $\mathbf{f}(\mathbf{x}, \mu_0)$ is not structurally stable is called a *bifurcation value*.

Four simple types of bifurcation, all at nonhyperbolic critical points, will be given in order to illustrate how the qualitative behavior of a structurally unstable system of differential equations can change with respect to a parameter value. Certain bifurcations can be classified by so-called *normal forms*. By finding suitable transformations, it is possible to reduce systems to a normal form. Schematic diagrams depicting four normal form bifurcations are illustrated below, and the theory of normal forms is introduced in the next section along with some simple examples.

7.1.I A Saddle-Node Bifurcation. Consider the system

$$\dot{x} = \mu - x^2, \quad \dot{y} = -y. \tag{7.2}$$

The critical points are found by solving the equations $\dot{x} = \dot{y} = 0$. There are (i) zero, (ii) one, or (iii) two critical points, depending on the value of μ. Consider the three cases separately.

Case (i). When $\mu < 0$, there are no critical points in the plane and the flow is from right to left since $\dot{x} < 0$. If $y > 0$, then $\dot{y} < 0$ and if $y < 0$, then $\dot{y} > 0$. A plot of the vector field is given in Figure 7.1(a). Note that the flow is invariant on the x-axis.

Case (ii). When $\mu = 0$, there is one critical point at the origin and it is nonhyperbolic. The solution curves may be found by solving the differential equation

$$\frac{dy}{dx} = \frac{\dot{y}}{\dot{x}} = \frac{y}{x^2}.$$

This is a separable differential equation (see Chapter 2) and the solution is given by $|y| = Ke^{-\frac{1}{x}}$, where K is a constant. Note that $\dot{x} < 0$ for all x. The vector field is plotted in Figure 7.1(b). Note that the flow is invariant along both the x-axis and the y-axis.

Case (iii). When $\mu > 0$, there are two critical points at $A = (\sqrt{\mu}, 0)$ and $B = (-\sqrt{\mu}, 0)$. Linearize in the usual way. The Jacobian matrix is given by

$$J = \begin{pmatrix} \frac{\partial P}{\partial x} & \frac{\partial P}{\partial y} \\ \frac{\partial Q}{\partial x} & \frac{\partial Q}{\partial y} \end{pmatrix} = \begin{pmatrix} -2x & 0 \\ 0 & -1 \end{pmatrix},$$

where $\dot{x} = P(x, y)$ and $\dot{y} = Q(x, y)$. Therefore,

$$J_A = \begin{pmatrix} -2\sqrt{\mu} & 0 \\ 0 & -1 \end{pmatrix}$$

and the eigenvalues and eigenvectors are given by $\lambda_1 = -2\sqrt{\mu}$, $(1, 0)^T$ and $\lambda_2 = -1$, $(0, 1)^T$. The critical point at A is thus a stable node and the stable manifolds are orthogonal to one another.

The Jordan matrix for the critical point at B is

$$J_B = \begin{pmatrix} 2\sqrt{\mu} & 0 \\ 0 & -1 \end{pmatrix}$$

and the eigenvalues and eigenvectors are $\lambda_1 = 2\sqrt{\mu}$, $(1, 0)^T$ and $\lambda_2 = -1$, $(0, 1)^T$. This critical point is a saddle point. The vector field and orthogonal stable and unstable manifolds are plotted in Figure 7.1(c).

In summary, there are no critical points if μ is negative; there is one nonhyperbolic critical point at the origin if $\mu = 0$; and there are two critical points—one a saddle and the other a node—when μ is positive. The qualitative behavior of the system changes as the parameter μ passes through the bifurcation value $\mu_0 = 0$. The behavior of the critical points can be summarized on a *bifurcation diagram* as depicted in Figure 7.2.

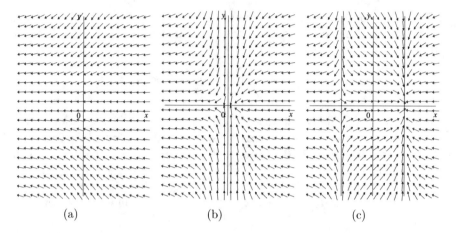

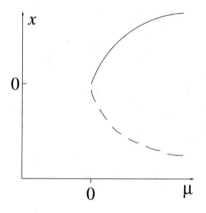

Figure 7.1: Vector field plots and manifolds when (a) $\mu < 0$, (b) $\mu = 0$, and (c) $\mu > 0$. There are no manifolds when $\mu < 0$.

Figure 7.2: A schematic of a bifurcation diagram for system (7.2) showing a saddle-node bifurcation. The solid curves depict stable behavior and the dashed curves depict unstable behavior.

When $\mu < 0$, there are no critical points, and as μ passes through zero the qualitative behavior changes and two critical points bifurcate from the origin. As μ increases, the critical points move farther and farther apart. Note that the critical points satisfy the equation $\mu = x^2$, hence the parabolic form of the bifurcation curve. More examples of saddle-node bifurcations are given in Section 7.3.

7.1.II A Transcritical Bifurcation. Consider the system

$$\dot{x} = \mu x - x^2, \quad \dot{y} = -y. \tag{7.3}$$

The critical points are found by solving the equations $\dot{x} = \dot{y} = 0$. There are either one or two critical points depending on the value of the parameter μ. The bifurcation value is again $\mu_0 = 0$. Consider the cases (i) $\mu < 0$, (ii) $\mu = 0$, and (iii) $\mu > 0$ separately.

Case (i). When $\mu < 0$, there are two critical points, one at $O = (0,0)$ and the other at $A = (\mu, 0)$. The origin is a stable node and A is a saddle point. A vector field and manifolds are plotted in Figure 7.3(a).

Case (ii). When $\mu = 0$, there is one nonhyperbolic critical point at the origin. The solution curves satisfy the differential equation

$$\frac{dy}{dx} = \frac{y}{x^2}$$

which has solutions $|y| = Ke^{-\frac{1}{x}}$, where K is a constant. A vector field and the manifolds through the origin are shown in Figure 7.3(b).

Case (iii). When $\mu > 0$, there are two critical points, one at $O = (0,0)$ and the other at $B = (\mu, 0)$. The origin is now a saddle point, and B is a stable node. A vector field and manifolds are plotted in Figure 7.3(c).

The behavior of the critical points can be summarized on a bifurcation diagram as depicted in Figure 7.4.

7.1.III A Pitchfork Bifurcation. Consider the system

$$\dot{x} = \mu x - x^3, \quad \dot{y} = -y. \tag{7.4}$$

The critical points are found by solving the equations $\dot{x} = \dot{y} = 0$. There are either one or three critical points depending on the value of the parameter μ. The bifurcation value is again $\mu_0 = 0$. Consider the cases (i) $\mu < 0$, (ii) $\mu = 0$, and (iii) $\mu > 0$, separately.

Case (i). When $\mu < 0$, there is one critical point at $O = (0,0)$. The origin is a stable node. A vector field and the manifolds at the origin are shown in Figure 7.5(a).

Case (ii). When $\mu = 0$, there is one nonhyperbolic critical point at the origin. The solution curves satisfy the differential equation

$$\frac{dy}{dx} = \frac{y}{x^3}$$

which has solutions $|y| = Ke^{-\frac{1}{2x^2}}$, where K is a constant. A vector field is plotted in Figure 7.5(a).

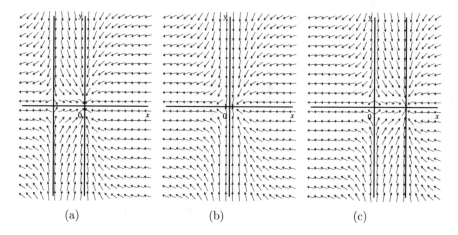

(a)   (b)   (c)

Figure 7.3: Vector field plots and manifolds when (a) $\mu < 0$, (b) $\mu = 0$, and (c) $\mu > 0$.

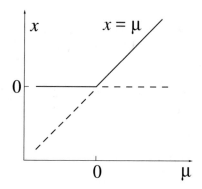

Figure 7.4: A bifurcation diagram for system (7.3) showing a transcritical bifurcation. The solid lines depict stable behavior and the dashed lines depict unstable behavior.

Case (iii). When $\mu > 0$, there are three critical points at $O = (0,0)$, $A = (\sqrt{\mu}, 0)$, and $B = (-\sqrt{\mu}, 0)$. The origin is now a saddle point and A and B are both stable nodes. A vector field and all of the stable and unstable manifolds are plotted in Figure 7.5(b).

The behavior of the critical points can be summarized on a bifurcation diagram as depicted in Figure 7.6.

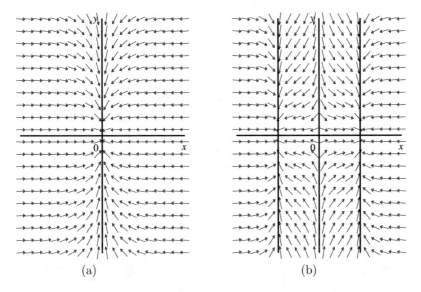

Figure 7.5: Vector field plots and manifolds when (a) $\mu \leq 0$ and (b) $\mu > 0$.

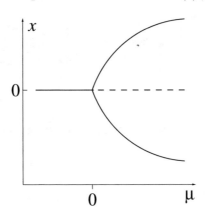

Figure 7.6: A schematic of a bifurcation diagram for system (7.4) showing a pitchfork bifurcation. The solid curves depict stable behavior and the dashed curves depict unstable behavior. Note the resemblance of the stable branches to a pitchfork.

7.1.IV A Hopf Bifurcation. Consider the system

$$\dot{r} = r(\mu - r^2), \quad \dot{\theta} = -1. \tag{7.5}$$

The origin is the only critical point since $\dot{\theta} \neq 0$. There are no limit cycles if (i) $\mu \leq 0$ and one if (ii) $\mu > 0$. Consider the two cases separately.

Case (i). When $\mu \leq 0$, the origin is a stable focus. Since $\dot{\theta} < 0$, the flow is clockwise. A phase portrait and vector field is shown in Figure 7.7(a).

Case (ii). When $\mu > 0$, there is an unstable focus at the origin and a stable limit cycle at $r = \sqrt{\mu}$ since $\dot{r} > 0$ if $0 < r < \sqrt{\mu}$ and $\dot{r} < 0$ if $r > \sqrt{\mu}$. A phase portrait is shown in Figure 7.7(b).

The qualitative behavior can be summarized on a bifurcation diagram as shown in Figure 7.8. As the parameter μ passes through the bifurcation value $\mu_0 = 0$, a limit cycle bifurcates from the origin. The amplitude of the limit cycle grows as μ increases. Think of the origin blowing a smoke ring. An animation of a Hopf bifurcation is given in the Mathematica notebook and the program is listed in Section 7.4.

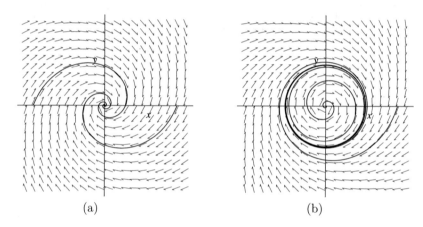

(a)   (b)

Figure 7.7: [Mathematica animation] Phase portraits when (a) $\mu \leq 0$ and (b) $\mu > 0$.

7.2 Normal Forms

This section introduces some basic theory of normal forms without any rigorous justification. To keep the theory simple, the author has decided to illustrate the method for planar systems only. Note that the theory can be applied to n-dimensional systems in general, see [2], [8], and [9]. The theory of normal forms began with Poincaré and Dulac and was later applied to Hamiltonian systems by Birkhoff.

The basic idea is to take nonlinear transformations of the nonlinear system $\dot{\mathbf{x}} = \mathbf{X}(\mathbf{x})$ to obtain a linear system $\dot{\mathbf{u}} = J\mathbf{u}$, where $\mathbf{X}(\mathbf{0}) = \mathbf{0}$, $(\mathbf{x}, \mathbf{X}, \mathbf{u} \in \Re^2)$ and J is a Jacobian matrix (see Section 3.4). The nonlinear terms are removed in a sequential manner starting with the quadratic terms. Of course, it is not always possible to obtain a linear system. In the majority of cases, one has to be satisfied with a "simplest" possible form, or normal form, which may not be

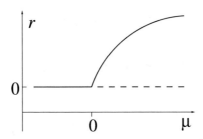

Figure 7.8: A schematic of a bifurcation diagram for system (7.5) showing a Hopf bifurcation. The solid curves depict stable behavior and the dashed curves depict unstable behavior.

unique. Normal forms are useful in the study of the local qualitative behavior of critical points and bifurcation problems.

In order to keep the explanations simple, we will start by trying to eliminate the quadratic terms of a planar system. Suppose that

$$\dot{\mathbf{w}} = A\mathbf{w} + \mathbf{H}_2(\mathbf{w}) + O\left(|\mathbf{w}|^3\right), \tag{7.6}$$

where $\mathbf{w} \in \Re^2$, $\mathbf{H}_2$ is a homogeneous polynomial vector of degree two, A is a 2×2 matrix, and $O\left(|\mathbf{w}|^3\right)$ denotes higher order terms. Let $\mathbf{w} = P\mathbf{x}$, then system (7.6) becomes

$$P\dot{\mathbf{x}} = AP\mathbf{x} + \mathbf{H}_2(P\mathbf{x}) + O\left(|\mathbf{x}|^3\right)$$

and multiplying by P^{-1}

$$\dot{\mathbf{x}} = J\mathbf{x} + \mathbf{h}_2(\mathbf{x}) + O\left(|\mathbf{x}|^3\right), \tag{7.7}$$

where P is such that $J = P^{-1}AP$ is a Jacobian matrix, and $\mathbf{h}_2(\mathbf{x}) = P^{-1}\mathbf{H}_2(P\mathbf{x})$ is a homogeneous vector of degree two. Take a transformation of the form

$$\mathbf{x} = \mathbf{u} + \mathbf{f}_2(\mathbf{u}) + O\left(|\mathbf{u}|^3\right), \tag{7.8}$$

Substitute (7.8) into (7.7). Thus,

$$\dot{\mathbf{u}} + D\mathbf{f}_2(\mathbf{u})\dot{\mathbf{u}} + O\left(|\mathbf{u}|^2\right)\dot{\mathbf{u}} = J\left(\mathbf{u} + \mathbf{f}_2(\mathbf{u}) + O\left(|\mathbf{u}|^3\right)\right) +$$

$$\mathbf{h}_2\left(\mathbf{u} + \mathbf{f}_2(\mathbf{u}) + O\left(|\mathbf{u}|^3\right)\right) + O\left(|\mathbf{u}|^3\right),$$

where D is the matrix of partial derivatives, an explicit example is given below. Now $\mathbf{h}_2\left(\mathbf{u} + \mathbf{f}_2(\mathbf{u})\right) = \mathbf{h}_2(\mathbf{u}) + O\left(|\mathbf{u}|^3\right)$ and $\dot{\mathbf{u}} = J\mathbf{u} + O\left(|\mathbf{u}|^2\right)$, therefore

$$\dot{\mathbf{u}} = J\mathbf{u} - (D\mathbf{f}_2(\mathbf{u})J\mathbf{u} - J\mathbf{f}_2(\mathbf{u})) + \mathbf{h}_2(\mathbf{u}) + O\left(|\mathbf{u}|^3\right). \qquad (7.9)$$

Equation (7.9) makes it clear how one may remove the quadratic terms by a suitable choice of the homogeneous quadratic polynomial $\mathbf{f}_2$. To eliminate the quadratic terms, one must find solutions to the equation

$$D\mathbf{f}_2(\mathbf{u})J\mathbf{u} - J\mathbf{f}_2(\mathbf{u}) = \mathbf{h}_2(\mathbf{u}). \qquad (7.10)$$

The method of normal forms will now be illustrated by means of simple examples.

Example 1. Determine the nonlinear transformation which eliminates terms of degree two from the planar system.

$$\dot{x} = \lambda_1 x + a_{20}x^2 + a_{11}xy + a_{02}y^2, \quad \dot{y} = \lambda_2 y + b_{20}x^2 + b_{11}xy + b_{02}y^2, \quad (7.11)$$

where $\lambda_{1,2} \neq 0$.

Solution. Now

$$J = \begin{pmatrix} \lambda_1 & 0 \\ 0 & \lambda_2 \end{pmatrix}$$

and

$$\mathbf{h}_2(\mathbf{x}) = \begin{pmatrix} h_{12} \\ h_{22} \end{pmatrix} = \begin{pmatrix} a_{20}x^2 + a_{11}xy + a_{02}y^2 \\ b_{20}x^2 + b_{11}xy + b_{02}y^2 \end{pmatrix}.$$

Equating coefficients of u^2, uv, and v^2, equation (7.10) can be written in the matrix form

$$MF = H$$

or more explicitly

$$\begin{pmatrix} \lambda_1 & 0 & 0 & 0 & 0 & 0 \\ 0 & \lambda_2 & 0 & 0 & 0 & 0 \\ 0 & 0 & 2\lambda_2 - \lambda_1 & 0 & 0 & 0 \\ 0 & 0 & 0 & 2\lambda_1 - \lambda_2 & 0 & 0 \\ 0 & 0 & 0 & 0 & \lambda_1 & 0 \\ 0 & 0 & 0 & 0 & 0 & \lambda_2 \end{pmatrix} \begin{pmatrix} f_{20} \\ f_{11} \\ f_{02} \\ g_{20} \\ g_{11} \\ g_{02} \end{pmatrix} = \begin{pmatrix} a_{20} \\ a_{11} \\ a_{02} \\ b_{20} \\ b_{11} \\ b_{02} \end{pmatrix}.$$

The inverse of matrix M exists if and only if all of the diagonal elements are nonzero. The computations above may be checked with Mathematica.

Definition 2. The 2-tuple of eigenvalues (λ_1, λ_2) is said to be resonant of order 2 if at least one of the diagonal elements of M is zero.

Therefore, if none of the diagonal elements of M are zero

$$f_{20} = \frac{a_{20}}{\lambda_1}, \ f_{11} = \frac{a_{11}}{\lambda_2}, \ f_{02} = \frac{a_{02}}{2\lambda_2 - \lambda_1}, \ g_{20} = \frac{b_{20}}{2\lambda_1 - \lambda_2}, \ g_{11} = \frac{b_{11}}{\lambda_1}, \ g_{02} = \frac{b_{02}}{\lambda_2},$$

and all of the quadratic terms can be eliminated from system (12.11) resulting in a linear normal form $\dot{\mathbf{u}} = J\mathbf{u}$.

Example 2. Find the change of coordinates of the form $\mathbf{x} = \mathbf{u} + \mathbf{f}_2(\mathbf{u})$ which transforms the system

$$\begin{pmatrix} \dot{x} \\ \dot{y} \end{pmatrix} = \begin{pmatrix} 5 & 0 \\ 0 & 3 \end{pmatrix} \begin{pmatrix} x \\ y \end{pmatrix} + \begin{pmatrix} 5x^2 \\ 0 \end{pmatrix}, \tag{7.12}$$

into the form

$$\begin{pmatrix} \dot{u} \\ \dot{v} \end{pmatrix} = \begin{pmatrix} 5 & 0 \\ 0 & 3 \end{pmatrix} \begin{pmatrix} u \\ v \end{pmatrix} + \mathrm{O}\left(|\mathbf{u}|^3\right), \tag{7.13}$$

Transform the system to verify the results.

Solution. Using the results from Example 1, $f_{20} = 1$, and

$$x = u + u^2, \quad y = v.$$

Differentiating with respect to time gives

$$\dot{x} = \dot{u} + 2u\dot{u}, \quad \dot{y} = \dot{v}.$$

Therefore,

$$\dot{u} = \frac{\dot{x}}{1 + 2u} = \frac{5x + 5x^2}{1 + 2u}, \quad \dot{v} = \dot{y} = 3y = 3v.$$

Now, taking a Taylor series expansion about $u = 0$,

$$\frac{1}{1 + 2u} = 1 - 2u + 4u^2 - 8u^3 + \mathrm{O}\left(u^4\right),$$

and

$$5x + 5x^2 = 5(u + u^2) + 5(u + u^2)^2 = 5u + 10u^2 + 10u^3 + \mathrm{O}\left(u^4\right).$$

Therefore,

$$\dot{u} = 5u\left(1 + 2u + 2u^2 + \mathrm{O}\left(u^3\right)\right)\left(1 - 2u + 4u^2 + \mathrm{O}\left(u^3\right)\right), \quad \dot{v} = 3v.$$

Finally, the linearized system is

$$\dot{u} = 5u + \mathrm{O}\left(u^3\right), \quad \dot{v} = 3v.$$

Note that in general, any terms that cannot be eliminated are called resonance terms, as the third example demonstrates.

Example 3. Determine the normal form of the following system with a non-hyperbolic critical point at the origin.

$$\begin{pmatrix} \dot{x} \\ \dot{y} \end{pmatrix} = \begin{pmatrix} \lambda_1 & 0 \\ 0 & 0 \end{pmatrix} \begin{pmatrix} x \\ y \end{pmatrix} + \begin{pmatrix} a_{20}x^2 + a_{11}xy + a_{02}y^2 \\ b_{20}x^2 + b_{11}xy + b_{02}y^2 \end{pmatrix} + \mathrm{O}\left(|\mathbf{x}|^3\right), \quad (7.14)$$

where $\lambda_1 \neq 0$.

Solution. Referring to Example 1, in this case $\lambda_2 = 0$, and the zero elements in matrix M are in the second and sixth rows. Therefore, there are resonance terms, auv and bv^2, and the normal form of equation (7.14) is given by

$$\begin{pmatrix} \dot{u} \\ \dot{v} \end{pmatrix} = \begin{pmatrix} \lambda_1 & 0 \\ 0 & 0 \end{pmatrix} \begin{pmatrix} u \\ v \end{pmatrix} + \begin{pmatrix} auv \\ bv^2 \end{pmatrix} + \mathrm{O}\left(|\mathbf{u}|^3\right).$$

7.3 Multistability and Bistability

There are two types of Hopf bifurcation, one in which stable limit cycles are created about an unstable critical point, called the *supercritical Hopf bifurcation* (see Figure 7.8), and the other in which an unstable limit cycle is created about a stable critical point, called the *subcritical Hopf bifurcation* (see Figure 7.9).

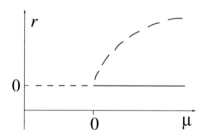

Figure 7.9:   A schematic of a bifurcation diagram showing a subcritical Hopf bifurcation. The solid curves depict stable behavior and the dashed curves depict unstable behavior.

In the engineering literature, supercritical bifurcations are sometimes called *soft* (or *safe*); the amplitude of the limit cycles builds up gradually as the parameter, μ in this case, is moved away from the bifurcation point. In contrast, subcritical bifurcations are *hard* (or *dangerous*). A steady state, say at the origin, could become unstable as a parameter varies and the nonzero solutions could tend to infinity. An example of this type of behavior can be found in Figure 7.9. As μ passes through zero from positive to negative values, the steady-state solution at the origin becomes unstable and trajectories starting anywhere other than the origin would tend to infinity.

It is also possible for limit cycles of finite amplitude to suddenly appear as the parameter μ is varied. These limit cycles are known as *large-amplitude limit cycles*. Examples of this type of behavior include surge oscillations in axial flow compressors and wing rock oscillations in aircraft flight dynamics, see [7] for examples. Generally, unstable limit cycles are not observed in physical applications, so it is only the stable large-amplitude limit cycles that are of interest. These limit cycles can appear in one of the two ways; either there is a jump from a stable critical point to a stable large-amplitude limit cycle or there is a jump from one stable limit cycle to another of larger amplitude. These bifurcations are illustrated in the following examples.

Large-Amplitude Limit Cycle Bifurcations.

Consider the system

$$\dot{r} = r(\mu + r^2 - r^4), \quad \dot{\theta} = -1. \qquad (7.15)$$

The origin is the only critical point since $\dot{\theta} \neq 0$. This critical point is stable if $\mu < 0$ and unstable if $\mu > 0$. The system undergoes a subcritical Hopf bifurcation at $\mu = 0$ as in Figure 7.9. However, the new feature here is the stable large-amplitude limit cycle which exists for, say, $\mu > \mu_S$. In the range $\mu_S < \mu < 0$, there exist two different steady-state solutions; hence system (7.15) is multistable in this range. The choice of initial conditions determines which steady state will be approached as $t \to \infty$.

Definition 3. A dynamical system, say (7.1), is said to be multistable if there is more than one possible steady-state solution for a fixed value of the parameter μ. The steady state obtained depends on the initial conditions.

The existence of multistable solutions allows for the possibility of *bistability* (or *hysteresis*) as a parameter is varied. The two essential ingredients for bistable behavior are nonlinearity and *feedback*. To create a bistable region, there must be some history in the system. Bistability is also discussed at some length in Chapter 16 when investigating nonlinear optical fiber resonators. Suppose that the parameter μ is increased from some value less than μ_S. The steady state remains at $r = 0$ until $\mu = 0$, where the origin loses stability. There is a sudden jump (a subcritical Hopf bifurcation) to the large-amplitude limit cycle, and the steady state remains on this cycle as μ is increased further. If the parameter μ is now decreased, then the steady state remains on the large-amplitude limit cycle until $\mu = \mu_S$, where the steady state suddenly jumps back to the origin (a saddle-node bifurcation) and remains there as μ is decreased further. In this way, a bistable region is obtained as depicted in Figure 7.10.

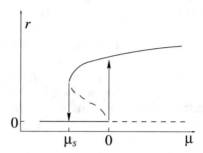

Figure 7.10:   A schematic of a bifurcation diagram depicting bistable behavior for system (7.15).

Definition 4. A dynamical system, say (7.1), has a bistable solution if there are two steady states for a fixed parameter μ and the steady state obtained depends on the history of the system.

Now consider the system

$$\dot{r} = r(\mu - 0.28r^6 + r^4 - r^2). \quad \dot{\theta} = -1. \tag{7.16}$$

A bistable region may be obtained by increasing and then decreasing the parameter μ as in the example above. A possible bifurcation diagram is given in Figure 7.11. In this case, there is a supercritical Hopf bifurcation at $\mu = 0$ and saddle-node bifurcations at μ_B and μ_A, respectively.

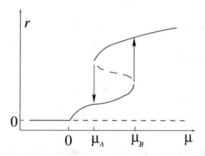

Figure 7.11:   A schematic of a bifurcation diagram depicting bistable behavior for system (7.16).

Jumps between different steady states have been observed in mechanical systems. Parameters need to be chosen which avoid such large-amplitude limit cycle bifurcations, and research is currently under way in this respect.

Bistability also has many positive applications in the real world; for example, *nonlinear bistable optical resonators* are investigated in Chapter 16. The author is also currently investigating multistability and bistability in a wide range of disciplines, including biology.

A Saddle-Node on an Invariant Cycle (SNIC) Bifurcation. In this case, a limit cycle bifurcates on an invariant cycle as a saddle point and a stable node collide. This type of bifurcation is best illustrated by means of an example.

Example 4. Consider the system:

$$\frac{dx}{dt} = x\left(1 - x^2 - y^2\right) - y(1 + \gamma + x), \quad \frac{dy}{dt} = y\left(1 - x^2 - y^2\right) + x(1 + \gamma + x),$$

where γ is a constant. Convert this system to polar form and show that the system has three critical points for $-2 < \gamma < 0$, and one critical point and one limit cycle, if either $\gamma > 0$ or if $\gamma < -2$. Determine the stability of the critical points, use the Manipulate command to produce an animation of the phase portrait and plot a bifurcation diagram as γ increases from $\gamma = -3$ to $\gamma = 1$.

Solution. Converting to polar coordinates one obtains:

$$\dot{r} = r\left(1 - r^2\right), \quad \dot{\theta} = 1 + \gamma + r\cos(\theta).$$

The origin $O = (0,0)$ is a critical point and there are additional two critical points at $A = \left(1, \cos^{-1}(-1 - \gamma)\right)$ and $B = \left(1, -\cos^{-1}(-1 - \gamma)\right)$. The Jacobian matrix is:

$$J = \begin{pmatrix} 1 - 3r^2 & 0 \\ \cos(\theta) & -r\sin(\theta) \end{pmatrix}.$$

When $-2 < \gamma < 0$, there is an invariant circle with a saddle point and a stable node which collide and vanish when either $\gamma = -2$ or $\gamma = 0$. The origin is unstable as, $\dot{r} > 0$, for $0 < r < 1$, the critical point at A is a saddle point as $\det(J_A) = -2\sqrt{1 - (\gamma + 1)^2} < 0$ and the critical point at B is stable as trace $(J_B) = -2 - 2\sqrt{1 - (\gamma + 1)^2} < 0, \det(J_B) = 2\sqrt{1 - (\gamma + 1)^2} > 0$. When $\gamma > 0$ or $\gamma < -2$, there is a stable limit cycle of radius one and one unstable critical point at the origin. Figure 7.12 shows a schematic of a saddle-node invariant cycle bifurcation diagram for Example 4 and a Mathematica program showing an animation of the bifurcation is listed in the next section.

7.4 Mathematica Commands

```
(* Animation of a simple curve. See Exercise 8(b). *)
Manipulate[Plot[r ((r-1)^2-mu r),{r,-3,3},PlotRange->{-3,3}],{mu,-1,1}]
```

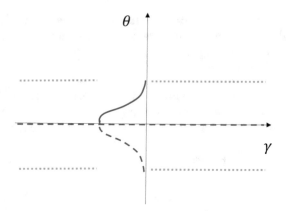

Figure 7.12: [Mathematica animation] A schematic of a bifurcation diagram showing a saddle-node invariant cycle (SNIC) bifurcation. The solid red curves depict stable critical points, the dashed red curves denote unstable critical points, and the blue dots represent a stable limit cycle.

```
(* Animation of a subcritical Hopf bifurcation. See Figure 7.7. *)
Manipulate[Module[{sol},sol=NDSolve[{x'[t]==y[t]+mu x[t]-x[t] y[t]^2,
y'[t]==mu y[t]-x[t]-y[t]^3,x[0]==.5,y[0]==.5},{x[t],y[t]},{t,0,80}];
ParametricPlot[Evaluate[Table[{x[t],y[t]}/.sol]],{t,0,80},
PlotRange->{{-1,1},{-1,1}}, PlotPoints->100]],{mu,-1,0.3}]
```

```
(* Saddle-Node on Invariant Circle (SNIC) bifurcation. *)
(* Animation displaying a SNIC bifurcation. See Figure 7.12. *)
Manipulate[Module[{sol},sol=NDSolve[{x'[t]==x[t]*(1-x[t]^2-y[t]^2)-
y[t]*(1+gamma+x[t]),
y'[t]==y[t]*(1-x[t]^2-y[t]^2)+x[t]*(1+gamma+x[t]),x[0]==.5,y[0]==.5},
{x[t],y[t]},{t,0,80}];ParametricPlot[Evaluate[{x[t],y[t]}/.sol],{t,0,80},
PlotRange->{{-2,2},{-2,2}},PlotPoints->100]],{gamma,-3,1}]
```

7.5 Exercises

1. Consider the following one-parameter families of first order differential equations defined on $\Re$:

   (a) $\dot{x} = \mu - x - e^{-x}$;

   (b) $\dot{x} = x(\mu + e^x)$;

   (c) $\dot{x} = x - \frac{\mu x}{1+x^2}$.

Determine the critical points and the bifurcation values, plot vector fields on the line, and draw a bifurcation diagram in each case.

Use the Manipulate command in Mathematica to show how $\dot{x}$ varies as μ increases from -4 to $+4$, for each of the differential equations in (a)-(c).

2. Construct first order ordinary differential equations having the following:

   (a) three critical points (one stable and two unstable) when $\mu < 0$, one critical point when $\mu = 0$, and three critical points (one unstable and two stable) when $\mu > 0$;

   (b) two critical points (one stable and one unstable) for $\mu \neq 0$ and one critical point when $\mu = 0$;

   (c) one critical point if $|\mu| \geq 1$ and three critical points if $|\mu| < 1$.

   Draw a bifurcation diagram in each case.

3. A certain species of fish in a large lake is harvested. The differential equation used to model the population, $x(t)$ in hundreds of thousands, is given by

$$\frac{dx}{dt} = x\left(1 - \frac{x}{5}\right) - \frac{hx}{0.2 + x}.$$

   Determine and classify the critical points and plot a bifurcation diagram. How can the model be interpreted in physical terms?

4. Consider the following one-parameter systems of differential equations:

   (a) $\dot{x} = x, \quad \dot{y} = \mu - y^4$;

   (b) $\dot{x} = x^2 - x\mu^2, \quad \dot{y} = -y$;

   (c) $\dot{x} = -x^4 + 5\mu x^2 - 4\mu^2, \quad \dot{y} = -y.$

   Find the critical points, plot phase portraits, and sketch a bifurcation diagram in each case.

5. Consider the following one-parameter systems of differential equations in polar form:

   (a) $\dot{r} = \mu r(r + \mu)^2, \quad \dot{\theta} = 1$;

   (b) $\dot{r} = r(\mu - r)(\mu - 2r), \quad \dot{\theta} = -1$;

   (c) $\dot{r} = r(\mu^2 - r^2), \quad \dot{\theta} = 1.$

   Plot phase portraits for $\mu < 0$, $\mu = 0$, and $\mu > 0$ in each case. Sketch the corresponding bifurcation diagrams.

6. Determine the nonlinear transformation which eliminates the terms of degree three from the planar system

$$\dot{x} = \lambda_1 x + a_{30}x^3 + a_{21}x^2y + a_{12}xy^2 + a_{03}y^3,$$
$$\dot{y} = \lambda_2 y + b_{30}x^3 + b_{21}x^2y + b_{12}xy^2 + b_{03}y^3,$$

where $\lambda_{1,2} \neq 0$.

7. Show that the normal form of a nondegenerate Hopf singularity is given by

$$\begin{pmatrix} \dot{u} \\ \dot{v} \end{pmatrix} = \begin{pmatrix} 0 & -\beta \\ \beta & 0 \end{pmatrix} \begin{pmatrix} u \\ v \end{pmatrix} + \begin{pmatrix} au\left(u^2 + v^2\right) - bv\left(u^2 + v^2\right) \\ av\left(u^2 + v^2\right) + bu\left(u^2 + v^2\right) \end{pmatrix} + \mathrm{O}\left(|\mathbf{u}|^5\right),$$

where $\beta > 0$ and $a \neq 0$.

8. Plot bifurcation diagrams for the planar systems

   (a) $\dot{r} = r\left(\mu - 0.2r^6 + r^4 - r^2\right), \quad \dot{\theta} = -1,$

   (b) $\dot{r} = r\left((r-1)^2 - \mu r\right), \quad \dot{\theta} = 1.$

   Give a possible explanation as to why the type of bifurcation in part (b) should be known as a *fold bifurcation*.

9. Show that the one-parameter system

$$\dot{x} = y + \mu x - xy^2, \quad \dot{y} = \mu y - x - y^3$$

   undergoes a Hopf bifurcation at $\mu_0 = 0$. Plot phase portraits and sketch a bifurcation diagram.

10. Thus far, the analysis has been restricted to bifurcations involving only one-parameter, and these are known as *codimension-1 bifurcations*. This example illustrates what can happen when two parameters are varied, allowing so-called *codimension-2 bifurcations*.
    The following two-parameter system of differential equations may be used to model a simple laser:

$$\dot{x} = x(y - 1), \quad \dot{y} = \alpha + \beta y - xy.$$

    Find and classify the critical points and sketch the phase portraits. Illustrate the different types of behavior in the (α, β) plane and determine whether or not any bifurcations occur.

Bibliography

[1] D.K. Arrowsmith and C.M. Place, *An Introduction to Dynamical Systems*, Cambridge University Press, Cambridge, UK, 1990.

[2] A.D. Bruno and V.F. Edneral, Normal forms and integrability of ODE systems, *Computer Algebra in Scientific Computing, Proceedings Lecture Notes in Computer Science* **3718**, (2005), 65–74.

[3] M. Demazure and D. Chillingworth (Translator), *Bifurcations and Catastrophes: Geometry of Solutions to Nonlinear Problems*, Springer-Verlag, New York, 2000.

[4] K. Ikeda and K. Murota, *Imperfect Bifurcations in Structures and Materials*, Springer-Verlag, New York, 2002.

[5] G. Iooss and D.D. Joseph, *Elementary Stability and Bifurcation Theory*, Springer-Verlag, New York, 1997.

[6] S. Liebscher, *Bifurcation without Parameters (Lecture Notes in Mathematics)*, Springer, New York, 2015.

[7] S. Lynch and C.J. Christopher, Limit cycles in highly nonlinear differential equations, *Journal of Sound and Vibration* **224**-3, (1999), 505–517.

[8] J. Murdock, *Normal Forms and Unfoldings for Local Dynamical Systems*, Springer-Verlag, New York, 2003.

[9] A.H. Nayfeh, *Method of Normal Forms*, Wiley Series in Nonlinear Science, New York, 1993.

[10] L.E. Reichl and W.C. Schieve, *Instabilities, Bifurcations, and Fluctuations in Chemical Systems*, University of Texas Press, Texas, 2015.

[11] M. Rocco ed., *Bifurcation and Chaos in Science and Engineering*, Clanrye International, New York, 2015.

[12] R. Seydel, *Practical Bifurcation and Stability Analysis, From Equilibrium to Chaos*, Springer-Verlag, New York, 1994.

[13] Shangjiang Guo and Jianhong Wu, *Bifurcation Theory of Functional Differential Equations*, Springer-Verlag, New York, 2013.

[14] S.H. Strogatz, *Nonlinear Dynamics and Chaos with Applications to Physics, Biology, Chemistry and Engineering*, Perseus Books, New York, 2001.

[15] J.M.T. Thompson and H.B. Stewart, *Nonlinear Dynamics and Chaos*, Second Edition, Wiley, New York, 2002.

Chapter 8
Three-Dimensional Autonomous Systems and Chaos

Aims and Objectives

- To introduce first-order ODEs in three variables.

- To plot phase portraits and chaotic attractors.

- To identify chaos.

On completion of this chapter the reader should be able to

- construct phase portraits for linear systems in three dimensions;

- use Mathematica to plot phase portraits and time series for nonlinear systems;

- identify chaotic solutions;

- interpret the solutions to modeling problems taken from various scientific disciplines, and in particular, chemical kinetics, electric circuits, and meteorology.

Three-dimensional autonomous systems of differential equations are considered. Critical points and stability are discussed, and the concept of chaos is introduced. Examples include the Lorenz equations, used as a simple meteorological model and in the theory of lasers; Chua's circuit, used in nonlinear electronics and radiophysics; and the Belousov-Zhabotinski reaction, used in chemistry and biophysics. All of these systems can display highly complex behavior that can be interpreted from phase portrait analysis or Poincaré maps (see Chapter 9).

Basic concepts are explained by means of example rather than mathematical rigor. Strange or chaotic attractors are constructed using Mathematica, and the reader is encouraged to investigate these systems through the exercises at the end of the chapter. Chaos will also be discussed in other chapters of the book.

© Springer International Publishing AG 2017   155
S. Lynch, *Dynamical Systems with Applications using Mathematica®*,
DOI 10.1007/978-3-319-61485-4_8

8.1 Linear Systems and Canonical Forms

Consider linear three-dimensional autonomous systems of the form

$$\dot{x} = a_{11}x + a_{12}y + a_{13}z,$$
$$\dot{y} = a_{21}x + a_{22}y + a_{23}z,$$
$$\dot{z} = a_{31}x + a_{32}y + a_{33}z, \tag{8.1}$$

where the a_{ij} are constants. The existence and uniqueness theorem (see Section 2.4) holds, which means that trajectories do not cross in three-dimensional space. The real canonical forms for 3×3 matrices are

$$J_1 = \begin{pmatrix} \lambda_1 & 0 & 0 \\ 0 & \lambda_2 & 0 \\ 0 & 0 & \lambda_3 \end{pmatrix}, \quad J_2 = \begin{pmatrix} \alpha & -\beta & 0 \\ \beta & \alpha & 0 \\ 0 & 0 & \lambda_3 \end{pmatrix},$$

$$J_3 = \begin{pmatrix} \lambda_1 & 1 & 0 \\ 0 & \lambda_1 & 0 \\ 0 & 0 & \lambda_2 \end{pmatrix}, \quad J_4 = \begin{pmatrix} \lambda_1 & 1 & 0 \\ 0 & \lambda_1 & 1 \\ 0 & 0 & \lambda_1 \end{pmatrix}.$$

Matrix J_1 has three real eigenvalues; matrix J_2 has a pair of complex eigenvalues; and matrices J_3 and J_4 have repeated eigenvalues. The type of phase portrait is determined from each of these canonical forms.

Definition 1. Suppose that $\mathbf{0} \in \Re^3$ is a critical point of the system (8.1). Then the stable and unstable manifolds of the critical point $\mathbf{0}$ are defined by

$$E_S(\mathbf{0}) = \{\mathbf{x} : \Lambda^+(\mathbf{x}) = \mathbf{0}\}, \quad E_U(\mathbf{0}) = \{\mathbf{x} : \Lambda^-(\mathbf{x}) = \mathbf{0}\}.$$

Example 1. Solve the following system of differential equations, sketch a phase portrait, and define the manifolds:

$$\dot{x} = x, \quad \dot{y} = y, \quad \dot{z} = -z. \tag{8.2}$$

Solution. There is one critical point at the origin. Each differential equation is integrable with solutions given by $x(t) = C_1 e^t$, $y(t) = C_2 e^t$, and $z(t) = C_3 e^{-t}$. The eigenvalues and corresponding eigenvectors are $\lambda_{1,2} = 1, (0,1,0)^T, (1,0,0)^T$ and $\lambda_3 = -1, (0,0,1)^T$. System (8.2) may be uncoupled in any of the xy, xz, or yz planes. Planar analysis gives an unstable singular node in the xy plane and cols in each of the xz and yz planes. The phase plane portraits for two of the uncoupled systems are given in Figure 8.1. If $z > 0$, $\dot{z} < 0$, and if $z < 0$, $\dot{z} > 0$. The z-axis is a one-dimensional stable manifold since trajectories on this line are attracted to the origin as $t \to +\infty$. The xy plane is a two-dimensional

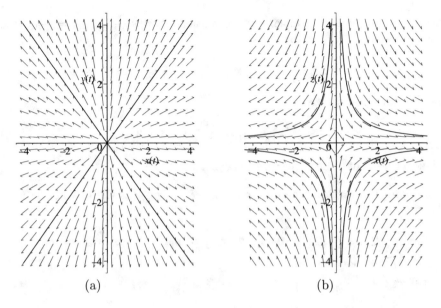

Figure 8.1: Phase plane portraits in the (a) xy and (b) xz planes. Note that (a) is an unstable planar manifold.

unstable manifold since all trajectories in this plane are attracted to the origin as $t \rightarrow -\infty$.

Putting all of this together, any trajectories not lying on the manifolds flow along "lampshades" in three-dimensional space, as depicted in Figure 8.2.

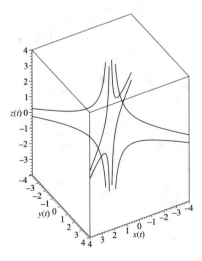

Figure 8.2: Phase portrait for system (8.2). The manifolds are not shown here.

Example 2. Given the linear transformations $x = x_1 - 2y_1$, $y = -y_1$, and $z = -y_1 + z_1$, show that the system

$$\dot{x}_1 = -3x_1 + 10y_1, \quad \dot{y}_1 = -2x_1 + 5y_1, \quad \dot{z}_1 = -2x_1 + 2y_1 + 3z_1$$

can be transformed into

$$\dot{x} = x - 2y, \quad \dot{y} = 2x + y, \quad \dot{z} = 3z. \tag{8.3}$$

Make a sketch of some trajectories in xyz space.

Solution. The origin is the only critical point. Consider the transformations. Then

$$\dot{x} = \dot{x}_1 - 2\dot{y}_1 = (-3x_1 + 10y_1) - 2(-2x_1 + 5y_1) = x_1 = x - 2y$$
$$\dot{y} = -\dot{y}_1 = -(-2x_1 + 5y_1) = 2x_1 - 5y_1 = 2x + y$$
$$\dot{z} = -\dot{y}_1 + \dot{z}_1 = -(-2x_1 + 5y_1) + (-2x_1 + 2y_1 + 3z_1) = 3(-y_1 + z_1) = 3z.$$

System (8.3) is already in canonical form, and the eigenvalues are $\lambda_{1,2} = 1 \pm i$ and $\lambda_3 = 3$; hence the critical point is hyperbolic. The system can be uncoupled; the critical point at the origin in the xy plane is an unstable focus. A phase plane portrait is given in Figure 8.3.

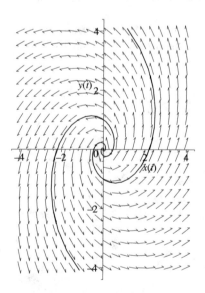

Figure 8.3: Some trajectories in the xy plane.

Note that all trajectories spiral away from the origin, as depicted in Figure 8.4. Since all trajectories tend to the origin as $t \to -\infty$, the whole phase space forms an unstable manifold.

Example 3. Solve the following initial value problem

$$\dot{x} = z - x, \quad \dot{y} = -y, \quad \dot{z} = z - 17x + 16, \tag{8.4}$$

with $x(0) = y(0) = z(0) = 0.8$ and plot the solution curve in three-dimensional space.

Solution. System (8.4) can be uncoupled. The differential equation $\dot{y} = -y$ has general solution $y(t) = y_0 e^{-t}$, and substituting $y_0 = 0.8$ gives $y(t) = 0.8e^{-t}$. Now $z = \dot{x} + x$, and therefore the equation $\dot{z} = z - 17x + 16$ becomes

$$(\ddot{x} + \dot{x}) = (\dot{x} + x) - 17x + 16,$$

which simplifies to

$$\ddot{x} + 16x = 16.$$

Take Laplace transforms of both sides and insert the initial conditions to obtain

$$\bar{x}(s) = \frac{1}{s} - \frac{0.2s}{s^2 + 16}.$$

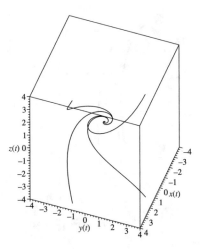

Figure 8.4: Phase portrait for system (8.3).

Take inverse transforms to get

$$x(t) = 1 - 0.2 \cos(4t),$$

and therefore
$$z(t) = 1 + 0.8\sin(4t) - 0.2\cos(4t).$$

The solution curve is plotted in Figure 8.5.

8.2 Nonlinear Systems and Stability

If the critical point of a three-dimensional autonomous system is hyperbolic, then the linearization methods of Hartman can be applied. If the critical point is not hyperbolic, then other methods need to be used.

Definition 2. Suppose that $\mathbf{p} \in \Re^3$ is a critical point of the nonlinear system $\dot{\mathbf{x}} = \mathbf{f}(\mathbf{x})$, where $\mathbf{x} \in \Re^3$. Then the stable and unstable manifolds of the critical point $\mathbf{p}$ are defined by

$$W_S(\mathbf{p}) = \{\mathbf{x} : \Lambda^+(\mathbf{x}) = \mathbf{p}\}, \quad W_U(\mathbf{p}) = \{\mathbf{x} : \Lambda^-(\mathbf{x}) = \mathbf{p}\}.$$

As for two-dimensional systems, three-dimensional systems can have stable and unstable manifolds. These manifolds can be convoluted surfaces in three-dimensional space. A survey of methods used for computing some manifolds is presented in [10].

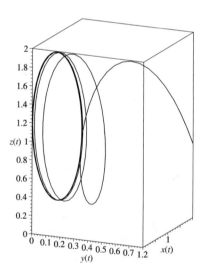

Figure 8.5:   The solution curve for the initial value problem in Example 3. The trajectory ends up on an ellipse in the $y = 0$ plane.

Theorem 1 *Consider the differential equation*

$$\dot{\mathbf{x}} = \mathbf{f}(\mathbf{x}), \quad \mathbf{x} \in \Re^n,$$

where $\mathbf{f} \in C^1(E)$ and E is an open subset of $\Re^n$ containing the origin. Suppose that $\mathbf{f}(\mathbf{0}) = \mathbf{0}$ and that the Jacobian matrix has n eigenvalues with nonzero real part. Then, in a small neighborhood of $\mathbf{x} = \mathbf{0}$, there exist stable and unstable manifolds W_S and W_U with the same dimensions n_S and n_U as the stable and unstable manifolds (E_S, E_U) of the linearized system

$$\dot{\mathbf{x}} = J\mathbf{x},$$

where W_S and W_U are tangent to E_S and E_U at $\mathbf{x} = \mathbf{0}$.

A proof to this theorem can be found in Hartman's book, see Chapter 3.

Definition 3. The *center eigenspace*, say, E_C, is defined by the eigenvectors corresponding to the eigenvalues with zero real part, and the *center manifold*, say, W_C, is the invariant subspace which is tangent to the center eigenspace E_C. In general, the center manifold is not unique.

Theorem 2 (The Center Manifold Theorem). *Let $\mathbf{f} \in C^r(E), (r \geq 1)$, where E is an open subset of $\Re^n$ containing the origin. If $\mathbf{f}(\mathbf{0}) = \mathbf{0}$ and the Jacobian matrix has n_S eigenvalues with negative real part, n_U eigenvalues with positive real part, and $n_C = n - n_S - n_U$ purely imaginary eigenvalues, then there exists an n_C-dimensional center manifold W_C of class C^r which is tangent to the center manifold E_C of the linearized system.*

To find out more about center manifolds, see Wiggins [18].

Example 4. Determine the stable, unstable, and center manifolds of the non-linear system

$$\dot{x} = x^2, \quad \dot{y} = -y, \quad \dot{z} = -2z.$$

Solution. There is a unique critical point at the origin. This system is easily solved, and it is not difficult to plot phase portraits for each of the uncoupled systems. The solutions are $x(t) = \frac{1}{C_1 - t}$, $y(t) = C_2 e^{-t}$, and $z(t) = C_3 e^{-2t}$. The eigenvalues and corresponding eigenvectors of the Jacobian matrix are $\lambda_1 = 0$, $(1, 0, 0)^T$, $\lambda_2 = -1$, $(0, 1, 0)^T$, and $\lambda_3 = -2$, $(0, 0, 1)^T$. In this case, $W_C = E_C$, the x-axis, and the yz plane forms a two-dimensional stable manifold, where $W_S = E_S$. Note that the center manifold is unique in this case, but it is not in general.

Example 5. Solve the following nonlinear differential system

$$\dot{x} = -x, \quad \dot{y} = -y + x^2, \quad \dot{z} = z + x^2,$$

and determine the stable and unstable manifolds.

Solution. The point $O = (0,0,0)$ is a unique critical point. Linearize by finding the Jacobian matrix. Hence

$$J = \begin{pmatrix} \frac{\partial P}{\partial x} & \frac{\partial P}{\partial y} & \frac{\partial P}{\partial z} \\ \frac{\partial Q}{\partial x} & \frac{\partial Q}{\partial y} & \frac{\partial Q}{\partial z} \\ \frac{\partial R}{\partial x} & \frac{\partial R}{\partial y} & \frac{\partial R}{\partial z} \end{pmatrix},$$

where $\dot{x} = P(x, y, z), \dot{y} = Q(x, y, z)$, and $\dot{z} = R(x, y, z)$. Therefore,

$$J_O = \begin{pmatrix} -1 & 0 & 0 \\ 0 & -1 & 0 \\ 0 & 0 & 1 \end{pmatrix},$$

and the origin is an unstable critical point. Note that two of the eigenvalues are negative. These give a two-dimensional stable manifold, which will now be defined.

The differential equation $\dot{x} = -x$ is integrable and has solution $x(t) = C_1 e^{-t}$. The other two differential equations are linear and have solutions $y(t) = C_2 e^{-t} + C_1^2(e^{-t} - e^{-2t})$ and $z(t) = C_3 e^t + \frac{C_1^2}{3}(e^t - e^{-2t})$. Now $\Lambda^+(\mathbf{x}) = \mathbf{0}$ if and only if $C_3 + \frac{C_1^2}{3} = 0$, where $\mathbf{x} \in \Re^3$, $C_1 = x(0), C_2 = y(0)$, and $C_3 = z(0)$. Therefore, the stable manifold is given by

$$W_S = \left\{ \mathbf{x} \in \Re^3 : z = -\frac{x^2}{3} \right\}.$$

Using similar arguments, $\Lambda^-(\mathbf{x}) = \mathbf{0}$ if and only if $C_1 = C_2 = 0$. Hence the unstable manifold is given by

$$W_U = \{ \mathbf{x} \in \Re^3 : x = y = 0 \}.$$

Note that the surface W_S is tangent to the xy plane at the origin.

Example 6. Sketch a phase portrait for the system

$$\dot{x} = x + y - x(x^2 + y^2), \quad \dot{y} = -x + y - y(x^2 + y^2), \quad \dot{z} = -z. \quad (8.5)$$

Solution. Convert to cylindrical polar coordinates by setting $x = r\cos\theta$ and $y = r\sin\theta$. System (8.5) then becomes

$$\dot{r} = r(1 - r^2), \quad \dot{\theta} = -1, \quad \dot{z} = -z.$$

The origin is the only critical point. The system uncouples; in the xy plane, the flow is clockwise and the origin is an unstable focus. If $z > 0$, then $\dot{z} < 0$, and if $z < 0$, then $\dot{z} > 0$. If $r = 1$, then $\dot{r} = 0$. Trajectories spiral toward the xy plane and onto the limit cycle, say, Γ_1, of radius 1 centered at the origin. Hence $\Lambda^+(\mathbf{x}) = \Gamma_1$ if $\mathbf{x} \neq \mathbf{0}$ and Γ_1 is a stable limit cycle. A phase portrait is shown in Figure 8.6.

Lyapunov functions were introduced in Chapter 6 and were used to determine the stability of critical points for certain planar systems. The theory is easily extended to the three-dimensional case as the following examples demonstrate. Once again, there is no systematic way to determine the Lyapunov functions, and they are given in the question.

Example 7. Prove that the origin of the system

$$\dot{x} = -2y + yz, \quad \dot{y} = x(1 - z), \quad \dot{z} = xy$$

is stable but not asymptotically stable by using the Lyapunov function $V(x, y, z) = ax^2 + by^2 + cz^2$.

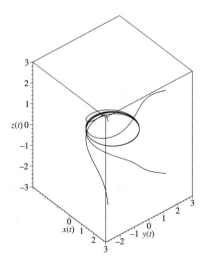

Figure 8.6:   Trajectories are attracted to a stable limit cycle in the xy plane.

Solution. Now

$$\frac{dV}{dt} = \frac{\partial V}{\partial x}\frac{dx}{dt} + \frac{\partial V}{\partial y}\frac{dy}{dt} + \frac{\partial V}{\partial z}\frac{dz}{dt} = 2(a - b + c)xyz + 2(b - 2a)xy.$$

If $b = 2a$ and $a = c > 0$, then $V(x, y, z) > 0$ for all $\mathbf{x} \neq 0$ and $\frac{dV}{dt} = 0$. Thus, the trajectories lie on the ellipsoids defined by $x^2 + 2y^2 + z^2 = r^2$. The origin is thus stable but not asymptotically stable.

Example 8. Prove that the origin of the system

$$\dot{x} = -y - xy^2 + z^2 - x^3, \quad \dot{y} = x + z^3 - y^3, \quad \dot{z} = -xz - x^2 z - yz^2 - z^5$$

is asymptotically stable by using the Lyapunov function $V(x, y, z) = x^2 + y^2 + z^2$.

Solution. Now

$$\frac{dV}{dt} = \frac{\partial V}{\partial x}\frac{dx}{dt} + \frac{\partial V}{\partial y}\frac{dy}{dt} + \frac{\partial V}{\partial z}\frac{dz}{dt} = -2(x^4 + y^4 + x^2 z^2 + x^2 y^2 + z^6).$$

Since $\frac{dV}{dt} < 0$ for $x, y, z \neq 0$, the origin is asymptotically stable. In fact, the origin is *globally asymptotically stable* since $\Lambda^+(\mathbf{x}) = (0, 0, 0)$ for all $\mathbf{x} \in \Re^3$.

8.3 The Rössler System and Chaos

8.3.I The Rössler Attractor. In 1976, Otto E. Rössler [15] constructed the following three-dimensional system of differential equations:

$$\dot{x} = -(y + z), \quad \dot{y} = x + ay, \quad \dot{z} = b + xz - cz, \tag{8.6}$$

where a, b, and c are all constants. Note that the only nonlinear term appears in the $\dot{z}$ equation and is quadratic. As the parameters vary, this simple system can display a wide range of behavior. Set $a = b = 0.2$, for example, and vary the parameter c. The dynamics of the system can be investigated using Mathematica. Four examples are considered here. Transitional trajectories have been omitted to avoid confusion. The initial conditions are $x(0) = y(0) = z(0) = 1$ in all cases.

Definition 4. A limit cycle in three-dimensional space is called a period-n cycle if $\mathbf{x}(t) = \mathbf{x}(t + nT)$ for some minimum constant T called the period. Note that n can be determined by the number of distinct amplitudes in a time series plot.

When $c = 2.3$, there is a period-one limit cycle which can be plotted in three-dimensional space. Figure 8.7(a) shows the limit cycle in phase space, and the periodic behavior with respect to $x(t)$ is shown in Figure 8.7(b).

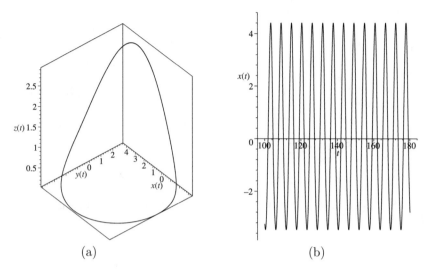

(a)   (b)

Figure 8.7: (a) A limit cycle for system (8.6) when $c = 2.3$. (b) Period-one behavior for $x(t)$.

When $c = 3.3$, there is period-two behavior. Figure 8.8(a) shows the closed orbit in phase space, and the periodic behavior is shown in Figure 8.8(b). Notice that there are two distinct amplitudes in Figure 8.8(b). This periodic behavior can be easily detected using Poincaré maps (see Chapter 9).

When $c = 5.3$, there is period-three behavior. Figure 8.9(a) shows the closed orbit in three-dimensional space, and the periodic behavior is shown in Figure 8.9(b). Note that there are three distinct amplitudes in Figure 8.9(b).

When $c = 6.3$, the system displays what appears to be *random behavior*. This type of behavior has been labeled *deterministic chaos*. A system is called *deterministic* if the behavior of the system is determined from the time evolution equations and the initial conditions alone, as in the case of the Rössler system. *Nondeterministic chaos* arises when there are no underlying equations, as in the United Kingdom national lottery, or there is noisy or random input. This text will be concerned with deterministic chaos only, and it will be referred to simply as chaos from now on.

8.3.II Chaos. Chaos is a multifaceted phenomenon that is not easily classified or identified. There is no universally accepted definition for chaos, but the following characteristics are nearly always displayed by the solutions of chaotic systems:

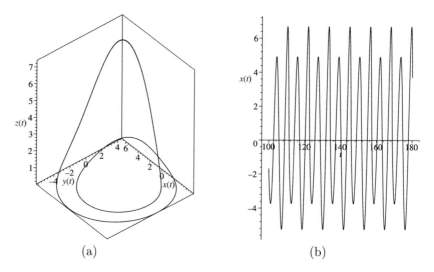

(a)   (b)

Figure 8.8: (a) A period-two limit cycle for system (8.6) when $c = 3.3$. (b) Period-two behavior for $x(t)$.

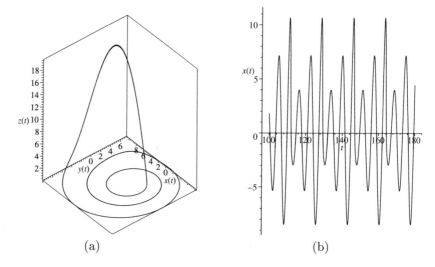

(a)   (b)

Figure 8.9: (a) A period-three limit cycle for system (8.6) when $c = 5.3$; (b) period-three behavior for $x(t)$.

1. long-term *aperiodic* (*nonperiodic*) *behavior*;
2. *sensitivity to initial conditions*;
3. *fractal structure*.

Consider each of these items independently. Note, however, that a chaotic system generally displays all three types of behavior listed above.

Case 1. It is very difficult to distinguish between aperiodic behavior and periodic behavior with a very long period. For example, it is possible for a chaotic system to have a periodic solution of period 10^{100}.

Case 2. A simple method used to test whether or not a system is chaotic is to check for sensitivity to initial conditions. Figure 8.10(a) shows the trajectory in phase space, and Figure 8.10(b) illustrates how the system is sensitive to the choice of initial conditions.

Definition 5. An *attractor* is a minimal closed invariant set that attracts nearby trajectories lying in the domain of stability (or basin of attraction) onto it.

Definition 6. A *strange attractor*, (*chaotic attractor*, *fractal attractor*) is an attractor that exhibits sensitivity to initial conditions.

Definition 7. The *spectrum of Lyapunov exponents* are quantities that characterize the rate of separation of infinitesimally close trajectories.

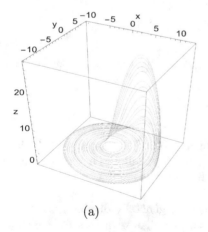
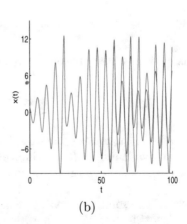

(a)   (b)

Figure 8.10: [Mathematica] (a) The chaotic attractor for system (8.6) when $c = 6.3$. (b) Time series plot of $x(t)$ showing sensitivity to initial conditions; the initial conditions for one time series are $x(0) = y(0) = z(0) = 1$ and for the other are $x(0) = 1.01, y(0) = z(0) = 1$. Use different colors when plotting in Mathematica.

An example of a strange attractor is shown in Figure 8.10(a). Another method for establishing whether or not a system is chaotic is to use the *Lyapunov exponents* (see Chapter 14 for examples in the discrete case). A system is chaotic if at least one of the Lyapunov exponents is positive. This implies that two trajectories that start close to each other on the strange attractor will diverge as time increases, as depicted in Figure 8.10(b). Note that an n-dimensional system will have n different Lyapunov exponents. Think of an infinitesimal sphere of perturbed initial conditions for a three-dimensional system. As time increases the sphere will evolve into an infinitesimal ellipsoid. If d_0 is the initial radius of the sphere, then $d_j = d_0 e^{\lambda_j}$ $(j = 1, 2, 3)$ define the axes of the ellipsoid. The following results are well known for three-dimensional systems. For chaotic attractors $\lambda_1 > 0$, $\lambda_2 = 0$, and $\lambda_3 < 0$; for single critical points $\lambda_1 < 0$, $\lambda_2 < 0$, and $\lambda_3 < 0$; for limit cycles $\lambda_1 = 0$, $\lambda_2 < 0$, and $\lambda_3 < 0$; and for a 2-torus $\lambda_1 = 0$, $\lambda_2 = 0$, and $\lambda_3 < 0$. A comparison of different methods for computing the Lyapunov exponents is given in [7]. A Mathematica program is listed in Section 8.5 for computing the Lyapunov exponents of the Lorenz attractor, and the method is described in [19]. Reference [13] provides a comprehensive description of Lyapunov exponents and gives applications to complex systems up to 2014. One interesting feature of strange attractors is that it is sometimes possible to reconstruct the attractor from time series data alone, see [21], for example. Many papers have also been published on the detection of chaos from time series data [4], [5], [8], [14], and [19], where the underlying equations may not be required.

Gottwald and Melbourne [8] describe a new test for deterministic chaos. Their diagnostic is the real-valued function

$$p(t) = \int_0^t \phi(\mathbf{x}(s)) \cos(\omega_0 s) ds,$$

where ϕ is an observable on the dynamics $\mathbf{x}(t)$ and $\omega_0 \neq 0$ is a constant. They set

$$K = \lim_{t \to \infty} \frac{\log \mathbf{M}(t)}{\log(t)},$$

where $\mathbf{M}$ is the mean-square displacement for $p(t)$. Typically, $K = 0$ signifying regular dynamics, or $K = 1$ indicating chaotic dynamics. They state that the test works well for both continuous and discrete systems.

Case 3. The solution curves to chaotic systems generally display fractal structure (see Chapter 17). The structure of the strange attractors for general n-dimensional systems may be complicated and difficult to observe clearly. To overcome these problems, Poincaré maps, which exist in lower-dimensional spaces, can be used, as in Chapter 9.

8.4 The Lorenz Equations, Chua's Circuit, and the Belousov-Zhabotinski Reaction

Note that most nonlinear systems display steady-state behavior most of the time, so it is possible to predict, for example, the weather, motion of the planets, spread of an epidemic, motion of a driven pendulum, or beat of the human heart. However, nonlinear systems can also display chaotic behavior where prediction becomes impossible.

There are many examples of applications of three-dimensional autonomous systems to the real world. These systems obey the existence and uniqueness theorem, but the dynamics can be much more complicated than in the two-dimensional case. The following examples taken from meteorology, electric circuit theory, and chemical kinetics have been widely investigated in recent years. There are more examples in the exercises at the end of the chapter.

8.4.I The Lorenz Equations. In 1963, the MIT meteorologist Edward Lorenz [11] constructed a highly simplified model of a convecting fluid. This simple model also displays a wide variety of behavior and for some parameter values is chaotic. The equations can be used to model convective flow up through the center and down on the sides of hexagonal columns. The system is given by

$$\dot{x} = \sigma(y - x), \quad \dot{y} = rx - y - xz, \quad \dot{z} = xy - bz, \qquad (8.7)$$

where x measures the rate of convective overturning, y measures the horizontal temperature variation, z measures the vertical temperature variation, σ is the Prandtl number, r is the Rayleigh number, and b is a scaling factor. The Prandtl number is related to the fluid viscosity, and the Rayleigh number is related to the temperature difference between the top and bottom of the column. Lorenz studied the system when $\sigma = 10$ and $b = \frac{8}{3}$.

The system can be considered to be a highly simplified model for the weather. Indeed, satellite photographs from space show hexagonal patterns on undisturbed desert floors. The astonishing conclusion derived by Lorenz is now widely labeled as the *butterfly effect*. Even this very simple model of the weather can display chaotic phenomena. Since the system is sensitive to initial conditions, small changes to wind speed (convective overturning), for example, generated by the flap of a butterfly's wings, can change the outcome of the results considerably. For example, a butterfly flapping its wings in Britain could cause or prevent a hurricane from occurring in the Bahamas in the not-so-distant future. Of course, there are many more variables that should be considered when trying to model weather systems, and this simplified model illustrates some of the problems meteorologists have to deal with.

Some simple properties of the Lorenz equations will now be listed, and all of these characteristics can be investigated with the aid of Mathematica:

1. System (8.7) has natural symmetry $(x, y, z) \rightarrow (-x, -y, z)$.

2. The z-axis is invariant.

3. The flow is volume contracting since $\mathrm{div}\mathbf{X} = -(\sigma + b + 1) < 0$, where $\mathbf{X}$ is the vector field.

4. If $0 < r < 1$, the origin is the only critical point, and it is a global attractor.

5. At $r = 1$, there is a bifurcation, and there are two more critical points at $C_1 = (\sqrt{b(r-1)}, \sqrt{b(r-1)}, r - 1)$ and $C_2 = (-\sqrt{b(r-1)}, -\sqrt{b(r-1)}, r - 1)$.

6. At $r = r_H \approx 13.93$, there is a homoclinic bifurcation (see Chapter 7) and the system enters a state of transient chaos.

7. At $r \approx 24.06$, a strange attractor is formed.

8. If $1 < r < r_O$, where $r_O \approx 24.74$, the origin is unstable and C_1 and C_2 are both stable.

9. At $r > r_O$, C_1 and C_2 lose their stability by absorbing an unstable limit cycle in a subcritical Hopf bifurcation.

For more details, see the work of Sparrow [17] or most textbooks on nonlinear dynamics. Most of the results above can be observed by plotting phase portraits or time series using Mathematica. A strange attractor is shown in Figure 8.11.

The trajectories wind around the two critical points C_1 and C_2 in an apparently random unpredictable manner. The strange attractor has the following properties:

- The trajectory is aperiodic (or not periodic).

- The trajectory remains on the attractor forever (the attractor is invariant).

- The general form is independent of initial conditions.

- The sequence of windings is sensitive to initial conditions.

- The attractor has fractal structure.

A variation on the Lorenz model has recently been discovered by Guanrong Chen and Tetsushi Ueta (see Figure 8.12). The equations are

$$\dot{x} = \sigma(y - x), \quad \dot{y} = (r - \sigma)x + ry - xz, \quad \dot{z} = xy - bz. \tag{8.8}$$

8.4.II Chua's Circuit Elementary electric circuit theory was introduced in Chapter 2. In the mid-1980s, Chua modeled a circuit that was a simple oscillator exhibiting a variety of bifurcation and chaotic phenomena. The circuit diagram is given in Figure 8.13. The circuit equations are given by

$$\frac{dv_1}{dt} = \frac{(G(v_2 - v_1) - f(v_1))}{C_1}, \frac{dv_2}{dt} = \frac{(G(v_1 - v_2) + i)}{C_2}, \frac{di}{dt} = -\frac{v_2}{L},$$

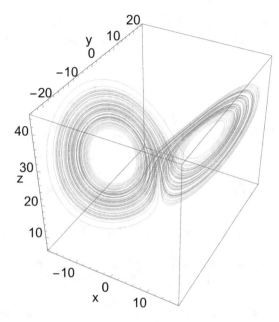

Figure 8.11: [Mathematica] A strange attractor for the Lorenz system when $\sigma = 10, b = \frac{8}{3}$, and $r = 28$.

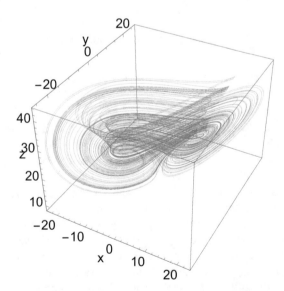

Figure 8.12: A strange attractor for system (8.8) when $\sigma = 35, b = 3$, and $r = 28$.

where v_1, v_2, and i are the voltages across C_1, C_2, and the current through L, respectively. The characteristic of the nonlinear resistor N_R is given by

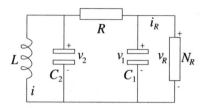

Figure 8.13:   Chua's electric circuit.

$$f(v_1) = G_b v_1 + 0.5(G_a - G_b)\left(|v_1 + B_p| - |v_1 - B_p|\right),$$

where $G = 1/R$. Typical parameters used are $C_1 = 10.1nF, C_2 = 101nF, L = 20.8mH, R = 1420\Omega, r = 63.8\Omega, G_a = -0.865mS, G_b = -0.519mS$, and $B_p = 1.85V$.

In the simple case, Chua's equations can be written in the following dimensionless form:

$$\dot{x} = a(y - x - g(x)), \quad \dot{y} = x - y + z, \quad \dot{z} = -by, \qquad (8.9)$$

where a and b are dimensionless parameters. The function $g(x)$ has the form

$$g(x) = cx + \frac{1}{2}(d - c)\left(|x + 1| - |x - 1|\right),$$

where c and d are constants.

Chua's circuit is investigated in some detail in [12] and exhibits many interesting phenomena including period-doubling cascades to chaos, intermittency routes to chaos, and *quasiperiodic* routes to chaos. For certain parameter values, the solutions lie on a *double-scroll attractor*, as shown in Figure 8.14.

The dynamics are more complicated than those appearing in either the Rössler or Lorenz attractors. Chua's circuit has proved to be a very suitable subject for study since laboratory experiments produce results which match very well with the results of the mathematical model. Recently, the author and Borresen [2] have shown the existence of a bistable cycle for Chua's electric circuit for the first time. Power spectra for Chua's circuit simulations are used to show how the qualitative nature of the solutions depends on the history of the system.

Zhou et al. [20] report on a new chaotic circuit that consists of only a few capacitors, operational amplifiers, and resistors and [3] provides a concise guide to chaotic electronic circuits up to 2014.

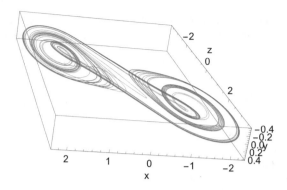

Figure 8.14: [Mathematica] Chua's double-scroll attractor: Phase portrait for system (8.9) when $a = 15, b = 25.58, c = -5/7$, and $d = -8/7$. The initial conditions are $x(0) = -1.6$, $y(0) = 0$, and $z(0) = 1.6$.

8.4.III The Belousov-Zhabotinski Reaction Periodic chemical reactions such as the Landolt clock and the Belousov-Zhabotinski reaction provide wonderful examples of relaxation oscillations in science (see [1], [6], [16]). They are often demonstrated in chemistry classes or used to astound the public at university open days. The first experiment was conducted by the Russian biochemist Boris Belousov in the 1950s, and the results were not confirmed until as late as 1968 by Zhabotinski.

Consider the following recipe for a Belousov periodic chemical reaction.

Ingredients.

- Solution A: Malonic acid, 15.6 gm/l.

- Solution B: Potassium bromate, 41.75 gm/l and potassium bromide, 0.006 gm/l.

- Solution C: Cerium IV sulfate, 3.23 gm/l in 6M sulfuric acid.

- Solution D: Ferroin indicator.

Procedure. Add 20 mls of solution A and 10 mls of solution B to a mixture of 10 mls of solution C and 1 ml of solution D. Stir continuously at room temperature. The mixture remains blue for about 10 minutes and then begins to oscillate blue-green-pink and back again with a period of approximately two minutes.

This reaction is often demonstrated by Chemistry Departments during university open days and is always a popular attraction.

Following the methods of Field and Noyes (see [6]), the chemical rate equations for an oscillating Belousov-Zhabotinski reaction are frequently written as

$$BrO_3^- + Br^- \rightarrow HBrO_2 + HOBr, \qquad \text{Rate} = k_1[BrO_3^-][Br^-]$$
$$HBrO_2 + Br^- \rightarrow 2HOBr, \qquad \text{Rate} = k_2[HBrO_2][Br^-]$$
$$BrO_3^- + HBrO_2 \rightarrow 2HBrO_2 + 2M_{OX}, \quad \text{Rate} = k_3[BrO_3^-][HBrO_2]$$
$$2HBrO_2 \rightarrow BrO_3^- + HOBr, \qquad \text{Rate} = k_4[HBrO_2]^2$$
$$OS + M_{OX} \rightarrow \tfrac{1}{2}C Br^-, \qquad \text{Rate} = k_5[OS][M_{OX}]$$

where OS represents all oxidizable organic species and C is a constant. Note that in the third equation, species $HBrO_2$ stimulates its own production, a process called *autocatalysis*. The reaction rate equations for the concentrations of intermediate species $x = [HBrO_2]$, $y = [Br^-]$, and $z = [M_{OX}]$ are

$$\dot{x} = k_1 a y - k_2 x y + k_3 a x - 2k_4 x^2,$$
$$\dot{y} = -k_1 a y - k_2 x y + \frac{1}{2} C k_5 b z,$$
$$\dot{z} = 2k_3 a x - k_5 b z, \qquad (8.10)$$

where $a = [BrO_3^-]$ and $b = [OS]$ are assumed to be constant. Taking the transformations

$$X = \frac{2k_4 x}{k_5 a}, \ Y = \frac{k_2 y}{k_3 a}, \ Z = \frac{k_5 k_4 b z}{(k_3 a)^2}, \ \tau = k_5 b t,$$

system (8.10) becomes

$$\frac{dX}{d\tau} = \frac{qY - XY + X(1 - X)}{\epsilon_1},$$
$$\frac{dY}{d\tau} = \frac{-qY - XY + CZ}{\epsilon_2},$$
$$\frac{dZ}{d\tau} = X - Z, \qquad (8.11)$$

where $\epsilon_1 = \frac{k_5 b}{k_3 a}$, $\epsilon_2 = \frac{2k_5 k_4 b}{k_2 k_3 a}$, and $q = \frac{2k_1 k_4}{k_2 k_3}$. Next, one assumes that $\epsilon_2 \ll 1$ so that $\frac{dY}{d\tau}$ is large unless the numerator $-qY - XY + CZ$ is also small. Assume that

$$Y = Y^* = \frac{CZ}{q + X}$$

at all times, so the bromide concentration $Y = [\mathrm{Br}^-]$ is in a steady state compared to X. In this way, a three-dimensional system of differential equations is reduced to a two-dimensional system of autonomous ODEs

$$\epsilon_1 \frac{dX}{d\tau} = X(1-X) - \frac{X-q}{X+q}CZ, \quad \frac{dZ}{d\tau} = X - Z. \tag{8.12}$$

For certain parameter values, system (8.12) has a limit cycle that represents an oscillating Belousov-Zhabotinski chemical reaction, as in Figure 8.15.

Example 9. Find and classify the critical points of system (8.12) when $\epsilon_1 = 0.05, q = 0.01$, and $C = 1$. Plot a phase portrait in the first quadrant.

Solution. There are two critical points, one at the origin and the other at $A \approx (0.1365, 0.1365)$. The Jacobian matrix is given by

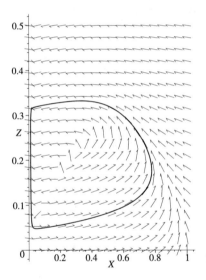

Figure 8.15: A limit cycle in the XZ plane for system (8.12) when $\epsilon_1 = 0.05, q = 0.01$, and $C = 1$.

$$J = \begin{pmatrix} \frac{1}{\epsilon_1}\left(1 - 2X - \frac{Z}{X+q} + \frac{(X-q)Z}{(X+q)^2}\right) & \frac{1}{\epsilon_1}\left(\frac{q-X}{X+q}\right) \\ 1 & -1 \end{pmatrix}.$$

It is not difficult to show that the origin is a saddle point and A is an unstable node. A phase portrait showing periodic oscillations is given in Figure 8.15.

The period of the limit cycle in Figure 8.15 is approximately 3.4. The trajectory moves quickly along the right and left branches of the limit cycle (up and down) and moves relatively slowly in the horizontal direction. This accounts for the rapid color changes and time spans between these changes.

It is important to note that chemical reactions are distinct from many other types of dynamical system in that closed chemical reactions cannot oscillate about their chemical equilibrium state. This problem is easily surmounted by exchanging mass or introducing a flow with the chemical reaction and its surroundings. For example, the Belousov-Zhabotinski reaction used during university open days is stirred constantly and mass is exchanged. It is also possible for the Belousov-Zhabotinski reaction to display chaotic phenomena; see [1], for example. Multistable and bistable chemical reactions are also discussed in [16]. In these cases, there is an inflow and outflow of certain species and more than one steady state can coexist.

Finally, consider the dimensionless system (8.11) when $\epsilon_1 = 0.0099$, $\epsilon_2 = 2.4802e\text{-}5$, $q = 3.1746e\text{-}5$, and $C = 1$. This is a stiff system of ODEs and is intractable to analytical approaches, instead we solve the system numerically using NDSolve. The Mathematica commands are listed in the next section and the relative concentrations are displayed in Figure 8.16.

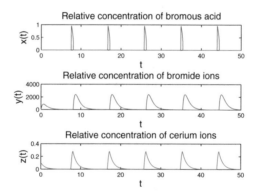

Figure 8.16: [Mathematica] Periodic behavior for the stiff ODE (8.11) when $X(0) = 0, Y(0) = 0, Z(0) = 0.1, \epsilon_1 = 0.0099, \epsilon_2 = 2.4802e\text{-}5, q = 3.1746e\text{-}5$, and $C = 1$. Note that the direct physical significance is lost and the graph shows relative concentrations of each of the concentrations of ions.

8.5 Mathematica Commands

Note that you can rotate the three-dimensional phase portraits by left clicking and moving the mouse.

```
(* Rossler chaotic attractor. See Fig. 8.10(a). *)
a=0.2;b=0.2;c=6.3;
```

```
Rossler=NDSolve[{x'[t]==-y[t]-z[t],y'[t]==x[t] +a y[t],
z'[t]==b+x[t]*z[t]-c*z[t],x[0]==1,y[0]==1,z[0]==1},{x,y,z},{t,0,200},
MaxSteps->Infinity];
ParametricPlot3D[Evaluate[{x[t],y[t],z[t]}/.Rossler],{t,0,200},
PlotPoints->10000,
PlotRange->All,PlotStyle->{Thickness[0.001]},AxesLabel->{"x","y","z"},
Axes->True,BaseStyle->{FontSize->18}]
```

```
(* Chua's double scroll attractor. See Fig. 8.14. *)
Chua=NDSolve[{x'[t]==15 (y[t]-x[t]-((-5/7) x[t]+(1/2) (-(8/7)-(-5/7))
(Abs[x[t]+1]-Abs[x[t]-1]))),y'[t]==x[t]-y[t]+z[t],z'[t]==-25.58 y[t],
x[0]==-1.6,y[0]==0,z[0]==1.6},{x,y,z},{t,0,200},MaxSteps->Infinity];
ParametricPlot3D[Evaluate[{x[t],y[t],z[t]}/.Chua],{t,50,200},
PlotPoints->10000,PlotRange->All,PlotStyle->{Thickness[0.001]},
AxesLabel->{"x","y","z"},Axes->True,BaseStyle->{FontSize->18},
BoxRatios->1]
```

```
(* The Chapman Cycle. See Exercise 6. *)
Chapman=NDSolve[{x'[t]==2 3 10^(-12) y[t]+5.5 10^(-4) z[t]-1.22 10^(-33)
x[t] y[t] 9 10^17-6.86 10^(-16) x[t] z[t],y'[t]==5.5 10^(-4) z[t]+
2 6.86 10^(-16) x[t] z[t]-3 10^(-12) y[t]-1.22 10^(-33) x[t] y[t] 9 10^(17),
z'[t]==1.22 10^(-33) x[t] y[t] 9 10^17-5.5 10^(-4) z[t]-6.86 10^(-16) x[t] z[t],
x[0]==4.6806 10^7,y[0]==6.999 10^16,z[0]==6.5396 10^12},{x,y,z},
{t,0,10^8},MaxSteps->Infinity];
Singlet=x[10^8] /. Chapman
Oxygen=y[10^8] /. Chapman
Ozone=z[10^8] /. Chapman
```

```
(* The Belousov-Zhabotinski Reaction. See Figure 8.16. *)
q=3.1746*10^(-5);f=1;\[Epsilon]=0.0099;\[Delta]=2.4802*10^(-5);Tmax=50;
BZ=NDSolve[{x'[t]==(q y[t]-x[t] y[t]+x[t](1-x[t]))/\[Epsilon],
y'[t]==(-q y[t]-x[t] y[t]+f z[t])/\[Delta],
z'[t]==x[t]-z[t],x[0]==0,y[0]==0,z[0]==0.1},{x,y,z},{t,0,Tmax},
Method -> {"StiffnessSwitching","NonstiffTest"->False}];
ParametricPlot[Evaluate[{t,x[t]}/.BZ],{t,0,Tmax},PlotPoints->1000,
PlotRange->{{0,50},{0,1}},AspectRatio->1/2,AxesLabel->{t,x}]
ParametricPlot[Evaluate[{t,y[t]}/.BZ],{t,0,Tmax},PlotPoints->1000,
PlotRange->{{0,50},{0,4000}},AspectRatio->1/2,AxesLabel->{t,y}]
ParametricPlot[Evaluate[{t,z[t]}/.BZ],{t,0,Tmax},PlotPoints->1000,
PlotRange->{{0,50},{0,0.4}},AspectRatio->1/2,AxesLabel->{t,z}]
```

```
(* Computing Lyapunov Exponents of Lorenz using the methods of *)
(* Wolf et al. in reference [19]. *)
Lorenz[{x_, y_, z_}] := {16 (y - x), x (45.92 - z) - y, x y - 4 z};
x0 = {19, 20, 50};
(* Runge-Kutta Method *)
RKStep[f_, y_, y0_, dt_] := Module[{k1, k2, k3, k4},
```

```
  k1 = dt N[f /. Thread[y -> y0]];
  k2 = dt N[f /. Thread[y -> y0 + k1/2]];
  k3 = dt N[f /. Thread[y -> y0 + k2/2]];
  k4 = dt N[f /. Thread[y -> y0 + k3]];
  y0 + (k1 + 2 k2 + 2 k3 + k4)/6]
IntVarEq[Lorenz_List, DPhi_List, x_List, Phi_List, x0_List,
         Phi0_List, {t1_, dt_}] :=
 Module[{n, f, y, y0, yt},
     n = Length[x0];
     f = Flatten[Join[Lorenz, DPhi]];
     y = Flatten[Join[x, Phi]];
     y0 = Flatten[Join[x0, Phi0]];
     yt = Nest[ RKStep[f, y, #, N[dt]] &, N[y0], Round[N[t1/dt]] ];
     {First[#], Rest[#]} & @ Partition[yt, n] ]
JacobianMatrix[funs_List, vars_List] := Outer[D, funs, vars]
T = 20; stepsize = 0.02; n = Length[x0]; x = Array[a, n];
Phi = Array[b, {n, n}];
DPhi = Phi.Transpose[JacobianMatrix[Lorenz[x], x]];
Phi0 = IdentityMatrix[n];
{xT, PhiT} =
  IntVarEq[Lorenz[x], DPhi, x, Phi, x0, Phi0, {T, stepsize}];
u = Table[Random[], {n}];
Log[Norm[PhiT.u]]/T;
T = 0.5; K = 1000; s = {};
PhiT = Phi0;
(* A Module for finding an orthogonal basis *)
OrthogonalBasis[w_?MatrixQ] :=
  Module[{v = ConstantArray[0, Length[w]]},
   Table[v[[n]] =
     w[[n]] -
     Sum[(v[[i]].w[[n]]/v[[i]].v[[i]])*v[[i]], {i, 1, n - 1}], {n, 1,
     Length[w]}];
   v];
Do[
       {xT, PhiT} =
   IntVarEq[Lorenz[x], DPhi, x, Phi, xT, PhiT, {T, stepsize}];
       W = OrthogonalBasis[PhiT];
       norms = Map[Norm, W];
       s = Append[s, norms];
       PhiT = W/norms,
       {K}
       ];
LyapunovExponents = Rest[FoldList[Plus, 0, Log[s]]]/(T Range[K]);
Last[LyapunovExponents]
ListPlot[Transpose[LyapunovExponents]]
```

8.6 Exercises

1. Find the eigenvalues and eigenvectors of the matrix

$$
A = \begin{pmatrix} 1 & 0 & -4 \\ 0 & 5 & 4 \\ -4 & 4 & 3 \end{pmatrix}.
$$

Hence show that the system $\dot{\mathbf{x}} = A\mathbf{x}$ can be transformed into $\dot{\mathbf{u}} = J\mathbf{u}$, where

$$
J = \begin{pmatrix} 3 & 0 & 0 \\ 0 & -3 & 0 \\ 0 & 0 & 9 \end{pmatrix}.
$$

Sketch a phase portrait for the system $\dot{\mathbf{u}} = J\mathbf{u}$.

2. Classify the critical point at the origin for the system

$$
\dot{x} = x + 2z, \quad \dot{y} = y - 3z, \quad \dot{z} = 2y + z.
$$

3. Find and classify the critical points of the system

$$
\dot{x} = x - y, \quad \dot{y} = y + y^2, \quad \dot{z} = x - z.
$$

4. Consider the system

$$
\dot{x} = -x + (\lambda - x)y, \quad \dot{y} = x - (\lambda - x)y - y + 2z, \quad \dot{z} = \frac{y}{2} - z,
$$

where $\lambda \geq 0$ is a constant. Show that the first quadrant is positively invariant and that the plane $x + y + 2z = $ constant is invariant. Find $\lambda^+(p)$ for p in the first quadrant given that there are no periodic orbits there.

5. (a) Prove that the origin of the system

$$
\dot{x} = -x - y^2 + xz - x^3, \quad \dot{y} = -y + z^2 + xy - y^3, \quad \dot{z} = -z + x^2 + yz - z^3
$$

is globally asymptotically stable.

   (b) Determine the domain of stability for the system

$$
\dot{x} = -ax + xyz, \quad \dot{y} = -by + xyz, \quad \dot{z} = -cz + xyz.
$$

6. The chemical rate equations for the Chapman cycle modeling the production of ozone are

$$
\begin{aligned}
O_2 + hv &\rightarrow O + O, & \text{Rate} &= k_1, \\
O_2 + O + M &\rightarrow O_3 + M, & \text{Rate} &= k_2, \\
O_3 + hv &\rightarrow O_2 + O, & \text{Rate} &= k_3, \\
O + O_3 &\rightarrow O_2 + O_2, & \text{Rate} &= k_4,
\end{aligned}
$$

where O is a singlet, O_2 is oxygen, and O_3 is ozone. The reaction rate equations for species $x = [O]$, $y = [O_2]$, and $z = [O_3]$ are

$$
\begin{aligned}
\dot{x} &= 2k_1 y + k_3 z - k_2 xy[M] - k_4 xz, \\
\dot{y} &= k_3 z + 2k_4 xz - k_1 y - k_2 xy[M], \\
\dot{z} &= k_2 xy[M] - k_3 z - k_4 xz.
\end{aligned}
$$

This is a stiff system of differential equations. Many differential equations applied in chemical kinetics are stiff. Given that $[M] = 9e17$, $k_1 = 3e{-}12$, $k_2 = 1.22e{-}33$, $k_3 = 5.5e{-}4$, $k_4 = 6.86e{-}16$, $x(0) = 4e16$, $y(0) = 2e16$, and $z(0) = 2e16$, show that the steady state reached is $[O] = 4.6806e7$, $[O_2] = 6.999e16$, and $[O_3] = 6.5396e12$.

7. A three-dimensional Lotka–Volterra model is given by

$$
\dot{x} = x(1 - 2x + y - 5z), \quad \dot{y} = y(1 - 5x - 2y - z), \quad \dot{z} = z(1 + x - 3y - 2z).
$$

Prove that there is a critical point in the first quadrant at $P(\frac{1}{14}, \frac{3}{14}, \frac{3}{14})$. Plot possible trajectories and show that there is a solution plane $x + y + z = \frac{1}{2}$. Interpret the results in terms of species behavior.

8. Assume that a given population consists of susceptibles (S), exposed (E), infectives (I), and recovered/immune (R) individuals. Suppose that $S + E + I + R = 1$ for all time. A seasonally driven epidemic model is given by

$$
\dot{S} = \mu(1 - S) - \beta SI, \quad \dot{E} = \beta SI - (\mu + \alpha)E, \quad \dot{I} = \alpha E - (\mu + \gamma)I,
$$

where $\beta = $ contact rate, $\alpha^{-1} = $ mean latency period, $\gamma^{-1} = $ mean infectivity period, and $\mu^{-1} = $ mean lifespan. The seasonality is introduced by assuming that $\beta = B(1 + A\cos(2\pi t))$, where $B \geq 0$ and $0 \leq A \leq 1$. Plot phase portraits when $A = 0.18$, $\alpha = 35.84$, $\gamma = 100$, $\mu = 0.02$, and $B = 1800$ for the initial conditions: (i) $S(0) = 0.065, E(0) = 0.00075, I(0) = 0.00025$, and (ii) $S(0) = 0.038, E(0) = 3.27 \times 10^{-8}, I(0) = 1.35 \times 10^{-8}$. Interpret the results for the populations.

9. Plot some time series data for the Lorenz system (8.7) when $\sigma = 10$, $b = \frac{8}{3}$, and $166 \leq r \leq 167$. When $r = 166.2$, the solution shows intermittent behavior, there are occasional chaotic bursts in between what looks like periodic behavior.

10. Consider system (8.12) given in the text to model the periodic behavior of the Belousov-Zhabotinski reaction. By considering the isoclines and gradients of the vector fields, explain what happens to the solution curves for $\epsilon_1 \ll 1$ and appropriate values of q and C.

Bibliography

[1] N. Arnold, *Chemical Chaos*, Hippo, London, 1997.

[2] J. Borresen and S. Lynch. Further investigation of hysteresis in Chua's circuit, *Int. J. Bifurcation and Chaos*, **12** (2002), 129–134.

[3] A. Buscarino, L. Fotuna, M. Frasca and G. Sciuto *A Concise Guide to Chaotic Electronic Circuits*, Springer, New York, 2014.

[4] J.P. Eckmann, S.O. Kamphorst, D. Ruelle and S. Ciliberto, Lyapunov exponents from time series, *Physical Review A*, **34**(6) (1986), 4971–4979.

[5] S. Ellner and P. Turchin, Chaos in a noisy world - new methods and evidence from time-series analysis, *American Naturalist* **145**(3) (1995), 343–375.

[6] R. Field and M. Burger, eds., *Oscillations and Travelling Waves in Chemical Systems*, Wiley, New York, 1985.

[7] K. Geist, U. Parlitz and W. Lauterborn, Comparison of different methods for computing Lyapunov exponents, *Progress of Theoretical Physics*, **83**-5 (1990), 875–893.

[8] G.A. Gottwald and I.Melbourne, A new test for chaos in deterministic systems, *Proc. Roy. Soc. Lond. A*, **460**-2042 (2004), 603–611.

[9] R.C. Hilborn, *Chaos and Nonlinear Dynamics - An Introduction for Scientists and Engineers*, Oxford University Press, Second ed., Oxford, UK, 2000.

[10] B. Krauskopf, H.M. Osinga, E.J. Doedel, M.E. Henderson, J.M. Guckenheimer, A. Vladimirsky, M. Dellnitz and O. Junge, A survey of methods for computing (un)stable manifolds of vector fields, *Int. J. Bifurcation and Chaos* **15**(3) (2005), 763–791.

[11] Lorenz E.N. Deterministic non-periodic flow, *J. Atmos. Sci.*, **20** (1963), 130–141.

[12] R.N. Madan, *Chua's Circuit: A Paradigm for Chaos*, Singapore, World Scientific, 1993.

[13] A. Pikovsky and A. Politi, *Lyapunov Exponents: A Tool to Explore Complex Dynamics*, Cambridge University Press, Cambridge, 2016.

[14] M.T. Rosenstein, J.J. Collins and C.J. Deluca, A practical method for calculating largest Lyapunov exponents from small data sets, *Physica D* **65** (1–2), (1993), 117–134.

[15] O.E. Rössler, An equation for continuous chaos, *Phys. Lett.*, **57**-A (1976), 397–398.

[16] S.K. Scott, *Oscillations, Waves, and Chaos in Chemical Kinetics*, Oxford Science Publications, Oxford, UK, 1994.

[17] C. Sparrow, *The Lorenz Equations: Bifurcations, Chaos and Strange Attractors*, Springer-Verlag, New York, 1982.

[18] S. Wiggins, *Introduction to Applied Nonlinear Dynamical Systems and Chaos*, 2nd ed., Springer-Verlag, New York, 2003.

[19] A. Wolf, J.B. Swift, H.L. Swinney and J.A. Vastano, Determining Lyapunov exponents from a time series, *Physica D* **16**, (1985), 285–317.

[20] P. Zhou, X.H. Luo and H.Y. Chen, A new chaotic circuit and its experimental results, *Acta Physica Sinica* **54**(11), (2005), 5048–5052.

[21] M. Zoltowski, An adaptive reconstruction of chaotic attractors out of their single trajectories, *Signal Process.*, **80**(6) (2000), 1099–1113.

Chapter 9
Poincaré Maps and Nonautonomous Systems in the Plane

Aims and Objectives

- To introduce the theory of Poincaré maps.

- To compare periodic and quasiperiodic behavior.

- To introduce Hamiltonian systems with two degrees of freedom.

- To use Poincaré maps to investigate a nonautonomous system of differential equations.

On completion of this chapter the reader should be able to

- understand the basic theory of Poincaré maps;

- plot return maps for certain systems;

- use the Poincaré map as a tool for studying stability and bifurcations.

Poincaré maps are introduced via example using two-dimensional autonomous systems of differential equations. They are used extensively to transform complicated behavior in the phase space to discrete maps in a lower-dimensional space. Unfortunately, this nearly always results in numerical work since analytic solutions can rarely be found.

A periodically forced nonautonomous system of differential equations is introduced, and Poincaré maps are used to determine stability and plot bifurcation diagrams.

Discrete maps have been dealt with in Chapters 13–18 of the book.

© Springer International Publishing AG 2017
S. Lynch, *Dynamical Systems with Applications using Mathematica®*,
DOI 10.1007/978-3-319-61485-4_9

9.1 Poincaré Maps

When plotting the solutions to some nonlinear problems, the phase space can become overcrowded and the underlying structure may become obscured. To overcome these difficulties, a basic tool was proposed by Henri Poincaré [10] at the end of the nineteenth century. An historical introduction to Poincaré maps is given in [4], and some mathematical applications are discussed in [6]. As a simple introduction to the theory of *Poincaré* (or *first return*) *maps*, consider two-dimensional autonomous systems of the form

$$\dot{x} = P(x, y), \quad \dot{y} = Q(x, y). \tag{9.1}$$

Suppose that there is a curve or straight line segment, say Σ, that is crossed *transversely* (no trajectories are tangential to Σ). Then Σ is called a *Poincaré section*. Consider a point r_0 lying on Σ. As shown in Figure 9.1, follow the flow of the trajectory until it next meets Σ at a point r_1. This point is known as the first return of the discrete Poincaré map $\mathbf{P} : \Sigma \to \Sigma$, defined by

$$r_{n+1} = \mathbf{P}(r_n),$$

where r_n maps to r_{n+1} and all points lie on Σ. Finding the function $\mathbf{P}$ is equivalent to solving the differential equations (9.1). Unfortunately, this is very seldom possible, and one must rely on numerical solvers to make any progress.

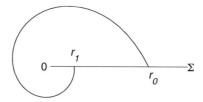

Figure 9.1:   A first return on a Poincaré section, Σ.

Definition 1. A point r^* that satisfies the equation $\mathbf{P}(r^*) = r^*$ is called a *fixed point of period one*.

To illustrate the method for finding Poincaré maps, consider the following two simple examples (Examples 1 and 2), for which $\mathbf{P}$ may be determined explicitly.

Example 1. By considering the line segment $\Sigma = \{(x, y) \in \Re^2 : 0 \le x \le 1, y = 0\}$, find the Poincaré map for the system

$$\dot{x} = -y - x\sqrt{x^2 + y^2}, \quad \dot{y} = x - y\sqrt{x^2 + y^2} \tag{9.2}$$

and list the first eight returns on Σ given that $r_0 = 1$.

Solution. Convert to polar coordinates. System (9.2) then becomes

$$\dot{r} = -r^2, \quad \dot{\theta} = 1. \tag{9.3}$$

The origin is a stable focus, and the flow is counterclockwise. A phase portrait showing the solution curve for this system is given in Figure 9.2.

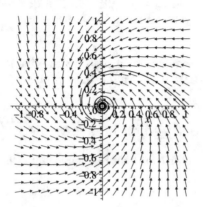

Figure 9.2:   A trajectory starting at $(1,0), (0 \le t \le 40)$ for system (9.3).

The set of equations (9.3) can be solved using the initial conditions $r(0) = 1$ and $\theta(0) = 0$. The solutions are given by

$$r(t) = \frac{1}{1+t}, \quad \theta(t) = t.$$

Trajectories flow around the origin with a period of 2π. Substituting for t, the flow is defined by

$$r(t) = \frac{1}{1 + \theta(t)}.$$

The flow is counterclockwise, and the required successive returns occur when $\theta = 2\pi, 4\pi, \dots$. A map defining these points is given by

$$r_n = \frac{1}{1 + 2n\pi}$$

on Σ, where $n = 1, 2, \dots$. As $n \to \infty$, the sequence of points moves toward the fixed point at the origin as expected. Now

$$r_{n+1} = \frac{1}{1 + 2(n+1)\pi}.$$

Elementary algebra is used to determine the Poincaré return map $\mathbf{P}$, which may be expressed as

$$r_{n+1} = \mathbf{P}(r_n) = \frac{r_n}{1 + 2\pi r_n}.$$

The first eight returns on the line segment Σ occur at the points $r_0 = 1, r_1 = 0.13730, r_2 = 0.07371, r_3 = 0.05038, r_4 = 0.03827, r_5 = 0.03085, r_6 = 0.02584, r_7 = 0.02223$, and $r_8 = 0.01951$, to five decimal places, respectively. Check these results for yourself using the Mathematica program at the end of the chapter.

Example 2. Use a one-dimensional map on the line segment $\Sigma = \{(x, y) \in \Re^2 : 0 \leq x < \infty, y = 0\}$ to determine the stability of the limit cycle in the following system:

$$\dot{x} = -y + x(1 - \sqrt{x^2 + y^2}), \quad \dot{y} = x + y(1 - \sqrt{x^2 + y^2}). \tag{9.4}$$

Solution. Convert to polar coordinates, then system (9.4) becomes

$$\dot{r} = r(1 - r), \quad \dot{\theta} = 1. \tag{9.5}$$

The origin is an unstable focus, and there is a limit cycle, say Γ, of radius 1 centered at the origin. A phase portrait showing two trajectories is given in Figure 9.3.

System (9.5) can be solved since both differential equations are separable. The solutions are given by

$$r(t) = \frac{1}{1 + Ce^{-t}}, \quad \theta(t) = t + \theta_0,$$

where C and θ_0 are constants. Trajectories flow around the origin with a period of 2π.

Suppose that a trajectory starts outside Γ on Σ, say at $r_0 = 2$. The solutions are then given by

$$r(t) = \frac{1}{1 - \frac{1}{2}e^{-t}}, \quad \theta(t) = t.$$

Therefore a return map can be expressed as

$$r_n = \frac{1}{1 - \frac{1}{2}e^{-2n\pi}},$$

where n is a natural number. If, however, a trajectory starts inside Γ at, say $r_0 = \frac{1}{2}$, then

$$r(t) = \frac{1}{1 + e^{-t}}, \quad \theta(t) = t,$$

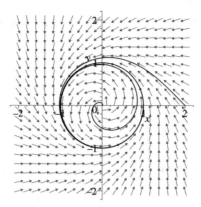

Figure 9.3: Two trajectories for system (9.5), one starting at $(2,0)$ and the other at $(0.01, 0)$.

and a return map is given by

$$r_n = \frac{1}{1 + e^{-2n\pi}}.$$

In both cases $r_n \to 1$ as $n \to \infty$. The limit cycle is stable on both sides, and the limit cycle Γ is hyperbolic stable since $r_n \to 1$ as $n \to \infty$ for any initial point apart from the origin. The next theorem gives a better method for determining the stability of a limit cycle.

Theorem 1. *Define the characteristic multiplier M to be*

$$M = \left. \frac{d\mathbf{P}}{dr} \right|_{r^*},$$

where r^* is a fixed point of the Poincaré map $\mathbf{P}$ corresponding to a limit cycle, say Γ. Then

1. *if $|M| < 1$, Γ is a hyperbolic stable limit cycle;*
2. *if $|M| > 1$, Γ is a hyperbolic unstable limit cycle;*
3. *if $|M| = 1$, and $\frac{d^2\mathbf{P}}{dr^2} \neq 0$, then the limit cycle is stable on one side and unstable on the other; in this case Γ is called a semistable limit cycle.*

Theorem 1 is sometimes referred to as the *derivative of the Poincaré map test*.

Definition 2. A fixed point of period one, say r^*, of a Poincaré map $\mathbf{P}$ is called *hyperbolic* if $|M| \neq 1$.

Example 3. Use Theorem 1 to determine the stability of the limit cycle in Example 2.

Solution. Consider system (9.5). The return map along Σ is given by

$$r_n = \frac{1}{1 + Ce^{-2n\pi}}, \tag{9.6}$$

where C is a constant. Therefore,

$$r_{n+1} = \frac{1}{1 + Ce^{-2(n+1)\pi}}. \tag{9.7}$$

Substituting $C = \frac{1-r_n}{r_n e^{2n\pi}}$ from equation (9.6) into (9.7) gives the Poincaré map

$$r_{n+1} = \mathbf{P}(r_n) = \frac{r_n}{r_n + (1 - r_n)e^{-2\pi}}.$$

The Poincaré map has two fixed points, one at zero (a trivial fixed point) and the other at $r^* = 1$, corresponding to the critical point at the origin and the limit cycle Γ, respectively. Now

$$\frac{d\mathbf{P}}{dr} = \frac{e^{-2\pi}}{(r + (1 - r)e^{-2\pi})^2},$$

using elementary calculus, and

$$\left.\frac{d\mathbf{P}}{dr}\right|_{r^*=1} = e^{-2\pi} \approx 0.00187 < 1,$$

and so the limit cycle Γ is hyperbolic attracting.

Definition 3. A point r^* that satisfies the equation $\mathbf{P}^m(r^*) = r^*$ is called a *fixed point of period m*.

Example 4. Consider the circle map $\mathbf{P}$ defined by

$$r_{n+1} = \mathbf{P}(r_n) = e^{i2\pi\frac{q_1}{q_2}} r_n,$$

which maps points on the unit circle to itself. Assuming that $r_0 = 1$, plot iterates when

(a) $q_1 = 0$, $q_2 = 1$,

(b) $q_1 = 1$, $q_2 = 2$,

(c) $q_1 = 2$, $q_2 = 3$, and

(d) $q_1 = 1$, $q_2 = \sqrt{2}$.

Explain the results displayed in Figures 9.4(a)–(d).

Solution. In Figure 9.4(a), there is a fixed point of period one since $r_{n+1} = \mathbf{P} = r_n$. Similarly, in Figures 9.4(b)–(c), there are fixed points of periods two and three since $r_{n+2} = \mathbf{P}^2 = r_n$ and $r_{n+3} = \mathbf{P}^3 = r_n$. For Figure 9.4(d), q_1 and q_2 are *rationally independent* since $c_1 q_1 + c_2 q_2 = 0$ with c_1 and c_2 integers is satisfied only by $c_1 = c_2 = 0$. This implies that the points on the circle map are never repeated and there is no periodic motion. (There is no integer c such that $r_{n+c} = \mathbf{P}^c = r_n$). Figure 9.4(d) shows the first 1000 iterates of this mapping. If one were to complete the number of iterations to infinity then a closed circle would be formed as new points approach other points arbitrarily closely an infinite number of times. This new type of qualitative behavior is known as *quasiperiodicity*. Note that one has to be careful when distinguishing between quasiperiodic points and points that have very high periods. For example, Figure 9.4(d) could be depicting a very high period trajectory. Systems displaying quasiperiodicity will be discussed in the next section.

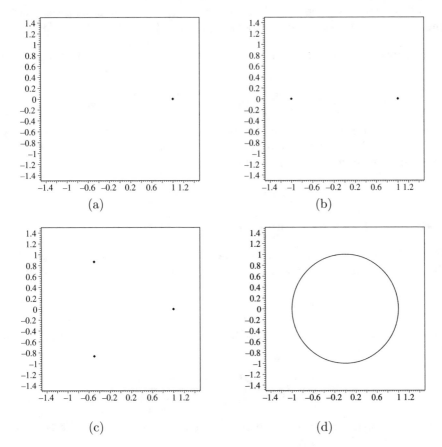

Figure 9.4: Fixed points of periods (a) one; (b) two; (c) three, and (d) quasi-periodic behavior, for the circle map $r_{n+1} = \mathbf{P}(r_n) = e^{i2\pi \frac{q_1}{q_2}} r_n$.

9.2 Hamiltonian Systems with Two Degrees of Freedom

Hamiltonian systems with one degree of freedom were introduced in Chapter 6. These systems can always be integrated completely. Hamiltonian (or conservative) systems with two degrees of freedom will be discussed briefly in this section, but the reader should note that it is possible to consider Hamiltonian systems with N—or even an infinite number of—degrees of freedom.

In general, the set of Hamiltonian systems with two degrees of freedom are not completely integrable, and those that are form a very restricted but important subset. The trajectories of these systems lie in four-dimensional space, but the overall structure can be determined by plotting Poincaré maps. It is known that completely integrable systems display remarkable smooth regular behavior in all parts of the phase space, which is in stark contrast to what happens with nonintegrable systems, which can display a wide variety of phenomena including chaotic behavior. A brief definition of integrability is given below, and Hamiltonian systems with two degrees of freedom will now be defined.

Definition 4. A Hamiltonian system with two degrees of freedom is defined by

$$\dot{p}_1 = -\frac{\partial H}{\partial q_1}, \quad \dot{q}_1 = \frac{\partial H}{\partial p_1}, \quad \dot{p}_2 = -\frac{\partial H}{\partial q_2}, \quad \dot{q}_2 = \frac{\partial H}{\partial p_2}, \tag{9.8}$$

where H is the Hamiltonian of the system. In physical applications, q_1 and q_2 are generalized coordinates and p_1 and p_2 represent a generalized momentum. The Hamiltonian may be expressed as

$$H(\mathbf{p}, \mathbf{q}) = K_E(\mathbf{p}, \mathbf{q}) + P_E(\mathbf{q}),$$

where K_E and P_E are the kinetic and potential energies, respectively.

Definition 5. The Hamiltonian system with two degrees of freedom given by (9.8) is *integrable* if the system has two integrals, say F_1 and F_2, such that

$$\{F_1, H\} = 0, \{F_2, H\} = 0, \{F_1, F_2\} = 0,$$

where F_1 and F_2 are functionally independent and $\{,\}$ are the so-called *Poisson brackets* defined by

$$\{F_1, F_2\} = \frac{\partial F_1}{\partial \mathbf{q}} \frac{\partial F_2}{\partial \mathbf{p}} - \frac{\partial F_1}{\partial \mathbf{p}} \frac{\partial F_2}{\partial \mathbf{q}}.$$

Some of the dynamics involved in these type of systems will now be described using some simple examples.

Example 5. Consider the Hamiltonian system with two degrees of freedom given by

$$H(\mathbf{p}, \mathbf{q}) = \frac{\omega_1}{2}(p_1^2 + q_1^2) + \frac{\omega_2}{2}(p_2^2 + q_2^2), \qquad (9.9)$$

which is integrable with integrals given by $F_1 = p_1^2 + q_1^2$ and $F_2 = p_2^2 + q_2^2$. This system can be used to model a linear harmonic oscillator with two degrees of freedom.

Plot three-dimensional and two-dimensional projections of the Poincaré surface-of-section for system (9.9) given the following set of initial conditions for p_1, p_2 and q_1, q_2:

(i) $\omega_1 = \omega_2 = 2$ with the initial conditions $t = 0, p_1 = 0.5, p_2 = 1.5, q_1 = 0.5, q_2 = 0$;

(ii) $\omega_1 = 8, \omega_2 = 3$ with the initial conditions $t = 0, p_1 = 0.5, p_2 = 1.5, q_1 = 0.3, q_2 = 0$;

(iii) $\omega_1 = \sqrt{2}, \omega_2 = 1$ with the initial conditions $t = 0, p_1 = 0.5, p_2 = 1.5, q_1 = 0.3, q_2 = 0$.

Solutions. A Mathematica program is listed in Section 9.4 (See Figure 9.5). The results may be interpreted as follows: In cases (i) and (ii) the solutions are periodic, and in case (iii) the solution is quasiperiodic. For the quasiperiodic solution, a closed curve will be formed in the $p_1 q_1$ plane as the number of iterations goes to infinity. The quasiperiodic cycle never closes on itself; however, the motion is not chaotic. Once more the trajectories are confined to invariant tori (see Figure 9.5(e), which shows a section of the torus).

Example 6. Consider the Hénon-Heiles Hamiltonian system (which may be used as a simple model of the motion of a star inside a galaxy) given by

$$H(\mathbf{p}, \mathbf{q}) = \frac{1}{2}(p_1^2 + q_1^2 + p_2^2 + q_2^2) + q_1^2 q_2 - \frac{q_2^3}{3}.$$

This Hamiltonian represents two simple harmonic oscillators (see Example 5(i)) coupled with a cubic term. The Hamiltonian in this case is nonintegrable. Plot three-dimensional and two-dimensional projections of the Poincaré surface-of-section of the Hamiltonian system for the set of initial conditions given by $t = 0, p_1 = 0.06, p_2 = 0.1, q_1 = -0.2, q_2 = -0.2$.

Solution. See Figure 9.6.
A rich variety of behavior is observed in the Poincaré section for the Hénon-Heiles system as the energy levels increase. For example, Figure 9.7 shows how the Poincaré section changes as the energy level increases from 0.041666 to 0.166666.

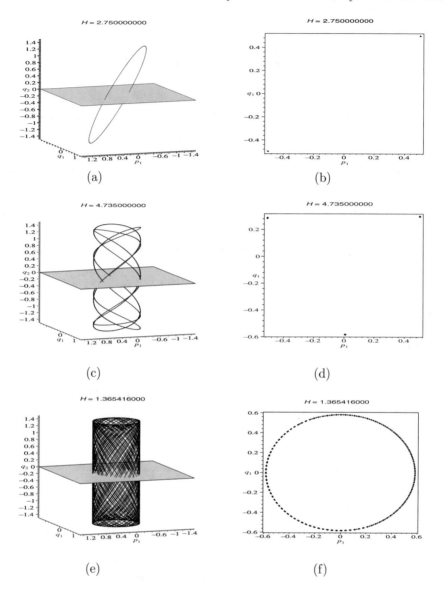

(a)   (b)

(c)   (d)

(e)   (f)

Figure 9.5:   Projections of the Poincaré surface-of-section for system (9.9) when
(a)–(b) $\omega_1 = \omega_2 = 2$, (c)–(d) $\omega_1 = 8$ and $\omega_2 = 3$, and (e)–(f) $\omega_1 = \sqrt{2}$ and
$\omega_2 = 1$. The initial conditions are listed in (i)–(iii) of Example 5.

As the energy levels increase the closed orbits, representing quasiperiodic
behavior, are replaced by irregular patterns, and eventually the Poincaré plane
seems to be swamped by chaos. In fact, there is a famous theorem due to

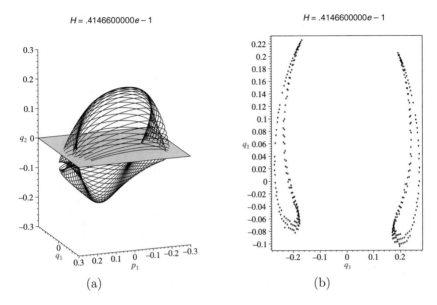

$H = .4146600000e - 1$ $H = .4146600000e - 1$

(a)   (b)

Figure 9.6: A three-dimensional and two-dimensional projection of the Poincaré surface-of-section for the Hénon-Heiles system with the initial conditions $t = 0, p_1 = 0.06, p_2 = 0.1, q_1 = -0.2, q_2 = -0.2$. Note that the energy level is equal to 0.041466 in this case.

Kolmogorov, Arnold, and Moser, now known as the KAM theorem. Interested readers are referred to the book of Guckenheimer and Holmes [5].

Theorem 2. *Suppose that a Hamiltonian system with two degrees of freedom is given by $H = H_0 + \epsilon H_1$, where ϵ is a small parameter, H_0 is integrable, and H_1 makes H nonintegrable. The quasiperiodic cycles (also known as KAM tori), which exist for $\epsilon = 0$, will also exist for $0 < \epsilon \ll 1$ but will be deformed by the perturbation. The KAM tori dissolve one by one as ϵ increases and points begin to scatter around the Poincaré plane. A similar pattern of behavior can be seen in Figure 9.7.*

9.3 Nonautonomous Systems in the Plane

The existence and uniqueness theorems introduced in Chapter 2 hold for autonomous systems of differential equations. This means that trajectories cannot cross, and the Poincaré-Bendixson Theorem implies that there is no chaos in two dimensions. However, chaos can be displayed in three-dimensional autonomous systems as shown in Chapter 8, where various strange attractors

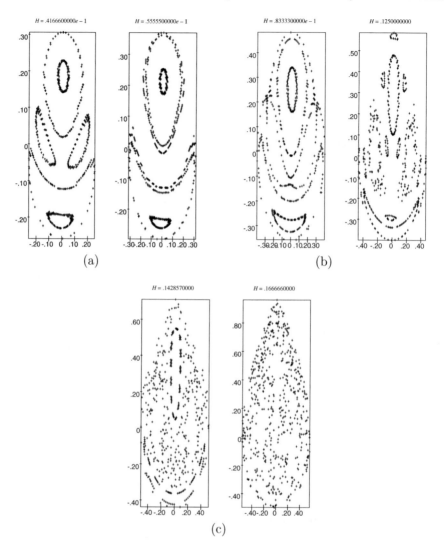

Figure 9.7: The Poincaré transversal plane for the Hénon-Heiles Hamiltonian system with different energy levels. The smoothness of the curves in both (p, q) planes is related to the integrability of the system.

were plotted using the Mathematica package. This section is concerned with *nonautonomous* (or *forced*) systems of differential equations of the form

$$\ddot{x} = f(x, \dot{x}, t),$$

where the function f depends explicitly on t. There is no longer uniqueness of the solutions, and trajectories can cross in the phase plane. For certain parameter values, the phase portrait can become entangled with trajectories crisscrossing one another. By introducing a Poincaré map, it becomes possible to observe the underlying structure of the complicated flow.

As a particular example, consider the *Duffing equation* given by

$$\ddot{x} + k\dot{x} + \beta x + \alpha x^3 = \Gamma \cos(\omega t),$$

where, in physical models, $k \geq 0$ is the damping coefficient, β is the stiffness, α is the nonlinear stiffness parameter, $\dot{x}$ is the speed of the mass, Γ is the amplitude of force vibration, and ω is the frequency of the driving force. Let $\dot{x} = y$; then the Duffing equation can be written as a system of the form

$$\dot{x} = y, \quad \dot{y} = -\beta x - ky - \alpha x^3 + \Gamma \cos(\omega t), \tag{9.10}$$

When $\beta < 0$, the Duffing equation models a periodically forced steel beam deflected between two magnets [7], see Figure 9.8(a). When $\beta > 0$, the Duffing equation models a periodically forced pendulum as depicted in Figure 9.8(b). When $\alpha > 0$, the spring is called a hardening spring and when $\alpha < 0$, the spring is called a softening spring. Consider the Poincaré map of system (9.10) as the amplitude Γ varies when k, β, α and ω are fixed. The radius of the limit cycle on the Poincaré section is given by r.

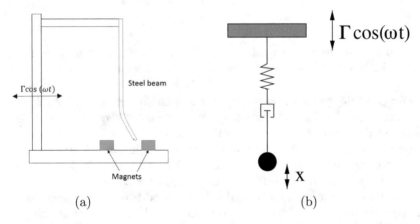

(a)   (b)

Figure 9.8: (a) A steel beam between two magnets. (b) A periodically driven pendulum.

Systems of the form (9.10) have been studied extensively in terms of, for example, stability, harmonic solutions, subharmonic solutions, transients, chaotic output, chaotic control, and Poincaré maps. The work here will be restricted to considering the Poincaré maps and bifurcation diagrams for system (9.10) as the driving amplitude Γ varies when $\alpha = 1, \beta = -1, k = 0.3$, and $\omega = 1.25$ are fixed.

It is interesting to apply quasiperiodic forcing to nonlinear systems, as in [9], where nonchaotic attractors appear for a quasiperiodically forced van der Pol system.

Any periodically forced nonautonomous differential equation can be represented in terms of an autonomous flow in a torus. To achieve this transformation, simply introduce a third variable $\theta = \omega t$. System (9.10) then becomes a three-dimensional autonomous system given by

$$\dot{x} = y, \quad \dot{y} = -\beta x - ky - \alpha x^3 + \Gamma \cos(\theta), \quad \dot{\theta} = \omega. \qquad (9.11)$$

A flow in this state space corresponds to a trajectory flowing around a torus with period $\frac{2\pi}{\omega}$. This naturally leads to a Poincaré mapping of a $\theta = \theta_0$ plane to itself as depicted in Figure 9.9.

When $\Gamma = 0$, system (9.10) has three critical points at $M = (-1, 0)$, $N = (1, 0)$, and $O = (0, 0)$. The points M and N are stable foci when $0 < k < 2\sqrt{2}$ and O is a saddle point. As Γ is increased from zero, stable periodic cycles appear from M and N and there are bifurcations of subharmonic oscillations. The system can also display chaotic behavior for certain values of Γ.

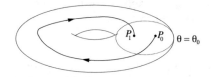

Figure 9.9: The first return of a point P_0 to P_1 in the plane $\theta = \theta_0$. The trajectories flow inside a torus in three-dimensional space.

Only periodic cycles initially appearing from the critical point N will be considered here. A gallery of phase portraits along with their respective Poincaré return maps are presented in Figures 9.10 and 9.11.

When $\Gamma = 0.2$, there is a period-one harmonic solution of period $\frac{2\pi}{\omega}$, which is depicted as a closed curve in the phase plane and as a single point in the $\theta = 0$ plane (see Figure 9.10(a)). When $\Gamma = 0.3$, a period-two cycle of period $\frac{4\pi}{\omega}$ appears; this is a subharmonic of order $\frac{1}{2}$. A period-two cycle is represented by two points in the Poincaré section (see Figure 9.10(b)); note that the trajectory crosses itself in this case. A period-four cycle of period $\frac{8\pi}{\omega}$ is present when $\Gamma = 0.31$ (see Figure 9.10(c)). When $\Gamma = 0.37$, there is a period-five cycle that is centered at O and also surrounds both M and N (see Figure 9.11(a)). When $\Gamma = 0.5$, the system becomes chaotic. A single trajectory plotted in the phase plane intersects itself many times, and the portrait soon becomes very messy. However, if one plots the first returns on the Poincaré section, then a strange attractor is formed that demonstrates some underlying structure (see Figure 9.11(b)). It must be noted that the chaotic attractor will have different forms on different Poincaré sections. This strange (or chaotic) attractor has fractal

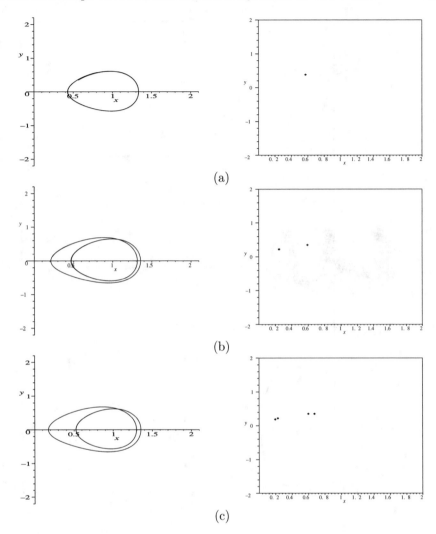

(a)

(b)

(c)

Figure 9.10: A gallery of phase portraits and Poincaré maps for system (9.10) when $\alpha = 1, \beta = -1, k = 0.3$, and $\omega = 1.25$: (a) $\Gamma = 0.2$ (forced period one), (b) $\Gamma = 0.3$ (a period-two subharmonic), and (c) $\Gamma = 0.31$ (a period-four subharmonic).

structure. At $\Gamma = 0.8$, there is once more a stable period-one solution. However, it is now centered at O (see Figure 9.11(c)).

Figures 9.10 and 9.11 display some of the behavior possible for the Duffing equation for specific values of the parameter Γ. Of course, it would be far better to summarize all of the possible behaviors as the parameter Γ varies on one diagram. To achieve this goal, one must plot bifurcation diagrams. There are

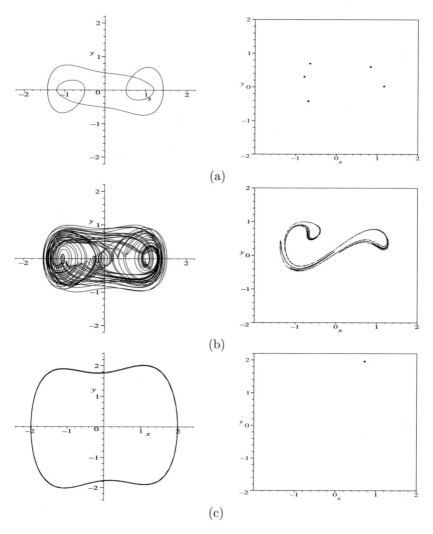

Figure 9.11: A gallery of phase portraits and Poincaré maps for system (9.10) when $\alpha = 1, \beta = -1, k = 0.3$, and $\omega = 1.25$: (a) $\Gamma = 0.37$ (a period-five subharmonic); (b) $\Gamma = 0.5$ (chaos), 4000 points are plotted; (c) $\Gamma = 0.8$ (forced period one).

basically two ways in which bifurcation diagrams may be produced; one involves a feedback mechanism, the other does not. The first and second iterative methods are described in Chapter 16.

Figure 9.12 shows a possible bifurcation diagram for system (9.10) for forcing amplitudes in the range $0 < \Gamma < 1$ near the critical point at N. The vertical axis labeled r represents the distance of the point in the Poincaré map from the origin ($r = \sqrt{x^2 + y^2}$). The first iterative method was employed in this case. For each value of Γ, the last 10 of 50 iterates were plotted, and the step length used in this case was 0.01. The initial values were chosen close to one of the existing periodic solutions. The diagram shows period-one behavior for $0 < \Gamma < 0.28$, approximately. For values of $\Gamma > 0.28$, there is period-two behavior, and then the results become a little obscure.

Figure 9.13 shows a possible bifurcation diagram produced using the second iterative method. The parameter Γ is increased from zero to 0.4 and then decreased from $\Gamma = 0.4$ back to zero. A similar study was carried out in Chapter 5. There were 2000 iterates used as Γ was increased and then decreased. The solid curve lying approximately between $0 \leq \Gamma < 0.32$ represents steady-state behavior. As Γ increases beyond 0.32, the system goes through a chaotic regime and returns to periodic behavior before $\Gamma = 0.4$. As the parameter Γ is decreased, the system returns through the periodic paths, enters a chaotic region, and period undoubles back to the steady-state solution at $\Gamma \approx 0.28$. Note that on the ramp-up part of the iterative scheme, the steady state overshoots into the region where the system is of period two, roughly where $0.28 < \Gamma < 0.32$.

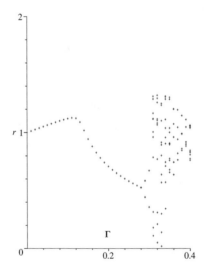

Figure 9.12:   A bifurcation diagram for system (9.10) when $\alpha = 1, \beta = -1, k = 0.3$, and $\omega = 1.25$, produced using the first iterative method.

Figure 9.14 shows a bifurcation diagram produced as Γ is increased from zero to 0.45 and then decreased back to zero. Once more as Γ is increased, there is steady-state behavior for Γ lying between zero and approximately 0.32. However, as the parameter is decreased a different steady state is produced and a large bistable region is present.

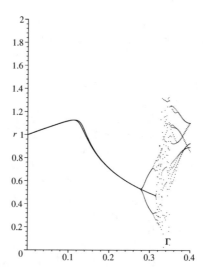

Figure 9.13:   A bifurcation diagram for system (9.10) when $\alpha = 1, \beta = -1, k = 0.3$, and $\omega = 1.25$, produced using the second iterative method.

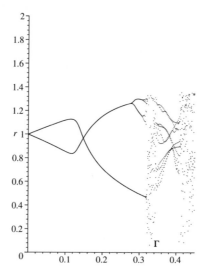

Figure 9.14:   A bifurcation diagram for system (9.10) when $\alpha = 1, \beta = -1, k = 0.3$, and $\omega = 1.25$, produced using the second iterative method. There is a large bistable region.

Note that there will also be steady-state behavior and bifurcations associated with the critical point at M. The flow near to saddle fixed points will now be considered.

Homoclinic and Heteroclinic Bifurcations. Some of the theory involved in the bifurcations to chaos for flows and maps is a result of the behavior of the stable and unstable manifolds of saddle points. Discrete maps have been discussed in some detail in earlier chapters. The stable and unstable manifolds can form homoclinic and heteroclinic orbits as a parameter is varied. Homoclinic and heteroclinic orbits were introduced in Chapter 6. It is also possible for the stable and unstable manifolds to approach one another and eventually intersect as a parameter varies. When this occurs, there is said to be a homoclinic (or heteroclinic) intersection. The intersection is homoclinic if a stable/unstable branch of a saddle point crosses the unstable/stable branch of the same saddle point, and it is heteroclinic if the stable/unstable branches of one saddle point cross the unstable/stable branches of a different saddle point. If the stable and unstable branches of saddle points intersect once, then it is known that there must be an infinite number of intersections, and a so-called *homoclinic (or heteroclinic) tangle* is formed. In 1967, Smale [11] provided an elegant geometric construction to describe this phenomenon. The mapping function used is now known as the *Smale horseshoe map*. Consider a small square, say S, of initial points surrounding a saddle point in the Poincaré section. Under the iterative scheme, this square of points will be stretched out in the direction of the unstable manifold and compressed along the stable branch of the saddle point. In Smale's construction, a square of initial points is stretched in one direction and then compressed in an orthogonal direction. Suppose that the map is given by $\mathbf{H}$: $S \to \Re^2$ and that $\mathbf{H}$ contracts S in the horizontal direction, expands S in the vertical direction, and then folds the rectangle back onto itself to form a horseshoe, as in Figure 9.15. Similarly, the action of $\mathbf{H}^{-1}$ on S is also given

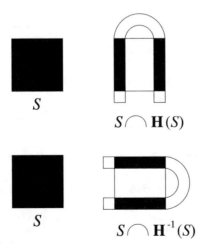

Figure 9.15:   The mappings $\mathbf{H}$ and $\mathbf{H}^{-1}$.

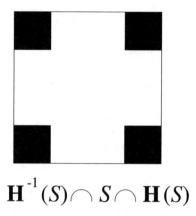

$$\mathbf{H}^{-1}(S) \cap S \cap \mathbf{H}(S)$$

Figure 9.16:   The first stage of the Smale horseshoe map.

in Figure 9.15. The result of the intersection of these first two sets is given in Figure 9.16.

As this process is iterated to infinity, points fall into the area contained by the original square in smaller and smaller subareas. The result is an *invariant Cantor set* that contains a countable set of periodic orbits and an uncountable set of bounded nonperiodic orbits.

The *Smale-Birkhoff Theorem* states that homoclinic tangles guarantee that a dynamical system will display *horseshoe dynamics*. For more details, the reader is directed once again to the excellent textbook of Guckenheimer and Holmes [5].

9.4 Mathematica Commands

```
(* Poincare returns. See Example 1. *)
returns=NDSolve[{r'[t]==-(r[t])^2,r[0]==1},r,{t,0,16 Pi}];
Table[Evaluate[r[2i Pi]],{i,8}] /. returns
```

```
(* Hamiltonian with two degrees of freedom. *)
(* Three-dimensional projection. See Example 5, Figure 9.5(e). *)
Ham1=NDSolve[{p1'[t]==-Sqrt[2] q1[t],q1'[t]==Sqrt[2] p1[t]
,p2'[t]==-q2[t],q2'[t]==p2[t],p1[0]==.5,p2[0]==1.5,q1[0]==.5,q2[0]==0},
{p1,p2,q1,q2},{t,0,1000},MaxSteps->Infinity];
ParametricPlot3D[Evaluate[{p1[t],q1[t],q2[t]} /.Ham1],{t,0,200},
AxesLabel->{"p1","q1","p2"},PlotPoints->10000]
(* Two-dimensional projection. See Figure 9.5(f). *)
ParametricPlot3D[Evaluate[{p1[t],q1[t],q2[t]} /.Ham1],{t,0,1000},
```

```
AxesLabel->{"p1","q1"," "},PlotRange->{{-1,1},{-1,1},{-0.01,0.01}},
PlotPoints->10000]
```

```
(* Phase portrait of a nonautonomous system. See Figure 9.11(b). *)
alpha=1;beta=-1;k=0.3;A=0.5;omega=1.25;
Duffing=NDSolve[{x'[t]==y[t],y'[t]==-beta*x[t]-k y[t]-alpha x[t]^3+A
Cos[omega t],x[0]==1,y[0]==0},{x[t],y[t]},{t,0,200},MaxSteps->Infinity];
ParametricPlot[Evaluate[{x[t],y[t]} /.Duffing],{t,0,200},PlotPoints->10000]
(* Poincare section. See Figure 9.11(b). *)
i=4000;
Poincare=NDSolve[{x'[t]==y[t],y'[t]==-beta x[t]-k y[t]-alpha x[t]^3+A
Cos[omega t],x[0]==1,y[0]==0},{x[t],y[t]},{t,0,2 i  Pi/omega},
MaxSteps->Infinity];
ListPlot[Table[Flatten[{x[t],y[t]} /. Poincare],{t,0,2 i Pi/omega,
2 Pi/omega}],PlotStyle->PointSize[0.001],AxesLabel->{"x","y"}]
```

```
(* Hysteresis loop. See Exercise 10. *)
k=0.1;omega=1.25;
M=1200;step=0.0001;MaxAmp=M*step;a=1;b=0;
For[i=1,i<=M,i++,A=step*i;pp=NDSolve[{x'[t]==y[t],
y'[t]==x[t]-k y[t]-(x[t])^3+A
Cos[omega t],x[0]==a,y[0]==b},{x,y},{t,0,2 Pi/omega},MaxSteps->100000];
a=Flatten[x[2 Pi/omega] /. pp];b=Flatten[y[2 Pi/omega] /. pp];
rampup[i]=Sqrt[a^2+b^2]];
p1=ListPlot[Table[Flatten[{m*step,rampup[m]}],{m,1,M}]];
For[i=1,i<=M,i++,A=MaxAmp-step*i;pp=NDSolve[{x'[t]==y[t],
y'[t]==x[t]-k y[t]-(x[t])^3+A Cos[omega t],x[0]==a,y[0]==b},{x,y},
{t,0,2 Pi/omega},MaxSteps->100000];a=Flatten[x[2 Pi/omega] /. pp];
b=Flatten[y[2 Pi/omega] /. pp];rampdown[i]=Sqrt[a^2+b^2]];
p2=ListPlot[Table[Flatten[{MaxAmp-m*step,rampdown[m]}],{m,1,M}],
PlotStyle->{Red,Thick}];
Show[{p1,p2},PlotRange->All,AxesLabel->{"\[CapitalGamma]","r"},
AxesOrigin->{0,0.4},TextStyle->{FontSize->15},
Ticks->{{0,0.04,0.08,0.12},Automatic}]
```

9.5 Exercises

1. Consider the system

$$\dot{x} = -y - 0.1x\sqrt{x^2 + y^2}, \quad \dot{y} = x - 0.1y\sqrt{x^2 + y^2}.$$

By considering the line segment $\Sigma = \{(x,y) \in \Re^2 : 0 \le x \le 4, y = 0\}$, list the first ten returns on Σ given that a trajectory starts at the point $(4,0)$.

2. Obtain a Poincaré map for the system

$$\dot{x} = \mu x + y - x\sqrt{x^2 + y^2}, \quad \dot{y} = -x + \mu y - y\sqrt{x^2 + y^2}$$

on the Poincaré section $\Sigma = \{(x, y) \in \Re^2 : 0 \le x < \infty, y = 0\}$.

3. Use the characteristic multiplier to determine the stability of the limit cycle in Example 2.

4. Solve the following differential equations:

$$\dot{r} = r(1 - r^2), \quad \dot{\theta} = 1.$$

Consider the line segment $\Sigma = \{(x, y) \in \Re^2 : 0 \le x \le \infty\}$ and find the Poincaré map for this system.

5. Use the characteristic multiplier to determine the stability of the limit cycle in Example 4.

6. Consider the two degrees of freedom Hamiltonian given by

$$H(\mathbf{p}, \mathbf{q}) = \frac{\omega_1}{2}(p_1^2 + q_1^2) + \frac{\omega_2}{2}(p_2^2 + q_2^2).$$

Plot three-dimensional and two-dimensional Poincaré sections when

(a) $\omega_1 = 3$ and $\omega_2 = 7$ for the set of initial conditions $t = 0, p_1 = 0.5, p_2 = 1.5, q_1 = 0.5, q_2 = 0$,

(b) $\omega_1 = \sqrt{2}$ and $\omega_2 = 3$ for the set of initial conditions $t = 0, p_1 = 0.5, p_2 = 1.5, q_1 = 0.5, q_2 = 0$.

7. Plot three-dimensional and two-dimensional Poincaré sections of the Toda Hamiltonian given by

$$H = \frac{p_1^2}{2} + \frac{p_2^2}{2} + \frac{e^{2q_2 + 2\sqrt{3}q_1}}{24} + \frac{e^{2q_2 - 2\sqrt{3}q_1}}{24} + \frac{e^{-4q_2}}{24} - \frac{1}{8},$$

for several different energy levels of your choice.

8. Plot the chaotic solution of the periodically driven Fitzhugh–Nagumo system

$$\dot{u} = 10\left(u - v - \frac{u^3}{3} + I(t)\right), \quad \dot{v} = u - 0.8v + 0.7,$$

where $I(t)$ is a periodic step function of period 2.025, amplitude 0.267, and width 0.3.

9. A damped driven pendulum may be modeled using the nonautonomous system of differential equations defined by

$$\frac{d^2\theta}{dt^2} + k\frac{d\theta}{dt} + \frac{g}{l}\sin(\theta) = \Gamma\cos(\omega t),$$

where k is a measure of the frictional force, Γ and ω are the amplitude and frequency of the driving force, g is the acceleration due to gravity, and l is the length of the pendulum. Plot a Poincaré map for this system when $k = 0.3$, $\Gamma = 4.5$, $\omega = 0.6$, and $\frac{g}{l} = 4$.

10. (a) Consider system (9.10) with $\alpha = 1, \beta = -1, k = 0.1$, and $\omega = 1.25$. Plot a bifurcation diagram for $0 \le \Gamma \le 0.1$ and show that there is a *clockwise hysteresis loop* at approximately $0.04 < \Gamma < 0.08$. Note that there is *ringing* (oscillation) at the ramp-up and ramp-down parts of the bistable region.

    (b) Plot the two stable limit cycles in the bistable region for Exercise 10(a) on one phase portrait. This shows that the system is multistable. For example, take $\Gamma = 0.07$. These limit cycles correspond to steady states on the upper and lower branches of the bistable cycle.

Bibliography

[1] S.S. Abdullaev, *Construction of Mappings for Hamiltonian Systems and their Applications (Lecture Notes in Physics)*, Springer-Verlag, New York, 2006.

[2] C-ODE-E (Consortium for ODE Experiments), *ODE Architect: The Ultimate ODE Power Tool*, John Wiley, New York, 1999.

[3] E.S. Cheb-Terrab and H. P. de Oliveira, Poincaré sections of Hamiltonian systems, *Comput. Phys. Comm.*, **95** (1996), 171.

[4] P. Galison, *Einstein's Clocks and Poincaré's Maps*, W.W. Norton and Company, London, 2004.

[5] J. Guckenheimer and P. Holmes, *Nonlinear Oscillations, Dynamical Systems, and Bifurcations of Vector Fields*, 3rd ed., Springer-Verlag, New York, 1990.

[6] J. Llibre and A.E. Teruel, *Introduction to the Qualitative Theory of Differential Systems: Planar, Symmetric and Continuous Piecewise Linear Systems*, Birkhäuser, Boston, 2013.

[7] F.C. Moon and P.J. Holmes, A magnetoelastic strange attractor, *Journal of Sound and Vibration*, **65** (1979), 276–296.

[8] M. Pettini, *Geometry and Topology in Hamiltonian Dynamics and Statistical Mechanics (Interdisciplinary Applied Mathematics)*, Springer-Verlag, New York, 2007.

[9] P. Pokorny, I. Schreiber and M. Marek, On the route to strangeness without chaos in the quasiperiodically forced van der Pol oscillator, *Chaos, Solitons and Fractals* **7** (1996), 409–424.

[10] H. Poincaré, Mémoire sur les courbes définies par une equation différentielle, *J. Math.*, **7** (1881), 375–422; Oeuvre, Gauthier-Villars, Paris, 1890.

[11] S. Smale, Differentiable dynamical systems, *Bull. Amer. Math. Soc.*, **73** (1967), 747–817.

[12] L.M. Surhone (Editor), M.T. Timpledon (Editor), S.F. Marseken (Editor), *Poincaré Map: Mathematics, Dynamical System, Henri Poincaré, Orbit, State Space, Dynamical System, Transversality, Flow, Recurrence Plot, Apsis*, Betascript Publishers, 2010.

Chapter 10
Local and Global Bifurcations

Aims and Objectives

- To introduce some local and global bifurcation theory in the plane.

- To bifurcate limit cycles in the plane.

- To introduce elementary theory of Gröbner bases.

On completion of this chapter, the reader should be able to

- bifurcate small-amplitude limit cycles from fine foci;

- solve systems of multivariate polynomial equations;

- bifurcate limit cycles from a center;

- investigate limit cycle bifurcation from homoclinic loops, numerically.

The problem of determining the maximum number of limit cycles for planar differential systems dates back more than 100 years and will be discussed in more detail in Chapter 11. Local limit cycles can be analyzed in terms of local behavior of the system near to a relevant critical point or limit cycle. The theory involved in global bifurcations is not so well developed and involves larger-scale behavior in the plane.

An algorithm is presented for bifurcating small-amplitude limit cycles out of a critical point. Gröbner bases are then introduced which can help with the reduction phase of the algorithm. The Melnikov function is used to determine the approximate location and number of limit cycles when a parameter is small. The limit cycles are bifurcated from a center. Bifurcations involving homoclinic loops are discussed in Section 10.4.

© Springer International Publishing AG 2017   209
S. Lynch, *Dynamical Systems with Applications using Mathematica®*,
DOI 10.1007/978-3-319-61485-4_10

10.1 Small-Amplitude Limit Cycle Bifurcations

The general problem of determining the maximum number and relative configurations of limit cycles in the plane has remained unresolved for over a century.
The problem will be stated in Chapter 11. Both local and global bifurcations
have been studied to create vector fields with as many limit cycles as possible.
All of these techniques rely heavily on symbolic manipulation packages such as
Mathematica. Unfortunately, the results in the global case number relatively
few. Only in recent years have many more results been found by restricting the
analysis to *small-amplitude limit cycle* bifurcations (see, e.g., Chapter 11 and
the references therein).

Consider systems of the form

$$\dot{x} = P(x, y), \quad \dot{y} = Q(x, y), \tag{10.1}$$

where P and Q are polynomials in x and y. It is well known that a nondegenerate
critical point, say $\mathbf{x_0}$, of center or focus type can be moved to the origin by a
linear change of coordinates to give

$$\dot{x} = \lambda x - y + p(x, y), \quad \dot{y} = x + \lambda y + q(x, y), \tag{10.2}$$

where p and q are at least quadratic in x and y. If $\lambda \neq 0$, then the origin is
structurally stable for all perturbations.

Definition 1. A critical point, say $\mathbf{x_0}$, is called a fine focus of system (10.1) if
it is a center for the linearized system at $\mathbf{x_0}$. Equivalently, if $\lambda = 0$ in system
(10.2), then the origin is a fine focus.

In the work to follow, assume that the unperturbed system does not have
a center at the origin. The technique used here is entirely local; limit cycles
bifurcate out of a fine focus when its stability is reversed by perturbing λ and
the coefficients arising in p and q. These are said to be local or small-amplitude
limit cycles. How close the origin is to being a center of the nonlinear system
determines the number of limit cycles that may be obtained from bifurcation.
The method for bifurcating limit cycles will now be summarized and is given in
detail in [2].

By a classical result, there exists a Lyapunov function, say $V(x, y) =
V_2(x, y) + V_3(x, y) + \ldots + V_k(x, y) + \ldots$, where V_k is a homogeneous polyno-
mial of degree k, such that

$$\frac{dV}{dt} = \eta_2 r^2 + \eta_4 r^4 + \ldots + \eta_{2i} r^{2i} + \ldots, \tag{10.3}$$

where $r^2 = x^2 + y^2$. The η_{2i} are polynomials in the coefficients of p and q and
are called the *focal values*. The origin is said to be a fine focus of order k if

$\eta_2 = \eta_4 = \ldots = \eta_{2k} = 0$ but $\eta_{2k+2} \neq 0$. Take an analytic transversal through the origin parameterized by some variable, say c. It is well known that the return map of (10.2), $c \mapsto h(c)$, is analytic if the critical point is nondegenerate. Limit cycles of system (10.2) then correspond to zeros of the *displacement function*, $d(c) = h(c) - c$ (see Chapter 9). Hence at most k limit cycles can bifurcate from the fine focus. The stability of the origin is clearly dependent on the sign of the first nonzero focal value, and the origin is a nonlinear center if and only if all of the focal values are zero. Consequently, it is the reduced values, or *Lyapunov quantities*, say $L(j)$, that are significant. One needs only consider the value η_{2k} reduced modulo the ideal $\langle \eta_2, \eta_4, \ldots, \eta_{2k-2} \rangle$ to obtain the Lyapunov quantity $L(k-1)$. To bifurcate limit cycles from the origin, select the coefficients in the Lyapunov quantities such that

$$|L(m)| \ll |L(m+1)| \quad \text{and} \quad L(m)L(m+1) < 0,$$

for $m = 0, 1, \ldots, k-1$. At each stage, the origin reverses stability and a limit cycle bifurcates in a small region of the critical point. If all of these conditions are satisfied, then there are exactly k small-amplitude limit cycles. Conversely, if $L(k) \neq 0$, then at most k limit cycles can bifurcate. Sometimes, it is not possible to bifurcate the full complement of limit cycles; an example is given in [9].

The algorithm for bifurcating small-amplitude limit cycles may be split into the following four steps:

1. computation of the focal values using a mathematical package;

2. reduction of the nth focal value modulo a Gröbner basis of the ideal generated by the first $n-1$ focal values (or the first $n-1$ Lyapunov quantities);

3. checking that the origin is a center when all of the relevant Lyapunov quantities are zero;

4. bifurcation of the limit cycles by suitable perturbations.

Dongming Wang [13] has recently developed software to deal with the reduction part of the algorithm for several differential systems, and Gröbner bases are introduced in the next section.

For some systems, the following theorems can be used to prove that the origin is a center.

The Divergence Test. Suppose that the origin of system (10.1) is a critical point of focus type. If

$$\mathrm{div}\,(\psi \mathbf{X}) = \frac{\partial(\psi P)}{\partial x} + \frac{\partial(\psi Q)}{\partial y} = 0,$$

where $\psi : \Re^2 \to \Re^2$, then the origin is a center.

The Classical Symmetry Argument. Suppose that $\lambda = 0$ in system (10.2) and that either

    (i) $p(x, y) = -p(x, -y)$ and $q(x, y) = q(x, -y)$ or
    (ii) $p(x, y) = p(-x, y)$ and $q(x, y) = -q(-x, y)$.

Then, the origin is a center.

Adapting the classical symmetry argument, it is also possible to prove the following theorem.

Theorem 1. *The origin of the system*

$$\dot{x} = y - F(G(x)), \quad \dot{y} = -\frac{G'(x)}{2}H(G(x)),$$

where F and H are polynomials, and $G(x) = \int_0^x g(s)ds$ with $g(x)sgn(x) > 0$ for $x \neq 0$, $g(0) = 0$, is a center.

The reader is asked to prove this theorem in the exercises at the end of this chapter.

To demonstrate the method for bifurcating small-amplitude limit cycles, consider Liénard equations

$$\dot{x} = y - F(x), \quad \dot{y} = -g(x), \tag{10.4}$$

where $F(x) = a_1 x + a_2 x^2 + \ldots + a_u x^u$ and $g(x) = x + b_2 x^2 + b_3 x^3 + \ldots + b_v x^v$. This system has proved very useful in the investigation of limit cycles when showing existence, uniqueness, and hyperbolicity of a limit cycle. In recent years, there have also been many local results (see, e.g., Table 11.1 in Chapter 11). Therefore, it seems sensible to use this class of system to illustrate the method.

The computation of the first three focal values will be given. Write $V_k(x, y) = \sum_{i+j=k} V_{i,j} x^i y^j$ and denote $V_{i,j}$ as being odd or even according to whether i is odd or even and that $V_{i,j}$ is 2-odd or 2-even according to whether j is odd or even, respectively. Solving equation (10.3), it is easily seen that $V_2 = \frac{1}{2}(x^2 + y^2)$ and $\eta_2 = -a_1$. Therefore, set $a_1 = 0$. The odd and even coefficients of V_3 are then given by the two pairs of equations

$$3V_{3,0} - 2V_{1,2} = b_2,$$
$$V_{1,2} = 0,$$

and

$$-V_{2,1} = a_2,$$
$$2V_{2,1} - 3V_{0,3} = 0,$$

respectively. Solve the equations to give

$$V_3 = \frac{1}{3}b_2 x^3 - a_2 x^2 y - \frac{2}{3}a_2 y^3.$$

Both η_4 and the odd coefficients of V_4 are determined by the equations

$$-\eta_4 - V_{3,1} = a_3,$$
$$-2\eta_4 + 3V_{3,1} - 3V_{1,3} = -2a_2 b_2,$$
$$-\eta_4 + V_{1,3} = 0.$$

The even coefficients are determined by the equations

$$4V_{4,0} - 2V_{2,2} = b_3 - 2a_2^2,$$
$$2V_{2,2} - 4V_{0,4} = 0$$

and the supplementary condition $V_{2,2} = 0$. In fact, when computing subsequent coefficients for V_{4m}, it is convenient to require that $V_{2m,2m} = 0$. This ensures that there will always be a solution. Solving these equations gives

$$V_4 = \frac{1}{4}(b_3 - 2a_2^2)x^4 - (\eta_4 + a_3)x^3 y + \eta_4 x y^3$$

and

$$\eta_4 = \frac{1}{8}(2a_2 b_2 - 3a_3).$$

Suppose that $\eta_4 = 0$ so that $a_3 = \frac{2}{3}a_2 b_2$. It can be checked that the two sets of equations for the coefficients of V_5 give

$$V_5 = \left(\frac{b_4}{5} - \frac{2a_2^2 b_2}{3}\right)x^5 + (2a_3^2 - a_4)x^4 y + \left(\frac{8a_2^3}{3} - \frac{4a_4}{3} + \frac{2a_2 b_3}{3}\right)x^2 y^3$$

$$+ \left(\frac{16a_2^3}{15} - \frac{8a_4}{15} - \frac{4a_2 b_3}{15}\right)y^5.$$

The coefficients of V_6 may be determined by inserting the extra condition $V_{4,2} + V_{2,4} = 0$. In fact, when computing subsequent even coefficients for V_{4m+2}, the extra condition, $V_{2m,2m+2} + V_{2m+2,2m} = 0$, is applied, which guarantees a solution. The polynomial V_6 contains 27 terms and will not be listed here. However, η_6 leads to the Lyapunov quantity

$$L(2) = 6a_2 b_4 - 10a_2 b_2 b_3 + 20a_4 b_2 - 15a_5.$$

Lemma 1. *The first three Lyapunov quantities for system (10.4) are* $L(0) = -a_1$, $L(1) = 2a_2b_2 - 3a_3$, *and* $L(2) = 6a_2b_4 - 10a_2b_2b_3 + 20a_4b_2 - 15a_5$.

Let $\hat{H}(u,v)$ denote the maximum number of small-amplitude limit cycles that can be bifurcated from the origin for system (10.4).

Example 1. Prove that

(i) $\hat{H}(3,2) = 1$ and

(ii) $\hat{H}(3,3) = 2$

for system (10.4).

Solutions. (i) Consider the case where $u = 3$ and $v = 2$. Now, $L(0)=0$ if $a_1 = 0$ and $L(1) = 0$ if $a_3 = \frac{2}{3}a_2b_2$. Thus, system (10.4) becomes

$$\dot{x} = y - a_2x^2 - \frac{2}{3}a_2b_2x^3, \quad \dot{y} = -x - b_2x^2,$$

and the origin is a center by Theorem 1. Therefore, the origin is a fine focus of order one if and only if $a_1 = 0$ and $2a_2b_2 - 3a_3 \neq 0$. The conditions are consistent. Select a_3 and a_1 such that

$$|L(0)| \ll |L(1)| \quad \text{and} \quad L(0)L(1) < 0.$$

The origin reverses stability once, and a limit cycle bifurcates. The perturbations are chosen such that the origin reverses stability once and the limit cycles that bifurcate persist. Thus, $\hat{H}(3,2) = 1$. Figure 10.1(a) shows a small-amplitude limit cycle for system (10.4) when $u = 3$ and $v = 2$.

(ii) Consider system (10.4) with $u = 3$ and $v = 3$. Now, $L(0) = 0$ if $a_1 = 0$, $L(1) = 0$ if $a_3 = \frac{2}{3}a_2b_2$, and $L(2) = 0$ if $a_2b_2b_3 = 0$. Thus, $L(2) = 0$ if

(a) $a_2 = 0$,

(b) $b_3 = 0$, or

(c) $b_2 = 0$.

If condition (a) holds, then $a_3 = 0$ and the origin is a center by the divergence test ($\text{div}\mathbf{X} = 0$). If condition (b) holds, then the origin is a center since $\hat{H}(3,2) = 1$. If condition (c) holds, then $a_3 = 0$ and system (10.3) becomes

$$\dot{x} = y - a_2x^2, \quad \dot{y} = -x - b_3x^3,$$

and the origin is a center by the classical symmetry argument. The origin is thus a fine focus of order two if and only if $a_1 = 0$ and $2a_2b_2 - 3a_3 = 0$ but $a_2b_2b_3 \neq 0$. The conditions are consistent. Select b_3, a_3, and a_1 such that

$$|L(1)| \ll |L(2)|, \ L(1)L(2) < 0 \quad \text{and} \quad |L(0)| \ll |L(1)|, \ L(0)L(1) < 0.$$

The origin has changed stability twice, and there are two small-amplitude limit cycles. The perturbations are chosen such that the origin reverses stability twice and the limit cycles that bifurcate persist. Thus $\hat{H}(3,3) = 2$. Figure 10.1(b) shows two small-amplitude limit cycles for system (10.4) when $u = 3$ and $v = 3$.

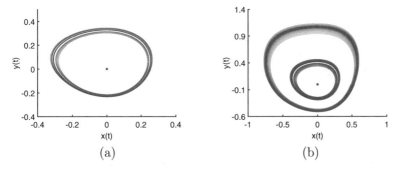

(a) (b)

Figure 10.1: Small-amplitude limit cycles (blue trajectories) for system (10.4): (a) one limit cycle when $u = 3$ and $v = 2$ and $a_1 = 0.01, a_2 = 1, b_2 = 1$, and $a_3 = \frac{1}{3}$; (b) two limit cycles when $u = 3$ and $v = 3$ and $a_1 = 0.01, a_2 = 1, b_2 = 1, a_3 = \frac{1}{3}$, and $b_3 = 2$.

10.2 Gröbner Bases

The field of *computer algebra* has expanded considerably in recent years and extends deeply into both mathematics and computer science. One fundamental tool in this new field is the theory of Gröbner bases [1, 6, 14]. In 1965, as part of his PhD research studies, Bruno Buchberger [5] devised an algorithm for computing Gröbner bases for a set of multivariate polynomials. The Gröbner bases algorithm was named in honor of his PhD supervisor Wolfgang Gröbner. The most common use of the Gröbner bases algorithm is in computing bases which can be related to operations for ideals in commutative polynomial rings. Most mathematical packages now have the Buchberger algorithm incorporated for computing Gröbner bases, and Mathematica is no exception. This section aims to give a brief overview of the method including some notation, definitions, and theorems without proof. Introductory theory on commutative rings and ideals and proofs to the theorems listed in this section can be found in most of the textbooks in the reference section of this chapter. There are a wide range of applications, see [4], for example; however, for this text, we will be interested in Gröbner bases in polynomial rings in several variables only. The theory of Gröbner bases originated with the desire to solve systems of nonlinear equations involving multivariate polynomial equations. Wang et al. [12, 13] have used Gröbner bases among other methods to test elimination algorithms when solving multivariate polynomial systems. One interesting unsolved example appears in

[12], when attempting to prove complete center conditions for a certain cubic system.

Recall some basic algebraic definitions:

Definition 2. A *ring*, say $(R, +, *)$ is a set R with two binary operations $+$ and $*$, satisfying the following conditions:

1. $(R, +)$ is an Abelian group;

2. $(R, *)$ is a semigroup, and

3. the *distributive laws* hold.

If $(R, +)$ is commutative, then $(R, +, *)$ is called a *commutative ring*.

Definition 3. A nonempty subset $I \subset (R, +, *)$ is called an *ideal* if for all $r \in R$ and $a \in I$, $r * a \in I$ and $a * r \in I$.

Notation. Let $\mathbb{N}$ denote the set of nonnegative integers $\mathbb{N} = \{0, 1, 2, \ldots\}$. Let $\alpha = (\alpha_1, \alpha_2, \ldots, \alpha_n)$ be a power vector in $\mathbb{N}^n$, and let $x_1, x_2, \ldots, x_n$ be any n variables. Write $\mathbf{x}^\alpha = x_1^{\alpha_1} x_2^{\alpha_2} \ldots x_n^{\alpha_n}$, where $|\alpha| = (\alpha_1 + \alpha_2 + \ldots + \alpha_n)$ is the *total degree* of the *monomial* $\mathbf{x}^\alpha$. Let $R = K[\mathbf{x}] = K[x_1, x_2, \ldots, x_n]$ be a commutative polynomial ring in n variables over an algebraically closed field K such as $\mathbb{C}, \mathbb{Q}$, or $\mathbb{R}$. Recall that a field is an algebraic structure in which the operations addition, subtraction, multiplication, and division (except by zero) may be performed.

Definition 4. Let $P = \{p_1, p_2, \ldots, p_s\}$ be a set of multivariate polynomials, then the ideal generated by P, denoted by $I = \langle P \rangle$, is given by:

$$\left\{ \sum_{i=1}^{s} f_i p_i : f_1, f_2, \ldots, f_s \in K[\mathbf{x}] \right\},$$

where the polynomials p_i form a *basis* for the ideal they generate.

In 1888, David Hilbert proved the following theorem:

Theorem 2. (Hilbert's Bases Theorem). *If K is a field, then every ideal in the polynomial ring $K[\mathbf{x}]$ is finitely generated.*

A proof to this theorem can be found in most textbooks in the reference section of this chapter.

An extremely useful basis of an ideal is the Gröbner basis, which will be defined after the notion of *monomial ordering* is introduced.

Definition 5. A monomial order, say $\succ$, is a total order on the monomials of R such that

1. for all $\alpha \in \mathbb{N}^n$, $\alpha \succ 0$;

2. for all $\alpha, \beta, \gamma \in \mathbb{N}^n$, $\alpha \succ \beta$ implies that $\alpha + \gamma \succ \beta + \gamma$.

The three most common monomial orderings are defined by the following:

Definition 6. Suppose that $\alpha, \beta \in \mathbb{N}^n$. Then the

1. *lexicographical order* is such that $\alpha \succ_{\text{lex}} \beta$ if and only if the leftmost nonzero entry in $\alpha - \beta$ is positive;

2. *degree lexicographical order* is such that $\alpha \succ_{\text{dlex}} \beta$ if and only if $|\alpha| \succ |\beta|$ or ($|\alpha| = |\beta|$ and $\alpha \succ_{\text{lex}} \beta$);

3. *degree reverse lexicographical order* is such that $\alpha \succ_{\text{drevlex}} \beta$ if and only if $|\alpha| \succ |\beta|$ or ($|\alpha| = |\beta|$ and the rightmost nonzero entry in $\alpha - \beta$ is negative.

Note that there are many other monomial orderings including weighted and grouped orders [7].

Example 2. Suppose that $\mathbf{x}^\alpha = x^3 y^3 z$, $\mathbf{x}^\beta = x^2 y^4 z^2$, and $\mathbf{x}^\gamma = x y^6 z$. Then

1. $(3, 3, 1) = \alpha \succ_{\text{lex}} \beta = (2, 4, 2)$ since in $(\alpha - \beta) = (1, -1, -1)$, the leftmost nonzero entry is positive. Hence, $x^3 y^3 z \succ_{\text{lex}} x^2 y^4 z^2$.

2. (i) $\beta = (2, 4, 2) \succ_{\text{dlex}} \alpha = (3, 3, 1)$ since $|\beta| = 8 > |\alpha| = 7$. Hence, $x^2 y^4 z^2 \succ_{\text{dlex}} x^3 y^3 z$. (ii) $\beta = (2, 4, 2) \succ_{\text{dlex}} \gamma = (1, 6, 1)$ since $|\beta| = |\gamma| = 8$ and in $(\beta - \gamma) = (1, -2, 1)$, the leftmost nonzero entry is positive. Hence, $x^2 y^4 z^2 \succ_{\text{dlex}} x y^6 z$.

3. (i) $\beta = (2, 4, 2) \succ_{\text{drevlex}} \alpha = (3, 3, 1)$ since $|\beta| = 8 > |\alpha| = 7$. Hence, $x^2 y^4 z^2 \succ_{\text{drevlex}} x^3 y^3 z$. (ii) $\gamma = (1, 6, 1) \succ_{\text{drevlex}} \beta = (2, 4, 2)$ since $|\gamma| = |\beta| = 8$ and in $(\gamma - \beta) = (-1, 2, -1)$, the rightmost nonzero entry is negative. Hence, $x y^6 z \succ_{\text{drevlex}} x^2 y^4 z^2$.

Definition 7. Assume that there is a fixed term order $\succ$ on a set of monomials that uniquely order the terms in a given nonzero polynomial $p = \sum_\alpha c_\alpha \mathbf{x}^\alpha \in K[\mathbf{x}]$. Define the

1. *multidegree* of p as $\text{multideg}(p) = \max(\alpha \in \mathbb{N}^n : c_\alpha \neq 0)$;

2. *leading coefficient* of p as $\text{LC}(p) = c_{\text{multideg}(p)}$;

3. *leading monomial* of p as $\text{LM}(p) = \mathbf{x}^{\text{multideg}(p)}$;

4. *leading term* of p as $\text{LT}(p) = \text{LC}(p)\text{LM}(p)$;

Example 3. Suppose that $p(x, y, z) = 2x^3 y^3 z + 3x^2 y^4 z^2 - 4x y^6 z$, then

- with respect to $\succ_{\text{lex}}$, $\text{multideg}(p) = (3, 3, 1)$, $\text{LC}(p) = 2$, $\text{LM}(p) = x^3 y^3 z$, $\text{LT}(p) = 2x^3 y^3 z$;

- with respect to $\succ_{\text{dlex}}$, $\text{multideg}(p) = (2, 4, 2)$, $\text{LC}(p) = 3$, $\text{LM}(p) = x^2 y^4 z^2$, $\text{LT}(p) = 3x^2 y^4 z^2$;

- with respect to $\succ_{\text{drevlex}}$, multideg$(p) = (1,6,1)$, LC$(p) = -4$, LM$(p) = xy^6z$, LT$(p) = -4xy^6z$.

Definition 8. A polynomial f is *reduced* with respect to $P = \{p_1, p_2, \ldots, p_s\}$ (or modulo P), $f \rightarrow_P h$, if and only if there exists $p_i \in P$, such that

$$h = f - \frac{\text{LT}(f)}{\text{LT}(p_i)}.$$

Furthermore, a polynomial g is *completely reduced* with respect to P if no monomial of g is divisible by any of the LM (p_i), for all $1 \le i \le s$.

Division Algorithm for Multivariate Polynomials Let $P = \{p_1, p_2, \ldots, p_s\}$ be an ordered set of polynomials in $K[\mathbf{x}]$, then there exist polynomials $q_1, q_2, \ldots, q_s, r \in K[\mathbf{x}]$ such that for $p \in K[\mathbf{x}]$

$$p = q_1 p_1 + q_2 p_2 + \ldots + q_s p_s + r,$$

and either $r = 0$ or r is completely reduced with respect to P. The algorithm is described briefly here and is a generalization of the division algorithm in $K[x_1]$. Perform the reduction of p modulo $p_1, p_2, \ldots, p_s$ by repeatedly applying the following procedure until doing so leaves p unchanged. Take the smallest i such that $a_i = \text{LT}(p_i)$ divides one of the terms of p. Let f be the largest (with respect to some monomial ordering $\succ$) term of p that is divisible by a_i, replace p by $p - \left(\frac{f}{a_i}\right) p_i$, and the process eventually terminates. For a more detailed explanation, see the textbooks at the end of the chapter.

When dealing with large ordered sets of polynomials with high total degrees, one must use computer algebra. There is a command in Mathematica for carrying out the division algorithm. The syntax is

$$\text{PolynomialReduce}\,[poly, \{poly_1, poly_2, \ldots\}, \{x_1, x_2, \ldots\}],$$

which gives a list representing a reduction of *poly* in terms of the *poly*$_i$ with $x_1 \succ x_2 \succ \ldots$.

Example 4. Fix a lexicographical order $x \succ_{\text{lex}} y \succ_{\text{lex}} z$. (i) Divide the polynomial $p = x^4 + y^4 + z^4$ by the ordered list of polynomials $\{x^2 + y, z^2y - 1, y - z^2\}$. (ii) Repeat the division with the divisors listed as $\{y - z^2, z^2y - 1, x^2 + y\}$.

Solution. Using the PolynomialReduce command in Mathematica:

(i) $x^4 + y^4 + z^4 = (x^2 - y)(x^2 + y) + (2 + y^2)(z^2y - 1) + (2y + y^3)(y - z^2) + 2 + z^4;$

(ii) $x^4 + y^4 + z^4 = (-x^2 + y^3 + z^2 + y^2z^2 + yz^4 + z^6)(y - z^2) + 0(z^2y - 1) + (x^2 - z^2)(x^2 + y) + 2z^4 + z^8.$

Note that the remainders are different. Unfortunately, the division algorithm for multivariate polynomials does not produce unique remainders. However, all is not lost and unique remainders exist when the basis of the ideal is a Gröbner basis.

Definition 9. The *lowest common multiple* (LCM) of two monomials $x_1^{\alpha_1} x_2^{\alpha_2} \ldots x_n^{\alpha_n}$ and $x_1^{\beta_1} x_2^{\beta_2} \ldots x_n^{\beta_n}$ is given by

$$\mathrm{LCM}\left(\mathbf{x}^\alpha, \mathbf{x}^\beta\right) = x_1^{\max(\alpha_1, \beta_1)} x_2^{\max(\alpha_2, \beta_2)} \ldots x_n^{\max(\alpha_n, \beta_n)}.$$

Definition 10. The S-polynomial of two nonzero ordered polynomials $p, \pi \in K[\mathbf{x}]$, is defined by

$$S\left(p, \pi\right) = \frac{\mathrm{LCM}(\mathrm{LM}(p), \mathrm{LM}(\pi))}{LT(p)} p - \frac{\mathrm{LCM}(\mathrm{LM}(p), \mathrm{LM}(\pi))}{LT(\pi)} \pi. \qquad (10.5)$$

The S-polynomials are constructed to cancel leading terms.

Example 5. Suppose that $p = x - 13y^2 - 12z^3$ and $\pi = x^2 - xy + 92z$, determine $S\left(p, \pi\right)$ with respect to the term order $x \succ_{\mathrm{lex}} y \succ_{\mathrm{lex}} z$.

Solution. Substituting into equation (10.5)

$$S\left(p, \pi\right) = \frac{x^2}{x}\left(x - 13y^2 - 12z^3\right) - \frac{x^2}{x^2}\left(x^2 - xy + 92z\right).$$

Hence,

$$S(p, \pi) = -13xy^2 - 12xz^3 + xy - 92z$$

and the leading terms of p and π have canceled.

The following theorem gives rise to Buchberger's algorithm:

Theorem 3. (Buchberger's Theorem) *Let $G = \{g_1, g_2, \ldots, g_s\}$ be a set of nonzero polynomials in $K[\mathbf{x}]$, then G is a Gröbner basis for the ideal $I = \langle G \rangle$ if and only if for all $i \neq j$,*

$$S\left(g_i, g_j\right) \to_G 0.$$

Buchberger's Algorithm to Compute Gröbner Bases The algorithm is used to transform a set of polynomial ideal generators into a Gröbner basis with respect to some monomial ordering. Suppose that $P = \{p_1, p_2, \ldots, p_s\}$ is a set of multivariate polynomials with a fixed term order $\succ$.

Step 1. Using the division algorithm for multivariate polynomials (PolynomialReduce in Mathematica), reduce all of the possible S-polynomial combinations modulo the set P.

Step 2. Add all nonzero polynomials resulting from Step 1 to P, and repeat Steps 1 and 2 until nothing new is added.

The Hilbert basis theorem guarantees that the algorithm eventually stops. Unfortunately, there are redundant polynomials in this Gröbner basis.

Definition 11. A Gröbner basis $G = \{g_1, g_2, \ldots g_s\}$ is *minimal* if for all $1 \leq i \leq s$, $\mathrm{LT}\,(g_i) \notin \langle \mathrm{LT}\,(g_1), \mathrm{LT}\,(g_2), \ldots, \mathrm{LT}\,(g_s) \rangle$.

Definition 12. A minimal Gröbner basis $G = \{g_1, g_2, \ldots g_s\}$ is *reduced* if for all pairs i, j, $i \neq j$, no term of g_i is divisible by $\mathrm{LT}\,(g_j)$.

Theorem 4. *Every polynomial ideal $I \subset K[\mathbf{x}]$ has a unique reduced Gröbner basis.*

A Gröbner basis for a polynomial ideal may be computed using the Mathematica command GroebnerBasis.

Example 6. Determine the critical points of the system

$$\dot{x} = x + y^2 - x^3, \quad \dot{y} = 4x^3 - 12xy^2 + x^4 + 2x^2y^2 + y^4. \qquad (10.6)$$

Solution. The critical points are found by solving the equations $\dot{x} = \dot{y} = 0$. Suppose that

$$I = \langle x + y^2 - x^3, 4x^3 - 12xy^2 + x^4 + 2x^2y^2 + y^4 \rangle,$$

then a reduced Gröbner basis for I with respect to $\succ_{\mathrm{lex}}$ may be computed using Mathematica. The command lines are given in Section 10.6. Note that a different reduced Gröbner basis might result if a different ordering is taken.

$$\{-195y^4 + 1278y^6 - 1037y^8 + 90y^{10} + y^{12}, 5970075x + 5970075y^2$$
$$+163845838y^4 - 162599547y^6 + 14472880y^8 + 160356y^{10}\}.$$

The first generator is expressed in terms of y alone, which can be determined from any one-variable technique. Back substitution is then used to determine the corresponding x values. There are seven critical points at

$$(0,0), (2.245, -3.011), (2.245, 3.011), (1.370, -1.097),$$
$$(1.370, 1.097), (-0.895, -0.422), (-0.895, 0.422).$$

Of course, the reader could also use the Solve command in Mathematica which is based on the Buchberger algorithm.

Example 7. The first five Lyapunov quantities for the Liénard system

$$\dot{x} = y - a_1x - a_2x^2 - a_3x^3 - a_4x^4, \quad \dot{y} = -x - b_2x^2 - b_3x^3,$$

are

$L(0) = -a_1;$

$L(1) = -3a_3 + 2b_2a_2;$

$L(2) = 5b_2(2a_4 - b_3a_2);$

$L(3) = -5b_2(92b_2^2a_4 - 99b_3^2a_2 + 1520a_2^2a_4 - 760a_2^3b_3 - 46b_2^2b_3a_2 + 198b_3a_4);$

$L(4) = -b_2(14546b_2^4a_4 + 105639a_2^3b_3^2 + 96664a_2^3b_2^2b_3 - 193328a_2^2b_2^2a_4 -$

$\qquad 891034a_2^4a_4 + 445517a_2^5b_3 + 211632a_2a_4^2 - 317094a_2^2b_3a_4 - 44190b_2^2b_3a_4 +$

$\qquad 22095b_2^2b_3^2a_2 - 7273b_2^4b_3a_2 + 5319b_3^3a_2 - 10638b_3^3a_4),$

where $a_3 = \frac{2}{3}a_2b_2$ was substituted from $L(1) = 0$. The polynomials can be reduced using a number of substitutions; however, the Gröbner basis is easily computed as:

$$GB = \{-4b_2a_4 + 3b_3a_3, -3a_3 + 2b_2a_2, a_1\},$$

under the ordering $a_1, a_2 \succ a_3 \succ a_4 \succ b_2 \succ b_3$. The Gröbner basis can then be used to help show that the origin is a center when all of the Lyapunov quantities are zero.

Note that there are specialist commutative algebraic packages, such as Singular and Macaulay, that use Gröbner bases intensely for really tough problems.

10.3 Melnikov Integrals and Bifurcating Limit Cycles from a Center

Consider perturbed two-dimensional differential systems of the form

$$\dot{\mathbf{x}} = \mathbf{f}(\mathbf{x}) + \epsilon\mathbf{g}(\mathbf{x}, \epsilon, \mu). \qquad (10.7)$$

Assume that the unperturbed system

$$\dot{\mathbf{x}} = \mathbf{f}(\mathbf{x}) \qquad (10.8)$$

has a one-parameter family of periodic orbits given by

$$\Gamma_r : \mathbf{x} = \gamma_r(t),$$

where the functions $\gamma_r(t)$ have minimum periods T_r and r belongs to an indexing set, say I, that is either a finite or semi-infinite open interval of $\Re$.

Definition 13. The *Melnikov function* for system (10.7) along the cycle Γ_r : $\mathbf{x} = \gamma_r(t)$, $0 \leq t \leq T_r$, of (10.8) is given by

$$M(r,\mu) = \int_0^{T_r} \exp\left(-\int_0^t \nabla.\mathbf{f}(\gamma_r(s))ds\right) \mathbf{f} \wedge \mathbf{g}(\gamma_r(t),0,\mu)dt.$$

Theorem 5. *Suppose that*

$$M(r_0,\mu_0) = 0 \quad and \quad \left.\frac{\partial M}{\partial r}\right|_{(r_0,\mu_0)} \neq 0,$$

where $r_0 \in I$. Then for $0 < \epsilon \ll 1$, system (10.7) has a unique hyperbolic limit cycle close to Γ_{r_0}. System (10.7) has no limit cycle close to Γ_{r_0} if $M(r_0,\mu_0) \neq 0$ and ϵ are small.

Theorem 6. *Suppose that $M(r,\mu_0) = 0$ has exactly k solutions $r_1, r_2, \ldots, r_k \in I$ with*

$$\left.\frac{\partial M}{\partial r}\right|_{(r_i,\mu_0)} \neq 0,$$

for some i from 1 to k. Then for $0 < \epsilon \ll 1$, exactly k one-parameter families of hyperbolic limit cycles bifurcate from the period annulus of (10.8) at the points $r_1, r_2, \ldots, r_k$. If $M(r,\mu_0) \neq 0$, then there are no limit cycles.

Melnikov-type integrals have been widely used since Poincaré's investigations at the end of the nineteenth century. It is well known that the Melnikov function for system (10.7) is proportional to the derivative of the Poincaré map for (10.7) with respect to ϵ. The interested reader may consult [2] for more details; the paper also deals with limit cycles of multiplicity greater than one and the bifurcation of limit cycles from separatrix cycles. To avoid elliptic integrals, only systems with $\gamma_r(t) = (x(t), y(t)) = (r\cos t, r\sin t)$ will be considered in this book.

Example 8. Consider the van der Pol system

$$\dot{x} = y, \quad \dot{y} = -x - \epsilon(1 - x^2)y.$$

Prove that there is a limit cycle asymptotic to the circle of radius two when ϵ is small.

Solution. In this case, $\mathbf{f}(\mathbf{x}) = (y, -x)^T$, $T_r = 2\pi$, $\mathbf{g}(\mathbf{x},\epsilon) = (0, -\epsilon y(1 - x^2))^T$, $x = r\cos(t)$, $y = r\sin(t)$ and $\nabla.\mathbf{f}(\mathbf{x}) = 0$. Therefore,

$$M(r,\mu) = \int_o^{T_r} \mathbf{f} \wedge \mathbf{g}(\gamma_r(t),0,\mu)\, dt.$$

Thus

$$M(r,\mu) = \int_0^{2\pi} -r^2\left(\sin^2 t(1 - r^2 \cos^2 t)\right)\, dt$$

and

$$M(r, \mu) = \frac{\pi}{4} r^2 (r^2 - 4).$$

Hence, $M(r_0, \mu) = 0$ when $r_0 = 2$ and $\frac{\partial M}{\partial r}\big|_{(r_0,0)} = \pi r_0(r_0^2 - 2) \neq 0$. Therefore, there exists a unique hyperbolic limit cycle asymptotic to a circle of radius two for the van der Pol system when ϵ is sufficiently small.

Example 9. Consider the Liénard system

$$\dot{x} = -y + \epsilon(a_1 x + a_3 x^3 + a_5 x^5), \quad \dot{y} = x. \tag{10.9}$$

Determine the maximum number and approximate location of the limit cycles when ϵ is sufficiently small.

Solution. Again, $\mathbf{f}(\mathbf{x}) = (-y, x)^T$, $T_r = 2\pi$, $\mathbf{g}(\mathbf{x}, \epsilon) = (\epsilon(a_1 x + a_3 x^3 + a_5 x^5), 0)^T$ and $\nabla . \mathbf{f}(\mathbf{x}) = 0$. Therefore,

$$M(r, \mu) = \int_0^{2\pi} -a_1 r^2 \cos^2 t - a_3 r^4 \cos^4 t - a_5 r^6 \cos^6 t \, dt$$

and

$$M(r, \mu) = -\pi r^2 \left(a_1 + \frac{3a_3}{4} r^2 + \frac{5a_5}{8} r^4 \right).$$

The polynomial $m(r) = a_1 + \frac{3a_3}{4} r^2 + \frac{5a_5}{8} r^4$ has at most two positive roots. Therefore, when ϵ is sufficiently small, system (10.9) has at most two hyperbolic limit cycles asymptotic to circles of radii r_j ($j = 1, 2$), where r_j are the positive roots of $m(r)$.

10.4 Bifurcations Involving Homoclinic Loops

Global bifurcations of limit cycles from centers were investigated in the previous section. Consider the following van der Pol type system

$$\dot{x} = y + 10x(0.1 - y^2), \quad \dot{y} = -x + C, \tag{10.10}$$

where C is a constant. If $C = 0$, the system has one critical point at the origin and a stable limit cycle surrounding it. However, if $C \neq 0$, there is a second critical point at $\left(C, \frac{1}{20C} + \sqrt{\left(\frac{1}{20C}\right)^2 + 0.1} \right)$, which is a saddle point. Figure 10.2 shows three possible phase portraits for varying values of the parameter C.

When C is large and negative, the saddle point is far from the origin. As C is increased and approaches the approximate value $C \approx -0.18$, one of the

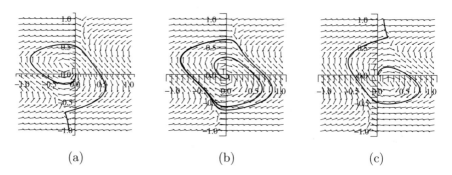

(a)   (b)   (c)

Figure 10.2: [Mathematica animation] Typical phase portraits for system (10.10) when (a) $C < -0.18$ (no limit cycle), (b) $-0.18 < C < 0.18$ (a stable limit cycle), and (c) $C > 0.18$ (no limit cycle).

stable and one of the unstable branches of the saddle point coalesce to form a homoclinic loop. As C is increased further toward $C = 0$, the saddle point moves away from the limit cycle (down the negative y axis). As C is increased through $C = 0$, the saddle point moves toward the limit cycle (down the positive y axis) and once more a homoclinic loop is formed at $C \approx 0.18$. As C passes through $C \approx 0.18$, the limit cycle vanishes.

Homoclinic Bifurcation The global bifurcation of limit cycles from homoclinic loops will now be discussed via example. The analysis involved in bifurcating limit cycles from separatrix cycles is beyond the scope of this book; however, interested readers are referred to [2]. Both homoclinic and heteroclinic bifurcations are used to obtain polynomial systems with a number of limit cycles (see Chapter 11). The Mathematica package can be used to investigate some of these systems numerically.

Example 10. Investigate the system

$$\dot{x} = y, \quad \dot{y} = x + x^2 - xy + \lambda y$$

as the parameter λ varies and plot possible phase portraits.

Solution. There are two critical points at $O = (0,0)$ and $P = (-1,0)$. The Jacobian is given by

$$J = \begin{pmatrix} 0 & 1 \\ 1 + 2x - y & -x + \lambda \end{pmatrix}.$$

The origin is a saddle point, and it can be shown that the point P is a node or focus. Since trace $J_P = 1 + \lambda$, it follows that P is stable if $\lambda < -1$ and unstable if $\lambda > -1$. The point P is also stable if $\lambda = -1$.

It can be shown that a limit cycle exists for $-1 < \lambda < \lambda_0$, where $\lambda_0 \approx -0.85$. Since the limit cycle appears from a homoclinic loop, which exists at a value, say λ_0, this is known as a homoclinic bifurcation. More details can be found

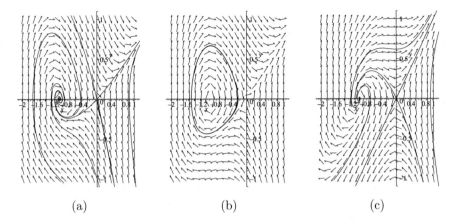

(a)   (b)   (c)

Figure 10.3: [Mathematica animation] Phase portraits for Example 10 when
(a) $\lambda = -1.5$, (b) $\lambda = -0.9$, and (c) $\lambda = 0$.

in [2]. Phase portraits for three values of λ are shown in Figure 10.3, and a
Mathematica program showing an animation is listed in the next section.

Another example is given in the exercises in Section 10.6.

10.5 Mathematica Commands

See the help menu for GroebnerBasis.

```
(*See Section 10.1 - Computation of focal values. See Example 1. *)
In[1]:=v[1,0]=0;v[0,1]=0;v[2,0]=1/2;v[1,1]=0;v[0,2]=1/2;v[0,0]=0;
In[2]:=vv=Sum[Sum[v[i,j-i]x^i y^(j-i),{i,0,j}],{j,0,20}];
In[3]:=xdot=y-a2 x^2-a3 x^3-a4 x^4;
In[4]:=dydt=-x-b2 x^2-b3 x^3;
In[5]:=dd =Expand[D[vv,x] xdot+D[vv,y] dydt-eta4 (x^2+y^2)^2-
         eta6(x^2+y^2)^3];
In[6]:=Do[eq[i]=Coefficient[Coefficient[dd,x,i],y,3-i],{i,0,3}];
In[7]:=SolsV3=Solve[{eq[0]==0,eq[1]==0,eq[2]==0,eq[3]==0},{v[3,0],
         v[2,1],v[1,2],v[0,3]}]
```

$\text{Out}[7] = \left\{ v[3,0] \to \frac{b_2}{3}, v[2,1] \to -a_2, v[1,2] \to 0, v[0,3] \to -\frac{2a_2}{3} \right\}$

```
In[8]:=Do[eq[i]=Coefficient[Coefficient[dd,x,i],y,4-i],{i,0,4}];
In[9]:=SolsV4=Solve[{eq[0]==0,eq[1]==0,eq[2]==0,eq[3]==0,eq[4]==0,
         v[2,2]== 0},{v[4,0],v[3,1],v[2,2],v[1,3],v[0,4],eta4}]
         /. SolsV3
```

$\text{Out}[9] = \{ v[4,0] \to \frac{1}{4}(-2a_2^2 + b_3), v[3,1] \to \frac{1}{8}(-5a_3 - 2a_2b_2), v[2,2] \to 0 ,$
$v[1,3] \to -\frac{1}{8}(-3a_3 + 2a_2b_2), v[0,4] \to 0, eta4 \to \frac{1}{8}(-3a_3 + 2a_2b_2) \}$

```
In[10]:=a3=2 a2 b2/3; (* Substitute from L(1)=0 *)
In[11]:=Do[eq[i]=Coefficient[Coefficient[dd,x,i],y,5-i],{i,0,5}];
In[12]:=SolsV5=Simplify[Solve[{eq[0]==0,eq[1]==0,eq[2]==0,eq[3]==0,
        eq[4]==0,eq[5]==0},{v[5,0],v[4,1],v[3,2],v[2,3],v[1,4],v[0,5]}]
        /. SolsV4 /. SolsV3]
```

$$\text{Out[12]} = \left\{ v[5,0] \rightarrow -\frac{2a_2^2 b_2}{3}, v[4,1] \rightarrow 2a_2^3 - a_4, v[3,2] \rightarrow 0 \, , \right.$$
$$\left. v[2,3] \rightarrow \tfrac{2}{3}(4a_2^3 - 2a_4 + a_2 b_3), v[1,4] \rightarrow 0, v[0,5] \rightarrow \tfrac{4}{15}(4a_2^3 - 2a_4 + a_2 b_3) \right\}$$

```
In[13]:=Do[eq[i]=Coefficient[Coefficient[dd,x,i],y,6-i],{i,0,6}];
In[14]:=SolsV6=Simplify[Solve[{eq[0]==0,eq[1]==0,eq[2]==0,eq[3]==0,
        eq[4]==0,eq[5]==0,eq[6]==0,v[4,2]+v[2,4]==0},{v[6,0],v[5,1],
        v[4,2],v[3,3],v[2,4],v[1,5],v[0,6],eta6}]
        /. SolsV3 /. SolsV4 /. SolsV5];
In[15]:=eta6 /. SolsV6
```

$$\text{Out[15]} = \text{eta6} := \left\{ -\tfrac{5}{24}(-2a4b2 + a2b2b3) \right\}$$

```
(*See Example 4 - Multivariate polynomial division.*)
In[16]:=PolynomialReduce[x^4+y^4+z^4,{x^2+y,z^2 y-1,y-z^2},{x,y,z}]
```

$$\text{Out[16]} = \left\{ \left\{ x^2 - y, 2 + y^2, 2y + y^3 \right\}, 2 + z^4 \right\}$$

```
In[17]:=PolynomialReduce[x^4+y^4+z^4,{y-z^2,z^2 y-1,x^2+y},{x,y,z}]
```

$$\text{Out[17]} = \left\{ \left\{ -x^2 + y^3 + z^2 + y^2 z^2 + yz^4 + z^6, 0, x^2 - z^2 \right\}, 2z^4 + z^8 \right\}$$

```
(*See Example 6 - Computing a reduced Groebner basis.*)
In[18]:=polys={y^2-x^3+x,2 x^2 y^2+x^4+4 x^3+y^4-12 x y^2};
In[19]:=GroebnerBasis[polys,{x,y},MonomialOrder->Lexicographic]
```

$$\text{Out[19]} = \left\{ -195y^4 + 1278y^6 - 1037y^8 + 90y^{10} + y^{12}, \right.$$
$$\left. 5970075x + 5970075y^2 + 163845838y^4 - 162599547y^6 + 14472880y^8 + 160356y^{10} \right\}$$

```
(*See Example 6 - Solving multivariate polynomial equations.*)
In[20]:=N[Solve[{x+y^2-x^3==0,2 x^2 y^2+x^4+4 x^3+y^4-12 x y^2==0},{x,y}]
        ,4]
```

$$\text{Out[20]} = \{ (y \rightarrow 0, x \rightarrow 0), (y \rightarrow -3.011, x \rightarrow 2.245), (y \rightarrow 3.011, x \rightarrow 2.245),$$
$$(y \rightarrow -1.097, x \rightarrow 1.370), (y \rightarrow 1.097, x \rightarrow 1.370), (y \rightarrow -0.4217, x \rightarrow -0.8952),$$
$$(y \rightarrow 0.4217, x \rightarrow -0.8952), (y \rightarrow -10.02i, x \rightarrow -4.720), (y \rightarrow 10.02i, x \rightarrow -4.720) \}$$

```
(*See Example 7 - Computing a Groebner basis.*)
In[21]:=polys={-a1,2 a2 b2-3a3,5 b2 (2a4-b3 a2),-5 b2 (92 b2^2 a4-
        99 b3^2 a2+1520 a2^2 a4-760 a2^3 b3-46 b2^2 b3 a2+198 b3 a4),
        -b2 (14546 b2^4 a4+105639 a2^3 b3^2+96664 a2^3 b2^2 b3-
        193328 a2^2 b2^2 a4-891034 a2^4 a4+445517 a2^5 b3+
        211632 a2 a4^2-317094 a2^2 b3 a4-44190 b2^2 b3 a4+
        22095 b2^2 b3^2 a2-7273 b2^4 b3 a2+5319 b3^3 a2-10638 b3^2 a4)};
In[22]:=GroebnerBasis[polys,{a1,a2,a3,a4,b2,b3}]
```

Out[22] = $\{-4a4b2 + 3a3b3, -3a3 + 2a2b2, a1\}$

```
(* Bifurcating limit cycles. See Figure 10.2. *)
Manipulate[Module[{sol},sol=NDSolve[{x'[t]==y[t]+10*x[t]*(0.1-y[t]^2),
y'[t]==-x[t]+Constant,x[0]==.01,y[0]==.01},{x[t],y[t]},{t,0,80}];
ParametricPlot[Evaluate[Table[{x[t],y[t]}/.sol]],{t,0,80},
PlotRange->{{-1.5,1.5},{-1,1}}, PlotPoints->100]],{Constant,-0.3,0.3}]
```

```
(* Bifurcating limit cycles. See Figure 10.3. *)
Manipulate[Module[{sol},sol=NDSolve[{x'[t]==y[t],
y'[t]==x[t]+x[t]^2-x[t]*y[t]+lambda*y[t],x[0]==-0.1,y[0]==-0.1},
{x[t],y[t]},{t,0,30},MaxSteps->Infinity];
ParametricPlot[Evaluate[Table[{x[t],y[t]}/.sol]],{t,0,30},
PlotRange->{{-2,1},{-1,1}}, PlotPoints->100]],{lambda,-1.2,-0.8,0.001}]
```

10.6 Exercises

1. Prove that the origin of the system

$$\dot{x} = y - F(G(x)), \quad \dot{y} = -\frac{G'(x)}{2}H(G(x))$$

   is a center using the transformation $u^2 = G(x)$ and the classical symmetry argument.

2. Fix a lexicographical order $x \succ y \succ z$. Divide the multivariate polynomial $p = x^3 + y^3 + z^3$ by the ordered list of polynomials $\{x + 3y, xy^2 - x, y - z\}$. Repeat the division with the divisors listed as $\{xy^2 - x, x + 3y, y - z\}$.

3. Use Mathematica to compute a Gröbner basis for the set of polynomials

$$\{y^2 - x^3 + x, y^3 - x^2\}$$

   under lexicographical, degree lexicographical, and degree reverse lexicographical ordering, respectively. Solve the simultaneous equations, $y^2 - x^3 + x = 0$, $y^3 - x^2 = 0$, for x and y.

4. Write a program to compute the first seven Lyapunov quantities of the Liénard system

$$\dot{x} = y - (a_1 x + a_2 x^2 + \ldots + a_{13} x^{13}), \quad \dot{y} = -x. \qquad (10.11)$$

   Prove that at most six small-amplitude limit cycles can be bifurcated from the origin of system (10.11).

5. Consider the system

$$\dot{x} = y - (a_1 x + a_3 x^3 + \ldots + a_{2n+1} x^{2n+1}), \quad \dot{y} = -x.$$

Prove by induction that at most n small-amplitude limit cycles can be bifurcated form the origin.

6. Write a program to compute the first five Lyapunov quantities for the Liénard system

$$\dot{x} = y - (a_1 x + a_2 x^2 + \ldots + a_7 x^7), \quad \dot{y} = -(x + b_2 x^2 + b_3 x^3 + \ldots + b_6 x^6).$$

Prove that $\hat{H}(4,2) = 2$, $\hat{H}(7,2) = 4$, and $\hat{H}(3,6) = 4$. Note that in $\hat{H}(u,v)$, u is the degree of F and v is the degree of g.

7. Consider the generalized mixed Rayleigh-Liénard oscillator equations given by

$$\dot{x} = y, \quad \dot{y} = -x - a_1 y - b_{30} x^3 - b_{21} x^2 y - b_{41} x^4 y - b_{03} y^3.$$

Prove that at most three small-amplitude limit cycles can be bifurcated from the origin.

8. Plot a phase portrait for the system

$$\dot{x} = y, \quad \dot{y} = x + x^2.$$

Determine an equation for the curve on which the homoclinic loop lies.

9. Consider the Liénard system given by

$$\dot{x} = y - \epsilon(a_1 x + a_2 x^2 + a_3 x^3), \quad \dot{y} = -x.$$

Prove that for sufficiently small ϵ, there is at most one limit cycle that is asymptotic to a circle of radius

$$r = \sqrt{\frac{4|a_1|}{3|a_3|}}.$$

10. Using the Mathematica package, investigate the system

$$\dot{x} = y, \quad \dot{y} = x - x^3 + \epsilon(\lambda y + x^2 y)$$

when $\epsilon = 0.1$ for values of λ from -1 to -0.5. How many limit cycles are there at most?

Bibliography

[1] T. Becker, V. Weispfenning and H. Kredel, *Gröbner Bases: A Computational Approach to Commutative Algebra (Graduate Texts in Mathematics)*, Springer, New York, 2012.

[2] T.R. Blows and L.M. Perko, Bifurcation of limit cycles from centers and separatrix cycles of planar analytic systems, *SIAM Review*, **36** (1994), 341–376.

[3] H. Broer, I. Hoveijn, G, Lunter, G. Vegter, *Bifurcations in Hamiltonian Systems: Computing Singularities by Gröbner Bases* (Lecture Notes in Mathematics), Springer-Verlag, New York, 2003.

[4] B. Buchberger Ed., *Gröbner Bases and Applications* (London Mathematical Society Lecture Note Series), Cambridge University Press, Cambridge, UK, 1998.

[5] B. Buchberger, *On Finding a Vector Space Basis of the Residue Class Ring Modulo a Zero Dimensional Polynomial Ideal*, PhD thesis, University of Innsbruck, Austria, 1965 (German).

[6] T. Hibi (Editor), *Gröbner Bases: Statistics and Software Systems*, Springer, New York, 2013.

[7] N. Lauritzen, *Concrete Abstract Algebra: From Numbers to Gröbner Bases*, Cambridge University Press, Cambridge, UK, 2003.

[8] N.G. Lloyd, *Limit cycles of polynomial systems, New Directions in Dynamical Systems* (ed. T. Bedford and J. Swift), L.M.S. Lecture Notes Series No. 127, Cambridge University Press, Cambridge, UK, 1988.

[9] N.G. Lloyd and S. Lynch, Small-amplitude limit cycles of certain Liénard systems, *Proc. Roy. Soc. Lond. Ser. A*, **418** (1988), 199–208.

[10] S. Lynch, Symbolic computation of Lyapunov quantities and the second part of Hilbert's sixteenth problem, in *Differential Equations with Symbolic Computation*, D.M. Wang and Z. Zheng Eds., Birkhäuser, Basel, Cambridge, MA, 2005.

[11] H. Maoan and Y. Pei, *Normal Forms, Melnikov Functions and Bifurcations of Limit Cycles (Applied Mathematical Sciences, Vol. 181)*, Springer, New York, 2012

[12] D.M. Wang, *Elimination Practice: Software Tools and Applications*, Imperial College Press, London, 2004.

[13] D.M. Wang, Polynomial systems from certain differential equations, *J. Symbolic Computation*, **28** (1999), 303–315.

[14] I. Yengui, *Constructive Commutative Algebra: Projective Modules Over Polynomial Rings and Dynamical Gröbner Bases (Lecture Notes in Mathematics)*, Springer, New York, 2015.

Chapter 11
The Second Part
of Hilbert's Sixteenth Problem

Aims and Objectives

- To describe the second part of Hilbert's sixteenth problem.

- To review the main results on the number of limit cycles of planar polynomial systems.

- To consider the flow at infinity after Poincaré compactification.

- To review the main results on the number of limit cycles of Liénard systems.

- To prove two theorems concerning limit cycles of certain Liénard systems.

On completion of this chapter the reader should be able to

- state the second part of Hilbert's sixteenth problem;

- describe the main results for this problem;

- compactify the plane and construct a global phase portrait which shows the behavior at infinity for some simple systems;

- compare local and global results;

- prove that certain systems have a unique limit cycle;

- prove that a limit cycle has a certain shape for a large parameter value.

The second part of Hilbert's sixteenth problem is stated, and the main results are listed. To understand these results, it is necessary to introduce Poincaré compactification, where the plane is mapped onto a sphere and the behavior on the equator of the sphere represents the behavior at infinity for planar systems.

© Springer International Publishing AG 2017   231
S. Lynch, *Dynamical Systems with Applications using Mathematica®*,
DOI 10.1007/978-3-319-61485-4_11

Many autonomous systems of two-dimensional differential equations can be transformed to systems of Liénard type. In recent years, there have been many results published associated with Liénard systems. The major results for both global and local bifurcations of limit cycles for these systems are listed.

A method for proving the existence, uniqueness, and hyperbolicity of a limit cycle is illustrated in this chapter, and the Poincaré-Bendixson theorem is applied to determine the shape of a limit cycle when a parameter is large.

11.1 Statement of Problem and Main Results

Poincaré began investigating isolated periodic cycles of planar polynomial vector fields in the 1880s. However, the general problem of determining the maximum number and relative configurations of limit cycles in the plane has remained unresolved for over a century. Recall that limit cycles in the plane can correspond to steady-state behavior for a physical system (see Chapter 7), so it is important to know how many possible steady states there are.

In 1900, David Hilbert presented a list of 23 problems to the International Congress of Mathematicians in Paris. Most of the problems have been solved, either completely or partially. However, the second part of the sixteenth problem remains unsolved. Il'yashenko [14] presents a centennial history of Hilbert's 16th problem, Jibin Li [17] has written a review article of the major results up to 2003, and more recently Han Maoan and Jibin Li [20] present some new lower bounds associated with the problem.

The Second Part of Hilbert's Sixteenth Problem. Consider planar polynomial systems of the form

$$\dot{x} = P(x,y), \quad \dot{y} = Q(x,y), \tag{11.1}$$

where P and Q are polynomials in x and y. The question is to estimate the maximal number and relative positions of the limit cycles of system (11.1). Let H_n denote the maximum possible number of limit cycles that system (16.1) can have when P and Q are of degree n. More formally, the Hilbert numbers H_n are given by

$$H_n = \sup\{\pi(P,Q) : \partial P, \partial Q \leq n\},$$

where ∂ denotes "the degree of" and $\pi(P,Q)$ is the number of limit cycles of system (11.1).

Dulac's Theorem states that a given polynomial system cannot have infinitely many limit cycles. This theorem has only recently been proved independently by Ecalle et al. [12] and Il'yashenko [15], respectively. Unfortunately, this does not imply that the Hilbert numbers are finite.

Of the many attempts to make progress in this question, one of the more fruitful approaches has been to create vector fields with as many isolated periodic orbits as possible using both local and global bifurcations. There are relatively few results in the case of general polynomial systems even when considering local bifurcations. Bautin [1] proved that no more than three small-amplitude limit cycles could bifurcate from a critical point for a quadratic system. For a homogeneous cubic system (no quadratic terms), Sibirskii [27] proved that no more than five small-amplitude limit cycles could be bifurcated from one critical point. Zoladek [31] recently found an example where 11 limit cycles could be bifurcated from the origin of a cubic system, but he was unable to prove that this was the maximum possible number.

Although easily stated, Hilbert's sixteenth problem remains almost completely unsolved. For quadratic systems, Shi Songling [26] has obtained a lower bound for the Hilbert number $H_2 \geq 4$. A possible global phase portrait showing the configuration of the limit cycles is given in Figure 11.1. The line at infinity is included and the properties on this line are determined using Poincaré compactification, which is described in Section 11.2. There are three small-amplitude limit cycles around the origin and at least one other surrounding another critical point. Some of the parameters used in this example are very small.

Blows and Rousseau [2] consider the bifurcation at infinity for polynomial vector fields and give examples of cubic systems having the following configurations:

$$\{(4), 1\}, \{(3), 2\}, \{(2), 5\}, \{(4), 2\}, \{(1), 5\} \text{ and } \{(2), 4\},$$

where $\{(l), L\}$ denotes the configuration of a vector field with l small-amplitude limit cycles bifurcated from a point in the plane and L large-amplitude limit cycles simultaneously bifurcated from infinity. There are many other configurations possible, some involving other critical points in the finite part of the plane as shown in Figure 11.2. Recall that a limit cycle must contain at least one critical point.

By considering cubic polynomial vector fields, in 1985, Li Jibin and Li Chunfu [18] produced an example with 11 limit cycles by bifurcating limit cycles out of homoclinic and heteroclinic orbits; see Figure 11.2. Yu Pei and Han Maoan [24] have more recently managed to show that $H_3 \geq 12$ by bifurcating exactly 12 small-amplitude limit cycles (two nests of six) from a cubic system with one saddle point at the origin and two focus points symmetric about the origin.

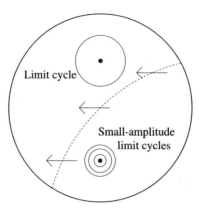

Figure 11.1: A possible configuration for a quadratic system with four limit cycles: one of large amplitude and three of small amplitude.

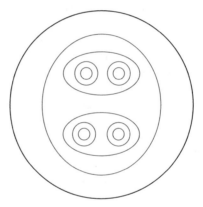

Figure 11.2: A possible configuration for a cubic system with 11 limit cycles.

Returning to the general problem, in 1995, Christopher and Lloyd [7] considered the rate of growth of H_n as n increases. They showed that H_n grows at least as rapidly as $n^2 \log n$. Other rates of growth of H_n with n are presented in [17] and [20].

In recent years, the focus of research in this area has been directed at a small number of classes of systems. Perhaps the most fruitful has been the Liénard system.

11.2 Poincaré Compactification

The method of compactification was introduced by Henri Poincaré at the end of the 19th century. By making a simple transformation, it is possible to map the phase plane onto a sphere. Note that the plane can be mapped to both the upper and lower hemispheres. In this way, the points at infinity are transformed to the points on the equator of the sphere. Suppose that a point (x, y) in the plane is mapped to a point (X, Y, Z) on the upper hemisphere of a sphere, say, $S^2 = \{(X, Y, Z) \in \Re^3 : X^2 + Y^2 + Z^2 = 1\}$. (Note that it is also possible to map onto the lower hemisphere). The equations defining (X, Y, Z) in terms of (x, y) are given by

$$X = \frac{x}{\sqrt{1+r^2}}, \quad Y = \frac{y}{\sqrt{1+r^2}}, \quad Z = \frac{1}{\sqrt{1+r^2}},$$

where $r^2 = x^2 + y^2$. A central projection is illustrated in Figure 11.3.

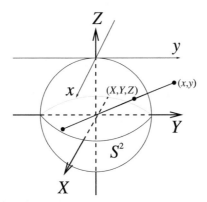

Figure 11.3: A mapping of (x, y) in the plane onto (X, Y, Z) on the upper part of the sphere.

Consider the autonomous system (11.1). Convert to polar coordinates. Thus system (11.1) transforms to

$$\dot{r} = r^n f_{n+1}(\theta) + r^{n-1} f_{n-1}(\theta) + \ldots + f_1(\theta)$$
$$\dot{\theta} = r^{n-1} g_{n+1}(\theta) + r^{n-2} g_{n-1}\theta + \ldots + r^{-1} g_1(\theta), \tag{11.2}$$

where f_m and g_m are polynomials of degree m in $\cos\theta$ and $\sin\theta$.

Let $\rho = \frac{1}{r}$. Hence $\dot{\rho} = -\frac{\dot{r}}{r^2}$, and system (11.2) becomes

$$\dot{\rho} = -\rho f_{n+1}(\theta) + O(\rho^2), \quad \dot{\theta} = g_{n+1}(\theta) + O(\rho).$$

Theorem 1. *The critical points at infinity are found by solving the equations $\dot{\rho} = \dot{\theta} = 0$ on $\rho = 0$, which is equivalent to solving*

$$g_{n+1}(\theta) = \cos\theta\, Q_n(\cos\theta, \sin\theta) - \sin\theta\, P_n(\cos\theta, \sin\theta) = 0,$$

where P_n and Q_n are homogeneous polynomials of degree n. Note that the solutions are given by the pairs θ_i and $\theta_i + \pi$. As long as $g_{n+1}(\theta)$ is nonzero, there are $n + 1$ pairs of roots and the flow is clockwise when $g_{n+1}(\theta) < 0$ and it is counterclockwise when $g_{n+1}(\theta) > 0$.

To determine the flow near the critical points at infinity, one must project the hemisphere with $X > 0$ onto the plane $X = 1$ with axes y and z or project the hemisphere with $Y > 0$ onto the plane $Y = 1$ with axes x and z. The projection of the sphere S^2 onto these planes is depicted in Figure 11.4.

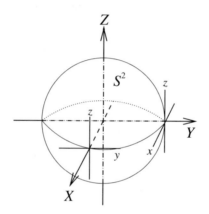

Figure 11.4:   The projections used to determine the behavior at infinity.

If n is odd, the antinodal points on S^2 are qualitatively equivalent. If n is even, the antinodal points are qualitatively equivalent but the direction of the flow is reversed.

The flow near a critical point at infinity can be determined using the following theorem.

Theorem 2. *The flow defined on the yz plane $(X = \pm 1)$, except the points $(0, \pm 1, 0)$, is qualitatively equivalent to the flow defined by*

$$\pm\dot{y} = y z^n P\left(\frac{1}{z}, \frac{y}{z}\right) - z^n Q\left(\frac{1}{z}, \frac{y}{z}\right), \qquad \pm\dot{z} = z^{n+1} P\left(\frac{1}{z}, \frac{y}{z}\right),$$

where the direction of the flow is determined from $g_{n+1}(\theta)$.

In a similar way, the flow defined on the xz plane ($Y = \pm 1$), except the points $(\pm 1, 0, 0)$, is qualitatively equivalent to the flow defined by

$$\pm \dot{x} = xz^n Q\left(\frac{x}{z}, \frac{1}{z}\right) - z^n P\left(\frac{x}{z}, \frac{1}{z}\right), \quad \pm \dot{z} = z^{n+1} Q\left(\frac{x}{z}, \frac{1}{z}\right),$$

where the direction of the flow is determined from $g_{n+1}(\theta)$.

Example 1. Construct global phase portraits, including the flow at infinity, for the following linear systems:
    (a) $\dot{x} = -x + 2y, \dot{y} = 2x + 2y$;
    (b) $\dot{x} = x + y, \dot{y} = -x + y$.

Solutions. (a) The origin is a saddle point with eigenvalues and corresponding eigenvectors are given by $\lambda_1 = 3$, $(1, 2)^T$ and $\lambda_2 = -2$, $(2, -1)^T$. The critical points at infinity satisfy the equation $g_2(\theta) = 0$, where

$$g_2(\theta) = \cos \theta Q_1(\cos \theta, \sin \theta) - \sin \theta P_1(\cos \theta, \sin \theta).$$

Now
$$g_2(\theta) = 2\cos^2 \theta + 3\cos \theta \sin \theta - 2\sin^2 \theta.$$

The roots are given by $\theta_1 = \tan^{-1}(2)$ radians, $\theta_2 = \tan^{-1}(2) + \pi$ radians, $\theta_3 = \tan^{-1}(-\frac{1}{2})$ radians, and $\theta_4 = \tan^{-1}(-\frac{1}{2}) + \pi$ radians.
    A plot of $g_2(\theta)$ is given in Figure 11.5.
    The flow near a critical point at infinity is qualitatively equivalent to the flow of the system

$$\pm \dot{y} = yz\left(-\frac{1}{2} + \frac{2y}{z}\right) - z\left(\frac{2}{z} - \frac{2y}{z}\right), \quad \pm \dot{z} = z^2\left(-\frac{1}{z} + \frac{2y}{z}\right).$$

From Figure 11.5, the flow is counterclockwise if $\tan^{-1}(-\frac{1}{2}) < \theta < \tan^{-1}(2)$. Therefore, the flow at infinity is determined by the system

$$-\dot{y} = -3y + 2y^2 - 2, \quad -\dot{z} = -z + 2yz.$$

There are critical points at $A = (2, 0)$ and $B = (-\frac{1}{2}, 0)$ in the yz plane. Point A is a stable node, and point B is an unstable node. A phase portrait is given in Figure 11.6.
    Since n is odd, the antinodal points are qualitatively equivalent. A global phase portrait is shown in Figure 11.7.

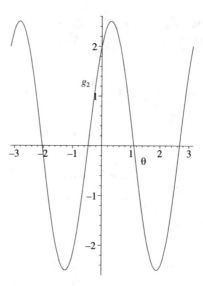

Figure 11.5:   The function $g_2(\theta)$.

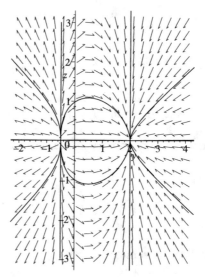

Figure 11.6:   Some trajectories in the yz plane (X=1) that define the flow at infinity.

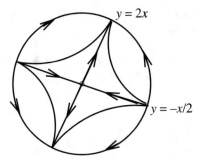

$y = 2x$

$y = -x/2$

Figure 11.7:   A global phase portrait for Example 1(a).

(b) The origin is an unstable focus and the flow is clockwise. The critical points at infinity satisfy the equation $g_2(\theta) = 0$, where

$$g_2(\theta) = \cos\theta Q_1(\cos\theta, \sin\theta) - \sin\theta P_1(\cos\theta, \sin\theta) = -(\cos^2\theta + \sin^2\theta).$$

There are no roots for $g_2(\theta)$, so there are no critical points at infinity. A global phase portrait is given in Figure 11.8.

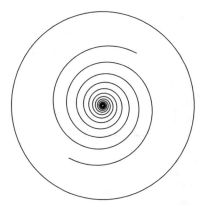

Figure 11.8:   A global phase portrait for Example 1(b). There are no critical points at infinity and the flow is clockwise.

Example 2. Show that the system given by

$$\dot{x} = -\frac{x}{2} - y - x^2 + xy + y^2, \quad \dot{y} = x(1 + x - 3y)$$

has at least two limit cycles.

Solution. There are two critical points at $O = (0,0)$ and $A = (0,1)$. The Jacobian matrix is given by

$$J = \begin{pmatrix} -\frac{1}{2} - 2x + y & -1 + x + 2y \\ 1 + 2x - 3y & -3x \end{pmatrix}.$$

Now

$$J_O = \begin{pmatrix} -\frac{1}{2} & -1 \\ 1 & 0 \end{pmatrix} \quad \text{and} \quad J_A = \begin{pmatrix} \frac{1}{2} & 1 \\ -2 & 0 \end{pmatrix}.$$

Therefore, O is a stable focus and A is an unstable focus. On the line L_1 : $1 + x - 3y = 0$, $\dot{y} = 0$ and $\dot{x} < 0$, so the flow is transverse to L_1.

The critical points at infinity satisfy the equation $g_3(\theta) = 0$, where

$$g_3(\theta) = \cos\theta Q_2(\cos\theta, \sin\theta) - \sin\theta P_2(\cos\theta, \sin\theta).$$

Now

$$g_3(\theta) = \cos^3\theta - 2\cos^2\theta \sin\theta - \cos\theta \sin^2\theta - \sin^3\theta.$$

A plot for $g_3(\theta)$ is given in Figure 11.9.

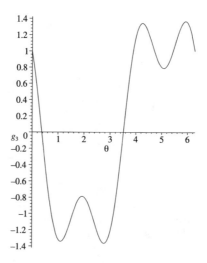

Figure 11.9:   The function $g_3(\theta)$.

There are two roots for $g_3(\theta)$: $\theta_1 = 0.37415$ radians and $\theta_2 = 3.51574$ radians. The flow near a critical point at infinity is qualitatively equivalent to the flow of the system

$$\pm\dot{y} = -\frac{yz}{2} - y^2z + 2y + y^2 + y^3 - z - 1$$

$$\pm\dot{z} = -\frac{z^2}{2} - yz^2 - z + yz + y^2z.$$

There is one critical point at $(y, z) = (0.39265, 0)$, which is a saddle point. Since n is even, the antinodal point is also a saddle point, but the direction of the flow is reversed. The direction of the flow may be established by inspecting $g_3(\theta)$ in Figure 11.9.

Part of the global phase portrait is shown in Figure 11.10, and from the corollary to the Poincaré-Bendixson Theorem, there are at least two limit cycles.

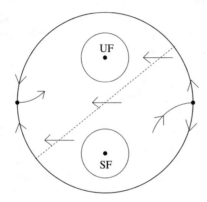

Figure 11.10:   A global phase portrait showing at least two limit cycles. UF and SF denote an unstable and stable focus, respectively.

If the system is nonlinear and there are no critical points at infinity, it is also possible to bifurcate limit cycles from infinity; see, for example, the work of Blows and Rousseau [2].

Marasco and Tenneriello [22] use Mathematica to propose methods that give the Fourier series of the periodic solutions and period of planar systems in the presence of isochronous centers and unstable limit cycles.

11.3 Global Results for Liénard Systems

Consider polynomial Liénard equations of the form

$$\ddot{x} + f(x)\dot{x} + g(x) = 0, \tag{11.3}$$

where $f(x)$ is known as the damping coefficient and $g(x)$ is called the restoring coefficient. Equation (11.3) corresponds to the class of systems

$$\dot{x} = y, \quad \dot{y} = -g(x) - f(x)y, \tag{11.4}$$

in the phase plane. Liénard applied the change of variable $Y = y + F(x)$, where $F(x) = \int_0^x f(s)ds$, to obtain an equivalent system in the so-called Liénard plane:

$$\dot{x} = Y - F(x), \quad \dot{Y} = -g(x). \tag{11.5}$$

For the critical point at the origin to be a nondegenerate focus or center, the conditions $g(0) = 0$ and $g'(0) > 0$ are imposed. Periodic solutions of (11.5) correspond to limit cycles of (11.2) and (11.5). There are many examples in both the natural sciences and technology where these and related systems are applied. The differential equation is often used to model either mechanical systems or electric circuits, and in the literature, many systems are transformed to Liénard type to aid in the investigations. For a list of applications to the real world, see, for example, Moreira [20]. In recent years, the number of results for this class of system has been phenomenal, and the allocation of this topic to a whole section of the book is well justified.

These systems have proved very useful in the investigation of multiple limit cycles and also when proving existence, uniqueness, and hyperbolicity of a limit cycle. Let ∂ denote the degree of a polynomial, and let $H(i, j)$ denote the maximum number of global limit cycles, where i is the degree of f and j is the degree of g. The main global results for systems (11.2) and (11.5) to date are listed below:

- In 1928, Liénard proved that when $\partial g = 1$ and F is a continuous odd function, which has a unique root at $x = a$ and is monotone increasing for $x \geq a$, then (11.5) has a unique limit cycle.

- In 1973, Rychkov [25] proved that if $\partial g = 1$ and F is an odd polynomial of degree five, then (11.5) has at most two limit cycles.

- In 1976, Cherkas [4] gave conditions in order for a Liénard equation to have a center.

- In 1977, Lins, de Melo, and Pugh [19] proved that $H(2, 1) = 1$. They also conjectured that $H(2m, 1) = H(2m + 1, 1) = m$, where m is a natural number.

- In 1988, Coppel [8] proved that $H(1, 2) = 1$.

- In 1992, Zhang Zhifen [35] proved that a certain generalized Liénard system has a unique limit cycle.

- In 1996, Dumortier and Chengzhi Li [9] proved that $H(1, 3) = 1$.

- In 1997, Dumortier and Chengzhi Li [10] proved that $H(2,2) = 1$.

- In 2005, Jiang et al. [16] proved that, when f and g are odd polynomials, $H(5,3) = 2$.

- In 2007, Dumortier et al. [11] proved that the conjecture by Lins, de Melo and Pugh from 1977 was incorrect.

- In 2014, Xiong and Han [29] obtain some new lower bounds for the Hilbert numbers of certain Liénard systems.

- In 2017, Sun and Huang [28] show that a Liénard system of type $(4,3)$ can have six limit cycles using an algorithm based on the Chebyshev criteria and the tools of regular chain theory in polynomial algebra.

Giacomini and Neukirch [13] introduced a new method to investigate the limit cycles of Liénard systems when $\partial g = 1$ and $F(x)$ is an odd polynomial. They are able to give algebraic approximations to the limit cycles and obtain information on the number and bifurcation sets of the periodic solutions even when the parameters are not small. Other work has been carried out on the algebraicity of limit cycles, but it is beyond the scope of this book.

Limit cycles were discussed in some detail in Chapter 5, and a method for proving the existence and uniqueness of a limit cycle was introduced. Another method for proving the existence, uniqueness, and hyperbolicity of a limit cycle is illustrated in Theorem 4.

Consider the general polynomial system

$$\dot{x} = P(x,y), \quad \dot{y} = Q(x,y),$$

where P and Q are polynomials in x and y, and define $\mathbf{X} = (P, Q)$ to be the vector field. Let a limit cycle, say, $\Gamma(t) = (x(t), y(t))$, have period T.

Definition 1. The quantity $\int_\Gamma \operatorname{div}(\mathbf{X})\,dt$ is known as the *characteristic exponent*.

Theorem 3. *Suppose that*

$$\int_\Gamma \operatorname{div}(\mathbf{X})\,dt = \int_0^T \left(\frac{\partial P}{\partial x} + \frac{\partial Q}{\partial y} \right)(x(t), y(t))\,dt.$$

Then
  (i) Γ is hyperbolic attracting if $\int_\Gamma \operatorname{div}(\mathbf{X})\,dt < 0$;
  (ii) Γ is hyperbolic repelling if $\int_\Gamma \operatorname{div}(\mathbf{X})\,dt > 0$.

Theorem 4. *Consider the Liénard system*

$$\dot{x} = y - (a_1 x + a_2 x^2 + a_3 x^3), \quad \dot{y} = -x. \tag{11.6}$$

There exists a unique hyperbolic limit cycle if $a_1 a_3 < 0$.

Proof. The method is taken from the paper of Lins, de Melo, and Pugh [19]. Note that the origin is the only critical point. The flow is horizontal on the line $x = 0$ and vertical on the curve $y = a_1 x + a_2 x^2 + a_3 x^3$. It is not difficult to prove that a trajectory starting on the positive (or negative) y-axis will meet the negative (or positive) y-axis. The solution may be divided into three stages:

I Every limit cycle of system (11.6) must cross both of the lines given by

$$L_1 : x_0 = -\sqrt{-\frac{a_1}{a_3}} \quad \text{and} \quad L_2 : x_1 = \sqrt{-\frac{a_1}{a_3}}.$$

II System (11.6) has at least one and at most two limit cycles; one of them is hyperbolic.

III System (11.6) has a unique hyperbolic limit cycle.

Stage I. Consider the Lyapunov function given by

$$V(x, y) = e^{-2a_2 y}\left(y - a_2 x^2 + \frac{1}{2a_2}\right).$$

Now

$$\frac{dV}{dt} = 2a_2 e^{-2a_2 y} x^2 (a_1 + a_3 x^2).$$

The Lyapunov function is symmetric with respect to the y-axis since $V(x, y) = V(-x, y)$, and there is a closed level curve $V(x, y) = C$ that is tangent to both L_1 and L_2. Since $\frac{dV}{dt}$ does not change sign inside the disc $V(x, y) = C$, no limit cycle can intersect the disk, which proves Stage I.

Stage II. Suppose that there are two limit cycles $\gamma_1 \subset \gamma_2$ surrounding the origin as in Figure 11.11.

Suppose that $a_1 < 0$ and $a_3 > 0$. Then the origin is unstable. Let γ_1 be the innermost periodic orbit, which must be attracting on the inside. Therefore,

$$\int_{\gamma_1} div(\mathbf{X})\, dt = \int_{\gamma_1} -(a_1 + 2a_2 x + 3a_3 x^2) \le 0.$$

Let P_i and Q_i, $i = 0, 1, 2, 3$, be the points of intersection of γ_1 and γ_2, respectively, with the lines L_1 and L_2. Now $\int_{\gamma_1} x \, dt = \int_{\gamma_1} -\frac{dy}{dt} \, dt = 0$, and similarly for the periodic orbit γ_2.

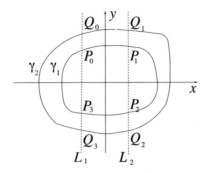

Figure 11.11:   Two limit cycles crossing the lines L_1 and L_2.

Consider the branches $P_0 P_1$ and $Q_0 Q_1$ on γ_1 and γ_2, respectively. The flow is never vertical on these branches. Hence one may parameterize the integrals by the variable x. Thus

$$\int_{P_0 P_1} -(a_1 + 3a_3 x^2) \, dt = \int_{x_0}^{x_1} \frac{-(a_1 + 3a_3 x^2)}{y_{\gamma_1}(x) - F(x)} \, dx$$

and

$$\int_{Q_0 Q_1} -(a_1 + 3a_3 x^2) \, dt = \int_{x_0}^{x_1} \frac{-(a_1 + 3a_3 x^2)}{y_{\gamma_2}(x) - F(x)} \, dx.$$

In the region $x_0 < x < x_1$, the quantity $-(a_1 + 3a_3 x^2) > 0$ and $y_{\gamma_2}(x) - F(x) > y_{\gamma_1}(x) - F(x) > 0$. It follows that

$$\int_{Q_0 Q_1} -(a_1 + 3a_3 x^2) \, dt < \int_{P_0 P_1} -(a_1 + 3a_3 x^2) \, dt.$$

Using similar arguments, it is not difficult to show that

$$\int_{Q_2 Q_3} -(a_1 + 3a_3 x^2) \, dt < \int_{P_2 P_3} -(a_1 + 3a_3 x^2) \, dt.$$

Consider the branches $P_1 P_2$ and $Q_1 Q_2$ on γ_1 and γ_2, respectively. The flow is never horizontal on these branches. Hence one may parameterize the integrals by the variable y. Thus

$$\int_{P_1 P_2} -(a_1 + 3a_3 x^2) \, dt = \int_{y_1}^{y_2} \frac{(a_1 + 3a_3 (x_{\gamma_1}(y))^2)}{x_{\gamma_1}} \, dy$$

and

$$\int_{Q_1 Q_2} -(a_1 + 3a_3 x^2)\, dt = \int_{y_1}^{y_2} \frac{-(a_1 + 3a_3 (x_{\gamma_2}(y))^2)}{x_{\gamma_2}}\, dy.$$

In the region $y_1 < y < y_2$, $x_{\gamma_2}(y) > x_{\gamma_1}(y)$. It follows that

$$\int_{Q_1 Q_2} -(a_1 + 3a_3 x^2)\, dt < \int_{P_1 P_2} -(a_1 + 3a_3 x^2)\, dt.$$

Using similar arguments, it is not difficult to show that

$$\int_{Q_3 Q_0} -(a_1 + 3a_3 x^2)\, dt < \int_{P_3 P_0} -(a_1 + 3a_3 x^2)\, dt.$$

Thus adding all of the branches together,

$$\int_{\gamma_2} \operatorname{div}(\mathbf{X})\, dt < \int_{\gamma_1} \operatorname{div}(\mathbf{X})\, dt \leq 0$$

which proves Stage II.

Stage III. Since the origin is unstable and $\int_{\gamma_2} \operatorname{div}(\mathbf{X}) dt < \int_{\gamma_1} \operatorname{div}(\mathbf{X}) dt \leq 0$, the limit cycle γ_2 is hyperbolic stable and the limit cycle γ_1 is semistable. By introducing a small perturbation such as $\dot{x} = y - F(x) - \epsilon x$, it is possible to bifurcate a limit cycle from γ_1 that lies between γ_2 and γ_1. Therefore, system (11.6) has at least three limit cycles, which contradicts the result at Stage II. Hence system (11.6) has a unique hyperbolic limit cycle. □

A Liénard System with a Large Parameter. Consider the parameterized cubic Liénard equation given by

$$\ddot{x} + \mu f(x)\dot{x} + g(x) = 0,$$

where $f(x) = -1 + 3x^2$ and $g(x) = x$, which becomes

$$\dot{x} = \mu y - \mu F(x), \quad \mu \dot{y} = -g(x), \tag{11.7}$$

where $F(x) = \int_0^x f(s)ds = -x + x^3$, in the Liénard plane. Liénard (see Chapter 10) proved that system (11.7) has a unique limit cycle. Systems containing small parameters were considered in Chapter 10 using Melnikov integrals.

The obvious question then is, what happens when μ is large? Figure 11.12 shows the limit cycle behavior in the Liénard and tx planes, when $\mu = 20$ for system (11.7).

Let $\mu = \frac{1}{\epsilon}$. Then system (11.7) can be written as an equivalent system in the form

$$\epsilon\dot{x} = y - F(x), \quad \dot{y} = -\epsilon g(x). \tag{11.8}$$

Theorem 5. *Consider system (11.8) and the Jordan curve J shown in Figure 11.13. As $\mu \to \infty$ or, alternatively, $\epsilon \to 0$, the limit cycle tends toward the piecewise analytic Jordan curve J.*

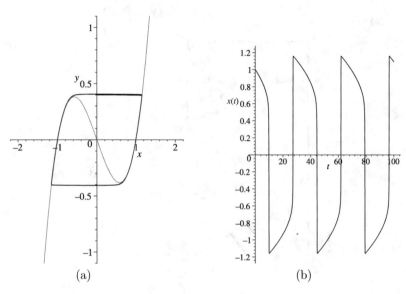

(a)   (b)

Figure 11.12: (a) A limit cycle for the cubic system when $F(x) = -x + x^3$; the function $y = F(x)$ is also shown. (b) Periodic behavior in the tx plane.

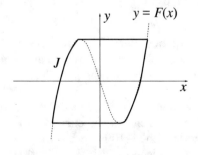

Figure 11.13: The Jordan curve and the function $y = F(x)$.

Proof. The method of proof involves the Poincaré-Bendixson Theorem from Chapter 5. Thus everything is reduced to the construction of an annular region A that is positively invariant and that contains no critical points. The construction is shown in Figure 11.14.

Note that system (11.8) is symmetric about the y-axis, so we need only consider one half of the plane.

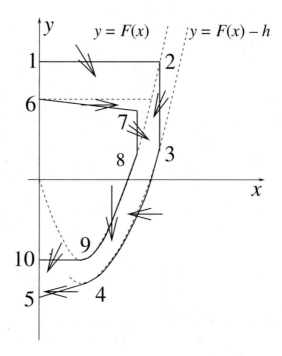

Figure 11.14: Construction of the inner and outer boundaries of the annular region that forms a positively invariant set in one half of the plane. A similar construction is used in the other half of the plane using symmetry arguments.

First, consider the outer boundary. The arc 1-2 is a horizontal line, and 2-3 is a vertical line from the graph $y = F(x)$ to the graph $y = F(x) - h$, where h is a small constant. The arc 3-4 follows the $y = F(x) - h$ curve, and the line 4-5 is a tangent.

Now consider the inner boundary. The line 6-7 is sloped below the horizontal, and the line 7-8 is vertical and meets the curve $y = F(x)$. The arc 8-9 follows the curve $y = F(x)$, and the line 9-10 is horizontal.

To prove that the region is positively invariant, one must show that the marked arrows point in the directions indicated in Figure 11.14. Consider each arc separately.

Notation. For any point n in Figure 11.14, let $F'(n)$ and $g(n)$ be the values of these functions at the abscissa of n.

Arc 1-2. On this line, $\dot{y} < 0$ since $\dot{y} = -x$ and $x > 0$.

Arc 2-3. On this line, $y \le F(x)$, so $\epsilon\dot{x} = y - F(x) \le 0$. Note that $\dot{y} < 0$ at point 2.

Arc 3-4. Suppose that p is a point on this arc. The slope of a trajectory crossing this arc is given by

$$\left.\frac{dy}{dx}\right|_p = \frac{-\epsilon^2 g(p)}{-h} < \frac{\epsilon^2 g(3)}{h},$$

and

$$\left.\frac{dy}{dx}\right|_p \to 0,$$

as $\epsilon \to 0$. Therefore, for ϵ small enough,

$$\left.\frac{dy}{dx}\right|_p < F'(4) < F'(p)$$

on the arc. Since $\dot{x} < 0$ along the arc, trajectories cross the boundary inwards.

Arc 4-5. Since $|y - F(x)| > h$, the slope of the curve 4-5 is

$$\left.\frac{dy}{dx}\right|_4 < \frac{\epsilon^2 g(4)}{h},$$

which tends to zero as $\epsilon \to 0$. Once more $\dot{x} < 0$ on this arc, for ϵ small enough, and the pointing is inward.

Arc 6-7. Let d_1 be the vertical distance of the line 7-8. For d_1 small enough, along the line 6-7, $|y - F(x)| > d_1$. Thus the slope of the curve at a point q say, on the line 6-7 is given by

$$\left.\frac{dy}{dx}\right|_q < \frac{\epsilon^2 g(q)}{d_1} < \frac{\epsilon^2 g(7)}{d_1},$$

which tends to zero as $\epsilon \to 0$. Since $\dot{x} > 0$ on this arc, for ϵ small enough, the pointing will be as indicated in Figure 11.14.

Arc 7-8. On this line, $y - F(x) > 0$, so $\dot{x} > 0$.

Arc 8-9. On the curve, $y = F(x)$ with $x > 0$, $\dot{y} < 0$, and $\dot{x} = 0$.

Arc 9-10. On this line, $y - F(x) < 0$ and $\dot{y} < 0$.

Using similar arguments on the left-hand side of the y-axis, a positively invariant annulus can be constructed. Since system (11.8) has a unique critical point at the origin, the Poincaré-Bendixson theorem can be applied to prove that there is a limit cycle in the annular region A. For suitably small values of h and d_1, the annular region will be arbitrarily near the Jordan curve J. Therefore, if $\Gamma(\epsilon)$ is the limit cycle, then $\Gamma(\epsilon) \to J$ as $\epsilon \to 0$. $\square$

11.4 Local Results for Liénard Systems

Although the Liénard equation (11.5) appears simple enough, the known global results on the maximum number of limit cycles are scant. By contrast, if the analysis is restricted to local bifurcations, then many more results may be obtained. The method for bifurcating small-amplitude limit cycles is given in Chapter 10. Consider the Liénard system

$$\dot{x} = y, \quad \dot{y} = -g(x) - f(x)y, \tag{11.9}$$

where $f(x) = a_0 + a_1 x + a_2 x^2 + \ldots + a_m x^m$ and $g(x) = x + b_2 x^2 + b_3 x^3 + \ldots + b_n x^n$; m and n are natural numbers. Let $\hat{H}(m, n)$ denote the maximum number of small-amplitude limit cycles that can be bifurcated from the origin for system (11.9), where m is the degree of f and n is the degree of g.

In 1984, Blows and Lloyd [3] proved the following results for system (11.9):

- If $\partial f = m = 2i$ or $2i + 1$, then $\hat{H}(m, 1) = i$.

- If g is odd and $\partial f = m = 2i$ or $2i + 1$, then $\hat{H}(m, n) = i$.

In addition to the above, the author has proved the following results by induction.

- If $\partial g = n = 2j$ or $2j + 1$, then $\hat{H}(1, n) = j$.

- If f is even, $\partial f = 2i$, then $\hat{H}(2i, n) = i$.

- If f is odd, $\partial f = 2i + 1$ and $\partial g = n = 2j + 2$ or $2j + 3$; then $\hat{H}(2i + 1, n) = i + j$.

- If $\partial f = 2$, $g(x) = x + g_e(x)$, where g_e is even and $\partial g = 2j$; then $\hat{H}(2, 2j) = j$.

Christopher and the author [5] have more recently developed a new algebraic method for determining the Lyapunov quantities, and this has allowed further computations. Let $\lfloor . \rfloor$ denote the integer part. Then the new results are listed below:

- $\hat{H}(2, n) = \left\lfloor \frac{2n+1}{3} \right\rfloor$.

- $\hat{H}(m, 2) = \left\lfloor \frac{2m+1}{3} \right\rfloor$.

- $\hat{H}(3, n) = 2 \left\lfloor \frac{3n+6}{8} \right\rfloor$, for all $1 < n \leq 50$.

Table 11.1: The values of $\hat{H}(m,n)$ for varying values of m and n

degree of f	1	2	3	4	5	6	7	8	9	10	11	12	13	⋯	48	49	50
50	↑	↑	38														
49	24	33	38														
48	24	32	36														
⋮	⋮	⋮	⋮														
13	6	9	10														
12	6	8	10														
11	5	7	8														
10	5	7	8														
9	4	6	8	9													
8	4	5	6	9													
7	3	5	6	8													
6	3	4	6	7													
5	2	3	4	6	6												
4	2	3	4	4	6	7	8	9	9								
3	1	2	2	4	4	6	6	6	8	8	8	10	10	⋯	36	38	38
2	1	1	2	3	3	4	5	5	6	7	7	8	9	⋯	32	33	→
1	0	1	1	2	2	3	3	4	4	5	5	6	6	⋯	24	24	→
	1	2	3	4	5	6	7	8	9	10	11	12	13	⋯	48	49	50

degree of g

- $\hat{H}(m,3) = 2 \left\lfloor \frac{3m+6}{8} \right\rfloor$, for all $1 < m \le 50$.

Complementing these results is the calculation of $\hat{H}(m,n)$ for specific values of m and n. The results are presented in Table 11.1.

The ultimate aim is to establish a general formula for $\hat{H}(m,n)$ as a function of the degrees of f and g. Christopher and Lloyd [6] have proven that Table 11.1 is symmetric but only in the restricted cases where the linear coefficient in $f(x)$ is nonzero. The author et al. [16] have recently started work on simultaneous bifurcations for symmetric Liénard systems and Maoan and Romanovski [21] have used a new method to give more results. Future work will concentrate on attempting to complete Table 11.1 and determining a relationship, if any, between global and local results.

It is important to note that programming with mathematical packages is a key tool that has to be used carefully. For example, it may be that two limit cycles bifurcating from a fine focus cannot be distinguished on a computer screen. There are always restrictions on how far a package can be used and this presents a good example of that fact. An example of a system with four limit cycles in close proximity is given in the Coursework examples in Chapter 23.

11.5 Mathematica Commands

```
(* Limit cycle of a Lienard system with changing parameter. *)
Manipulate[Module[{p1,sol,p2},p1=Plot[-x+x^3,{x,-2,2},PlotStyle->{Red}];
sol=NDSolve[{x'[t]==mu*y[t]-mu*(-x[t]+x[t]^3),y'[t]==-x[t]/mu,x[0]==1,
y[0]==0},{x[t],y[t]},{t,50,100}];
p2=ParametricPlot[Evaluate[Table[{x[t],y[t]}/.sol]],{t,50,100},
```

```
PlotRange->{{-2,2},{-2.2,2.2}},PlotPoints->100,AxesLabel->{"x","y"}];
Show[{p1,p2},PlotRange->{{-2,2},{-2.2,2.2}},AxesLabel->{"x","y"},
Axes->True]],{mu,0.5, 10}]
```

11.6 Exercises

1. Draw a global phase portrait for the linear system

$$\dot{x} = y, \quad \dot{x} = -4x - 5y$$

   including the flow at infinity.

2. Draw a global phase portrait for the system

$$\dot{x} = -3x + 4y, \quad \dot{y} = -2x + 3y$$

   and give the equations defining the flow near critical points at infinity.

3. Determine a global phase portrait for the quadratic system given by

$$\dot{x} = x^2 + y^2 - 1, \quad \dot{y} = 5xy - 5.$$

4. Draw a global phase portrait for the Liénard system

$$\dot{x} = y - x^3 - x, \quad \dot{y} = -y.$$

5. Draw a global phase portrait for the Liénard system

$$\dot{x} = y - x^3 + x, \quad \dot{y} = -y.$$

6. Edit the Mathematica files in Chapter 5 to compare the limit cycles for Liénard systems in the phase plane and in the Liénard plane. Plot the periodic orbits in the xt plane.

7. Use Mathematica to investigate the system

$$\dot{x} = y - (a_1x + a_2x^2 + a_3x^3), \quad \dot{y} = -x$$

   for varying values of the parameters a_1, a_2, and a_3.

8. Use Mathematica to investigate the limit cycles, if they exist, of the system

$$\dot{x} = y - \epsilon(a_1x + a_2x^2 + \ldots + a_Mx^M), \quad \dot{y} = -x,$$

as the parameter ϵ varies from zero to infinity.

9. Prove Liénard's theorem, that when $\partial g = 1$ and $F(x)$ is a continuous odd function that has a unique root at $x = a$ and is monotone increasing for $x \geq a$, (11.5) has a unique limit cycle.

10. This is quite a difficult question. Consider the Liénard system

$$\dot{x} = y - F(x), \quad \dot{y} = -x, \qquad (11.10)$$

where $F(x) = (a_1 x + a_3 x^3 + a_5 x^5)$ is odd. Prove that system (11.10) has at most two limit cycles.

Bibliography

[1] N. Bautin, On the number of limit cycles which appear with the variation of the coefficients from an equilibrium point of focus or center type, *Amer. Math. Soc. Trans.* **5**, (1962), 396–414.

[2] T.R. Blows and C. Rousseau, Bifurcation at infinity in polynomial vector fields, *J. Differential Equations* **104**, (1993), 215–242.

[3] T.R. Blows and N.G. Lloyd, The number of small-amplitude limit cycles of Liénard equations, *Math. Proc. Camb. Philos. Soc.* **95**, (1984), 359–366.

[4] L.A. Cherkas, Conditions for a Liénard equation to have a center, *Differentsial'nye Uravneniya* **12**, (1976), 201–206.

[5] C.J. Christopher and S. Lynch, Small-amplitude limit cycle bifurcations for Liénard systems with quadratic or cubic damping or restoring forces, *Nonlinearity* **12**, (1999), 1099–1112.

[6] C.J. Christopher and N.G. Lloyd, Small-amplitude limit cycles in polynomial Liénard systems, *Nonlinear Diff. Eqns. Appl.* **3**, (1996), 183–190.

[7] C.J. Christopher and N.G. Lloyd, Polynomial systems: a lower bound for the Hilbert numbers, *Proc. Roy. Soc. Lond. A* **450**, (1995), 219–224.

[8] W.A. Coppel, Some quadratic systems with at most one limit cycle, *Dynamics Reported, New York* **2**, (1988), 61–68.

[9] F. Dumortier, D. Panazzolo and R. Roussarie, More limit cycles than expected in Liénard equations *Proceedings of the American Mathematical Society* **135**(6), (2007), 1895–1904.

[10] F. Dumortier and L. Chengzhi, Quadratic Liénard equations with quadratic damping, *J. Diff. Eqns* **139**, (1997), 41–59.

[11] F. Dumortier and L. Chengzhi, On the uniqueness of limit cycles surrounding one or more singularities for Liénard equations, *Nonlinearity* **9**, (1996), 1489–1500.

[12] J. Ecalle, J. Martinet, J. Moussu and J.P. Ramis, Non-accumulation des cycles-limites I, *C.R. Acad. Sci. Paris Sér. I Math.* **304**, (1987), 375–377.

[13] H. Giacomini and S. Neukirch, Improving a method for the study of limit cycles of the Liénard equation, *Phys. Rev. E* **57**, (1998), 6573–6576.

[14] Y. Ilyashenko, Centennial history of Hilbert's 16th problem, *Bull. Amer. Math. Soc.* **39**-3, (2002), 301–354.

[15] Y. Ilyashenko, *Finiteness Theorems for Limit Cycles*, Translations of Mathematical Monographs 94, American Mathematical Society, Providence, RI, 1991.

[16] J. Jiang, H. Maoan, Y. Pei and S. Lynch, Small-amplitude limit cycles of two types of symmetric Liénard systems, *Int. J. of Bifurcation and Chaos*, **17**(6), (2007), 2169–2174.

[17] Li Jibin, Hilbert's sixteenth problem and bifurcations of planar polynomial vector fields, *Int, J. of Bifurcation and Chaos* **13**, (2003), 47–106.

[18] Li Jibin and Li Chunfu, Global bifurcation of planar disturbed Hamiltonian systems and distributions of limit cycles of cubic systems, *Acta. Math. Sinica* **28**, (1985), 509–521.

[19] A. Lins, W. de Melo and C. Pugh, On Liénards equation with linear damping, *Lecture Notes in Mathematics*, **597**, (Edited by J. Palis and M. do Carno), 335-357, Springer-Verlag, Berlin, 1977, 335–357.

[20] Han Maoan and Jibin Li, Lower bounds for the Hilbert number of polynomial systems, *J. of Differential Equations* **252**(4), (2012) 3278–3304.

[21] Han Maoan and V.G. Romanovski, On the number of limit cycles of polynomial Lienard systems, *Nonlinear Analysis - Real World Applications*, **14**(3), (2013), 1655–1668.

[22] A. Marasco and C. Tenneriello, Periodic solutions of a 2D-autonomous system using Mathematica, *Mathematical and Computer Modelling* **45**(5-6), (2007), 681–693.

[23] H.N. Moreira, Liénard-type equations and the epidemiology of maleria, *Ecological Modelling* **60**, (1992), 139–150.

[24] Y. Pei and H. Maoan, Twelve limit cycles in a cubic case of the 16'th Hilbert problem, *Int. J. of Bifurcation and Chaos* **15**, (2005), 2191–2205.

[25] G.S. Rychkov, The maximum number of limit cycles of the system $\dot{x} = y - a_0x - a_1x^3 - a_2x^5, \dot{y} = -x$ is two, *Differentsial'nye Uravneniya* **11**, (1973), 380–391.

[26] Shi Songling, A concrete example of the existence of four limit cycles for plane quadratic systems, *Sci. Sinica A* **23**, (1980), 153–158.

[27] K.S. Sibirskii, The number of limit cycles in the neighbourhood of a critical point, *Differential Equations* **1**, (1965), 36–47.

[28] X.B. Sun and W.T. Huang, Bounding the number of limit cycles for a polynomial Lienard system by using regular chains, *J. of Symbolic Computation* **79**, (2017), 197–210.

[29] Y.Q. Xiong, and M.A. Han, New lower bounds for the Hilbert number of polynomial systems of Liénard type, *J. of Diff Eqns.* **257**, (2014), 2565–2590.

[30] Zhang Zhifen, Ding Tongren, Huang Wenzao, and Dong Zhenxi, *Qualitative Theory of Differential Equations*, Translation of Mathematical Monographs 102, American Mathematical Society, Providence, RI, 1992.

[31] H. Zoladek, Eleven small limit cycles in a cubic vector field, *Nonlinearity* **8**, (1995), 843–860.

Chapter 12
Delay Differential Equations

Aims and Objectives

- To introduce the method of steps for Delay Differential Equations (DDEs).

- To investigate stability of a simple DDE.

- To use the Manipulate command to produce animations.

- To investigate applications in dynamical systems.

On completion of this chapter the reader should be able to

- use the method of steps to solve simple DDEs;

- determine the stability of simple DDEs;

- apply the theory of DDEs to examples from biology, economics, environmental science, mechanical systems, neural networks, and nonlinear optics and interpret the numerical solutions physically.

Almost all dynamical systems can be subject to some sort of feedback control, where a time delay arises due to a finite time interval being required for the system to sense a change and react to it. Also, many dynamical systems, especially in biology, have the delays inherently built in. Seeking solutions to these type of problems has led to the field of mathematics known as Delay Differential Equations, abbreviated to DDEs in most of the literature.

The chapter begins with an introduction and an outline of the method of steps used to solve certain DDEs. The following sections highlight applications in biology, nonlinear optics, and other dynamical systems.

© Springer International Publishing AG 2017   257
S. Lynch, *Dynamical Systems with Applications using Mathematica*®,
DOI 10.1007/978-3-319-61485-4_12

12.1 Introduction and the Method of Steps

Dynamical systems subject to some sort of delay have been studied for over two hundred years, the paper of Schmitt [21] provides references and lists some properties of simple linear DDEs. DDEs were studied more extensively after the second world war with the need for control engineering in technology but it is only in the last few decades that DDEs have become extensively studied with the development of mathematics packages such as Mathematica. An introduction to the theory of DDEs is given in [7], applications of DDEs are discussed in [8], DDEs applied to the life sciences are covered in [25] and a nice introduction to the dynamics of nonlinear time delay systems is provided in [16]. DDEs differ from ODEs in that the derivative at any time depends on the solution at prior times. These systems are infinite-dimensional; it is necessary to provide a so-called initial *history function* to specify the value of the solution set before time $t = 0$, and the time delays could be constants or state-dependent. Currently, Mathematica is only able to solve DDEs with constant time delays and those are the type of system investigated in this book.

Definition 1. A DDE subject to constant time delays is of the form:

$$\frac{d\mathbf{x}}{dt} = \mathbf{f}\left(\mathbf{x}(t), \mathbf{x}(t - \tau_1), \mathbf{x}(t - \tau_2), \dots, \mathbf{x}(t - \tau_n)\right), \tag{12.1}$$

where $\mathbf{x} \in \Re^n$ and the delays τ_i are positive constants.

In order to solve DDEs, it is necessary to define an initial history function which determines the behavior of the dynamical system $\mathbf{x}(t)$ defined on the interval $[-\tau, 0]$, assuming that the systems start at $t = 0$.

The simplest method for solving some systems of DDEs has been labeled as the *method of steps*. DDEs differ from ODEs in that the solution for the DDE can be thought of as a mapping from functions on an interval $[t - \tau, t]$ on to functions on an interval $[t, \tau + t]$. In some very simple cases, it is possible to work out an analytical solution to this problem as the following example demonstrates.

Example 1. Solve the simple linear DDE given by

$$\frac{dx}{dt} = -x(t - 1), \tag{12.2}$$

with initial history function $x(t) = 1$, on $[-1, 0]$.

Solution. Suppose that $x(t) = \phi_{i-1}(t)$ on the interval $[t_i - 1, t_i]$. Then, using separation of variables, on the interval $[t_i, t_i + 1]$:

$$\int_{\phi_{i-1}}^{x(t)} dx' = -\int_{t_i}^{t} \phi_{i-1}\left(t'-1\right) dt'$$

and

$$x(t) = \phi_i(t) = \phi_{i-1}\left(t_i\right) - \int_{t_i}^{t} \phi_{i-1}\left(t'-1\right) dt'. \tag{12.3}$$

Therefore, in the interval $[0,1]$, equation (12.3) gives

$$x(t) = 1 - \int_0^t 1 dt' = 1 - t$$

and in the interval $[1,2]$:

$$x(t) = 0 - \int_1^t 1 - (t'-1)dt' = -\left[2t - \frac{t^2}{2}\right]_1^t = -2t + \frac{t^2}{2} + \frac{3}{2}.$$

One could continue to calculate the solution on further intervals but the process can be easily implemented in Mathematica.

```
(* The method of steps. See Figure 12.1. *)
In[1]:=tmax=4;
phi[i_,t_]:=If[i==0,1,phi[i-1,i-1]-Integrate[phi[i-1,xi-1],{xi,i-1,t}]];
Do[p[i]=phi[i,t],{i,0,tmax}];Table[p[i],{i,0,tmax}];

In[2]:=tmax=10;
phi[i_,t_]:=If[i==0,1,phi[i-1,i-1]-Integrate[phi[i-1,ti-1],{ti,i-1,t}]];
Do[p[i]=phi[i,t],{i,0,tmax}];
Do[plots[i]=Flatten[List[p[i],{i-1<t<i}]],{i,0,tmax}];
p1=Plot[Piecewise[Table[plots[i],{i,0,tmax}]],{t,-1,tmax}];
Show[{p1},AxesLabel->{"t","x(t)"},LabelStyle->{FontSize->20}]
```

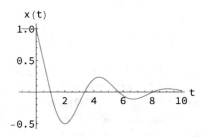

Figure 12.1: [Mathematica] The solution $x(t)$ for the DDE equation (12.2) for $-1 < t < 10$. The initial history function in this case was $x(t) = 1$ on $[-1, 0]$.

Figure 12.1 shows the solution obtained by the method of steps for $-1 < t < 10$. Using Mathematica (see the commands above), the analytical solutions on the intervals $[t_{i-1}, t_i]$ for $i = 0$ to $i = 4$ and t in the range $[-1, 4]$ are output as a list of the form:

$$Out[1] = \left\{ 1, 1 - t, \frac{3}{2} - 2t + \frac{t^2}{2}, \frac{17}{6} - 4t + \frac{3t^2}{2} - \frac{t^3}{6}, \frac{149}{24} - \frac{17t}{2} + \frac{15t^2}{4} - \frac{2t^3}{3} + \frac{t^4}{24} \right\}.$$

Of course, one can also solve the DDE (12.2) directly using the NDSolve command in Mathematica. Readers can check the syntax of the NDSolve command by clicking on the Help tab and typing in "Solving Delay Differential Equations" in Wolfram Documentation.

```
(* Using NDSolve to solve a DDE. See Figure 12.1. *)
In[3]:=sol=NDSolve[{x'[t]==-x[t-1],x[t/;t<=0]==1},x,{t,-1,10},
MaxSteps->Infinity]

In[4]:=p1=ParametricPlot[Evaluate[{t,x[t]}/.sol],{t,-1,10}];
Show[{p1},AxesLabel->{"t","x(t)"},LabelStyle->{FontSize->20}]
```

The next example shows how the initial history function affects the solution for $t > 0$.

Example 2. Solve the simple linear DDE (12.2) with initial history functions on $[-1, 0]$ given by: (a) $x(t) = e^t$; (b) $x(t) = t^2$; (c) $x(t) = t$; (d) $x(t) = \sin(t)$.

Solution. In the Mathematica notebook, the Manipulate command is used, and the reader can move the slider to obtain the graphs displayed in Figure 12.2.

Linear Stability Analysis. As with ODEs, it is important to establish the location and stability of critical points of DDEs. A critical point of a DDE of the form (12.1) satisfies the equation

$$\mathbf{f}(\mathbf{x}^*, \mathbf{x}^*, \ldots, \mathbf{x}^*) = 0,$$

where $\mathbf{x}^*$ is a critical point of system (12.1). The methods to determine the location and stability of critical points of ODEs are covered in other chapters of the book. The process is similar with DDEs except that the solution space is an infinite-dimensional function space. Consider small perturbations from equilibrium in this space, then the displacements are time-dependent functions, $\delta\mathbf{x}(t)$, say, that can persist for an interval at least the maximum value of τ_i. In order to simplify the notation, it is usual to write $\mathbf{x}_\tau = \mathbf{x}(t - \tau)$,

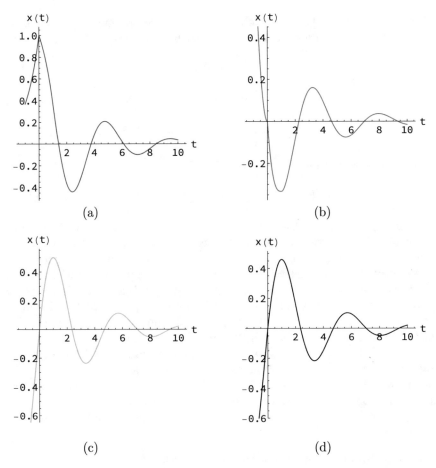

Figure 12.2: [Mathematica animation] Solutions to the DDE (12.2) with initial history functions on $[-1,0]$ are defined by: (a) $x(t) = e^t$; (b) $x(t) = t^2$; (c) $x(t) = t$; (d) $x(t) = \sin(t)$. Note that in the notebook, the Manipulate command is used to produce a single animation. Make sure that under the Evaluation tab you tick Dynamic Updating Enabled.

which is what we will do here. Using methods similar to those described in Chapters 2 and 3, suppose that

$$\mathbf{x} = \mathbf{x}^* + \delta\mathbf{x},$$

take a Taylor series expansion and linearize to obtain

$$\delta\dot{\mathbf{x}} \approx \mathbf{J}_0\delta\mathbf{x} + \mathbf{J}_{\tau_1}\delta\mathbf{x}_{\tau_1} + \ldots + \mathbf{J}_{\tau_n}\delta\mathbf{x}_{\tau_n}, \tag{12.4}$$

where the matrices $\mathbf{J}_{\tau_i}$ are Jacobians of the corresponding $\mathbf{x}_{\tau_i}$. Supposing that DDEs have exponential solutions as with ODEs, then

$$\delta\mathbf{x}(t) = \mathbf{A}e^{\lambda t},$$

where λ is an eigenvalue of the Jacobian matrix. Substituting into (12.4) gives

$$\lambda\mathbf{A} = \mathbf{A}\left(\mathbf{J}_0 + e^{-\lambda\tau_1}\mathbf{J}_{\tau_1} + \ldots + e^{-\lambda\tau_n}\mathbf{J}_{\tau_n}\right).$$

The characteristic equation is then given by

$$\left|\mathbf{J}_0 + e^{-\lambda\tau_1}\mathbf{J}_{\tau_1} + \ldots + e^{-\lambda\tau_n}\mathbf{J}_{\tau_n} - \lambda\mathbf{I}\right| = 0. \tag{12.5}$$

Expanding out the determinant leads to polynomials which include some terms in $e^{\lambda\tau_i}$ and these are called *quasi-polynomials*. As with ODEs, if all of the solutions of equation (12.5) have negative real part, then the critical point is stable, otherwise it is unstable, or if some of the leading characteristic values are zero, the critical point is non-hyperbolic. Quasi-polynomials generally have an infinite number of roots in the complex plane, and this is where Mathematica can help in determining the stability of critical points. The method will now be illustrated by means of an example.

In the next example, a type of logistic equation subject to delay is investigated using the **Manipulate** command. It is shown that there are two critical points and the stability of one of these points will be investigated. From this simple logistic DDE, it is shown that a Hopf bifurcation takes place, where a stable critical point loses its stability and a stable limit cycle bifurcates from the point.

Example 3. Investigate the logistic DDE given by

$$\frac{dx}{dt} = \mu x(t)(1 - x(t - \tau_1)), \tag{12.6}$$

with initial history function $x(t) = 0.1$, on $[-1, 0]$, $\tau_1 = 1$, as the parameter μ varies.

Solution. The Mathematica commands for producing an animation, snapshots of which are shown in Figure 12.3, are listed below:

```
(* Using NDSolve to solve a logistic DDE. See Figure 12.3. *)
Manipulate[Module[{T=50,sol,x,t},
sol=First[x/.NDSolve[{x'[t]==mu x[t] (1-x[t-1]),x[t/;t<=0]==0.1},x,
{t,0,T}]];
If[pp,ParametricPlot[{sol[t],sol[t-1]},{t,1,T},PlotRange->{{0,3},{0,3}}],
Plot[sol[t],{t,0,T},PlotRange->{{0,50},{0,3}}]]],
{{pp,False,"Phase Portrait"},{False,True}},{{mu,1},0,2}]
```

For system (12.6), there are two critical points at $x^* = 0$ and $x^* = 1$. One can show that the trivial critical point at $\mathbf{x}^* = 0$ is unstable. Consider the critical point at $x^* = 1$, where more interesting behavior is present. The characteristic equation is given by

$$\mu e^{-\lambda \tau_1} + \lambda = 0. \tag{12.7}$$

For the critical point $\mathbf{x}^* = 1$ to be stable, the complex roots of equation (12.7) must lie on the left-half of the λ plane. A Hopf bifurcation is a local bifurcation in which a critical point loses stability as a pair of complex con- jugate eigenvalues cross the imaginary axis. Therefore, a bifurcation takes place when the roots lie on the imaginary axis. Suppose that $\lambda = 0 + iy$, then equation(12.7) becomes

$$\mu e^{-iy\tau_1} + iy = 0$$

or

$$\mu \left(\cos\left(y\tau_1\right) - i\sin\left(y\tau_1\right) \right) + iy = 0.$$

Equating real and imaginary parts to zero, one obtains

$$\mu \cos\left(y\tau_1\right) = 0, \quad y - \mu\sin\left(y\tau_1\right) = 0.$$

The first equation has solution $y\tau_1 = \frac{(2n+1)\pi}{2}, n = 0, 1, 2, \ldots$, and the mini- mum order solution is found when $n = 0$, giving $\mu\tau_1 = \frac{\pi}{2}$. Hence, a necessary and sufficient condition for the critical point at $\mathbf{x}^* = 1$ to be stable is $\mu\tau_1 < \frac{\pi}{2}$, and when $\mu\tau_1 \geq \frac{\pi}{2}$, the critical point goes through a bifurcation and becomes unstable and a small-amplitude limit cycle bifurcates from the critical point. Readers can check the bifurcation value by running the animation program in the Mathematica notebook, see Figure 12.3.

The logistic DDE was first investigated by Hutchinson [15] in 1948 who devised a more realistic single species model subject to time delays. These time delays could be a result of density dependent population growth, discrete breeding seasons, disease, food supply, and seasonal effects, for example. Most processes in biology and especially in the human body are subject to some kind of time delay, and this naturally leads to systems displaying oscillatory behavior. The next section concentrates on examples of DDEs in the field of biology.

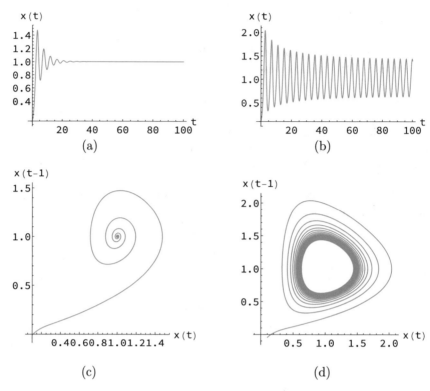

Figure 12.3: [Mathematica animation] Solutions to the DDE (12.4) with initial history function on $[-1, 0]$ are defined by $x(t) = 0.1$ for varying values of the parameter μ. (a) Time series showing that when $\mu = 1.2$, $x(t)$ approaches the stable critical point at $x^* = 1$. (b) Time series showing that when $\mu = 1.6$, $x(t)$ approaches a stable limit cycle. (c) Phase portrait showing that when $\mu = 1.2$, $x(t)$ approaches the stable critical point at $x^* = 1$. (d) Phase portrait showing that when $\mu = 1.6$, $x(t)$ approaches a stable limit cycle. Note that the Manipulate command is used to produce an animation in the notebook for values of μ from 0 to 2. A linear stability analysis shows that the system goes through a Hopf bifurcation when $\mu = \frac{\pi}{2}$.

12.2 Applications in Biology

In 2016, Agarwal [1] published a book on logistic DDEs and considers oscillation of delay logistic models and chapters also cover stability, piecewise arguments, food-limited populations, and diffusion. Predator-prey DDE models with disease in prey are investigated in [26], the effect of diffusion in predator-prey DDE models is covered in [10] and predator-prey DDE models with multiple delays are investigated in [5]. When considering DDE models in epidemiology, a pest management strategy is considered in [29], an analysis of an SEIRS model with two delays is investigated in [6] and global stability for DDE SIR and SEIR models with nonlinear incidence rate are discussed in [14].

The human body is predominantly composed of periodic processes from the cellular and molecular level up to the organ and inter-organ system level. Many cells, such as neurons, heart cells, muscle cells, retinal cells and blood cells oscillate, [20] and periodic processes in the human body, encompass phenomena such as heartbeat rhythms, bacterial oscillations, cytoskeletal structures, genetic interactions, rhythmic behavior in growth and development and cancer, for instance, see [3] and [27].

In this section, the author has decided to concentrate on two physiological processes, namely haematopoiesis (the formation of blood cellular components) and angiogenesis (the development of new blood vessels) which include the dynamics of tumour growth from a dormant to a malignant state. More detailed information can be found in [17].

A DDE Model of Haematopoiesis. Consider the one-dimensional Mackey-Glass model related to haematopoiesis [19] defined by

$$\frac{dx}{dt} = \frac{\beta x(t-\tau)}{1 + x(t-\tau)^n} - \delta x(t), \qquad (12.8)$$

where $x(t)$ is the blood cell population at time t, τ is a constant time lag, and n is a constant. The first term in the right-hand side of equation (12.8) models the delayed production rate of the blood cells, and δ represents a death rate of blood cells. The DDE can be numerically solved with Mathematica, and an animation of the phase space solution can be generated using the Manipulate command. Figure 12.4(a) shows a phase portrait of equation (12.8), and Figure 12.4(b) shows the corresponding power spectrum for parameter values $\beta = 2$, $n = 10$, $\tau = 2$, and $\delta = 0.8$.

The Mathematica program for plotting an animated phase portrait for the Mackey-Glass model is listed below. Within the Mathematica notebook, users can move the slider left and right (decreasing and increasing δ), and the phase portrait updates automatically.

```
(* Animated phase portrait of the Mackey-Glass model of haematopoeisis. *)
beta=2;tau=2;n=10;
Manipulate[sol=First[NDSolve[{x'[t]==beta x[t-tau]/(1+x[t-tau]^n)-
delta x[t],x[t/;t<=0]==1/2},x,{t,0,200},MaxSteps->Infinity]];
p1=ParametricPlot[Evaluate[{x[t],x[t-tau]}/.sol],{t,100,200}];
Show[{p1},AxesLabel->{"x(t)","x(t-\[Tau])"},
LabelStyle->{FontSize->20}],{\[Delta],0.5,1.5}]
```

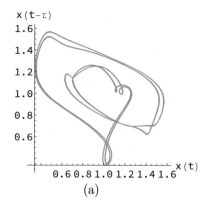

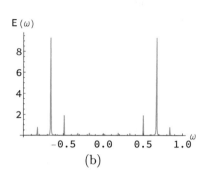

(a)   (b)

Figure 12.4: [Mathematica] Periodic behavior in the blood cell population model for equation (12.8) when $\beta = 2$, $n = 10$, $\tau = 2$, and $\delta = 0.8$. (a) Phase portrait. (b) Power spectrum.

The next programs show how to plot a phase portrait and the corresponding power spectrum. See Chapter 18 for more details on plotting power spectra using Fourier transforms.

```
(* Phase portrait and power spectrum of the Mackey-Glass model of *)
(* haematopoeisis. *)
(* See Figure 12.4. *)
beta=2;tau=2;n=10;delta=0.8;
sol=First[NDSolve[{x'[t]==beta x[t-tau]/(1+x[t-tau]^n)-delta x[t],
x[t/;t<=0]==1/2},x,{t,0,500},MaxSteps->Infinity]];
p1=ParametricPlot[Evaluate[{x[t],x[t-tau]}/.sol],{t,100,500}];
Show[{p1},AxesLabel->{"x(t)","x(t-\[Tau])"},
LabelStyle->{FontSize->20}]

TS=Table[x[i],{i,101,500,1}]/.sol;Nits=400;
p2=ListLinePlot[Rest[Transpose[{Range[-Nits/2+1,Nits/2]*(2/Nits),
Abs[Fourier[TS]]^2}]],PlotRange->All,
AxesOrigin->{-1,0},AxesLabel->{\[Omega],E(\[Omega])}];
Show[{p2},LabelStyle->20,AxesLabel->{\[Omega],E(\[Omega])}]
```

Figure 12.5(a) shows a phase portrait of equation (12.8), and Figure 12.5(b) shows the corresponding power spectrum for parameter values $\beta = 2$, $n = 10$, $\tau = 2$, and $\delta = 1$.

The next Mathematica program plots a bifurcation diagram for the Mackey-Glass model using the second iterative method, where the parameters β, τ, and n are kept constant and the parameter δ is increased and then decreased again.

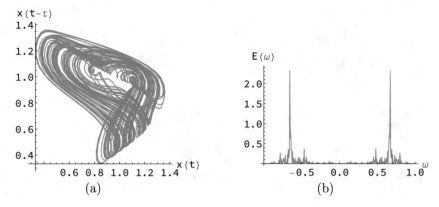

Figure 12.5: [Mathematica] Chaotic behavior in the blood cell population model for equation (12.8) when $\beta = 2$, $n = 10$, $\tau = 2$, and $\delta = 1$. (a) Phase portrait. (b) Power spectrum.

```
(* Bifurcation diagram of the Mackey-Glass model of haematopoeisis. *)
Init\[Delta]=0;M=18000;step=0.0001;Max\[Delta]=Init\[Delta]+M*step;
\[Tau]=2;\[Beta]=2;n=10;Tmax=\[Tau];x0=1/2;
(* Ramp parameter delta up *)
For[i=1,i<=M,i++,\[Delta]=Init\[Delta]+step*i;
MG=NDSolve[{x'[t]==\[Beta] x[t-\[Tau]]/(1+x[t-\[Tau]]^n)-\[Delta] x[t],
x[t/;t<=0]==x0},x,{t,0,Tmax},MaxSteps->Infinity];
x0=Flatten[x[Tmax] /. MG];rampup[i]=x0];
p1=ListPlot[Table[Flatten[{Init\[Delta]+m*step,rampup[m]}],
{m,1,M}],PlotRange->All];
(* Ramp parameter delta down *)
For[i=1,i<=M,i++,\[Delta]=Max\[Delta]-step*i;
MG=NDSolve[{x'[t]==\[Beta] x[t-\[Tau]]/(1+x[t-\[Tau]]^n)-\[Delta] x[t],
x[t/;t<=0]==x0},x,{t,0,Tmax},MaxSteps->Infinity];
x0=Flatten[x[Tmax] /. MG];rampdown[i]=x0];
p2=ListPlot[Table[Flatten[{Max\[Delta]-m*step,rampdown[m]}],
{m,1,M}],PlotRange->All];
Show[{p1,p2},PlotRange->All,AxesLabel->{\[Delta],"x(t)"},
AxesOrigin->{0,0},LabelStyle->{FontSize->15}]
```

Figure 12.6 shows the bifurcation diagram for system (12.8) obtained using the second iterative method with feedback. There are clearly regions of periodic behavior and regions of period doubling and un-doubling to and from chaos.

To complete the study of the Mackey-Glass system, consider the following modified model with two constant time delays:

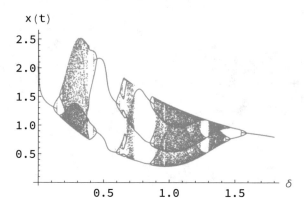

Figure 12.6: [Mathematica] Bifurcation diagram for system (12.8) as the parameter δ is increased linearly from $\delta = 0$ to $\delta = 1.8$, and then ramped back down again. Note that the single branches are depicting periodic solutions.

$$\frac{dx}{dt} = \frac{\beta x\,(t - \tau_1) + \beta x\,(t - \tau_2)}{1 + (x\,(t - \tau_1) + \beta x\,(t - \tau_2))^n} - \delta x(t), \qquad (12.9)$$

where the time delays are τ_1 and τ_2. Fix the parameters $\beta = 2.4$, $n = 10$, $\tau_1 = 2.4$, $\tau_2 = 6.2$, and $\delta = 2$. Figure 12.7 shows the phase portrait and corresponding power spectrum, and the system is clearly demonstrating quasiperiodic behavior. A *double Hopf bifurcation* is a local bifurcation in which a critical point has two pairs of purely imaginary eigenvalues. Generically, two branches of torus bifurcations evolve leading to periodic, quasiperiodic and chaotic behaviors. Equation (12.9) allows double Hopf bifurcations. The Mathematica program for producing Figure 12.7 is listed below.

```
(* Phase portrait of the modified Mackey-Glass model of haematopoeisis. *)
Clear[\[Tau]];
\[Tau]1=2.4;\[Tau]2=6.2;\[Beta]=2.4;\[Delta]=2;n=10;w1=1;w2=1;
sol=First[NDSolve[{x'[t]==\[Beta] *(w1*x[t-\[Tau]1]+
w2*x[t-\[Tau]2])/(1+(w1*x[t-\[Tau]1]+w2*x[t-\[Tau]2])^n)-
\[Delta]*x[t],x[t/;t<=0]==0.2},x,{t,0,500},MaxSteps->Infinity,
Method -> {"StiffnessSwitching"}]];
p1=ParametricPlot[Evaluate[{x[t],x[t-\[Tau]1]}/.sol],{t,200,500}];
Show[{p1},LabelStyle->20,AxesLabel->{"x(t)",x(t-Subscript[\[Tau], 1])}]
```

DDE Model of Angiogenesis. The final biological example presents a mathematical model of angiogenesis. Angiogenesis is the process where new blood vessels are formed from pre-existing vessels and is vital in growth and development and is part of wound healing and the formation of granulation tissue. Unfortunately, there are also negative effects of angiogenesis particularly in the promotion of cancer growth. Following the work of Hahnfeltd

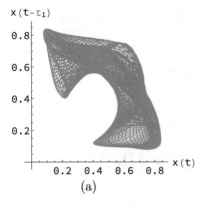

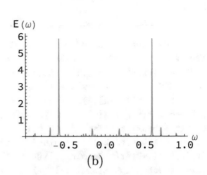

(a)   (b)

Figure 12.7: [Mathematica] Quasiperiodic behavior in the blood cell population model for equation (12.9) when $\beta = 2.4$, $n = 10$, $\tau_1 = 2.4$, $\tau_2 = 6.2$ and $\delta = 2$. (a) Phase portrait. (b) Power spectrum.

[13], Agur et al. [2], and Bodnar et al. [4], the following DDEs can be used to model cancer growth:

$$\frac{dN}{dt} = \alpha N(t)\left(1 - \frac{N(t)}{1 + f_1(E(t - \tau_1))}\right),$$

$$\frac{dP}{dt} = f_2(E(t))N(t) - \delta P(t),$$

$$\frac{dE}{dt} = f_3(P(t - \tau_2)) - \alpha\left(1 - \frac{N(t)E(t)}{1 + f_1(E(t - \tau_1))}\right), \qquad (12.10)$$

where N is the number of tumour cells or tumour size, P is the quantity growth factors known to be involved in supplying the tumour, and E represents the vessel density, where $E = \frac{V}{N}$, and V is the volume of blood vessels feeding the tumour. The functions f_1, f_2, f_3, model tumour cell proliferation rate, the protein production rate, and the vessel growth rate, respectively, and are given by:

$$f_1(E) = \frac{b_1 E^n}{c_1 + E^n}, \quad f_2(E) = \frac{a_2 c_2}{c_2 + E}, \quad f_3(P) = \frac{b_3\left(P^2 - m(t)^2\right)}{\frac{m(t)^2 b_3}{a_3} + P^2}. \qquad (12.11)$$

Take $\alpha = 1, a_2 = 0.4, a_3 = 1, b_1 = 2.3, b_3 = 1, c_1 = 1.5, c_2 = 1$, and $\delta = 0.34$. The parameter m alters the stimulation of tumour vessel production and is taken to be:

$$m(t) = 1.05 + 0.1\cos(t/120).$$

The Mathematica program used to produce the plots in Figure 12.8 is listed below:

```
(* Mathematical model of angiogenesis. See Figure 12.8. *)
a2=0.4;c1=1.5;b1=2.3; \[Delta] = 0.34; Tmin=2000; Tmax = \
5000; Subscript[\[Tau],1]=1;Subscript[\[Tau], 2] = 1;
Angio=NDSolve[{x'[t]==x[t]*(1-(x[t]/(1+b1
z[t-Subscript[\[Tau],1]]^2/(c1+z[t-Subscript[\[Tau],1]]^2)))),
y'[t]==a2*x[t]/(1+z[t])-\[Delta]*y[t],z'[t]==
(((y[t-Subscript[\[Tau],2]]^2 -(1.05+0.1*Cos[t/120])^2)/
((1.05+0.1*Cos[t/120])^2 + y[t-Subscript[\[Tau], 2]]^2))-
(1-(x[t]/(1+b1*z[t-Subscript[\[Tau],1]]^2/
(c1+z[t-Subscript[\[Tau],1]]^2)))))*z[t],
x[t /; t <= 0] == 0.4, y[t /; t <= 0] == 0,
z[t /; t <= 0] == 0.1}, {x[t], y[t], z[t]}, {t, 0, Tmax},
MaxSteps -> Infinity, Method->{"StiffnessSwitching"}];
p1 = ParametricPlot[Evaluate[{t, x[t]} /. Angio], {t, Tmin, Tmax},
PlotPoints -> 10000, PlotRange -> All,
AspectRatio -> 1/2, PlotLegends -> {"N(t)"}];
p2 = ParametricPlot[Evaluate[{t, y[t]} /. Angio], {t, Tmin, Tmax},
PlotPoints -> 10000, PlotRange -> All,
AspectRatio -> 1/2, PlotStyle -> Green, PlotLegends -> {"P(t)"}];
p3 = ParametricPlot[Evaluate[{t, z[t]} /. Angio], {t, Tmin, Tmax},
PlotPoints -> 10000, PlotRange -> All,
AspectRatio -> 1/2, PlotStyle -> Red, PlotLegends -> {"E(t)"}];
Show[{p1, p2, p3}, PlotRange -> All, AxesLabel -> {t, " "},
AxesOrigin -> {Tmin, 0}, LabelStyle -> {FontSize -> 15}]
```

Figure 12.8 shows a gallery of time series plots for the quantities $N(t)$, $P(t)$, and $E(t)$ for system (12.10) subject to conditions (12.11) for varying parameter values of τ_1 and τ_2 when driven by a periodic stimulation of tumour vessel production, $m(t) = 1.05 + 0.1\cos(t/120)$. In Figure 12.8(a), there is regular oscillatory behavior; in Figure 12.8(b), one can see that bursting is starting to form and becomes well established in Figure 12.8(c). In Figure 12.8(d), there is large amplitude modulation in all $N(t)$, $P(t)$, and $E(t)$. In Figure 12.8(e), there is intermitent fast spiking and finally, in Figure 12.8(f), there is large amplitude modulation in $E(t)$ but not in $P(t)$ or $N(t)$. The reader is encouraged to investigate the system as other parameters vary using the Mathematica program listed above.

12.3 Applications in Nonlinear Optics

Electromagnetic waves and optical resonators are discussed in detail in Chapter 16. Here, we present two DDE models from nonlinear optics. The first example is the Ikeda DDE modelling a nonlinear passive cavity in a ring

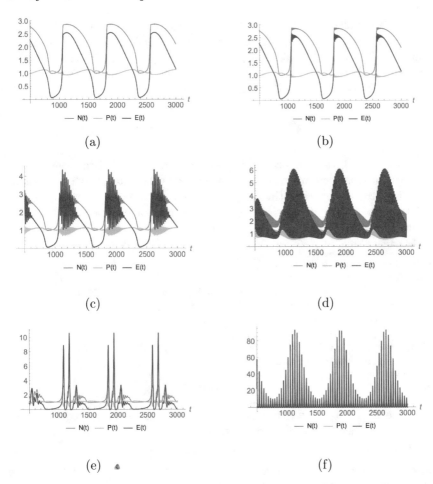

Figure 12.8: [Mathematica] A periodically forced angiogenesis model (12.10) with parameter values is defined by (12.11) and $m = 1.05 + 0.1\cos(t/120)$: (a) $\tau_1 = 1, \tau_2 = 1$, periodic; (b) $\tau_1 = 1, \tau_2 = 3$, bursting; (c) $\tau_1 = 1, \tau_2 = 5$, bursting; (d) $\tau_1 = 3, \tau_2 = 3$, amplitude modulation; (e) $\tau_1 = 1, \tau_2 = 15$, fast spiking; (f) $\tau_1 = 10, \tau_2 = 10$, large amplitude modulation in $E(t)$.

resonator, derived from the Maxwell-Debye equations for highly dispersive media. Readers should look at Chapter 16 for more details, and a figure of a ring cavity is given there. The coupled DDEs are given by:

$$E = A + BE\left(t - \tau_C\right)e^{i(\phi - \phi_0)}, \quad \tau\frac{d\phi}{dt} = -\phi + |E\left(t - \tau_C\right)|^2,$$

where E is the electric field strength in a ring cavity, ϕ is a phase shift as light propagates around the ring, ϕ_0 is a linear phase shift, A is related to input intensity, B is related to dissipation in the cavity, τ is the Debye relaxation time, and τ_C is the delay time of light in the ring. The coupled DDE may be simplified [9] to obtain the Ikeda DDE:

$$\tau \frac{d\phi}{dt} = -\phi + A^2 \left[1 + 2B \cos\left(\phi\left(t - \tau_C\right) - \phi_0\right)\right]. \qquad (12.12)$$

It is shown in [9] that as well as displaying chaos, Hopf bifurcations to sine wave oscillations occur for small time delays and that square wave oscillations occur for large delays. Fix the parameters $A = 2$, $B = 0.4$, $\tau = 0.8$, $\phi_0 = 0$ and vary the cavity delay time τ_C. A Mathematica program used to plot Figure 12.9 is listed below:

```
(* The Ikeda DDE. See Figure 12.9. *)
Manipulate[
Module[{tmax = 500, \[Tau]=0.8,A = 2,B=0.4,\[Phi]0=0,sol,\[Phi], t},
sol=First[\[Phi] /. NDSolve[{\[Tau]*\[Phi]'[t]==-\[Phi][t]+
A^2*(1+2*B*Cos[\[Phi][t-\[Tau]C]]-\[Phi]0),\[Phi][
t /; t <= 0]==1},\[Phi], {t,0,tmax}]];
If[pp,ParametricPlot[{sol[t],sol[t-1]},{t,1,tmax},PlotRange-> All],
Plot[sol[t],{t,0,tmax},PlotRange-> All,
AxesLabel-> {"t", "\[Phi]"}]]], {{pp, False,
"Phase Portrait"}, {False, True}}, {{\[Tau]C, 0.1}, 0, 50}]
```

Figure 12.9 shows two time series displaying chaos and a square wave oscillation. Readers will be asked to find a Hopf bifurcation to a limit cycle in the Exercises at the end of the chapter.

Delayed-Dynamical Optical Bistability Within and Without the Rotating Wave Approximation. The work to follow here is based on our recent papers [18] and [23]. We consider a bistable model of homogeneously broadened two-level atomic medium of length l placed inside a bulk ring cavity of length L. In the paper, it is shown that the DDEs for the field components $x_{0,\pm}$, are given by:

$$\frac{dx_0}{d\tau} = |y| - (1 + i\theta)x_0\left(\tau - \tau_0\right) - \frac{2Cx_0(\tau)(1 - i\delta)}{1 + \delta^2 + |x_0(\tau)|^2}$$

$$\frac{dx_+}{d\tau} = -(1 + i\theta)x_+\left(\tau - \tau_0\right) - \frac{i\eta}{\kappa}x_+(\tau) - \frac{2Cx_0^*(\tau)(1 + \delta^2)}{\left(1 + \delta^2 + |x_0(\tau)|^2\right)\left(1 + \frac{i\delta + 2i}{\lambda}\right)}$$

$$\frac{dx_-}{d\tau} = -(1 + i\theta)x_-\left(\tau - \tau_0\right) + \frac{i\eta}{\kappa}x_-(\tau), \qquad (12.13)$$

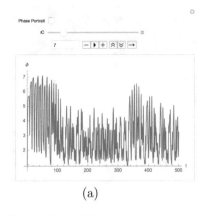
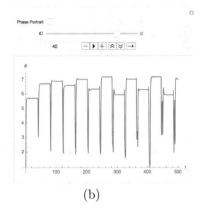

(a)   (b)

Figure 12.9: [Mathematica animation] The Ikeda DDE (12.12) for parameter values $A = 2$, $B = 0.4$, $\tau = 0.8$, $\phi_0 = 0$ and vary the cavity delay time τ_C. (a) $\tau_C = 7$; (b) $\tau_C = 40$.

where $|y|$ is the normalized input field amplitude; $\theta = \frac{\omega_c - \omega_L}{\kappa}$ is the normalized cavity detuning with cavity mode frequency ω_c, input field frequency ω_L, and κ is the cavity decay constant; $\tau = \frac{L-l}{c}$ is the cavity round trip delay, where c is the velocity of light in a vacuum, $C = \frac{g^2}{\gamma\kappa}$ is the cooperative parameter, where g is the coupling constant between the cavity field and the atoms, and γ is the A-coefficient; $\delta = \frac{2(\omega_0 - \omega_L)}{\gamma}$ is the normalized atomic detuning with the atomic transition frequency ω_0, and finally $\eta = 2\omega_L$. Note that $x_0(\tau)$ is the fundamental field component (of the same form within the Rotating Wave Approximation (RWA)) and the nonlinear source term for the first harmonic field component x_+ in equation (12.13) depends on x_0 and is of $O\left(\lambda^2\right)$, to the same order.

When the delay is zero $\tau = 0$, the system can display three different forms of hysteresis as shown if Figure 12.10.

Readers are directed to the research papers [18] and [23] to see how time delays affect the bistable operations depicted in Figure 12.11, however, in the main, time delays cause instabilities to encroach upon the bistable regions. There are some exceptions to the rule, however. Figure 12.11 shows a large butterfly bistable region for transverse Gaussian field effects within RWA when subject to a time delay. Readers can find further details in our paper [23].

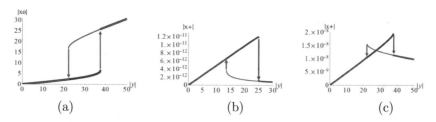

Figure 12.10: Bistable regions for system (12.13) when $\tau = 0$. (a) A counterclockwise bistable region in the fundamental field component when $\theta = 2$, $\delta = 8$, $C = 50$, $\gamma = 1$, $\kappa = 0.1$, and $\lambda = 10^{-9}$. (b) A clockwise bistable cycle for the first harmonic field component when $\theta = 0$, $\delta = 0$, $C = 24$, $\kappa = 0.1$, $\gamma = 1$, and $\lambda = 10^{-7}$. (c) A butterfly bistable region for the first harmonic field component when $\theta = 2$, $\delta = 8$, $C = 50$, $\gamma = 1$, $\kappa = 0.1$, and $\lambda = 10^{-9}$.

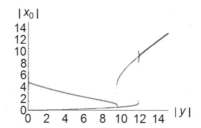

Figure 12.11: For certain parameter values with transverse Gaussian field effects within the RWA and with a non-zero time delay τ, a large butterfly bistable region, isolated from any instabilities, is clearly evident. Please see [23] for more details.

12.4 Other Applications

The ENSO Environmental Model. The first example presented is a simple nonlinear model for the El Niño and La Niña phenomena in the Eastern Pacific Ocean. The El Niño Southern Oscillation (ENSO) refers to the warming and cooling cycles of sea surface temperatures — the warm phase of the cycle is called El Niño and the cool phase is called La Niña. The ENSO events have been occurring for thousands of years and with the development of global warming and climate change over past decades the mathematical models have become increasingly important and heavily investigated. In 1988, Schopf and Suarez [22] devised a simple nonlinear DDE model for ENSO. Suppose that $T(t)$ represents the temperature anomaly, which represents a small perturbation from the long-term temperature average. The DDE model is given by:

$$\frac{dT}{dt} = T - T^3 - rT(t - \tau_w), \qquad (12.14)$$

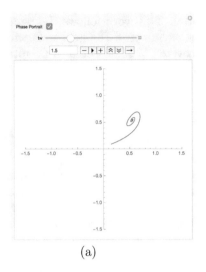
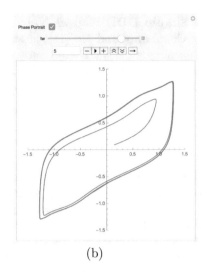

(a)   (b)

Figure 12.12: [Mathematica animation] Phase portraits of the ENSO model (12.14) when $r = 0.7$ and the initial history function is $T(t) = 0.1$. (a) When $\tau_w = 1.5$, there is no oscillation. (b) When $\tau_w = 5$, there is an oscillation. Note that readers can move the slider in the interactive Mathematica notebook.

where r represents the influence of the returning signal relative to local feedback, and τ_w is the nondimensional delay representing the wave transit time. Figure 12.12 shows that when $r = 0.7$, there is a bifurcation point where a limit cycle bifurcates (the bifurcation occurs at $\tau_w \approx 1.84$). The delay time in this model represents the time taken for equatorially trapped oceanic waves to propagate across the Pacific Ocean. When the delay time is below a certain threshold $\tau_w \approx 1.84$, in this case, then there is no oscillation, this could explain the lack of ENSO phenomena in smaller bodies of water such as the Atlantic and Indian Oceans. At the current time, the most cited paper on ENSO was published in 1997 [30] and there is also a lot of information published on the web. The Mathematica commands used to plot the phase portraits in Figure 12.12 is listed below.

```
(* The ENSO model. See Figure 12.12. *)
Manipulate[
Module[{tmax=100,r=0.7,sol,T,t},
sol = First[T /. NDSolve[{T'[t]==T[t]-T[t]^3-r*T[t-tw],
T[t /;t <= 0]==0.1},T,{t,0,tmax}]];
If[pp,ParametricPlot[{sol[t],sol[t-1]},{t,1,tmax},
PlotRange->{{-1.5,1.5}, {-1.5,1.5}}],
Plot[sol[t], {t,0,tmax},
PlotRange->{{0,tmax},{-1.5,1.5}}]]], {{pp,False,
"Phase Portrait"}, {False,True}},{{tw,0.1},0,6}]
```

A Simple DDE Neural Network Model. This work follows the paper by Gopalsamy and He [11] on stability in asymmetric Hopfield nets with transmission delays. Neural networks are covered in some detail in Chapter 20. A simple two-neuron coupled DDE network is given by:

$$\frac{dx}{dt} = -x(t) + a_{11} \tanh\left(x\left(t - \tau_1\right)\right) + a_{12} \tanh\left(y\left(t - \tau_2\right)\right)$$

$$\frac{dy}{dt} = -y(t) + a_{21} \tanh\left(x\left(t - \tau_1\right)\right) + a_{22} \tanh\left(y\left(t - \tau_2\right)\right), \quad (12.15)$$

where x, y are activation levels, a_{ij} denote couplings between the two neurons and τ_1, τ_2 are transmission delays. A linear stability analysis shows that the system can go through a Hopf bifurcation and a pitchfork bifurcation simultaneously. This is demonstrated in Figure 12.13, where the different dynamics of the system are shown in four plots. The Mathematica program for producing Figure 12.13(d) is listed below:

```
(* A two-neuron neural network DDE. See Figure 12.13(d). *)
a11=-1;a12=2;a21=3;a22=-1;t1=1;t2=1;tmax=10;
ode1[x0_, y0_]:=
NDSolve[{x'[t]==-x[t]+a11*Tanh[x[t-t1]]+a12*Tanh[y[t-t2]],
y'[t]==-y[t]+a21*Tanh[x[t-t1]]+a22*Tanh[y[t-t2]],
x[t /; t <= 0]==x0,y[t /; t <=0]==y0},{x[t],y[t]},{t,0,tmax}];
sol[1]=ode1[2,1];sol[2]=ode1[-1,1];sol[3]=ode1[-1,-1] ;
ParametricPlot[Evaluate[Table[{t,x[t]} /. sol[i], {i,3}]], {t,0,tmax},
PlotRange->All,PlotPoints->100,AxesLabel->{x, y},
LabelStyle->Directive[Black,Large]]
```

Figure 12.13 shows solutions to the two-neuron neural network DDE (12.15) with varying parameters and initial history functions. (a) Phase portrait showing that when $a_{11} = a_{12} = -1, a_{21} = 2, a_{22} = 1, \tau_1 = \tau_2 = 1$, and the initial history functions, $x(t) = y(t) = 0.5$, $(x(t), y(t))$ approaches a stable critical point at the origin. (b) Phase portrait showing that when $a_{11} = -1, a_{12} = a_{21} = -2, a_{22} = -3, \tau_1 = \tau_2 = 1$, and the initial history functions, $x(t) = y(t) = 0.5$, $(x(t), y(t))$ approaches in-phase oscillation. (c) Phase portrait showing that when $a_{11} = 0.33, a_{12} = -1, a_{21} = 2, a_{22} = 0.34, \tau_1 = \tau_2 = 1$, and the initial history functions, $x(t) = y(t) = 0.5$, $(x(t), y(t))$ approaches out of phase oscillation. (d) Three time series showing that when $a_{11} = -1, a_{12} = 2, a_{21} = 3, a_{22} = -1, \tau_1 = \tau_2 = 1$, and the initial history functions (i) $x(t) = 2, y(t) = 1$, (ii) $x(t) = -1, y(t) = 1$ and (ii) $x(t) = -1, y(t) = -1$, there are two stable critical points and one stable oscillation in $x(t)$.

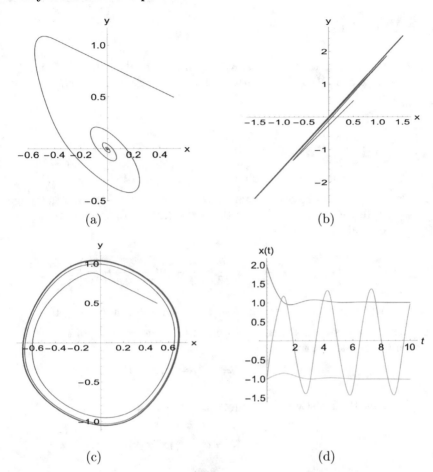

Figure 12.13: [Mathematica] A gallery of plots for system (12.12) for varying parameter values and initial history functions. (a) The solution approaches the stable critical point at the origin; (b) the system settles on to two oscillating solutions which are in-phase; (c) the system has settled on to an oscillatory solution where $x(t)$ and $y(t)$ are now out of phase; (d) the system has co-existing stable critical points and oscillatory solutions which depend upon the initial history functions.

12.5 Exercises

1. Use the method of steps to show that an analytical solution in the range $[-1, 2]$ to the DDE (12.2) with the initial history function $x(t) = e^t$ on $[-1, 0]$ is given by:

$$\left\{ e^t, 1 - \frac{e^t - 1}{e}, 2 - \frac{e - 1}{e} + e^{t-2} - t - \frac{t}{e} \right\}.$$

   Use Mathematica to determine an analytical solution on $[-1, 4]$.

2. Use the method of steps to show that an analytical solution in the range $[-1, 2]$ to the DDE (12.2) with the initial history function $x(t) = t^2$ on $[-1, 0]$ is given by:

$$\left\{ t^2, -t + t^2 - t^3/3, \frac{7}{12} - \frac{7t}{3} + 2t^2 - \frac{2t^3}{3} + \frac{t^4}{12} \right\}.$$

   Use Mathematica to determine an analytical solution on $[-1, 4]$.

3. Use a linear stability analysis to prove that the critical point $\mathbf{x}^* = 0$ is stable for the system

$$\frac{dx}{dt} = -x(t - 1).$$

4. Consider the logistic DDE subject to two delays:

$$\frac{dx}{dt} = -\mu x(t) \left(1 - x(t - \tau_1) - x(t - \tau_2) \right).$$

   Show that a necessary and sufficient condition for the critical point at $\mathbf{x}^* = \frac{1}{2}$ to be stable is

$$\mu (\tau_1 + \tau_2) \cos \left(\frac{\pi (\tau_1 - \tau_2)}{2 (\tau_1 + \tau_2)} \right) < \pi.$$

   Use Mathematica to show a Hopf bifurcation of a limit cycle for suitable parameter values.

5. Plot a bifurcation diagram for the modified Mackey-Glass system (12.9) when $\beta = 2.4$, $\delta = 2$, $n = 10$, $\tau_1 = 2.4$, as the parameter τ_2 is linearly ramped up from $\tau_2 = 0$ to $\tau_2 = 10$, and then ramped down again.

6. Perform a linear stability analysis on the Ikeda DDE (12.12) and show that there exists both stable and unstable Hopf bifurcation points.

7. A simple ENSO environmental model with a global warming effect, W, say, is given by:

$$\frac{dT}{dt} = T(t) - T(t)^3 - rT(t - \tau_w) + W.$$

Using the same parameters as those given in equation (12.14), investigate how the global warming term W affects the model.

8. Consider the periodically forced mechanical oscillator described by the equations:

$$\frac{d^2x}{dt^2} + a\left(\frac{dx}{dt}\right)^3 - b\frac{dx}{dt} + cx = \Gamma\cos(\omega t) + A\left(\dot{x}_\tau\right) + B\left(\dot{x}_\tau\right)^3,$$

where constants a, b, c, and Γ are positive, $x_\tau = x(t - \tau)$, and τ is a time delay. The feedback is negative if $A, B < 0$, and positive if $A, B > 0$. Suppose that $a = 0.3, b = 2, c = 2, \omega = 2, A = 6, B = 2, \Gamma = 9$, and vary the delay τ. Plot phase portraits of $x(t)$ versus $\dot{x}(t)$ when (i) $\tau = 0.4$; (ii) $\tau = 0.8$; (iii) $\tau = 1.2$. Describe the behavior in each case. Read paper [28], where a study of double Hopf bifurcation and chaos for this system is analyzed.

9. Consider the simple mathematical model of an inverted pendulum that is balanced using linear feedback control [24]. Think of an inverted pendulum that is pivoted on a cart and the cart can move horizontally left and right. The system is modeled by a pair of DDEs:

$$\frac{d^2\theta}{dt^2} = \sin(\theta) - F\cos(\theta), \quad \frac{d^2\delta}{dt^2} = \frac{2}{3}LF,$$

$$F = \left(c_1\theta(t - \tau) + c_2\frac{d\theta}{dt}(t - \tau)\right),$$

where θ is an angular displacement of the inverted pendulum, c_1, c_2 are feedback control gains, δ is the horizontal displacement from the pivot point, F is a delayed feedback control force, L is the length of the pendulum, and τ is a control loop latency. Use Mathematica to investigate the system and show that there are parameter regions which show (i) stabilization to $\theta = 0$; (ii) small oscillations about $\theta = 0$; (iii) runaway oscillations, where δ gets large, and (iv) chaotic solutions.

10. Consider the economic model of neoclassical growth with both workers and capitalists [12], using the Cobb-Douglas production function, $f(k_t) = k_t^\alpha$, where $\alpha \in (0, 1)$, given by:

$$\frac{dk}{dt} = (s_W + \alpha(s_C - s_W))k(t - \tau)^\alpha - nk(t - \tau),$$

where $k(t)$ denotes capital per worker, s_W and s_C are the propensities for workers and capitalists to save, respectively, $n > 0$ is the constant labor force growth rate and τ represents a time lag in the production technology. Prove that there is a unique critical point at $k^* = \left(\frac{(s_W + \alpha(s_C - s_W))}{n} \right)^{\frac{1}{1-\alpha}}$, and prove that there is a Hopf bifurcation point at $\tau = \frac{\pi}{2n(1-\alpha)}$.

Bibliography

[1] R.P. Agarwal, D. O'Regan and S.H. Saker, *Oscillation and Stability of Delay Models in Biology*, Springer, New York, 2016.

[2] Z. Agur, L. Arakelyan, P. Daugulis and Y. Dinosaur, Hopf point analysis for angiogenesis models, *Discrete and Continuous Dynamical Systems-Series B* **4**, (2004), 29–38.

[3] B.J. Altman, Cancer Clocks Out for Lunch: Disruption of Circadian Rhythm and Metabolic Oscillation in Cancer, *Front. Cell Dev. Bill.* 4:62. doi:10.3389/fcell.2016.00062.eCollection (2016).

[4] M. Bodnar, M.J. Piotrowska, U. Forys and E. Nizinska, Model of tumour angiogenesis - analysis of stability with respect to delays, *Math. Biosciences and Eng*, **10**, (2013), 19–35.

[5] K. Chakraborty, M. Chakraborty and T.K. Kar, Bifurcation and control of a bioeconomic model of a prey-predator system with a time delay, *Nonlinear Analysis-Hybrid Systems* **5**, (2011), 613–625.

[6] K.L. Cooke and P. van den Driessche, Analysis of an SEIRS epidemic model with two delays, *J. Math. Biol.* **35** (1996), 240–260.

[7] O. Diekmann, S. A. van Gils, S. M. V. Lunel and H. O Walther, *Delay Equations: Functional-, Complex-, and Nonlinear Analysis*, Springer, New York, 1995.

[8] T. Erneux, *Applied Delay Differential Equations*, Springer, New York, 2009.

[9] T. Erneux, L. Larger, M.W. Lee and J.P. Goedgebuer, Ikeda Hopf bifurcation revisited, *Physica D* **194**, (2004), 49–64.

[10] T. Faria, Stability and bifurcation for a delayed predator-prey model and the effect of diffusion, *J. of Math. Anal. and Appl.* **254** (2001), 433–463.

[11] K. Gopalsamy and X.Z.He, Stability in asymmetric Hopfield nets with transmission delays, *Physica D* **76**, (1994), 344–358.

[12] L. Guerrini, Time delay in a neoclassical growth model with differential saving, *Int. J. of Pure and Applied Mathematics* **78**, (2012), 999–1003.

[13] P. Hahnfeltd, D. Paigrahy, J. Folkman and L. Hlatky, Tumor development under angiogenic signaling: a dynamical theory of tumor growth, treatment response, and postvascular dormancy, *Cancer Res.* **59**, (1999), 4770–4775.

[14] G. Huang, Y. Takeuchi, W. Ma et al., Global Stability for Delay SIR and SEIR Epidemic Models with Nonlinear Incidence Rate, *Bulletin Math. Biol.* **72** (2010), 1192–1207.

[15] G.E. Hutchinson, Circular causal system in ecology, *Ann. New York Acad. Sci.* **50**, (1948), 221–246.

[16] M. Lakshmanan and D.V. Senthilkumar, *Dynamics of Nonlinear Time-Delay Systems*, Springer, New York, 2011.

[17] S. Lynch and J. Borresen J, Oscillations, feedback and bifurcations in mathematical models of angiogenesis and haematopoiesis, in *Handbook of Vascular Biology Techniques*, Slevin M, McDowell G, Cao Y, Kitajewski J eds., Springer, New York, (2015), 373–390.

[18] S. Lynch, R.A. Alharbey, S.S. Hassan and H.A. Batarfi, Delayed-dynamical bistability within and without rotating wave approximation, *J. of Nonlinear Optical Physics and Materials* **24**, (2015), 1550037.

[19] M.C. Mackey and L. Glass, Oscillation and chaos in physiological control systems, *Science* **197** (1977) 287–289.

[20] P.E. Rap, An atlas of cellular oscillators, *J. Exp. Biol.* **81** (1979), 281–306.

[21] E. Schmitt, Uber eine Klasse linearer funktionaler Differentialgleichungen, *Math. Ann.* **70**, (1911), 499–524.

[22] P.S. Schopf and M.J. Suarez (1988) Vacillations in a coupled ocean-atmosphere model, *J. Atmos. Sci.* 45 (1988), 549–566.

[23] Y.A. Sharaby, S. Lynch and S.S. Hassan, Inhomogeneous and transverse field effects on time delayed optical bistability inside and outside the rotating wave approximation, *J. of Nonlinear Optical Physics and Materials* **25**, (2016), 1650021.

[24] J. Sieber and B. Krauskopf, Complex balancing motions of an inverted pendulum subject to delayed feedback control, *Physica D* **197**, (2004), 332–345.

[25] H. Smith, *An Introduction to Delay Differential Equations with Applications to the Life Sciences*, Springer, New York, 2010.

[26] Y.N. Xiao and L.S. Chen, Modeling and analysis of a predator-prey model with disease in the prey, *Math. Biosciences* **171** (2001), 59–82.

[27] S. Yom-Tov and I. Golani, Oscillators in the human body and circular-muscle gymnastics, *Med. Hypothesis* **41** (1993), 118–122.

[28] P. Yu, Y. Yuan and J. Xu, Study of double Hopf bifurcation and chaos for an oscillator with time delayed feedback, *Communications in Nonlinear Science and Numerical Simulation* **7**, (2002), 69–91.

[29] H. Zhang, L. Chen and J.J. Nieto, A delayed epidemic model with stage-structure and pulses for pest management strategy, *Nonlin. Anal. Real-World Appl.* **9** (2008), 1714-1726.

[30] Y. Zhang, J.M. Wallace and D.S. Battisti, DS, ENSO-like interdecadal variability: 1900-93, *J. of Climate* **10** (1997), 1004–1020.

Chapter 13
Linear Discrete Dynamical Systems

Aims and Objectives

- To introduce recurrence relations for first- and second-order difference equations.

- To introduce the theory of the Leslie model.

- To apply the theory to modeling the population of a single species.

On completion of this chapter, the reader should be able to

- solve first- and second-order homogeneous linear difference equations;

- find eigenvalues and eigenvectors of matrices;

- model a single population with different age classes;

- predict the long-term rate of growth/decline of the population;

- investigate how harvesting and culling policies affect the model.

This chapter deals with linear discrete dynamical systems, where time is measured by the number of iterations carried out and the dynamics are not continuous. In applications this would imply that the solutions are observed at discrete time intervals.

Recurrence relations can be used to construct mathematical models of discrete systems. They are also used extensively to solve many differential equations which do not have an analytic solution; the differential equations are represented by recurrence relations (or difference equations) that can be solved numerically on a computer. Of course one has to be careful when considering the accuracy of the numerical solutions. Ordinary differential equations are used to model continuous dynamical systems in the first part of the book. More information on discrete systems can be found in the textbooks [1], [2] and [5].

The bulk of this chapter is concerned with a linear discrete dynamical system that can be used to model the population of a single species. As with continuous systems, in applications to the real world, linear models generally produce good results over only a limited range of time. The Leslie model introduced here is

© Springer International Publishing AG 2017   285
S. Lynch, *Dynamical Systems with Applications using Mathematica®*,
DOI 10.1007/978-3-319-61485-4_13

useful when establishing harvesting and culling policies. References [3], [4], [6], [7], [8], [9], and [10] are concerned with the Leslie model.

Nonlinear discrete dynamical systems will be discussed in the next chapter, and the Poincaré maps introduced in Chapter 9, for example, illustrate how discrete systems can be used to help in the understanding of how continuous systems behave.

13.1 Recurrence Relations

This section is intended to give the reader a brief introduction to *difference equations* and illustrate the theory with some simple models.

First-Order Difference Equations

A *recurrence relation* can be defined by a difference equation of the form

$$x_{n+1} = f(x_n), \tag{13.1}$$

where x_{n+1} is derived from x_n and $n = 0, 1, 2, 3, \ldots$. If one starts with an initial value, say, x_0, then *iteration* of equation (13.1) leads to a sequence of the form

$$\{x_i : i = 0 \text{ to } \infty\} = \{x_0, x_1, x_2, \ldots, x_n, x_{n+1}, \ldots\}.$$

In applications, one would like to know how this sequence can be interpreted in physical terms. Equations of the form (13.1) are called *first-order difference equations* because the suffices differ by one. Consider the following simple example.

Example 1. The difference equation used to model the interest in a bank account compounded once per year is given by

$$x_{n+1} = \left(1 + \frac{3}{100}\right) x_n, \quad n = 0, 1, 2, 3, \ldots.$$

Find a general solution and determine the balance in the account after five years given that the initial deposit is 10,000 dollars and the interest is compounded annually.

Solution. Using the recurrence relation

$$x_1 = \left(1 + \frac{3}{100}\right) \times 10,000,$$

$$x_2 = \left(1 + \frac{3}{100}\right) \times x_1 = \left(1 + \frac{3}{100}\right)^2 \times 10,000,$$

and, in general,

$$x_n = \left(1 + \frac{3}{100}\right)^n \times 10,000,$$

where $n = 0, 1, 2, 3, \ldots$. Given that $x_0 = 10,000$ and $n = 5$, the balance after five years will be $x_5 = 11,592.74$ dollars.

Theorem 1. *The general solution of the first-order linear difference equation*

$$x_{n+1} = mx_n + c, \quad n = 0, 1, 2, 3, \ldots, \tag{13.2}$$

is given by

$$x_n = m^n x_0 + \begin{cases} \frac{m^n - 1}{m-1} c & \text{if } m \neq 1 \\ nc & \text{if } m = 1. \end{cases}$$

Proof. Applying the recurrence relation given in (13.2)

$$x_1 = mx_0 + c,$$

$$x_2 = mx_1 + c = m^2 x_0 + mc + c,$$

$$x_3 = mx_2 + c = m^3 x_0 + m^2 c + mc + c,$$

and the pattern in general is

$$x_n = m^n x_0 + (m^{n-1} + m^{n-2} + \ldots + m + 1)c.$$

Using geometric series, $m^{n-1} + m^{n-2} + \ldots + m + 1 = \frac{m^n - 1}{m-1}$, provided that $m \neq 1$. If $m = 1$, then the sum of the geometric sequence is n. This concludes the proof of Theorem 1. Note that, if $|m| < 1$ then $x_n \to \frac{c}{1-m}$ as $n \to \infty$. □

Second-Order Linear Difference Equations
Recurrence relations involving terms whose suffices differ by two are known as *second-order linear difference equations*. The general form of these equations with constant coefficients is

$$ax_{n+2} = bx_{n+1} + cx_n. \tag{13.3}$$

Theorem 2. *The general solution of the second-order recurrence relation (13.3) is*

$$x_n = k_1 \lambda_1^n + k_2 \lambda_2^n,$$

where k_1, k_2 are constants and $\lambda_1 \neq \lambda_2$ are the roots of the quadratic equation $a\lambda^2 - b\lambda - c = 0$. If $\lambda_1 = \lambda_2$, then the general solution is of the form

$$x_n = (k_3 + nk_4)\lambda_1^n.$$

Note that when λ_1 and λ_2 are complex, the general solution can be expressed as

$$x_n = k_1\lambda_1^n + k_2\lambda_2^n = k_1(re^{i\theta})^n + k_2(re^{-i\theta})^n = r^n \left(A\cos(n\theta) + B\sin(n\theta) \right),$$

where A and B are constants. When the eigenvalues are complex, the solution oscillates and is real.

Proof. The solution of system (13.2) gives us a clue where to start. Assume that $x_n = \lambda^n k$ is a solution, where λ and k are to be found. Substituting, equation (13.3) becomes

$$a\lambda^{n+2}k = b\lambda^{n+1}k + c\lambda^n k$$

or

$$\lambda^n k(a\lambda^2 - b\lambda - c) = 0.$$

Assuming that $\lambda^n k \neq 0$, this equation has solutions if

$$a\lambda^2 - b\lambda - c = 0. \tag{13.4}$$

Equation (13.4) is called the *characteristic equation*. The difference equation (13.3) has two solutions and because the equation is linear, a solution is given by

$$x_n = k_1\lambda_1^n + k_2\lambda_2^n,$$

where $\lambda_1 \neq \lambda_2$, are the roots of the characteristic equation.

If $\lambda_1 = \lambda_2$, then the characteristic equation can be written as

$$a\lambda^2 - b\lambda - c = a(\lambda - \lambda_1)^2 = a\lambda^2 - 2a\lambda_1\lambda + a\lambda_1^2.$$

Therefore, $b = 2a\lambda_1$ and $c = -a\lambda_1^2$. Now assume that another solution is of the form $kn\lambda^n$. Substituting, equation (13.3) becomes

$$ax_{n+2} - bx_{n+1} - cx_n = a(n+2)k\lambda_1^{n+2} - b(n+1)k\lambda_1^{n+1} - cnk\lambda_1^n,$$

therefore

$$ax_{n+2} - bx_{n+1} - cx_n = kn\lambda_1^n(a\lambda_1^2 - b\lambda_1 - c) + k\lambda_1(2a\lambda_1 - b),$$

which equates to zero from the above. This confirms that $kn\lambda_n$ is a solution to equation (13.3). Since the system is linear, the general solution is thus of the form

$$x_n = (k_3 + nk_4)\lambda_1^n.$$ □

The values of k_j can be determined if x_0 and x_1 are given. Consider the following simple examples.

Example 2. Solve the following second-order linear difference equations:

(i) $x_{n+2} = x_{n+1} + 6x_n, n = 0, 1, 2, 3, \ldots$, given that $x_0 = 1$ and $x_1 = 2$;

(ii) $x_{n+2} = 4x_{n+1} - 4x_n, n = 0, 1, 2, 3, \ldots$, given that $x_0 = 1$ and $x_1 = 3$;

(iii) $x_{n+2} = x_{n+1} - x_n, n = 0, 1, 2, 3, \ldots$, given that $x_0 = 1$ and $x_1 = 2$.

Solution. (i) The characteristic equation is

$$\lambda^2 - \lambda - 6 = 0,$$

which has roots at $\lambda_1 = 3$ and $\lambda_2 = -2$. The general solution is therefore

$$x_n = k_1 3^n + k_2(-2)^n, \quad n = 0, 1, 2, 3, \ldots.$$

The constants k_1 and k_2 can be found by setting $n = 0$ and $n = 1$. The final solution is

$$x_n = \frac{4}{5}3^n + \frac{1}{5}(-2)^n, \quad n = 0, 1, 2, 3, \ldots.$$

(ii) The characteristic equation is

$$\lambda^2 - 4\lambda + 4 = 0,$$

which has a repeated root at $\lambda_1 = 2$. The general solution is

$$x_n = (k_3 + k_4 n)2^n, \quad n = 0, 1, 2, 3, \ldots.$$

Substituting for x_0 and x_1 gives the solution

$$x_n = \left(1 + \frac{n}{2}\right)2^n, \quad n = 0, 1, 2, 3, \ldots.$$

(iii) The characteristic equation is

$$\lambda^2 - \lambda + 1 = 0,$$

which has complex roots $\lambda_1 = \frac{1}{2} + i\frac{\sqrt{3}}{2} = e^{\frac{i\pi}{3}}$ and $\lambda_2 = \frac{1}{2} - i\frac{\sqrt{3}}{2} = e^{\frac{-i\pi}{3}}$. The general solution is

$$x_n = k_1\lambda_1^n + k_2\lambda_2^n, \quad n = 0, 1, 2, 3, \ldots.$$

Substituting for λ_1 and λ_2, the general solution becomes

$$x_n = (k_1 + k_2)\cos\left(\frac{n\pi}{3}\right) + i(k_1 - k_2)\sin\left(\frac{n\pi}{3}\right), \quad n = 0, 1, 2, 3, \ldots.$$

Substituting for x_0 and x_1, gives $k_1 = \frac{1}{2} - \frac{i}{2\sqrt{3}}$ and $k_2 = \frac{1}{2} + \frac{i}{2\sqrt{3}}$, and so

$$x_n = \cos\left(\frac{n\pi}{3}\right) + \sqrt{3}\sin\left(\frac{n\pi}{3}\right), \quad n = 0, 1, 2, 3, \ldots.$$

Example 3. Suppose that the national income of a small country in year n is given by $I_n = S_n + P_n + G_n$, where S_n, P_n, and G_n represent national spending by the populous, private investment, and government spending, respectively. If the national income increases from one year to the next, then assume that consumers will spend more the following year; in this case, suppose that consumers spend $\frac{1}{6}$ of the previous year's income, then $S_{n+1} = \frac{1}{6}I_n$. An increase in consumer spending should also lead to increased investment the following year, assume that $P_{n+1} = S_{n+1} - S_n$. Substitution for S_n then gives $P_{n+1} = \frac{1}{6}(I_n - I_{n-1})$. Finally, assume that the government spending is kept constant. Simple manipulation then leads to the following economic model

$$I_{n+2} = \frac{5}{6}I_{n+1} - \frac{1}{6}I_n + G, \tag{13.5}$$

where I_n is the national income in year n, and G is a constant. If the initial national income is G dollars and one year later is $\frac{3}{2}G$ dollars, determine

(i) a general solution to this model;

(ii) the national income after 5 years; and

(iii) the long-term state of the economy.

Solution. (i) The characteristic equation is given by

$$\lambda^2 - \frac{5}{6}\lambda + \frac{1}{6} = 0,$$

which has solutions $\lambda_1 = \frac{1}{2}$ and $\lambda_2 = \frac{1}{3}$. Equation (13.5) also has a constant term G. Assume that the solution involves a constant term also; try $I_n = k_3 G$, then from equation (13.5)

$$k_3 G = \frac{5}{6} k_3 G - \frac{1}{6} k_3 G + G,$$

and so $k_3 = \frac{1}{1 - \frac{5}{6} + \frac{1}{6}} = 3$. Therefore, a general solution is of the form

$$I_n = k_1 \lambda_1^n + k_2 \lambda_2^n + 3G.$$

(ii) Given that $I_0 = G$ and $I_1 = \frac{3}{2}G$, simple algebra gives $k_1 = -5$ and $k_2 = 3$. When $n = 5$, $I_5 = 2.856G$, to three decimal places.

(iii) As $n \to \infty$, $I_n \to 3G$, since $|\lambda_1| < 1$ and $|\lambda_2| < 1$. Therefore, the economy stabilizes in the long term to a constant value of $3G$. This is obviously a very crude model.

A general n-dimensional linear discrete population model is discussed in the following sections using matrix algebra.

13.2 The Leslie Model

The Leslie model was developed around 1940 to describe the population dynamics of the female portion of a species. For most species, the number of females is equal to the number of males, and this assumption is made here. The model can be applied to human populations, insect populations, and animal and fish populations. The model is an example of a discrete dynamical system. As explained throughout the text, we live in a nonlinear world and universe; since this model is linear, one would expect the results to be inaccurate in the long term. However, the model can give some interesting results, and it incorporates some features not discussed in later chapters. The following characteristics are ignored—diseases, environmental effects, and seasonal effects. The book [9] provides an extension of the Leslie model, where individuals exhibit migration characteristics. A nonlinear Leslie matrix model for predicting the dynamics of biological populations in polluted environments is discussed in [7].

Assumptions: The females are divided into n age classes; thus, if N is the theoretical maximum age attainable by a female of the species, then each age class will span a period of $\frac{N}{n}$ equally spaced, days, weeks, months, years, etc. The population is observed at regular discrete time intervals which are each equal to the length of one age class. Thus, the kth time period will be given by $t_k = \frac{kN}{n}$. Define $x_i^{(k)}$ to be the number of females in the ith age class after the kth time period. Let b_i denote the number of female offspring born to one female during the ith age class, and let c_i be the proportion of females which continue to survive from the ith to the $(i+1)$st age class.

In order for this to be a realistic model, the following conditions must be satisfied:

$$(i) \quad b_i \geq 0, \qquad 1 \leq i \leq n;$$
$$(ii) \ 0 < c_i \leq 1, \qquad 1 \leq i < n.$$

Obviously, some b_i have to be positive in order to ensure that some births do occur and no c_i are zero, otherwise there would be no females in the $(i+1)$st age class.

Working with the female population as a whole, the following sets of linear equations can be derived. The number of females in the first age class after the kth time period is equal to the number of females born to females in all n age classes between the time t_{k-1} and t_k; thus

$$x_1^{(k)} = b_1 x_1^{(k-1)} + b_2 x_2^{(k-1)} + \ldots + b_n x_n^{(k-1)}.$$

The number of females in the $(i+1)$st age class at time t_k is equal to the number of females in the ith age class at time t_{k-1} who continue to survive to enter the $(i+1)$st age class, hence

$$x_{i+1}^{(k)} = c_i x_i^{(k-1)}.$$

Equations of the above form can be written in matrix form, and so

$$
\begin{pmatrix}
x_1^{(k)} \\
x_2^{(k)} \\
x_3^{(k)} \\
\vdots \\
x_n^{(k)}
\end{pmatrix}
=
\begin{pmatrix}
b_1 & b_2 & b_3 & \cdots & b_{n-1} & b_n \\
c_1 & 0 & 0 & \cdots & 0 & 0 \\
0 & c_2 & 0 & \cdots & 0 & 0 \\
\vdots & \vdots & \vdots & \ddots & \vdots & \vdots \\
0 & 0 & 0 & \cdots & c_{n-1} & 0
\end{pmatrix}
\begin{pmatrix}
x_1^{(k-1)} \\
x_2^{(k-2)} \\
x_3^{(k-1)} \\
\vdots \\
x_n^{(k-1)}
\end{pmatrix},
$$

or

$$X^{(k)} = L X^{(k-1)}, \qquad k = 1, 2, \ldots,$$

where $X \in \Re^n$ and the matrix L is called the *Leslie matrix*.

Suppose that $X^{(0)}$ is a vector giving the initial number of females in each of the n age classes, then

$$X^{(1)} = L X^{(0)},$$

$$X^{(2)} = L X^{(1)} = L^2 X^{(0)},$$

$$\vdots$$

$$X^{(k)} = L X^{(k-1)} = L^k X^{(0)}.$$

Therefore, given the initial age distribution and the Leslie matrix L, it is possible to determine the female age distribution at any later time interval.

Example 4. Consider a species of bird that can be split into three age-groupings: those aged 0–1 year, those aged 1–2 years, and those aged 2–3 years. The population is observed once a year. Given that the Leslie matrix is equal to

$$L = \begin{pmatrix} 0 & 3 & 1 \\ 0.3 & 0 & 0 \\ 0 & 0.5 & 0 \end{pmatrix},$$

and the initial population distribution of females is $x_1^{(0)} = 1000$, $x_2^{(0)} = 2000$, and $x_3^{(0)} = 3000$, compute the number of females in each age-group after

(a) 10 years;

(b) 20 years;

(c) 50 years.

Solution. Using the above,

$$(a)\ X^{(10)} = L^{10}X^{(0)} = \begin{pmatrix} 5383 \\ 2177 \\ 712 \end{pmatrix},$$

$$(b)\ X^{(20)} = L^{20}X^{(0)} = \begin{pmatrix} 7740 \\ 2388 \\ 1097 \end{pmatrix},$$

$$(c)\ X^{(50)} = L^{50}X^{(0)} = \begin{pmatrix} 15695 \\ 4603 \\ 2249 \end{pmatrix}.$$

The numbers are rounded down to whole numbers since it is not possible to have a fraction of a living bird. Obviously, the populations cannot keep on growing indefinitely. However, the model does give useful results for some species when the time periods are relatively short.

In order to investigate the limiting behavior of the system, it is necessary to consider the eigenvalues and eigenvectors of the matrix L. These can be used to determine the eventual population distribution with respect to the age classes.

Theorem 3. *Let the Leslie matrix L be as defined above and assume that*

(a) $b_i \geq 0$ *for* $1 \leq i \leq n$;

(b) *at least two succesive b_i are strictly positive; and*

(c) $0 < c_i \leq 1$ *for* $1 \leq i < n.$

Then,

(i) *matrix* L *has a unique positive eigenvalue, say,* λ_1;

(ii) λ_1 *is simple, or has algebraic multiplicity one;*

(iii) *the eigenvector—X_1, say—corresponding to λ_1, has positive components;*

(iv) *any other eigenvalue,* $\lambda_i \neq \lambda_1$, *of L satisfies*
$$|\lambda_i| < \lambda_1,$$

and the positive eigenvalue λ_1 is called strictly dominant.

The reader will be asked to prove part (i) in the exercises at the end of the chapter.

If the Leslie matrix L has a unique positive strictly dominant eigenvalue, then an eigenvector corresponding to λ_1 is a nonzero vector solution of

$$LX = \lambda_1 X.$$

Assume that $x_1 = 1$, then a possible eigenvector corresponding to λ_1 is given by

$$X_1 = \begin{pmatrix} 1 \\ \frac{c_1}{\lambda_1} \\ \frac{c_1 c_2}{\lambda_1^2} \\ \vdots \\ \frac{c_1 c_2 \ldots c_{n-1}}{\lambda_1^{n-1}} \end{pmatrix}.$$

Assume that L has n linearly independent eigenvectors, say, $X_1, X_2, \ldots, X_n$. Therefore, L is diagonalizable. If the initial population distribution is given by $X^{(0)} = X_0$, then there exist constants $b_1, b_2, \ldots, b_n$, such that

$$X_0 = b_1 X_1 + b_2 X_2 + \ldots + b_n X_n.$$

Since
$$X^{(k)} = L^k X_0 \quad \text{and} \quad L^k X_i = \lambda_i^k X_i,$$

then

$$X^{(k)} = L^k (b_1 X_1 + b_2 X_2 + \ldots + b_n X_n) = b_1 \lambda_1^k X_1 + b_2 \lambda_2^k X_2 + \ldots b_n \lambda_n^k X_n.$$

Therefore,

$$X^{(k)} = \lambda_1^k \left(b_1 X_1 + b_2 \left(\frac{\lambda_2}{\lambda_1} \right)^k X_2 + \ldots + b_n \left(\frac{\lambda_n}{\lambda_1} \right)^k X_n \right).$$

Since λ_1 is dominant, $\left| \frac{\lambda_i}{\lambda_1} \right| < 1$ for $\lambda_i \neq \lambda_1$, and $\left(\frac{\lambda_i}{\lambda_1} \right)^k \to 0$ as $k \to \infty$. Thus for large k,

$$X^{(k)} \approx b_1 \lambda_1^k X_1.$$

In the long run, the age distribution stabilizes and is proportional to the vector X_1. Each age-group will change by a factor of λ_1 in each time period. The vector X_1 can be normalized so that its components sum to one, and the normalized vector then gives the eventual proportions of females in each of the n age-groupings.

Note that if $\lambda_1 > 1$, the population eventually increases; if $\lambda_1 = 1$, the population stabilizes, and if $\lambda_1 < 1$, the population eventually decreases.

Example 5. Determine the eventual distribution of the age classes for Example 4.

Solution. The characteristic equation is given by

$$\det(L - \lambda I) = \begin{vmatrix} -\lambda & 3 & 1 \\ 0.3 & -\lambda & 0 \\ 0 & 0.5 & -\lambda \end{vmatrix} = -\lambda^3 + 0.9\lambda + 0.15 = 0.$$

The roots of the characteristic equation are:

$$\lambda_1 = 1.023, \quad \lambda_2 = -0.851, \quad \lambda_3 = -0.172,$$

to three decimal places. Note that λ_1 is the dominant eigenvalue.

To find the eigenvector corresponding to λ_1, solve

$$\begin{pmatrix} -1.023 & 3 & 1 \\ 0.3 & -1.023 & 0 \\ 0 & 0.5 & -1.023 \end{pmatrix} \begin{pmatrix} x_1 \\ x_2 \\ x_3 \end{pmatrix} = \begin{pmatrix} 0 \\ 0 \\ 0 \end{pmatrix}.$$

One solution is $x_1 = 2.929, x_2 = 0.855$, and $x_3 = 0.420$. Divide each term by the sum to obtain the normalized eigenvector

$$\hat{X}_1 = \begin{pmatrix} 0.696 \\ 0.204 \\ 0.1 \end{pmatrix}.$$

Hence, after a number of years, the population will increase by approximately 2.3% every year. The percentage of females aged 0–1 year will be 69.6%; aged 1–2 years will be 20.4%, and aged 2–3 years will be 10%.

13.3 Harvesting and Culling Policies

This section will be concerned with insect and fish populations only since they tend to be very large. The model has applications when considering insect species which survive on crops, for example. An insect population can be culled each year by applying either an insecticide or a predator species. Harvesting of fish populations is particularly important nowadays; certain policies have to be employed to avoid depletion and extinction of the fish species. Harvesting indiscriminately could cause extinction of certain species of fish from our oceans. References [8] and [10] provide examples for the populations of yellow-legged gulls in the Mediterranean and roach in rivers in Belgium, respectively.

A harvesting or culling policy should only be used if the population is increasing.

Definition 1. A harvesting or culling policy is said to be *sustainable* if the number of fish, or insects, killed and the age distribution of the population remaining are the same after each time period.

Assume that the fish or insects are killed in short sharp bursts at the end of each time period. Let X be the population distribution vector for the species just before the harvesting or culling is applied. Suppose that a fraction of the females about to enter the $(i+1)$st class are killed, giving a matrix

$$D = \begin{pmatrix} d_1 & 0 & 0 & \cdots & 0 \\ 0 & d_2 & 0 & \cdots & 0 \\ 0 & 0 & d_3 & \cdots & 0 \\ \vdots & \vdots & \vdots & \ddots & \vdots \\ 0 & 0 & 0 & \cdots & d_n \end{pmatrix}.$$

By definition, $0 \le d_i \le 1$, where $1 \le i \le n$. The numbers killed will be given by DLX, and the population distribution of those remaining will be

$$LX - DLX = (I - D)LX.$$

In order for the policy to be sustainable, one must have

$$(I - D)LX = X. \tag{13.6}$$

If the dominant eigenvalue of $(I - D)L$ is one, then X will be an eigenvector for this eigenvalue and the population will stabilize. This will impose certain conditions on the matrix D. Hence

$$I - D = \begin{pmatrix} (1-d_1) & 0 & 0 & \cdots & 0 \\ 0 & (1-d_2) & 0 & \cdots & 0 \\ 0 & 0 & (1-d_3) & \cdots & 0 \\ \vdots & \vdots & \vdots & \ddots & \vdots \\ 0 & 0 & 0 & \cdots & (1-d_n) \end{pmatrix}$$

and the matrix, say, $M = (I - D)L$, is easily computed. The matrix M is also a Leslie matrix and hence has an eigenvalue $\lambda_1 = 1$ if and only if

$$(1 - d_1)(b_1 + b_2 c_1 (1 - d_1) + b_3 c_1 c_2 (1 - d_2)(1 - d_3) + \ldots$$
$$+ b_n c_1 \ldots c_{n-1} (1 - d_1) \ldots (1 - d_n)) = 1. \qquad (13.7)$$

Only values of $0 \le d_i \le 1$, which satisfy equation (13.7), can produce a sustainable policy.

A possible eigenvector corresponding to $\lambda_1 = 1$ is given by

$$X_1 = \begin{pmatrix} 1 \\ (1 - d_2)c_1 \\ (1 - d_2)(1 - d_3)c_1 c_2 \\ \vdots \\ (1 - d_2) \ldots (1 - d_n)c_1 c_2 \ldots c_{n-1} \end{pmatrix}.$$

The sustainable population will be $C_1 X_1$, where C_1 is a constant. Consider the following policies:

Sustainable Uniform Harvesting or Culling. Let $d = d_1 = d_2 = \ldots = d_n$, then (13.6) becomes

$$(1 - d)LX = X,$$

which means that $\lambda_1 = \frac{1}{1-d}$. Hence a possible eigenvector corresponding to λ_1 is given by

$$X_1 = \begin{pmatrix} 1 \\ \frac{c_1}{\lambda_1} \\ \frac{c_1 c_2}{\lambda_1^2} \\ \vdots \\ \frac{c_1 c_2 \ldots c_{n-1}}{\lambda_1^{n-1}} \end{pmatrix}.$$

Sustainable Harvesting or Culling of the Youngest Class. Let $d_1 = d$ and $d_2 = d_3 = \ldots = d_n = 0$, therefore equation (13.7) becomes

$$(1 - d)(b_1 + b_2 c_1 + b_3 c_1 c_2 + \ldots + b_n c_1 c_2 \ldots c_{n-1}) = 1,$$

or, equivalently,

$$(1 - d)R = 1,$$

where R is known as the *net reproduction rate*. Harvesting or culling is only viable if $R > 1$, unless you wish to eliminate an insect species. The age distribution after each harvest or cull is then given by

$$X_1 = \begin{pmatrix} 1 \\ c_1 \\ c_1 c_2 \\ \vdots \\ c_1 c_2 \dots c_{n-1} \end{pmatrix}.$$

Definition 2. An *optimal sustainable* harvesting or culling policy is one in which either one or two age classes are killed. If two classes are killed, then the older age class is completely killed.

Example 6. A certain species of fish can be divided into three six-month age classes and has Leslie matrix

$$L = \begin{pmatrix} 0 & 4 & 3 \\ 0.5 & 0 & 0 \\ 0 & 0.25 & 0 \end{pmatrix}.$$

The species of fish is to be harvested by fishermen using one of four different policies which are uniform harvesting or harvesting one of the three age classes, respectively. Which of these four policies are sustainable? Decide which of the sustainable policies the fishermen should use.

Solution. The characteristic equation is given by

$$\det(L - \lambda I) = \begin{vmatrix} -\lambda & 4 & 3 \\ 0.5 & -\lambda & 0 \\ 0 & 0.25 & -\lambda \end{vmatrix} = -\lambda^3 + 2\lambda + 0.375 = 0.$$

The eigenvalues are given by $\lambda_1 = 1.5$, $\lambda_2 = -0.191$, and $\lambda_3 = -1.309$, to three decimal places. The eigenvalue λ_1 is dominant, and the population will eventually increase by 50% every six months. The normalized eigenvector corresponding to λ_1 is given by

$$\hat{X}_1 = \begin{pmatrix} 0.529 \\ 0.177 \\ 0.294 \end{pmatrix}.$$

So, after a number of years, there will be 52.9% of females aged 0–6 months; 17.7% of females aged 6–12 months, and 29.4% of females aged 12–18 months.

If the harvesting policy is to be sustainable, then equation (13.7) becomes

$$(1 - d_1)(b_1 + b_2 c_1(1 - d_2) + b_3 c_1 c_2(1 - d_2)(1 - d_3)) = 1.$$

Suppose that $h_i = (1 - d_i)$, then

$$h_1 h_2 (2 + 0.375 h_3) = 1. \tag{13.8}$$

Consider the four policies separately.
(i) *Uniform harvesting:* let $\mathbf{h} = (h, h, h)$. Equation (13.8) becomes

$$h^2(2 + 0.375h) = 1,$$

which has solutions $h = 0.667$ and $d = 0.333$. The normalized eigenvector is given by

$$\hat{X}_U = \begin{pmatrix} 0.720 \\ 0.240 \\ 0.040 \end{pmatrix}.$$

(ii) *Harvesting the youngest age class:* let $\mathbf{h} = (h_1, 1, 1)$. Equation (13.8) becomes

$$h_1(2 + 0.375) = 1,$$

which has solutions $h_1 = 0.421$ and $d_1 = 0.579$. The normalized eigenvector is given by

$$\hat{X}_{A_1} = \begin{pmatrix} 0.615 \\ 0.308 \\ 0.077 \end{pmatrix}.$$

(iii) *Harvesting the middle age class:* let $\mathbf{h} = (1, h_2, 1)$. Equation (13.8) becomes

$$h_2(2 + 0.375) = 1,$$

which has solutions $h_2 = 0.421$ and $d_2 = 0.579$. The normalized eigenvector is given by

$$\hat{X}_{A_2} = \begin{pmatrix} 0.791 \\ 0.167 \\ 0.042 \end{pmatrix}.$$

(iv) Harvesting the oldest age class: let $\mathbf{h} = (1, 1, h_3)$. Equation (13.8) becomes

$$1(2 + 0.375 h_3) = 1,$$

which has no solutions if $0 \leq h_3 \leq 1$.

Therefore, harvesting policies (i)–(iii) are sustainable and policy (iv) is not. The long-term distributions of the populations of fish are determined by the normalized eigenvectors $\hat{X}_U, \hat{X}_{A_1}$, and $\hat{X}_{A_2}$, given above. If, for example, the fishermen wanted to leave as many fish as possible in the youngest age class, then the policy which should be adopted is the second age class harvesting. Then 79.1% of the females would be in the youngest age class after a number of years.

13.4 Mathematica Commands

```
(* See Example 1. Solving recurrence relations. *)
In[1]:=Savings=RSolve[{x[n+1]==(1+(3/100)) x[n],x[0]==10000},x[n],n]
```
$\mathrm{Out}[1]=x[n] \rightarrow 100^{(2-n)}103^n$
```
In[2]:=n=5;
In[3]:=N[Savings,7]
```
$\mathrm{Out}[3]=x[5.000000] \rightarrow 11592.74$
```
In[4]:=Clear[n];
(* See Example 2(i).*)
In[5]:=RSolve[{x[n+2]==x[n+1]+6 x[n],x[0]==1,x[1]==2},x[n],n]
```
$\mathrm{Out}[5]=x[n] \rightarrow \frac{1}{5}((-2)^n + 4 \times 3^n)$

```
(* See Example 4. Leslie matrices. *)
In[6]:=L={{0,3,1},{.3,0,0},{0,.5,0}};
In[7]:=X0={{1000},{2000},{3000}};
In[8]:=X10=MatrixPower[L,10].X0
Out[8]={{5383.66},{2177.08},{712.327}}
In[9]:=Eigenvalues[L]
Out[9]={1.02305,-0.850689,-0.172356}
In[10]:=Eigenvectors[L]
Out[10]={{0.950645,0.278769,0.136245},{-0.925557,0.326403,-0.191846},
         {0.184033,-0.320326,0.929259}}
```

13.5 Exercises

1. The difference equation used to model the length of a carpet, say, l_n, rolled n times is given by

$$l_{n+1} = l_n + \pi(4 + 2cn), \quad n = 0, 1, 2, 3, \ldots,$$

where c is the thickness of the carpet. Solve this recurrence relation.

2. Solve the following second-order linear difference equations:

   (a) $x_{n+2} = 5x_{n+1} - 6x_n$, $n = 0, 1, 2, 3, \ldots$, if $x_0 = 1$, $x_1 = 4$;

   (b) $x_{n+2} = x_{n+1} - \frac{1}{4}x_n$, $n = 0, 1, 2, 3, \ldots$, if $x_0 = 1$, $x_1 = 2$;

   (c) $x_{n+2} = 2x_{n+1} - 2x_n$, $n = 0, 1, 2, 3, \ldots$, if $x_0 = 1$, $x_1 = 2$;

   (d) $F_{n+2} = F_{n+1} + F_n$, $n = 0, 1, 2, 3, \ldots$, if $F_1 = 1$ and $F_2 = 1$ (the sequence of numbers is known as the Fibonacci sequence);

   (e) $x_{n+2} = x_{n+1} + 2x_n - f(n)$, $n = 0, 1, 2, \ldots$, given that $x_0 = 2$ and $x_1 = 3$, when (i) $f(n) = 2$, (ii) $f(n) = 2n$ and (iii) $f(n) = e^n$ (use Mathematica for part (iii) only).

3. Consider a human population that is divided into three age classes; those aged 0–15 years, those aged 15–30 years, and those aged 30–45 years. The Leslie matrix for the female population is given by

$$L = \begin{pmatrix} 0 & 1 & 0.5 \\ 0.9 & 0 & 0 \\ 0 & 0.8 & 0 \end{pmatrix}.$$

Given that the initial population distribution of females is $x_1^{(0)} = 10000$, $x_2^{(0)} = 15000$, and $x_3^{(0)} = 8000$, compute the number of females in each of these groupings after

   (a) 225 years;

   (b) 750 years;

   (c) 1500 years.

4. Consider the following Leslie matrix used to model the female portion of a species

$$L = \begin{pmatrix} 0 & 0 & 6 \\ \frac{1}{2} & 0 & 0 \\ 0 & \frac{1}{3} & 0 \end{pmatrix}.$$

Determine the eigenvalues and eigenvectors of L. Show that there is no dominant eigenvalue and describe how the population would develop in the long term.

5. Consider a human population that is divided into five age classes: those aged 0–15 years, those aged 15–30 years, those aged 30–45 years, those aged 45–60 years, and those aged 60–75 years. The Leslie matrix for the female population is given by

$$L = \begin{pmatrix} 0 & 1 & 1.5 & 0 & 0 \\ 0.9 & 0 & 0 & 0 & 0 \\ 0 & 0.8 & 0 & 0 & 0 \\ 0 & 0 & 0.7 & 0 & 0 \\ 0 & 0 & 0 & 0.5 & 0 \end{pmatrix}.$$

Determine the eigenvalues and eigenvectors of L and describe how the population distribution develops.

6. Given that

$$L = \begin{pmatrix} b_1 & b_2 & b_3 & \cdots & b_{n-1} & b_n \\ c_1 & 0 & 0 & \cdots & 0 & 0 \\ 0 & c_2 & 0 & \cdots & 0 & 0 \\ \vdots & \vdots & \vdots & \ddots & \vdots & \vdots \\ 0 & 0 & 0 & \cdots & c_{n-1} & 0 \end{pmatrix},$$

where $b_i \geq 0, 0 < c_i \leq 1$, and at least two successive b_i are strictly positive, prove that $p(\lambda) = 1$, if λ is an eigenvalue of L, where

$$p(\lambda) = \frac{b_1}{\lambda} + \frac{b_2 c_1}{\lambda^2} + \ldots + \frac{b_n c_1 c_2 \ldots c_{n-1}}{\lambda^n}.$$

Show the following:

    (a) $p(\lambda)$ is strictly decreasing;

    (b) $p(\lambda)$ has a vertical asymptote at $\lambda = 0$;

    (c) $p(\lambda) \to 0$ as $\lambda \to \infty$.

Prove that a general Leslie matrix has a unique positive eigenvalue.

7. A certain species of insect can be divided into three age classes: 0–6 months, 6–12 months, and 12–18 months. A Leslie matrix for the female population is given by

$$L = \begin{pmatrix} 0 & 4 & 10 \\ 0.4 & 0 & 0 \\ 0 & 0.2 & 0 \end{pmatrix}.$$

Determine the long-term distribution of the insect population. An insecticide is applied which kills off 50% of the youngest age class. Determine the long-term distribution if the insecticide is applied every six months.

8. Assuming the same model for the insects as in Exercise 7, determine the long-term distribution if an insecticide is applied every six months which kills 10% of the youngest age class, 40% of the middle age class, and 60% of the oldest age class.

9. In a fishery, a certain species of fish can be divided into three age-groups each one year long. The Leslie matrix for the female portion of the population is given by

$$L = \begin{pmatrix} 0 & 3 & 36 \\ \frac{1}{3} & 0 & 0 \\ 0 & \frac{1}{2} & 0 \end{pmatrix}.$$

Show that, without harvesting, the fish population would double each year. Describe the long-term behavior of the system if the following policies are applied:

(a) harvest 50% from each age class;

(b) harvest the youngest fish only, using a sustainable policy;

(c) harvest 50% of the youngest fish;

(d) harvest 50% of the whole population from the youngest class only;

(e) harvest 50% of the oldest fish.

10. Determine an optimal sustainable harvesting policy for the system given in Exercise 9 if the youngest age class is left untouched.

Bibliography

[1] H. Anton and C. Rorres, *Elementary Linear Algebra, 11th Ed.*, Wiley, New York, 2013.

[2] S. Barnet, *Discrete Mathematics*, Addison-Wesley, New York, 1998.

[3] E. Chambon-Dubreuil, P. Auger, J.M. Gaillard and M. Khaladi, Effect of aggressive behaviour on age-structured population dynamics, *Ecological Modelling* **193**, (2006), 777–786.

[4] S. Jal & T.G. Hallam, Effects of delay, truncations and density dependence in reproduction schedules on stability of nonlinear Leslie matrix models, *Journal of Mathematical Biology* **31**, (1993), 367–395.

[5] M.R.S Kulenovic and O. Merino, *Discrete Dynamical Systems and Difference Equations with Mathematica*, Chapman & Hall, London, 2002.

[6] V.N. Lopatin and S.V. Rosolovsky, *Evaluation of the State and Productivity of Moose Populations using Leslie Matrix Analyses. An article from: Alces [HTML] (Digital)*, Alces, 2002.

[7] L. Monte, Characterisation of a nonlinear Leslie matrix model for predicting the dynamics of biological populations in polluted environments: Applications to radioecology, *Ecological modeling* **248**, (2013), 174–183.

[8] W. Otjacques, F. De Laender, and P. Kestemont, P., Discerning the causes of a decline in a common European fish, the roach (Rutilus rutilus L.): A modelling approach, *Ecological Modelling*, **322**, (2016), 92–100.

[9] E. Otumba, *The Extended Leslie Matrix Model: Estimation of Population Sizes of Migrating Sizes*, LAP LAMBERT Academic Publishing, Saarbrücken, 2011.

[10] S. Roy, R. Ridley, J. Sandon et al., Adapting strategies to maintain efficiency during a cull of yellow-legged gulls, *Human-Wildlife Interactions*, **10**, (2016), 83–90.

Chapter 14
Nonlinear Discrete Dynamical Systems

Aims and Objectives

- To introduce nonlinear one- and two-dimensional iterated maps.

- To investigate period-doubling bifurcations to chaos.

- To introduce the notion of universality.

On completion of this chapter, the reader should be able to

- produce graphical iterations of one-dimensional iterated maps;

- test whether or not certain systems are chaotic;

- plot bifurcation diagrams;

- apply some of the theory to model simple problems from biology, economics, neural networks, nonlinear optics, and population dynamics.

Most of the dynamics displayed by highly complicated nonlinear systems also appear for simple nonlinear systems. The reader is first introduced to the tent function, which is composed of two straight lines. The graphical method of iteration is introduced using this simple function since the constructions may be easily carried out with graph paper, rule, and pencil. The reader is also shown how to graph composite functions. The system can display periodicity, mixing, and sensitivity to initial conditions, the essential ingredients for chaos.

The logistic map is used as a simple model for the population growth of an insect species. Bifurcation diagrams are plotted and period-doubling bifurcations to chaos are displayed.

Bifurcation diagrams are plotted for the Gaussian map. Two-dimensional Hénon maps are investigated, periodic points are found, and chaotic (or strange) attractors are produced.

© Springer International Publishing AG 2017   305
S. Lynch, *Dynamical Systems with Applications using Mathematica*®,
DOI 10.1007/978-3-319-61485-4_14

The chapter ends with some applications from biology, economics, nonlinear optics and neural networks. There are a number of textbooks available on discrete dynamical systems, e.g., see [4], [7], [8] and [14]. Recent conferences have concentrated on applications of discrete dynamical systems, see [2] and [16], for example.

14.1 The Tent Map and Graphical Iterations

As a simple introduction to one-dimensional nonlinear discrete dynamical systems, consider the *tent map*, $T : [0, 1] \rightarrow [0, 1]$ defined by

$$T(x) = \begin{cases} \mu x & 0 \leq x < \frac{1}{2} \\ \mu(1 - x) & \frac{1}{2} \leq x \leq 1, \end{cases}$$

where $0 \leq \mu \leq 2$. The tent map is constructed from two straight lines, which makes the analysis simpler than for truly nonlinear systems. The graph of the T function may be plotted by hand and is given in Figure 14.1.

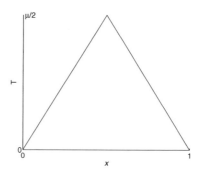

Figure 14.1:   A graph of the tent function.

Define an iterative map by

$$x_{n+1} = T(x_n), \tag{14.1}$$

where $x_n \in [0, 1]$. Although the form of the tent map is simple and the equations involved are linear, for certain parameter values, this system can display highly complex behavior and even chaotic phenomena. In fact, most of the features discussed in other chapters of this text are displayed by this relatively simple system. For certain parameter values, the mapping undergoes stretching and folding transformations and displays sensitivity to initial conditions and periodicity. Fortunately, it is not difficult to carry out simple iterations for system (14.1) as the following examples demonstrate.

Example 1. Iterate the tent function numerically for the following μ and x_0 values:

(I) $\mu = \frac{1}{2}$:

(i) $x_0 = \frac{1}{4}$, (ii) $x_0 = \frac{1}{2}$, (iii) $x_0 = \frac{3}{4}$;

(II) $\mu = 1$:

(i) $x_0 = \frac{1}{3}$, (ii) $x_0 = \frac{2}{3}$;

(III) $\mu = \frac{3}{2}$:

(i) $x_0 = \frac{3}{5}$, (ii) $x_0 = \frac{6}{13}$, (iii) $x_0 = \frac{1}{3}$;

(IV) $\mu = 2$:

(i) $x_0 = \frac{1}{3}$, (ii) $x_0 = \frac{1}{5}$, (iii) $x_0 = \frac{1}{7}$, (iv) $x_0 = \frac{1}{11}$.

Solution. A calculator or computer is not needed here. It is very easy to carry out the iterations by hand. For the sake of simplicity, the iterates will be listed as $\{x_0, x_1, x_2, \ldots, x_n, \ldots\}$. The solutions are as follows:

(I) $\mu = \frac{1}{2}$:

(i) $\{\frac{1}{4}, \frac{1}{8}, \frac{1}{16}, \ldots, \frac{1}{4 \times 2^n}, \ldots\}$;

(ii) $\{\frac{1}{2}, \frac{1}{4}, \frac{1}{8}, \ldots, \frac{1}{2^{n+1}}, \ldots\}$;

(iii) $\{\frac{3}{4}, \frac{3}{8}, \frac{3}{16}, \ldots, \frac{3}{4 \times 2^n}, \ldots\}$.

In each case, $x_n \to 0$ as $n \to \infty$.

(II) $\mu = 1$:

(i) $\{\frac{1}{3}, \frac{1}{3}, \frac{1}{3}, \ldots, \frac{1}{3}, \ldots\}$;

(ii) $\{\frac{2}{3}, \frac{1}{3}, \frac{1}{3}, \ldots, \frac{1}{3}, \ldots\}$.

The orbits tend to points of period one in the range $\left[0, \frac{1}{2}\right]$.

(III) $\mu = \frac{3}{2}$:

(i) $\{\frac{3}{5}, \frac{3}{5}, \frac{3}{5}, \ldots, \frac{3}{5}, \ldots\}$;

(ii) $\{\frac{6}{13}, \frac{9}{13}, \frac{6}{13}, \frac{9}{13} \ldots, \frac{6}{13}, \frac{9}{13}, \ldots\}$;

(iii) $\{\frac{1}{3}, \frac{1}{2}, \frac{3}{4}, \frac{3}{8}, \frac{9}{16}, \frac{21}{32}, \frac{33}{64}, \frac{93}{128}, \frac{105}{256}, \frac{315}{512}, \frac{591}{1024}, \ldots\}$.

In case (i), the iterate x_{n+1} is equal to x_n for all n. This type of sequence displays *period-one* behavior. In case (ii), the iterate x_{n+2} is equal to x_n for all n, and the result is *period-two* behavior. In case (iii), the first 11 iterates are listed but other methods need to be used in order to establish the long-term behavior of the sequence.

(IV) $\mu = 2$:

(i) $\{\frac{1}{3}, \frac{2}{3}, \frac{2}{3}, \ldots, \frac{2}{3}, \ldots\}$;

(ii) $\{\frac{1}{5}, \frac{2}{5}, \frac{4}{5}, \frac{2}{5}, \frac{4}{5} \ldots, \frac{2}{5}, \frac{4}{5}, \ldots\}$;

(iii) $\{\frac{1}{7}, \frac{2}{7}, \frac{4}{7}, \frac{6}{7}, \frac{2}{7}, \frac{4}{7}, \frac{6}{7}, \ldots, \frac{2}{7}, \frac{4}{7}, \frac{6}{7}, \ldots\}$;

(iv) $\{\frac{1}{11}, \frac{2}{11}, \frac{4}{11}, \frac{8}{11}, \frac{6}{11}, \frac{10}{11}, \frac{2}{11}, \ldots, \frac{2}{11}, \frac{4}{11}, \frac{8}{11}, \frac{6}{11}, \frac{10}{11}, \ldots\}$.

The sequences behave as follows: (i) there is period-one behavior, (ii) there is period-two behavior, (iii) there is a period-three sequence, and (iv) there is a period-five sequence.

Example 2. Using the tent map defined by equation (14.1) when $\mu = 2$, compute the first 20 iterates for the following two initial conditions:

(i) $x_0 = 0.2$;

(ii) $x_0 = 0.2001 = 0.2 + \epsilon$.

Solution. The iterates may be computed using the Mathematica package. The first 20 iterates for both initial conditions are listed side-by-side in Table 14.1.

The system clearly shows sensitivity to initial conditions for the parameter value $\mu = 2$. Comparing the numerical values in the second and third columns, it is not difficult to see that the sequences diverge markedly when $n > 9$. This test for sensitivity to initial conditions gives researchers a simple tool to determine whether or not a system is chaotic. A more in-depth description of chaos is given in Chapter 8.

The results of Examples 1 and 2 show that there is a rich variety of dynamics which system (14.1) can display. Indeed, a now famous result due to Li and Yorke [17] states that if a system displays period-three behavior, then the system can display periodic behavior of any period, and they go on to prove that the system can display chaotic phenomena. Hence, when $\mu = 2$, system (14.1) is chaotic since it has a period-three sequence (Example 1(IV)(iii)).

Unfortunately, numerical iterations do not always give a clear insight into how the sequences develop as n gets large. Another popular method used to display the sequence of iterations more clearly is the so-called *graphical method*.

The Graphical Method. From an initial point x_0, draw a vertical line up to the function, in this case, $T(x)$. From this point, draw a horizontal line either left or right to join the diagonal $y = x$. The x-ordinate corresponds to the iterate $x_1 = T(x_0)$. From the point $(x_1, T(x_0))$, draw a vertical line up or down to join the function $T(x)$. Draw a horizontal line from this point to the diagonal at the point $(x_2, T(x_1))$. The first two iterates are shown in Figure 14.2.

The iterative procedure may be summarized as a simple repeated two-step algorithm.

Table 14.1: The first 20 iterates for both initial conditions in Example 2

n	x_n	x_n
0	$x_0 = 0.2$	$x_0 = 0.2001$
1	0.4	0.4002
2	0.8	0.8004
3	0.4	0.3992
4	0.8	0.7984
5	0.4	0.4032
6	0.8	0.8064
7	0.4	0.3872
8	0.8	0.7744
9	0.4	0.4512
10	0.8	0.9024
11	0.4	0.1952
12	0.8	0.3904
13	0.4	0.7808
14	0.8	0.4384
15	0.4	0.8768
16	0.8	0.2464
17	0.4	0.4928
18	0.8	0.9856
19	0.4	0.0288
20	0.8	0.0576

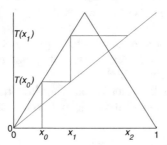

Figure 14.2: A possible graphical iteration when $n = 2$.

1. Draw a vertical line to the function (evaluation).

2. Draw a horizontal line to the diagonal (feedback); go back to 1.

The algorithm generates successive iterates along the x-axis, corresponding to the sequence of points $\{x_0, x_1, x_2, \ldots, x_n, \ldots\}$.

To demonstrate the method, the numerical iterations carried out in Examples 1 and 2 will now be repeated using the graphical method.

Example 3. Iterate the tent function graphically for the following μ and x_0 values:

(I) $\mu = \frac{1}{2}$:

(i) $x_0 = \frac{1}{4}$, (ii) $x_0 = \frac{1}{2}$, (iii) $x_0 = \frac{3}{4}$;

(II) $\mu = 1$:

(i) $x_0 = \frac{1}{3}$, (ii) $x_0 = \frac{2}{3}$;

(III) $\mu = \frac{3}{2}$:

(i) $x_0 = \frac{3}{5}$, (ii) $x_0 = \frac{6}{13}$, (iii) $x_0 = \frac{1}{3}$;

(IV) $\mu = 2$:

(i) $x_0 = \frac{1}{3}$, (ii) $x_0 = \frac{1}{5}$, (iii) $x_0 = \frac{1}{7}$, (iv) $x_0 = \frac{1}{11}$.

(V) $\mu = 2$:

(i) $x_0 = 0.2$, (ii) $x_0 = 0.2001$.

Solution. Each of the diagrams (Figures 14.3, 14.4, 14.5 14.6 and 14.7) can be reproduced using the Mathematica package. Most of the graphical iterations are self-explanatory; however, Figures 14.5(c) and 14.7 warrant further explanation. When $\mu = \frac{3}{2}$, the tent map displays sensitivity to initial conditions and can be described as being chaotic. The iterative path plotted in Figure 14.5(c) appears to wander randomly. It is still not clear whether the path is chaotic or whether the path is periodic of a very high period. Figure 14.7 clearly shows the sensitivity to initial conditions. Again, it is not clear in case (ii) whether the path is chaotic or of a very high period.

What is clear from the diagrams is that the three basic properties of chaos—mixing, periodicity, and sensitivity to initial conditions—are all exhibited for certain parameter values.

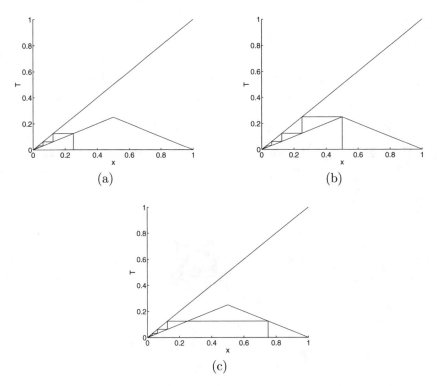

Figure 14.3: Graphical iterations when $\mu = \frac{1}{2}$: (a) $x_0 = \frac{1}{4}$; (b) $x_0 = \frac{1}{2}$; and (c) $x_0 = \frac{3}{4}$.

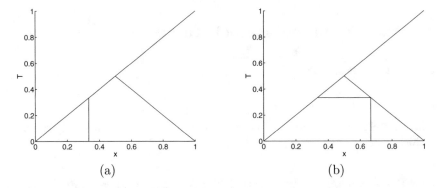

Figure 14.4: Graphical iterations when $\mu = 1$: (a) $x_0 = \frac{1}{3}$ and (b) $x_0 = \frac{2}{3}$.

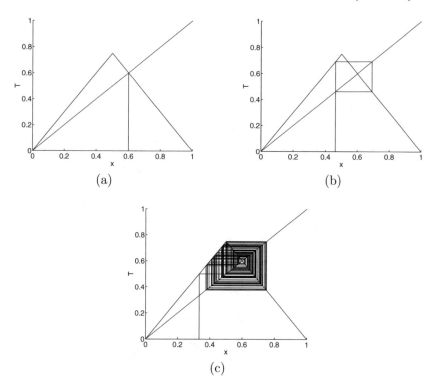

(a)

(b)

(c)

Figure 14.5: Graphical iterations when $\mu = \frac{3}{2}$: (a) $x_0 = \frac{3}{5}$; (b) $x_0 = \frac{6}{13}$; and (c) $x_0 = \frac{1}{3}$, for 200 iterations.

14.2 Fixed Points and Periodic Orbits

Consider the general map

$$x_{n+1} = f(x_n). \tag{14.2}$$

Definition 1. A *fixed point*, or point of period one, of system (14.2) is a point at which $x_{n+1} = f(x_n) = x_n$, for all n.

For the tent map, this implies that $T(x_n) = x_n$, for all n. Graphically, the fixed points can be found by identifying intersections of the function $T(x)$ with the diagonal.

As with other dynamical systems, the fixed points of period one can be attracting, repelling, or indifferent. The type of fixed point is determined from the gradient of the tangent to the function, $T(x)$ in this case, at the fixed point. For straight line segments with equation $y = mx + c$, it can be easily shown that if

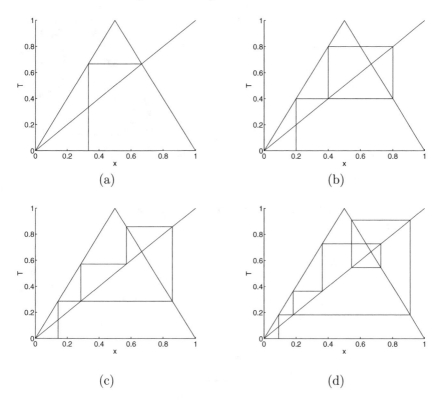

Figure 14.6: [Mathematica] Graphical iterations when $\mu = 2$: (a) $x_0 = \frac{1}{3}$; (b) $x_0 = \frac{1}{5}$; (c) $x_0 = \frac{1}{7}$; and (d) $x_0 = \frac{1}{11}$.

- $m < -1$, the iterative path is repelled and spirals away from the fixed point;

- $-1 < m < 0$, the iterative path is attracted and spirals into the fixed point;

- $0 < m < 1$, the iterative path is attracted and staircases into the fixed point;

- $m > 1$, the iterative path is repelled and staircases away from the critical. point.

When $|m| = 1$, the fixed point is neither repelling nor attracting and $m = 0$ is a trivial case. A test for stability of fixed points for nonlinear iterative maps will be given in Section 14.2.

Using Definition 1, it is possible to determine the fixed points of period one for the tent map (14.1). If $0 < \mu < 1$, the only fixed point is at $x = 0$ (see Figure 14.8) and since the gradient at $x = 0$ is less than one, the fixed point is stable. Note that the origin is called the *trivial fixed point*.

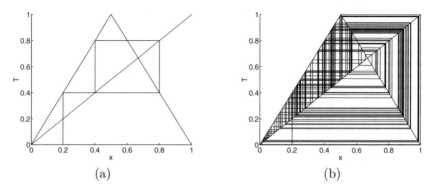

(a)   (b)

Figure 14.7:   Graphical iterations when $\mu = 2$: (a) $x_0 = 0.2$ and (b) $x_0 = 0.2001$, for 200 iterations.

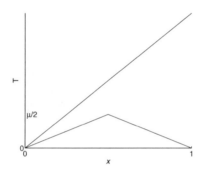

Figure 14.8:   The intersection $T(x) = x$ when $0 < \mu < 1$.

When $\mu = 1$, the branch μx of $T(x)$ coincides with the diagonal and all points lying in the interval $0 \leq x \leq 1/2$ are of period one. Once the tent function crosses the diagonal the origin becomes unstable since the gradient of the tent map at this point now exceeds one.

When $1 < \mu \leq 2$, there are two fixed points of period one, $x_{1,1} = 0$ and $x_{1,2} = \frac{\mu}{1+\mu}$ (see Figure 14.9).

Notation. Throughout this text, the periodic point given by $x_{i,j}$ will denote the jth point of period i. This notation is useful when determining the number of points of period i. For example, $x_{1,1}$ and $x_{1,2}$ above are the two fixed points of period one.

The gradient of the function $T(x)$ is greater than one at $x_{1,1}$, so this point is unstable; the gradient of $T(x)$ at the point $x_{1,2}$ is less than -1. Therefore, this point is also unstable.

In summary, when $0 \leq \mu < 1$, there is one stable period-one point at $x = 0$; when $\mu = 1$, there are an infinite number of period-one points in the interval

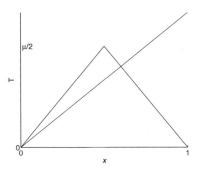

Figure 14.9: The intersections $T(x) = x$ when $1 < \mu \leq 2$. There are two intersections.

$0 \leq x \leq 1/2$; and when $1 < \mu \leq 2$, there are two unstable period-one points at $x_{1,1}$ and $x_{1,2}$. The obvious question then is, where do the paths go if not to these two points of period one? The answer to this question will be given later (see Exercise 3 in Section 14.7).

Definition 2. For system (14.2), a *fixed point of period N* is a point at which $x_{n+N} = f^N(x_n) = x_n$, for all n.

In order to determine the fixed points of period two for the tent map, it is necessary to find the points of intersection of $T^2(x)$ with the diagonal. Consider the case where $\mu = 2$, the methods below can be applied for any value of μ in the interval $[0, 2]$.

The function of the function $T(T(x)) = T^2(x)$ is determined by replacing x with $T(x)$ in the mapping

$$T(x) = \begin{cases} 2x & 0 \leq x < \frac{1}{2} \\ 2(1-x) & \frac{1}{2} \leq x \leq 1. \end{cases}$$

Hence

$$T^2(x) = \begin{cases} 2T(x) & 0 \leq T(x) < \frac{1}{2} \\ 2(1-T(x)) & \frac{1}{2} \leq T(x) \leq 1. \end{cases}$$

The interval $0 \leq T(x) < \frac{1}{2}$ on the vertical axis corresponds to two intervals, namely $0 \leq x < T^{-1}\left(\frac{1}{2}\right)$ and $T^{-1}\left(\frac{1}{2}\right) \leq x \leq 1$ on the horizontal axis. When $\mu = 2$, it is not difficult to show that $T^{-1}\left(\frac{1}{2}\right) = \frac{1}{4}$ or $\frac{3}{4}$, depending on the branch of $T(x)$. The process may be repeated for $T(x)$ lying in the interval $\left[\frac{1}{2}, 1\right]$. Therefore, $T^2(x)$ becomes

$$T^2(x) = \begin{cases} 4x & 0 \leq x < \frac{1}{4} \\ 2 - 4x & \frac{1}{4} \leq x < \frac{1}{2} \\ 4x - 2 & \frac{1}{2} \leq x < \frac{3}{4} \\ 4 - 4x & \frac{3}{4} \leq x \leq 1. \end{cases}$$

This function intersects the diagonal at four points corresponding to $x = 0, 2/5, 2/3$, and $4/5$ as shown in Figure 14.10.

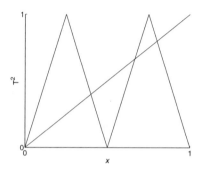

Figure 14.10:   The graphs of $T^2(x)$ and $y = x$ when $\mu = 2$.

The fixed points at $x = 0$ and $x = 2/3$ are of period one; therefore, there are two points of period two given by $x_{2,1} = \frac{2}{5}$ and $x_{2,2} = \frac{4}{5}$. Since the gradient of $|T^2(x)|$ is greater than one at these points, $x_{2,1}$ and $x_{2,2}$ are unstable.

It is not difficult to show that there are no period-two points for $0 \leq \mu \leq 1$ and there are two points of period two for $1 < \mu \leq 2$.

To determine the fixed points of period three, it is necessary to find the points of intersection of $T^3(x)$ with the diagonal. Consider the case where $\mu = 2$. The methods below can be applied for any value of μ in the interval $[0, 2]$.

The function $T(T(T(x))) = T^3(x)$ is determined by replacing x with $T(x)$ in the mapping for $T^2(x)$. Hence,

$$T^3(x) = \begin{cases} 4T(x) & 0 \leq T(x) < \frac{1}{4} \\ 2 - 4T(x) & \frac{1}{4} \leq T(x) < \frac{1}{2} \\ 4T(x) - 2 & \frac{1}{2} \leq T(x) < \frac{3}{4} \\ 4 - 4T(x) & \frac{3}{4} \leq T(x) \leq 1. \end{cases}$$

The interval $0 \leq T(x) < \frac{1}{4}$ on the vertical axis corresponds to two intervals, namely $0 \leq x < T^{-1}\left(\frac{1}{4}\right)$ and $T^{-1}\left(\frac{1}{4}\right) \leq x \leq 1$ on the horizontal axis. When $\mu = 2$, it is not difficult to show that $T^{-1}\left(\frac{1}{4}\right) = \frac{1}{8}$ or $\frac{7}{8}$, depending on the branch of $T(x)$. The process may be repeated for $T(x)$ lying in the other intervals. Therefore, $T^3(x)$ becomes

$$T^3(x) = \begin{cases} 8x & 0 \le x < \frac{1}{8} \\ 2 - 8x & \frac{1}{8} \le x < \frac{1}{4} \\ 8x - 2 & \frac{1}{4} \le x < \frac{3}{8} \\ 4 - 8x & \frac{3}{8} \le x < \frac{1}{2} \\ 8x - 4 & \frac{1}{2} \le x < \frac{5}{8} \\ 6 - 8x & \frac{5}{8} \le x < \frac{3}{4} \\ 8x - 6 & \frac{3}{4} \le x < \frac{7}{8} \\ 8 - 8x & \frac{7}{8} \le x \le 1. \end{cases}$$

This function intersects the diagonal at eight points corresponding to $x = 0, \frac{2}{9}, \frac{2}{7}, \frac{4}{9}, \frac{4}{7}, \frac{2}{3}, \frac{6}{7}$, and $\frac{8}{9}$ as shown in Figure 14.11. Note that points of period two do not repeat on every third cycle and hence do not appear here.

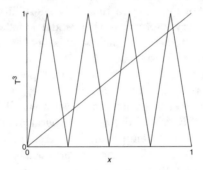

Figure 14.11: The graphs of $T^3(x)$ and $y = x$ when $\mu = 2$.

The fixed points at $x = 0$ and $x = 2/3$ are of period one; therefore, there are six points of period three given by $x_{3,1} = \frac{2}{9}, x_{3,2} = \frac{4}{9}, x_{3,3} = \frac{8}{9}, x_{3,4} = \frac{2}{7}, x_{3,5} = \frac{4}{7}$, and $x_{3,6} = \frac{6}{7}$. Since the gradient of $|T^3(x)|$ is greater than one at these points, all six points are unstable. Thus, an initial point close to the periodic orbit, but not on it, will move away and the orbits will diverge.

This process may be repeated to determine points of any period for the tent map. Recall that the results due to Li and Yorke imply that the map contains periodic points of all periods. It is therefore possible to find points of period $10, 10^6$, or even 10^{100}, for example. There are also aperiodic (or nonperiodic) orbits and the system is sensitive to initial conditions. Similar phenomena are observed for three-dimensional autonomous systems in Chapter 8, in fact, most of the dynamics exhibited there appear for this much simpler system.

14.3 The Logistic Map, Bifurcation Diagram, and Feigenbaum Number

In the early 1970s, May [13] and others began to investigate the equations used by fish biologists and entomologists to model the fluctuations in certain species. Simple population models have been discussed in other chapters using continuous dynamical models, but the analysis here will be restricted to simple nonlinear discrete systems. Perhaps the most famous system used to model a single species is that known as the *logistic map* given by

$$x_{n+1} = f_\mu(x_n) = \mu x_n(1 - x_n), \tag{14.3}$$

where μ is a parameter and $0 \leq x_n \leq 1$ represents the scaled population size. Consider the case where μ is related to the reproduction rate and x_n represents the population of blowflies at time n, which can be measured in hours, days, weeks, months, etc. Blowflies have a relatively short lifespan and are easy to monitor in the laboratory. Note that this model is extremely simple but as with the tent map a rich variety of behavior is displayed as the parameter μ is varied. We note that scientists would find it difficult to change reproduction rates of individual flies directly; however, for many species the reproduction rate depends on other factors such as temperature. Hence, imagine a tank containing a large number of blowflies. Experimentally, we would like to observe how the population fluctuates, if at all, at different temperatures. A population of zero would imply that the tank is empty and a scaled population of one would indicate that the tank is full. The numbers produced in this model would be rounded down to guarantee that fractions would be ignored as in the continuous case.

It must be pointed out that this model does not take into account many features which would influence a population in real applications. For example, age classes, diseases, other species interactions, and environmental effects are all ignored. Even though many factors are left out of the equation, the results show a wide range of dynamical behavior which has been observed both experimentally and in the field.

Consider the logistic map $f_\mu : [0, 1] \to [0, 1]$ given by

$$x_{n+1} = f_\mu(x_n),$$

where $f_\mu(x) = \mu x(1 - x)$. The parameter μ lies in the interval $[0, 4]$. The graph of f_μ is given in Figure 14.12.

As with the tent map, simple numerical and graphical iterations may be carried out for varying values of the parameter μ. To avoid repetition, these tasks will be left as exercises at the end of the chapter. Instead, the analysis will be restricted to finding periodic points and plotting a bifurcation diagram.

To find points of period one, it is necessary to solve the equation given by

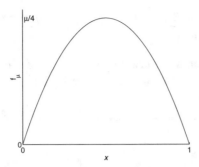

Figure 14.12: A graph of the logistic map function.

$$f_\mu(x) = \mu x(1-x) = x,$$

which gives the points which satisfy the condition $x_{n+1} = x_n$ for all n. There are two solutions given by $x_{1,1} = 0$ and $x_{1,2} = 1 - \frac{1}{\mu}$. The stability of the critical points may be determined using the following theorem.

Theorem 1. *Suppose that the map $f_\mu(x)$ has a fixed point at x^*. Then the fixed point is stable if*

$$\left| \frac{d}{dx} f_\mu(x^*) \right| < 1$$

and it is unstable if

$$\left| \frac{d}{dx} f_\mu(x^*) \right| > 1.$$

Using Theorem 1, $\left| \frac{df_\mu(0)}{dx} \right| = \mu$. Thus, the point $x_{1,1}$ is stable for $0 < \mu < 1$ and unstable if $\mu > 1$. Since $\left| \frac{df_\mu(x_{1,2})}{dx} \right| = 2 - \mu$, this fixed point is stable for $1 < \mu < 3$ and is unstable when $\mu < 1$ or $\mu > 3$.

To find points of period two, it is necessary to solve the equation given by

$$f_\mu^2(x) = \mu(\mu x(1-x)(1 - \mu x(1-x))) = x, \tag{14.4}$$

which gives the points which satisfy the condition $x_{n+2} = x_n$ for all n. Two solutions for equation (14.4) are known, namely $x_{1,1}$ and $x_{1,2}$, since points of period one repeat on every second iterate. Therefore, equation (14.4) factorizes as follows:

$$x \left(x - \left(1 - \frac{1}{\mu} \right) \right) \left(-\mu^3 x^2 + (\mu^2 + \mu^3)x - (\mu^2 + \mu) \right) = 0.$$

The equation $-\mu^3 x^2 + (\mu^2 + \mu^3)x - (\mu^2 + \mu) = 0$ has roots at

$$x_{2,1} = \frac{\mu + 1 + \sqrt{(\mu - 3)(\mu + 1)}}{2\mu} \quad \text{and} \quad x_{2,2} = \frac{\mu + 1 - \sqrt{(\mu - 3)(\mu + 1)}}{2\mu}.$$

Thus there are two points of period two when $\mu > 3$. Let $b_1 = 3$ correspond to the first bifurcation point for the logistic map. Now

$$\frac{d}{dx} f_\mu^2(x_{2,1}) = -4\mu^3 x^3 + 6\mu^3 x^2 - 2(\mu^2 + \mu^3)x + \mu^2$$

and

$$\left| \frac{d}{dx} f_\mu^2(x_{2,1}) \right| = 1,$$

when $\mu = b_2 = 1 + \sqrt{6}$. The value b_2 corresponds to the second bifurcation point for the logistic map. Hence, $x_{2,1}$ and $x_{2,2}$ lose their stability at $\mu = b_2$ (check this using Mathematica).

In summary, for $0 < \mu < 1$, the fixed point at $x = 0$ is stable and iterative paths will be attracted to that point. Physically, this would mean that the population of blowflies would die away to zero. One can think of the temperature of the tank being too low to sustain life. As μ passes through one, the trivial fixed point becomes unstable and the iterative paths are attracted to the fixed point at $x_{1,2} = 1 - \frac{1}{\mu}$. For $1 < \mu < b_1$, the fixed point of period one is stable which means that the population stabilizes to a constant value after a sufficiently long time. As μ passes through b_1, the fixed point of period one becomes unstable and a fixed point of period two is created. For $b_1 < \mu < b_2$, the population of blowflies will alternate between two values on each iterative step after a sufficient amount of time. As μ passes through b_2, the fixed point of period two loses its stability and a fixed point of period four is created.

As with other dynamical systems, all of the information gained so far can be summarized on a bifurcation diagram. Figure 14.13 shows a bifurcation diagram for the logistic map when $0 \leq \mu \leq 3.5$. The first two bifurcation points are labeled b_1 and b_2.

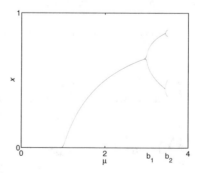

Figure 14.13:   The first two bifurcations for the logistic map.

For other values of μ, it is interesting to plot time series data obtained from the logistic map. Figure 14.14 shows iterative paths and time series data when $x_0 = 0.2$ (assuming the tank is initially, say, $\frac{1}{5}$ full) for the following four cases: (i) $\mu = 2$, (ii) $\mu = 3.2$, (iii) $\mu = 3.5$, and (iv) $\mu = 4$.

It is not too difficult to extend the diagram to cover the whole range of values for μ, namely $0 \le \mu \le 4$. The bifurcation diagram given in Figure 14.15 was produced using the Mathematica package. Thus, even the simple quadratic function $f_\mu(x) = \mu x(1 - x)$ exhibits an extraordinary variety of behaviors as μ varies from one to four. In the past, scientists believed that in order to model complicated behavior, one must have complicated or many equations. One of the most exciting developments to emerge from the realm of nonlinear dynamical systems was the realization that simple equations can lead to extremely complex seemingly random behavior.

Figure 14.15 shows *period-doubling bifurcations to chaos*. This means that as μ increases beyond b_1, points of period one become period two; at b_2 points of period two become period four, and so on. The sequence of period-doublings ends at about $\mu = 3.569945\ldots$, where the system becomes chaotic. This is not the end of the story, however, Figure 14.16 clearly shows regions where the system returns to periodic behavior, even if for only a small range of μ values. These regions are called *periodic windows*.

Near to the period-three window, the logistic map can display a new type of behavior known as *intermittency*, which is almost periodic behavior interrupted by occasional chaotic bursts. A graphical iteration and time series plot are shown in Figure 14.17. The intermittent nature becomes more evident as more points are plotted.

The geometry underlying this behavior can be seen by plotting a graphical iteration for f_μ^3 when $\mu = 3.8282$, for example. This is left as an exercise for the reader. As the parameter μ is increased, the length of the intervals of chaotic bursts become larger and larger until the system becomes fully chaotic. This phenomenon is known as an *intermittency route to chaos* and appears in many other physical examples.

An even more remarkable discovery was made by Mitchell J. Feigenbaum in the mid-1970s and involves the concept of *universality*. The first seven bifurcation points computed numerically, are given by $b_1 = 3.0, b_2 = 3.449490\ldots, b_3 = 3.544090\ldots, b_4 = 3.564407\ldots, b_5 = 3.568759\ldots, b_6 = 3.569692\ldots$, and $b_7 = 3.569891\ldots$. Feigenbaum discovered that if d_k is defined by $d_k = b_{k+1} - b_k$, then

$$\delta = \lim_{k \to \infty} \frac{d_k}{d_{k+1}} = 4.669202\ldots.$$

The number δ, known as the *Feigenbaum constant*, is much like the numbers π and e in that it appears throughout the realms of science. The constant δ can be found, not only in iterative maps but also in certain differential equations and even in physical experiments exhibiting period-doubling cascades to chaos. Hence, the Feigenbaum constant is called a universal constant.

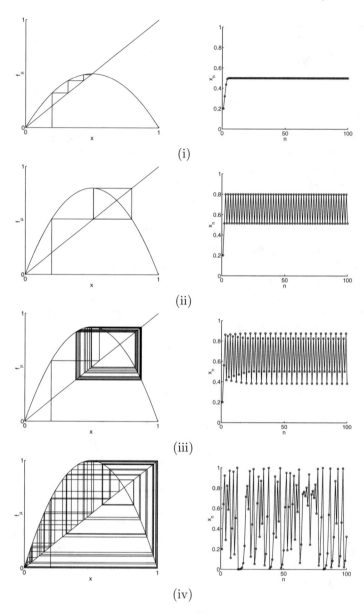

Figure 14.14:   Iterative paths and time series data representing the population of blowflies at time n. The population can vary periodically or in an erratic unpredictable manner.

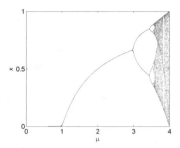

Figure 14.15: [Mathematica] The bifurcation diagram of the logistic map produced using the first iterative method (see Chapter 16).

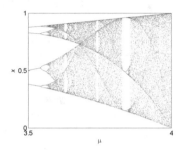

Figure 14.16: A magnification of the bifurcation diagram for the logistic map in the range $3.5 \le \mu \le 4$.

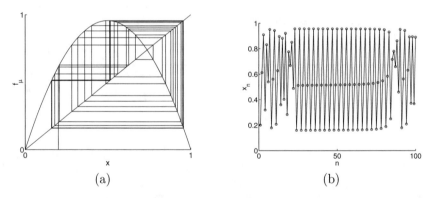

Figure 14.17: (a) Iterative paths when $\mu = 3.8282$ and $x_0 = 0.2$. (b) Time series data.

Figure 14.15 also has fractal structure, one may see similar patterns as you zoom into the picture. Fractals will be discussed in detail in Chapter 17.

Another method often used to determine whether or not a system is chaotic, is to use the *Lyapunov exponent*. One of the properties of chaos is the sensitivity to

initial conditions. However, it is known that an orbit on a chaotic attractor for a
bounded system also returns to all accessible states with equal probability. This
property is known as *ergodicity*. Thus, iterates return infinitely closely, infinitely
often to any previous point on the chaotic attractor. The formula below may
be applied to compute the Lyapunov exponent for iterates in the logistic map.
It gives an indication as to whether two orbits starting close together diverge
or converge.

Definition 3. The Lyapunov exponent L computed using the derivative
method is defined by

$$L = \frac{1}{n}\left(\ln|f'_\mu(x_1)| + \ln|f'_\mu(x_2)| + \ldots + \ln|f'_\mu(x_n)|\right),$$

where f'_μ represents differentiation with respect to x and $x_0, x_1, x_2, \ldots, x_n$ are
successive iterates. The Lyapunov exponent may be computed for a sample of
points near the attractor to obtain an *average Lyapunov exponent*.

Theorem 2. *If at least one of the average Lyapunov exponents is positive, then
the system is chaotic; if the average Lyapunov exponent is negative, then the
orbit is periodic and when the average Lyapunov exponent is zero, a bifurcation
occurs*

Table 14.2: The Lyapunov exponents computed to 4 decimal places using the
first derivative method for the logistic map. A total of 50000 iterates was used
in each case

μ	0.5	1	2.1	3	3.5	3.8282	4
Average L	-0.6932	-0.0003	-2.3025	-0.0002	-0.8720	0.2632	0.6932

Table 14.2 lists Lyapunov exponents computed for the logistic map (14.3) for
several values of the parameter μ. Note that there are other methods available
for determining Lyapunov exponents (see Chapter 8).

The numerical results agree quite well with Theorem 2. In fact, the more
chaotic a system the higher the value of the Lyapunov exponent, as can be seen
in Table 14.2. The Mathematica program is given in Section 14.6. In order to
find a better approximation of the Lyapunov exponent a much larger number
of iterates would be required.

Let us return briefly to the tent map (14.1). The Lyapunov exponent of the tent map can be found exactly since $T'(x) = \pm\mu$ for all values of x. Hence,

$$L = \lim_{n\to\infty} \left(\frac{1}{n} \sum_{i=1}^{n} \ln |T'(x_i)| \right) = \ln\mu.$$

Problem. Show that for the logistic map with $\mu = 4$, the Lyapunov exponent is in fact $L = \ln(2)$.

14.4 Gaussian and Hénon Maps

The Gaussian Map. Another type of nonlinear one-dimensional iterative map is the Gaussian map $G : \Re \to \Re$ defined by

$$G(x) = e^{-\alpha x^2} + \beta,$$

where α and β are constants. The graph of the Gaussian function has a general form as depicted in Figure 14.18. The parameters α and β are related to the width and height of the Gaussian curve, respectively.

Define an iterative map by

$$x_{n+1} = G(x_n).$$

Since there are two parameters associated with this map, one would expect the dynamics to be more complicated than for the logistic map. All of the features which appear in the logistic map are also present for the Gaussian map. However, certain features of the latter map are not exhibited at all by the logistic map. Some of these additional phenomena may be described as *period*

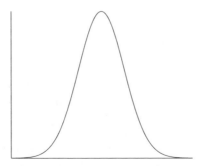

Figure 14.18: The Gaussian map function.

bubblings, period undoublings and *bistability*. These features can appear in the bifurcation diagrams.

Simple numerical and graphical iterations may be carried out as for the tent and logistic maps (see the exercises at the end of the chapter). The fixed points of period one may be found by solving the iterative equation $x_{n+1} = x_n$ for all n, which is equivalent to finding the intersection points of the function $G(x)$ with the diagonal. It is not difficult to see that there can be one, two or three intersections as shown in Figure 14.19. For certain parameter values, it is possible to have two stable fixed points of period one.

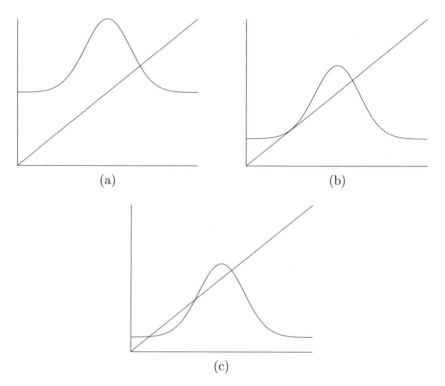

(a)   (b)

(c)

Figure 14.19:   Possible intersections of the Gaussian function with the diagonal.

The Gaussian map has two points of inflection at $x = \pm \frac{1}{\sqrt{2\alpha}}$. This implies that period-one behavior can exist for two ranges of the parameters. This in turn means that a period-one point can make a transition from being stable to unstable and back to stable again, as depicted in Figure 14.20.

As the parameter β is increased from $\beta = -1$, a fixed point of period one becomes unstable and a sequence of period bubbling occurs through period-two,

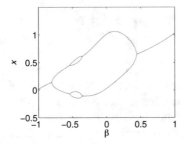

Figure 14.20: A bifurcation diagram for the Gaussian map when $\alpha = 4$ produced using the first iterative method.

period-four, and back to period-two behavior. As the parameter is increased still further, the unstable fixed point of period one becomes stable again and a single branch appears once more. For higher values of the parameter α, the system can display more complex dynamics. An example is shown in Figure 14.21.

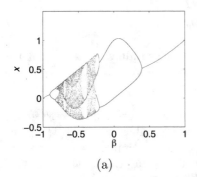

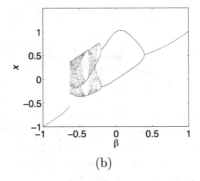

(a)   (b)

Figure 14.21: Bifurcation diagrams for the Gaussian map when $\alpha = 8$ produced using the first iterative method. In (a) $x_0 = 0$ and (b) $x_0 = -1$ for each value of β.

Figure 14.21 displays period-doubling and period-undoubling bifurcations and multistability. For example, when $\beta = -1$, there are two possible steady-state solutions. It is possible for these systems to display bistable phenomena as explained in other chapters of the book. The tent and logistic maps cannot display bistability.

The Hénon Map. Consider the two-dimensional iterated map function given by

$$x_{n+1} = 1 + y_n - \alpha x_n^2$$
$$y_{n+1} = \beta x_n, \tag{14.5}$$

where $\alpha > 0$ and $|\beta| < 1$. The map was first discussed by Hénon [6] in 1976 who used it as a simple model for the Poincaré map of the Lorenz system. The Hénon map displays periodicity, mixing, and sensitivity to initial conditions. The system can also display hysteresis and bistability can be observed in the bifurcation diagrams. Each of these phenomena will now be discussed briefly in turn.

Suppose that the discrete nonlinear system

$$x_{n+1} = P(x_n, y_n), \quad y_{n+1} = Q(x_n, y_n),$$

has a fixed point at (x_1, y_1), where P and Q are at least quadratic in x_n and y_n. The fixed point can be transformed to the origin and the nonlinear terms can be discarded after taking a Taylor series expansion. The Jacobian matrix is given by

$$J(x_1, y_1) = \left. \begin{pmatrix} \frac{\partial P}{\partial x} & \frac{\partial P}{\partial y} \\ \frac{\partial Q}{\partial x} & \frac{\partial Q}{\partial y} \end{pmatrix} \right|_{(x_1, y_1)}$$

Definition 4. Suppose that the Jacobian has eigenvalues λ_1 and λ_2. A fixed point is called *hyperbolic* if both $|\lambda_1| \neq 1$ and $|\lambda_2| \neq 1$. If either $|\lambda_1| = 1$ or $|\lambda_2| = 1$, then the fixed point is called nonhyperbolic.

The type of fixed point is determined using arguments similar to those used in Chapter 3. In the discrete case, the fixed point is stable as long as $|\lambda_1| < 1$ and $|\lambda_2| < 1$; otherwise, the fixed point is unstable. For example, the fixed points of period one for the Hénon map can be found by solving the equations given by $x_{n+1} = x_n$ and $y_{n+1} = y_n$ simultaneously. Therefore, period-one points satisfy the equations

$$x = 1 - \alpha x^2 + y, \quad y = \beta x.$$

The solutions are given by

$$x = \frac{(\beta - 1) \pm \sqrt{(1 - \beta)^2 + 4\alpha}}{2\alpha}, \quad y = \beta \left(\frac{(\beta - 1) \pm \sqrt{(1 - \beta)^2 + 4\alpha}}{2\alpha} \right).$$

Thus the Hénon map has two fixed points of period one if and only if $(1 - \beta)^2 + 4\alpha > 0$. As a particular example, consider system (14.5) with $\alpha = \frac{3}{16}$ and $\beta = \frac{1}{2}$. There are two fixed points of period one given by $A = (-4, -2)$ and $B = \left(\frac{4}{3}, \frac{2}{3} \right)$. The Jacobian is given by

$$J = \begin{pmatrix} -2\alpha x & 1 \\ \beta & 0 \end{pmatrix}.$$

The eigenvalues for the fixed point A are $\lambda_1 \approx -0.28$ and $\lambda_2 \approx 1.78$; therefore, A is a saddle point. The eigenvalues for the fixed point B are $\lambda_1 = -1$ and $\lambda_2 = 0.5$. Thus, this critical point is nonhyperbolic.

Fix the parameter $\beta = 0.4$ in the Hénon map (14.5). There are points of periods one (when $\alpha = 0.2$), two (when $\alpha = 0.5$), and four (when $\alpha = 0.9$), for example. The reader can verify these results using the Mathematica program in Section 14.6. Some iterative plots are given in Figure 14.22.

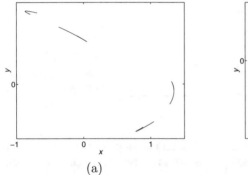
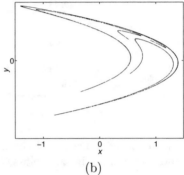

(a)   (b)

Figure 14.22: Iterative plots for system (14.5) when $\beta = 0.4$ and (a) $\alpha = 1$ and (b) $\alpha = 1.2$. In each case, the initial point was $(0.1, 0)$.

The choice of initial conditions is important in these cases as some orbits are unbounded and move off to infinity. One must start with points that are within the *basin of attraction* for this map. Basins of attraction are discussed in other chapters of this book. Of course, all of this information can be summarized on a bifurcation diagram, and this will be left as an exercise for the reader. There are the usual phenomena associated with bifurcation diagrams. However, for the Hénon map, different chaotic attractors can exist simultaneously for a range of parameter values of α. This system also displays hysteresis for certain parameter values.

To demonstrate the stretching and folding associated with this map, consider a set of initial points lying on the square of length two centered at the origin. Figure 14.23 shows how the square is stretched and folded after only two iterations. This stretching and folding is reminiscent of the Smale horseshoe discussed in Chapter 9.

The chaotic attractor formed is an invariant set and has fractal structure. Note that $\det(J)$ for the Hénon map is equal to $|\beta|$. This implies that a small area is reduced by a factor of β on each iteration since $|\beta| < 1$.

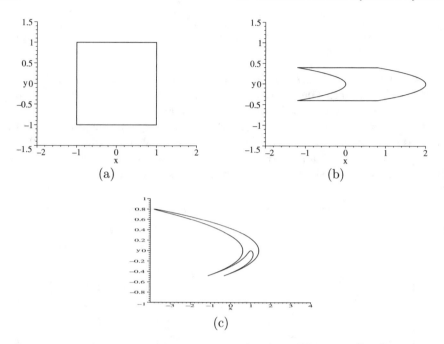

Figure 14.23:   Application of the Hénon transformation to a square when $\alpha = 1.2$ and $\beta = 0.4$: (a) initial points, (b) first iterates, and (c) second iterates.

14.5 Applications

This section introduces four discrete dynamical systems taken from biology, economics, nonlinear optics, and neural networks. The reader can investigate these systems via the exercises in Section 14.7.

Biology. The average human's 70 liter body contains five liters of blood, a small amount of which consists of *erythrocytes* or red blood cells. These cells, which are structurally the simplest in the body, can be counted to measure hematologic conditions such as *anemia*. Anemia is any condition resulting in a significant decrease in total body erythrocyte mass. The population of red blood cells oscillates in a healthy human with the average woman having 4.2-5.4 per μL, and the average man having 4.7-6.1 per μL. A simple blood cell population model was investigated by Lasota [9] in 1977. Let c_n denote the red cell count per unit volume in the nth time interval, then

$$c_{n+1} = c_n - d_n + p_n,$$

where d_n and p_n are the number of cells destroyed and produced in one time interval, respectively. In the model considered by Lasota, $d_n = ac_n$ and $p_n =$

$bc_n^r e^{-sc_n}$, where $0 < a \le 1$ and $b, r, s > 0$. Hence,

$$c_{n+1} = (1-a)c_n + bc_n^r e^{-sc_n}. \tag{14.6}$$

Typical parameters used in the model are $b = 1.1 \times 10^6$, $r = 8$ and $s = 16$. Clinical examples are cited in the author's paper [11], where a model is investigated in which production and destruction rates vary.

Economics. The gross national product (GNP) measures economic activity based on labor and production output within a country. Consider the following simple growth model investigated by Day [3] in 1982:

$$k_{t+1} = \frac{s(k_t)f(k_t)}{1+\lambda},$$

where k_t is the capital–labor ratio, s is the savings ratio function, f is the per capita production function, and λ is the natural rate of population growth. In one case considered by Day,

$$s(k) = \sigma, \quad f(k) = \frac{Bk^\beta (m-k)^\gamma}{(1+\lambda)},$$

where $\beta, \gamma, m > 0$. This leads to the following discrete dynamical system:

$$k_{t+1} = \sigma \frac{Bk_t^\beta (m-k_t)^\gamma}{(1+\lambda)}, \tag{14.7}$$

which can be thought of as a highly simplified model for the GNP of a country.

Nonlinear Optics. When modeling the intracavity field of a laser in a bulk cavity ring under the assumption that saturable absorption can be ignored, Hammel, Jones, and Moloney [5] obtained the following complex one-dimensional difference equation relating the field amplitude, say, E_{n+1}, at the $(n+1)$st cavity pass to that of a round trip earlier:

$$E_{n+1} = A + BE_n \exp\left[i\left(\phi - \frac{C}{1+|E_n|^2}\right)\right], \tag{14.8}$$

where ϕ is a phase angle, and A, B, C are all constant. This mapping can also be thought of as two-dimensional (one-dimensional complex). Splitting E_n into its real and imaginary parts, equation (14.8) becomes

$$\begin{aligned}
x_{n+1} &= A + B\left[x_n \cos(\theta) - y_n \sin(\theta)\right] \\
y_{n+1} &= B\left[x_n \sin(\theta) + y_n \cos(\theta)\right],
\end{aligned} \tag{14.9}$$

where $\theta = \left(\phi - \frac{C}{1+|E_n|^2}\right)$. Equations (14.8) and (14.9) are known as *Ikeda mappings*. Electromagnetic waves and optical resonators are dealt with in some detail in Chapter 16.

Neural Networks. According to Pasemann and Stollenwerk [15], the activity of a recurrent two-neuron module shown in Figure 14.24 at time n is given by the vector $\mathbf{x_n} = (x_n, y_n)^T$. The discrete dynamical system used to model the neuromodule is given by

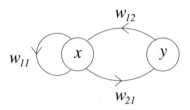

Figure 14.24: A recurrent two-neuron module with an excitory neuron with activity y_n, and a self-connected inhibitory neuron with activity x_n.

$$x_{n+1} = \theta_1 + w_{11}\sigma(x_n) + w_{12}\sigma(y_n)$$
$$y_{n+1} = \theta_2 + w_{21}\sigma(x_n), \qquad (14.10)$$

where σ defines the sigmoidal transfer function defining the output of a neuron

$$\sigma(x) = \frac{1}{1+e^{-x}};$$

θ_1 and θ_2 are the neuron biases; the w_{ij} are weights; the index i indicates the neuron destination for that weight, and the index j represents the source of the signal fed to the neuron. The author and Bandar [12] consider a simple neuromodule subject to feedback. Neural networks are dealt with in some detail in Chapter 20.

14.6 Mathematica Commands

Type NestList, Prepend, Partition, Drop, Compile, Map, and Union in the Mathematica Help Browser, for explanations of what these commands do in the following programs.

   The character # is equivalent to the command Slot[1]. *Body* & is equivalent to Function[*body*].

```
(* Iteration of the tent map. See Example 1 and Table 14.1. *)
In[1]:=mu=2;
In[2]:=T[x_]=Which[0<=x<=1/2,mu x,1/2<=x<=1,mu (1-x)];
In[3]:=NestList[T,1/11,10]
```

$$\text{Out}[3]=\{\tfrac{1}{11},\tfrac{2}{11},\tfrac{4}{11},\tfrac{8}{11},\tfrac{6}{11},\tfrac{10}{11},\tfrac{2}{11},\tfrac{4}{11},\tfrac{8}{11},\tfrac{6}{11}\}$$

```
In[4]:=NestList[T, 0.2, 20]
Out[4]={0.2,0.4,0.8,0.4,0.8,0.4,0.8,0.4,0.8,0.4,0.8,0.4,0.8,0.4,0.8,0.4,
        0.8,0.4,0.8,0.4,0.8}
In[5]:=NestList[T,0.2001,20]
Out[5]={0.2001,0.4002,0.8004,0.3992,0.7984,0.4032,0.8064,0.3872,0.7744,
        0.4512,0.9024,0.1952,0.3904,0.7808,0.4384,0.8768,0.2464,0.4928,
        0.9856,0.0288,0.0576}
```

```
(* Graphical iteration of the tent map. *)
In[6]:=Iterations=20;x0=1/11;
In[7]:=NL=NestList[T,x0,Iterations];
In[8]:=x=Drop[Prepend[Flatten[Partition[NL,2,1]],x0],-Iterations];
In[9]:=y=Drop[Prepend[Drop[Flatten[Partition[NL,2,1]],1],0],
        -(Iterations-1)];
In[10]:=p1=ListPlot[Table[{x[[i]], y[[i]]},{i,1,Iterations+1}],
        Joined->True,PlotStyle->RGBColor[1,0,0]];
In[11]:=p2=Plot[{x,T[x]},{x,0,1}];
In[12]:=Show[{p1,p2},PlotRange->All,AxesLabel->{"x","T"},Axes->True,
        AspectRatio->Automatic]
(* See Figure 14.6(d). *)
```

```
(* Computing the Lyapunov exponent for the logistic map. *)
In[13]:=Clear[mu,x,i];mu=4;
In[14]:=fmu[x_]=N[mu x (1-x)];
In[15]:=Itermax=50000;
In[16]:=p=NestList[fmu,.1,Itermax];
In[17]:=pp=Log[Abs[mu (1-2 p)]];
In[18]:=Total[pp/(Itermax+1)]
Out[18]=0.693146
```

```
(* Bifurcation diagram of the logistic map using the first iterative *)
(* method. *)
In[19]:=x0=0.4;stepsize=0.001;
In[20]:=Itermax=200;
In[21]:=iterate=Compile[{mu},Map[{mu, #}&,Union[Drop[NestList
        [mu # (1-#)&,x0,Itermax],100]]]];
In[22]:=pts=Flatten[Table[iterate[mu],{mu,0,4,stepsize}],1];
In[23]:=ListPlot[pts,PlotStyle->{PointSize[0.001],RGBColor[1,0,0]},
```

```
        AxesLabel->{"mu", "x"}]
(*See Figure 14.15.*)
```

```
(* Iteration of the Henon map. *)
In[24]:=Quit[ ]
In[1]:=a=1.2;b=0.4;Itermax=10000;
In[2]:=Henon[{x_,y_}]={1+y-a x^2,b x};
In[3]:=HenonMap=Drop[NestList[Henon,{0.1,0},Itermax],100];
In[4]:=ListPlot[HenonMap,AxesLabel->{"x","y"},Axes->None,
        PlotStyle->{PointSize[0.001],RGBColor[0,1,0]},Frame->True]
(*See Figure 14.22(b).*)
```

```
(* Compute the Lyapunov exponents of the Henon map. See Exercise 8(c). *)
MaxIter=500;a=1.2;b=0.4;x=0;y=0;
vec1={1,0};vec2={0,1};
Do[x1=1-a*x^2+y;y1=b*x;x=x1;y=y1;
J={{-2*a*x,1},{b,0}};vec1=J.vec1;vec2=J.vec2;
dotprod1=vec1.vec1;
dotprod2=vec1.vec2;
vec2=vec2-(dotprod2/dotprod1)*vec1;
lengthv1=Sqrt[dotprod1];
area=Abs[vec1[[1]]*vec2[[2]]-vec1[[2]]*vec2[[1]]];
h1=Log[lengthv1]/i;h2=Log[area]/i-h1,{i,1,MaxIter}];
Print["h1=",h1]
Print["h2=",h2]
```

14.7 Exercises

1. Consider the tent map defined by

$$T(x) = \begin{cases} 2x & 0 \le x < \frac{1}{2} \\ 2(1-x) & \frac{1}{2} \le x \le 1. \end{cases}$$

Sketch graphical iterations for the initial conditions (i) $x_0 = \frac{1}{4}$, (ii) $x_0 = \frac{1}{6}$, (iii) $x_0 = \frac{5}{7}$, and (iv) $x_0 = \frac{1}{19}$. Find the points of periods one, two, three, and four. Give a formula for the number of points of period N.

2. (a) Let T be the function $T : [0, 1] \to [0, 1]$ defined by

$$T(x) = \begin{cases} \frac{3}{2}x & 0 \le x < \frac{1}{2} \\ \frac{3}{2}(1-x) & \frac{1}{2} \le x \le 1. \end{cases}$$

Sketch the graphs of $T(x)$, $T^2(x)$, and $T^3(x)$. How many points are there of periods one, two, and three, respectively?

(b) Let T be the function $T : [0,1] \to [0,1]$ defined by

$$T(x) = \begin{cases} \frac{9}{5}x & 0 \le x < \frac{1}{2} \\ \frac{9}{5}(1-x) & \frac{1}{2} \le x \le 1. \end{cases}$$

Determine the fixed points of periods one, two, and three.

3. By editing the Mathematica program given in Section 14.6, plot a bifurcation diagram for the tent map.

4. Consider the logistic map function defined by $f_\mu(x) = \mu x(1-x)$. Determine the functions $f_\mu(x)$, $f_\mu^2(x)$, $f_\mu^3(x)$, and $f_\mu^4(x)$, and plot the graphs when $\mu = 4.0$. How many points are there of periods one, two, three, and four?

5. Consider the iterative equation

$$x_{n+1} = \mu x_n (100 - x_n),$$

which may be used to model the population of a certain species of insect. Given that the population size periodically alternates between two distinct values, determine a value of μ that would be consistent with this behavior. Determine an equation that gives the points of period two for a general μ value.

6. Plot bifurcation diagrams for

(a) the Gaussian map when $\alpha = 20$ for $-1 \le \beta \le 1$;

(b) the Gaussian map when $\beta = -0.5$ for $0 \le \alpha \le 20$.

7. Find the fixed points of periods one and two for the Hénon map given by

$$x_{n+1} = \frac{3}{50} + \frac{9}{10}y_n - x_n^2, \quad y_{n+1} = x_n.$$

Derive the inverse map.

8. (a) Show that the Hénon map given by

$$x_{n+1} = 1 - \alpha x_n^2 + y_n, \quad y_{n+1} = \beta x_n,$$

where $\alpha > 0$ and $|\beta| < 1$ undergoes a bifurcation from period-one to period-two behavior exactly when $\alpha = \frac{3(\beta-1)^2}{4}$ for fixed β.

(b) Investigate the bifurcation diagrams for the Hénon map by plotting the x_n values as a function of α for $\beta = 0.4$.

(c) Derive the Lyapunov exponents of the Hénon map when $\alpha = 1.2$ and $\beta = 0.4$.

9. (a) Consider the blood cell iterative equation (14.6). Assuming that $b = 1.1 \times 10^6$, $r = 8$ and $s = 16$, show that there are (i) two stable and one unstable fixed points of period one when $a = 0.2$, and (ii) two unstable and one stable fixed point of period one when $a = 0.3$.

(b) Assume that $\sigma = 0.5$, $\beta = 0.3$, $\gamma = 0.2$, $\lambda = 0.2$, $m = 1$ in the economic model (14.7). Show that there is a stable fixed point of period one at $x_{1,2} = 0.263$ when $B = 1$, and an unstable fixed point of period one at $x_{1,2} = 0.873$ when $B = 3.3$.

(c) Show that the inverse map of equation (14.8) is given by

$$E_{n+1} = \frac{(E_n - A)}{B} \exp\left[-i\left(\phi - \frac{CB^2}{(B^2 + |E_n - A|^2)}\right)\right].$$

(d) Consider the neuromodule model (14.10). Assume that $\theta_1 = -2$, $\theta_2 = 3$, $w_{11} = -20$, $w_{12} = 6$, and $w_{21} = -6$. Show that there is one fixed point of period one approximately at $(-1.280, 1.695)$ and that it is a saddle point.

10. According to Ahmed et al. [1], an inflation-unemployment model is given by

$$U_{n+1} = U_n - b(m - I_n), \quad I_{n+1} = I_n - (1 - c)f(U_n) + f(U_n - b(m - I_n)),$$

where $f(U) = \beta_1 + \beta_2 e^{-U}$, U_n and I_n are measures of unemployment and inflation at time n, respectively, and b, c, β_1, and β_2 are constants. Show that the system has a unique fixed point of period one at

$$\left(\ln\left(\frac{-\beta_2}{\beta_1}\right), m\right).$$

Given that $m = 2$, $\beta_1 = -2.5$, $\beta_2 = 20$ and $c = 0.18$, show that the eigenvalues of the Jacobian matrix are given by

$$\lambda_{1,2} = 1 - \frac{5b}{4} \pm \frac{\sqrt{25b^2 - 40bc}}{4}.$$

Bibliography

[1] E. Ahmed, A. El-Misiery and H.N. Agiza, On controlling chaos in an inflation-unemployment dynamical system, *Chaos, Solitons and Fractals*, **10** (1999), 1567–1570.

[2] Z. AlSharawi, J.M. Cushing and S. Elaydi (Editors), *Theory and Applications of Difference Equations and Discrete Dynamical Systems: ICDEA, Muscat, Oman, May 26–30, 2013*, Springer, New York, 2016.

[3] R.H. Day, Irregular growth cycles, *The American Economic Review*, **72** (1982), 406–414.

[4] O. Galor, *Discrete Dynamical Systems*, Springer, 2010.

[5] S.M. Hammel, C.K.R.T. Jones, and J.V. Maloney, Global dynamical behaviour of the optical field in a ring cavity, *J. Opt. Soc. Am. B*, **2** (1985), 552–564.

[6] M. Hénon, Numerical study of quadratic area-preserving mappings, *Quart. Appl. Math.*, **27** (1969), 291–311.

[7] R.A. Holmgrem, *A First Course in Discrete Dynamical Systems, 2nd ed.*, Springer-Verlag, New York, 2013.

[8] D. Kaplan and L. Glass, *Understanding Nonlinear Dynamics*, Springer-Verlag, New York, 1995.

[9] A. Lasota, Ergodic problems in biology, *Astérisque*, **50** (1977), 239–250.

[10] T.Y. Li and J.A. Yorke, Period three implies chaos, *Amer. Math. Monthly*, **82** (1975), 985–992.

[11] S. Lynch, Analysis of a blood cell population model, *Int. J. of Bifurcation and Chaos*, **15** (2005), 2311–2316.

[12] S. Lynch and Z.G. Bandar, Bistable neuromodules, *Nonlinear Anal. Theory, Meth. & Appl.*, **63** (2005), 669–677.

[13] R.M. May, *Stability and Complexity in Model Ecosystems*, Princeton University Press, Princeton, NJ, 1974.

[14] H. Nagashima and Y. Baba, *Introduction to Chaos, Physics and Mathematics of Chaotic Phenomena*, Institute of Physics, London, UK, 1998.

[15] F. Pasemann and N. Stollenwerk, Attractor switching by neural control of chaotic neurodynamics, *Computer Neural Systems*, **9** (1998), 549–561.

[16] L.A. Soler, J.M. Cushing, S. Elaydi and A.A. Pinto (Editors), *Difference Equations, Discrete Dynamical Systems and Applications: ICDEA, Barcelona, Spain, July 2012*, Springer, New York, 2016.

Chapter 15
Complex Iterative Maps

Aims and Objectives

- To introduce simple complex iterative maps;

- To introduce Julia sets and the Mandelbrot set;

- To carry out some analysis on these sets;

- To introduce Newton fractals;

On completion of this chapter, the reader should be able to

- carry out simple complex iterations;

- plot Julia sets and the Mandelbrot set using simple Mathematica programs;

- determine boundaries of points with low periods;

- find basins of attraction (or domains of stability);

- plot Newton fractals;

It is assumed that the reader is familiar with complex numbers and the Argand diagram. Julia sets are defined, and Mathematica is used to plot approximations of these sets.

There are an infinite number of Julia sets associated with one mapping. In one particular case, these sets are categorized by plotting the so-called Mandelbrot set. A Mathematica program for plotting a color version of the Mandelbrot set is listed.

Applications of complex iterative maps to the real world are presented in Chapter 16, and generalizations of Julia and Mandelbrot set are discussed in [4].

© Springer International Publishing AG 2017   339
S. Lynch, *Dynamical Systems with Applications using Mathematica®*,
DOI 10.1007/978-3-319-61485-4_15

15.1 Julia Sets and the Mandelbrot Set

As a simple introduction to one-dimensional nonlinear complex iterative maps, consider the quadratic map

$$z_{n+1} = f_c(z_n) = z_n^2 + c, \tag{15.1}$$

where z_n and c are complex numbers. Although equation (15.1) is as simple as the equation of a real circle, the dynamics displayed are highly complicated. In 1919, Gaston Julia published a prize-winning lengthy article on certain types of conformal complex mappings, the images of which would not appear until the advent of computer graphics many years later. Recall that a conformal mapping preserves both the size and the sign of angles.

Definition 1. Consider a complex polynomial mapping of the form $z_{n+1} = f(z_n)$. The points that lie on the boundary between points that orbit under f and are bounded and those that orbit under f and are unbounded are collectively referred to as the *Julia set*.

The following properties of a Julia set, say, J, are well known:

- The set J is a repellor.

- The set J is invariant.

- An orbit on J is either periodic or chaotic.

- All unstable periodic points are on J.

- The set J is either wholly connected or wholly disconnected.

- The set J nearly always has fractal structure (see Chapter 17).

As a gentle introduction to Julia sets and the Mandelbrot set, the reader is directed to the book entitled "Fractals for the Classroom" [7], and to see the true beauty and some detail of the Julia sets and the Mandelbrot set, the author would encourage the reader to watch the video [8]. There are also numerous videos on You Tube, where the viewer can take a virtual journey in to the Mandelbrot set.

To generate Julia sets, some of the properties listed above are utilized. For example, if the set J is a repellor under the forward iterative map (15.1), then the Julia set will become an attractor under an inverse mapping. For computational reasons, it is best to work with the real and imaginary parts of the complex numbers separately. For equation (15.1), it is not difficult to determine the inverse map. Now

$$z_{n+1} = z_n^2 + c,$$

and thus

$$x_{n+1} = x_n^2 - y_n^2 + a, \quad \text{and} \quad y_{n+1} = 2x_n y_n + b,$$

where $z_n = x_n + iy_n$ and $c = a + ib$. To find the inverse map, one must find expressions for x_n and y_n in terms of x_{n+1} and y_{n+1}. Now

$$x_n^2 - y_n^2 = x_{n+1} - a,$$

and note that

$$(x_n^2 + y_n^2)^2 = (x_n^2 - y_n^2)^2 + 4x_n^2 y_n^2 = (x_{n+1} - a)^2 + (y_{n+1} - b)^2.$$

Hence

$$x_n^2 + y_n^2 = +\sqrt{(x_{n+1} - a)^2 + (y_{n+1} - b)^2},$$

since $x_n^2 + y_n^2 > 0$. Suppose that

$$u = \frac{\sqrt{(x_{n+1} - a)^2 + (y_{n+1} - b)^2}}{2}, \quad \text{and} \quad v = \frac{x_{n+1} - a}{2}.$$

Then

$$x_n = \pm\sqrt{u + v} \quad \text{and} \quad y_n = \frac{y_{n+1} - b}{2x_n}. \tag{15.2}$$

In terms of the computation, there will be a problem if $x_n = 0$. To overcome this difficulty, the following simple algorithm is applied. Suppose that the two roots of equation (15.2) are given by $x_1 + iy_1$ and $x_2 + iy_2$. If $x_1 = \sqrt{u + v}$, then $y_1 = \sqrt{u - v}$ if $y > b$, or $y_1 = -\sqrt{u - v}$ if $y < b$. The other root is then given by $x_2 = -\sqrt{u + v}$ and $y_2 = -y_1$.

This transformation has a two-valued inverse, and twice as many predecessors are generated on each iteration. One of these points is chosen randomly in the computer program. Recall that all unstable periodic points are on J. It is not difficult to determine the fixed points of period one for mapping (15.1). Suppose that z is a fixed point of period one. Then $z_{n+1} = z_n = z$, and

$$z^2 - z + c = 0,$$

which gives two solutions, either

$$z_{1,1} = \frac{1 + \sqrt{1 - 4c}}{2} \quad \text{or} \quad z_{1,2} = \frac{1 - \sqrt{1 - 4c}}{2}.$$

The stability of these fixed points can be determined in the usual way. Hence the fixed point is stable if

$$\left|\frac{df_c}{dz}\right| < 1$$

and, it is unstable if

$$\left|\frac{df_c}{dz}\right| > 1.$$

By selecting an unstable fixed point of period one as an initial point, it is possible to generate a Julia set using the so-called *backward training iterative process*.

Julia sets define the border between bounded and unbounded orbits. Suppose that the Julia set associated with the point $c = a + ib$ is denoted by $J(a, b)$. As a simple example, consider the mapping

$$z_{n+1} = z_n^2. \tag{15.3}$$

One of two fixed points of equation (15.3) lies at the origin, say, z^*. There is also a fixed point at $z = 1$. Initial points that start wholly inside the circle of radius one are attracted to z^*. An initial point starting on $|z| = 1$ will generate points that again lie on the unit circle $|z| = 1$. Initial points starting outside the unit circle will be repelled to infinity, since $|z| > 1$. Therefore, the circle $|z| = 1$ defines the Julia set $J(0, 0)$ that is a repellor (points starting near to but not on the circle are repelled), invariant (orbits that start on the circle are mapped to other points on the unit circle), and wholly connected. The interior of the unit circle defines the *basin of attraction* (or domain of stability) for the fixed point at z^*. In other words, any point starting inside the unit circle is attracted to z^*.

Suppose that $c = -0.5 + 0.3i$, in equation (15.1). Figure 15.1(a) shows a picture of the Julia set $J(-0.5, 0.3)$ containing 2^{15} points. The Julia set $J(-0.5, 0.3)$ defines the border between bounded and unbounded orbits. For example, an orbit starting inside the set $J(-0.5, 0.3)$ at $z_0 = 0 + 0i$ remains bounded, whereas an orbit starting outside the set $J(-0.5, 0.3)$ at $z = -1 - i$, for instance, is unbounded. The reader will be asked to demonstrate this in the exercises at the end of the chapter.

Four of an infinite number of Julia sets are plotted in Figure 15.1. The first three are totally connected, but $J(0, 1.1)$ is totally disconnected. A program for plotting Julia sets is listed in Section 15.3. Note that there may be regions where the Julia set is sparsely populated (see Figure 15.1(c)). You can of course increase the number of iterations to try to close these gaps, but other improved methods are available.

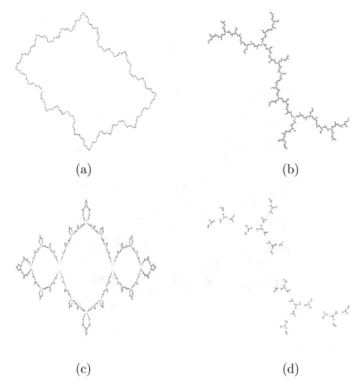

(a)   (b)

(c)   (d)

Figure 15.1: [Mathematica] Four Julia sets for the mapping (15.1), where $J(a,b)$ denotes the Julia set associated with the point $c = a+ib$: (a) $J(-0.5, 0.3)$, (b) $J(0,1)$, (c) $J(-1,0)$, and (d) $J(0,1.1)$.

In 1979, Mandelbrot devised a way of distinguishing those Julia sets that is wholly connected from those that are wholly disconnected. He used the fact that $J(a,b)$ is connected if and only if the orbit generated by $z \rightarrow z^2 + c$ is bounded. In this way, it is not difficult to generate the now famous *Mandelbrot set*.

Assign a point on a computer screen to a coordinate position $c = (a,b)$, in the Argand plane. The point $z = 0 + 0i$ is then iterated under the mapping (15.1) to give an orbit

$$0 + 0i, c, c^2 + c, (c^2 + c)^2 + c, \ldots .$$

If after 50 iterations, the orbit remains bounded (within a circle of radius 4 in the program used here), then the point is colored black. If the orbit leaves the circle of radius 4 after m iterations, where $1 < m < 50$, then the point is colored black if m is even and white if m is odd. In this way a black-and-white picture of the Mandelbrot set can be plotted. A color version of the program that gives Figure 15.2 is listed in Section 15.4.

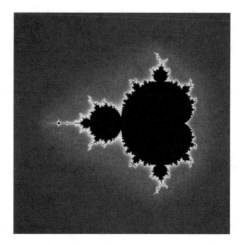

Figure 15.2: [Mathematica] The color Mandelbrot set (central black figure) produced using a personal computer.

Unfortunately, Figure 15.2 does no justice to the beauty and intricacy of the Mandelbrot set. This figure is a theoretical object that can be generated to an infinite amount of detail, and the set is a kind of fractal displaying self-similarity in certain parts and scaling behavior. One has to try to imagine a whole new universe that can be seen by zooming into the picture. For a video journey into the Mandelbrot set, the reader is once more directed to the video [8] and You Tube. In 2016, the current record for an animated zoom consisted of 750 million iterations.

It has been found that this remarkable figure is a universal "constant" much like the Feigenbaum number introduced in Chapter 14. Some simple properties of the Mandelbrot set will be investigated in the next section.

15.2 Boundaries of Periodic Orbits

For the Mandelbrot set, the fixed points of period one may be found by solving the equation $z_{n+1} = z_n$ for all n, or equivalently,

$$f_c(z) = z^2 + c = z,$$

which is a quadratic equation of the form

$$z^2 - z + c = 0. \tag{15.4}$$

The solutions occur at

$$z_{1,1} = \frac{1 + \sqrt{1 - 4c}}{2} \quad \text{and} \quad z_{1,2} = \frac{1 - \sqrt{1 - 4c}}{2},$$

where $z_{1,1}$ is the first fixed point of period one, and $z_{1,2}$ is the second fixed point of period one using the notation introduced in Chapter 14. As with other discrete systems, the stability of each period-one point is determined from the derivative of the map at the point. Now

$$\frac{df_c}{dz} = 2z = re^{i\theta}, \tag{15.5}$$

where $r \geq 0$ and $0 \leq \theta < 2\pi$. Substituting from equation (15.5), equation (15.4) then becomes

$$\left(\frac{re^{i\theta}}{2}\right)^2 - \frac{re^{i\theta}}{2} + c = 0.$$

The solution for c is

$$c = \frac{re^{i\theta}}{2} - \frac{r^2 e^{i2\theta}}{4}. \tag{15.6}$$

One of the fixed points, say, $z_{1,1}$, is stable as long as

$$\left| \frac{df_c}{dz}(z_{1,1}) \right| < 1.$$

Therefore, using equation (15.5), the boundary of the points of period one is given by

$$\left| \frac{df_c}{dz}(z_{1,1}) \right| = |2z_{1,1}| = r = 1$$

in this particular case. Let $c = x + iy$. Then from equation (15.6), the boundary is given by the following parametric equations:

$$x = \frac{1}{2}\cos\theta - \frac{1}{4}\cos(2\theta), \quad y = \frac{1}{2}\sin\theta - \frac{1}{4}\sin(2\theta).$$

The parametric curve is plotted in Figure 15.3 and forms a cardioid that lies at the heart of the Mandelbrot set.

Using similar arguments to those above, it is not difficult to extend the analysis to determine the boundary for the fixed points of period two. Fixed points of period two satisfy the equation $z_{n+2} = z_n$ for all n. Therefore,

$$f_c^2(z) = (z^2 + c)^2 + c = z,$$

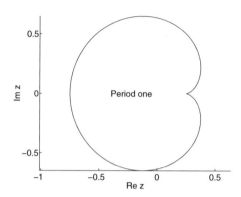

Figure 15.3: The boundary of fixed points of period one for the Mandelbrot set.

or, equivalently,

$$z^4 + 2cz^2 - z + c^2 + c = 0. \tag{15.7}$$

However, since points of period one repeat on every second iterate, the points $z_{1,1}$ and $z_{1,2}$ satisfy equation (15.7). Therefore, equation (15.7) factorizes into

$$(z^2 - z + c)(z^2 + z + c + 1) = 0.$$

Hence the fixed points of period two satisfy the quadratic equation

$$z^2 + z + c + 1 = 0, \tag{15.8}$$

which has roots at

$$z_{2,1} = \frac{-1 + \sqrt{-3 - 4c}}{2} \quad \text{and} \quad z_{2,2} = \frac{-1 - \sqrt{-3 - 4c}}{2}.$$

Once more the stability of each critical point is determined from the derivative of the map at the point, now

$$\frac{df_c^2}{dz} = 4z^3 + 4cz = 4z(z^2 + c).$$

Thus,

$$\left| \frac{df_c^2}{dz}(z_{2,1}) \right| = |4 + 4c|,$$

and the boundary is given by

$$|c + 1| = \frac{1}{4}.$$

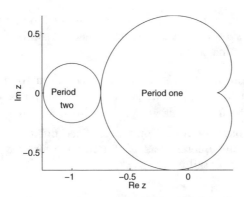

Figure 15.4: The boundary of fixed points of periods one and two for the Mandelbrot set.

The parametric curve is plotted in Figure 15.4 and forms a circle centered at $(-1, 0)$ of radius $1/4$ in the Argand plane. This circle forms the "head" of the Mandelbrot set, sometimes referred to as the *potato man*.

The Mandelbrot set for the nonlinear complex iterative map $z_{n+1} = z_n^2 - 2z_n + c$, is plotted in Figure 15.5.

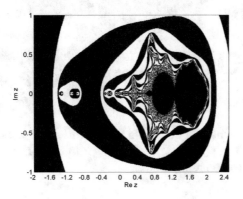

Figure 15.5: The Mandelbrot set for the mapping $z_{n+1} = z_n^2 - 2z_n + c$.

15.3 The Newton Fractal

It is well known that in numerical analysis [5], Newton's method, or the Newton-Raphson method, can be used to find the roots of the equation $f(z) = 0$ using the iterative formula

$$z_{n+1} = z_n - \frac{f(z_n)}{f'(z_n)}.$$

Definition 2. A Newton fractal is the Julia set of the meromorphic function $z_{n+1} = f(z_n)$ and shows that the numerical method can be very sensitive to its choice of initial starting point.

In Figure 15.6, all points in the blue regions are in the basin of attraction for the fixed point at $z_{11} = 1$; all points in the red regions are in the basin of attraction for the fixed point at $z_{12} = \frac{-1+\sqrt{3}i}{2}$; and all points in the green regions are in the basin of attraction for the fixed point at $z_{13} = \frac{-1-\sqrt{3}i}{2}$. The boundary between the different basins of attraction forms a Julia set. Fractals are discussed in more detail in Chapter 17.

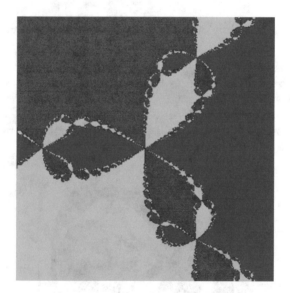

Figure 15.6: [Mathematica] Julia sets for the rational function associated with Newton's method for $f(z) = z^3 - 1$. Using three colors to identify basins of attraction for the three fixed points.

Note that other fractals may be constructed using different numerical techniques such as the Halley, Householder, Secant and Schröder methods [10].

Mandelbrot and Hudson [6] provide a fractal view of the stock markets, and Chapter 16 illustrates how nonlinear complex iterative maps are being applied in physical applications when modeling lasers and the propagation of light through optical fibers.

15.4 Mathematica Commands

```
(* Plotting Julia sets. See Figure 15.1. *)
In[1]:=Coordinates={Complex[x_,y_]->{x,y}};
In[2]:=JuliaSet[c_,z0_,Npoints_]:=Module[{Nc=N[c],Nz0=N[z0]},
       NestList[(1-2 Random[Integer]) Sqrt[#-Nc] &,Nz0,Npoints]
       /. Coordinates];
In[3]:=JuliaData=Drop[Evaluate[JuliaSet[-1+0 I,0+5 I,2^18]],100];
In[4]:=ListPlot[JuliaData,Axes->False,AspectRatio->Automatic,
       PlotStyle->PointSize[0.001]]
```

```
(* A color Mandelbrot set. See Figure 15.2. *)
In[5]:=xmin =-2.2;xmax=0.6;ymin=-1.2;ymax=1.2;
In[6]:=Itermax=50;stepsize=0.005;
In[7]:=Counter=Compile[{x,y,{Itermax,_Integer}},
       Module[{w,z,count=0},w=x+I y;z=w;
       While[(Abs[z]<4.0) && (count<=Itermax),++count;z=z^2+w;];
       Return[count]]];
In[8]:=MandelbrotData=Table[Counter[x,y,Itermax],{y,ymin,ymax,stepsize},
       {x,xmin,xmax,stepsize}];
In[9]:=ListDensityPlot[MandelbrotData,Frame->False,Mesh->False,
       DataRange->{{xmin,xmax},{ymin,ymax}},AspectRatio->Automatic,
       ColorFunction->(Hue[Log[1+#]]&)]
```

```
(* Plot a Newton Fractal. See Figure 15.6. *)
In[10]:=f=#^3-1 &;
In[11]:=NewtonMethod=Compile[{{z,_Complex}},
        Arg[FixedPoint[#-f[#]/f'[#]&,N[z],50]]/(2*Pi)];
In[12]:=NewtonFractal[NumPoints_,xmin_,xmax_,ymin_,ymax_]:=
        DensityPlot[NewtonMethod[x+I*y],{x,xmin,xmax},{y,ymin,ymax},
        Mesh->False,PlotPoints->NumPoints,Frame->False,
        ColorFunctionScaling->False,ColorFunction->(Hue[#]&)];
In[13]:=NewtonFractal[250,-2.5,2.5,-2.5,2.5]
```

15.5 Exercises

1. Consider the Julia set given in Figure 15.1(a). Take the mapping $z_{n+1} = z_n^2 + c$, where $c = -0.5 + 0.3i$.

   (a) Iterate the initial point $z_0 = 0 + 0i$ for 500 iterations and list the final 100. Increase the number of iterations, what can you deduce about the orbit?

   (b) Iterate the initial point $z_0 = -1 - i$ and list z_1 to z_{10}. What can you deduce about this orbit?

2. Given that $c = -1 + i$, determine the fixed points of periods one and two for the mapping $z_{n+1} = z_n^2 + c$.

3. Consider equation (15.1); plot the Julia sets $J(0,0)$, $J(-0.5, 0)$, $J(-0.7, 0)$, and $J(-2, 0)$.

4. Compute the fixed points of period one for the complex mapping

$$z_{n+1} = 2 + \frac{z_n e^{i|z_n|^2}}{10}.$$

5. Determine the boundaries of points of periods one and two for the mapping

$$z_{n+1} = c - z_n^2.$$

6. Plot the Mandelbrot set for the mapping

$$z_{n+1} = c - z_n^2.$$

7. Determine the fixed points of periods one and two for the mapping $z_{n+1} = z_n^2 - 2z_n + c$.

8. Modify the Mathematica program in Section 15.4 to plot a Mandelbrot set for the mappings (i) $z_{n+1} = z_n^4 + c$ and (ii) $z_{n+1} = z_n^3 + c$.

9. Determine the periods of the points (i) $c = -1.3$ and (ii) $c = -0.1 + 0.8i$ for the mapping $z_{n+1} = z_n^2 + c$.

10. Plot a Newton fractal (of the same format to that shown in Figure 15.6) for the function $f(z) = z^3 - 2z + 2$.

Bibliography

[1] R.L. Devaney and L. Keen (eds.), *Complex Dynamics: Twenty-five Years After the Appearance of the Mandelbrot Set (Contemporary Mathematics)*, American Mathematical Society, Providence, RI, 2005.

[2] R.L. Devaney *Complex Dynamical Systems: The Mathematics Behind the Mandelbrot and Julia Sets (Proceedings of Symposia in Applied Mathematics)*, American Mathematical Society, 1995.

[3] P.W. Jones, B. Mandelbrot, C.J.G. Evertsz and M.C. Gutzwiller, *Fractals and Chaos: The Mandelbrot Set and Beyond*, Springer, 2004.

[4] A. Katunin, *A Concise Introduction to Hypercomplex Fractals*, CRC Press, Florida, 2017.

[5] A. Kharab and R.B. Guenther, *An Introduction to Numerical Methods: A MATLAB Approach, 3rd Ed.*, CRC Press, Florida, 2011.

[6] B.B. Mandelbrot and R.L. Hudson, *The (Mis)Behavior of the Markets: A Fractal View of Risk, Ruin and Reward*, Perseus Books Group, New York, 2006.

[7] H-O. Peitgen (ed.), E.M. Maletsky, H. Jürgens, T. Perciante, D. Saupe, and L. Yunker, *Fractals for the Classroom: Strategic Activities Volume 2*, Springer-Verlag, New York, 1994.

[8] H-O. Peitgen, H. Jürgens, D. Saupe, and C. Zahlten, *Fractals: An Animated Discussion*, SpektrumAkademischer Verlag, Heidelberg, 1989; W.H. Freeman, New York, 1990.

[9] T. Rashid, *Make Your Own Mandelbrot: A gentle journey through the mathematics of the of the Mandelbrot and Julia fractals, and making your own using the Python computer language*, CreateSpace Independent Publishing Platform, 2014.

[10] Root-Finding Fractals, Softology's Blog, (Jan 20, 2011), web pages last accessed 16th May 2016. https://softologyblog.wordpress.com/2011/01/20/root-finding-fractals.

Chapter 16
Electromagnetic Waves and Optical Resonators

Aims and Objectives

- To introduce some theory of electromagnetic waves.

- To introduce optical bistability and show some related devices.

- To discuss possible future applications.

- To apply some of the theory of nonlinear dynamical systems to model a real physical system.

On completion of this chapter, the reader should be able to

- understand the basic theory of Maxwell's equations;

- derive the equations to model a nonlinear simple fiber ring (SFR) resonator;

- investigate some of the dynamics displayed by these devices and plot chaotic attractors;

- use a linear stability analysis to predict regions of instability and bistability;

- plot bifurcation diagrams using the first and second iterative methods;

- compare the results from four different methods of analysis.

As an introduction to optics, electromagnetic waves are discussed via Maxwell's equations.

The reader is briefly introduced to a range of bistable optical resonators including the nonlinear Fabry–Perot interferometer, the cavity ring, the SFR, the double-coupler fiber ring, the fiber double-ring, and a nonlinear optical loop mirror (NOLM) with feedback. All of these devices can display hysteresis, and

© Springer International Publishing AG 2017
S. Lynch, *Dynamical Systems with Applications using Mathematica®*,
DOI 10.1007/978-3-319-61485-4_16

all can be affected by instabilities. Possible applications are discussed in the physical world.

Linear stability analysis is applied to the nonlinear SFR resonator. The analysis gives intervals where the system is bistable and unstable but does not give any information on the dynamics involved in these regions. To use optical resonators as bistable devices, the bistable region must be isolated from any instabilities. To supplement the linear stability analysis, iterative methods are used to plot bifurcation diagrams.

For a small range of parameter values, the resonator can be used as a bistable device. Investigations are carried out to see how the bistable region is affected by the linear phase shift due to propagation of the electric field through the fiber loop.

16.1 Maxwell's Equations and Electromagnetic Waves

This section is intended to give the reader a simple general introduction to optics. Most undergraduate physics textbooks discuss *Maxwell's electromagnetic equations* in some detail. The aim of this section is to list the equations and show that Maxwell's equations can be expressed as *wave equations*. Maxwell was able to show conclusively that just four equations could be used to interpret and explain a great deal of electromagnetic phenomena.

The four equations, collectively referred to as Maxwell's equations, did not originate entirely with him but with Ampère, Coulomb, Faraday, Gauss, and others. First, consider Faraday's law of induction, which describes how electric fields are produced from changing magnetic fields. This equation can be written as

$$\oint_C \mathbf{E} \cdot d\mathbf{r} = -\frac{\partial \phi}{\partial t},$$

where $\mathbf{E}$ is the electric field strength, $\mathbf{r}$ is a spatial vector, and ϕ is the magnetic flux. This equation may be written as

$$\oint_C \mathbf{E} \cdot d\mathbf{r} = -\frac{\partial}{\partial t} \iint_S \mathbf{B} \cdot d\mathbf{S},$$

where $\mathbf{B}$ is a magnetic field vector. Applying Stokes's theorem,

$$\iint_S \nabla \wedge \mathbf{E} \cdot d\mathbf{S} = -\frac{\partial}{\partial t} \iint_S \mathbf{B} \cdot d\mathbf{S}.$$

Therefore,

$$\nabla \wedge \mathbf{E} = -\frac{\partial \mathbf{B}}{\partial t}, \tag{16.1}$$

which is the point form of Faraday's law of induction.

Ampère's law describes the production of magnetic fields by electric currents. Now

$$\oint_C \mathbf{H} \cdot d\mathbf{r} = \iint_S \mathbf{J} \cdot d\mathbf{S},$$

where $\mathbf{H}$ is another magnetic field vector $(\mathbf{B} = \mu\mathbf{H})$ and $\mathbf{J}$ is the current density. By Stokes's theorem

$$\oint_C \mathbf{H} \cdot d\mathbf{r} = \iint_S \nabla \wedge \mathbf{H} \cdot d\mathbf{S} = \iint_S \mathbf{J} \cdot d\mathbf{S}.$$

Therefore,

$$\nabla \wedge \mathbf{H} = \mathbf{J}.$$

Maxwell modified this equation by adding the time rate of change of the electric flux density (electric displacement) to obtain

$$\nabla \wedge \mathbf{H} = \mathbf{J} + \frac{\partial \mathbf{D}}{\partial t}, \qquad (16.2)$$

where $\mathbf{D}$ is the electric displacement vector.

Gauss's law for electricity describes the electric field for electric charges, and Gauss's law for magnetism shows that magnetic field lines are continuous without end. The equations are

$$\nabla \cdot \mathbf{E} = \frac{\rho}{\epsilon_0}, \qquad (16.3)$$

where ρ is the charge density and ϵ_0 is the permittivity of free space (a vacuum), and

$$\nabla \cdot \mathbf{B} = 0. \qquad (16.4)$$

In using Maxwell's equations, (16.1) to (16.4), above and solving problems in electromagnetism, the three so-called constitutive relations are also used. These are

$$\mathbf{B} = \mu\mathbf{H} = \mu_r\mu_0\mathbf{H}; \quad \mathbf{D} = \epsilon\mathbf{E} = \epsilon_r\epsilon_0\mathbf{E} \quad \text{and} \quad \mathbf{J} = \sigma\mathbf{E},$$

where μ_r, μ_0 are the relative permeabilities of a material and free space, respectively; ϵ_r, ϵ_0 are the relative permittivities of a material and free space, respectively; and σ is conductivity.

If $\mathbf{E}$ and $\mathbf{H}$ are sinusoidally varying functions of time, then in a region of free space, Maxwells' equations become

$$\nabla \cdot \mathbf{E} = 0; \quad \nabla \cdot \mathbf{H} = 0; \quad \nabla \wedge \mathbf{E} + i\omega\mu_0\mathbf{H} = 0 \quad \text{and} \quad \nabla \wedge \mathbf{H} - i\omega\epsilon_0\mathbf{E} = 0.$$

The wave equations are obtained by taking the curls of the last two equations; thus

$$\nabla^2 \mathbf{E} + \epsilon_0 \mu_0 \omega^2 \mathbf{E} = 0 \quad \text{and} \quad \nabla^2 \mathbf{H} + \epsilon_0 \mu_0 \omega^2 \mathbf{H} = 0,$$

where ω is the angular frequency of the wave. These differential equations model an unattenuated wave traveling with velocity

$$c = \frac{1}{\sqrt{\epsilon_0 \mu_0}},$$

where c is the speed of light in a vacuum. The field equation

$$\mathbf{E}(\mathbf{r}, t) = \mathbf{E_0} \exp\left[i(\omega t - \mathbf{k} \mathbf{r})\right]$$

satisfies the wave equation, where $|\mathbf{k}| = 2\pi/\lambda$ is the modulus of the wave vector and λ is the wavelength of the wave. The remarkable conclusion drawn by Maxwell is that light is an electromagnetic wave and that its properties can all be deduced from his equations. The electric fields propagating through an optical fiber loop will be investigated in this chapter.

Similar equations are used to model the propagation of light waves through different media including a dielectric (a nonconducting material whose properties are isotropic); see the next section. In applications to nonlinear optics, the Maxwell–Debye or Maxwell–Bloch equations are usually used, but the theory is beyond the scope of this book. Interested readers are referred to [9], [15], and the research papers listed at the end of this chapter.

16.2 Historical Background

In recent years, there has been a great deal of interest in optical bistability because of its potential applications in high-speed all-optical signal processing and all-optical computing. Indeed, in 1984, Smith [21] published an article in *Nature* with the enthralling title "Towards the Optical Computer," and in 1999, Matthews [17] reported on work carried out by A. Wixforth and his group on the possibility of optical memories. Bistable devices can be used as logic gates, memory devices, switches, and differential amplifiers. The electronic components used nowadays can interfere with one another, need wires to guide the electronic signals, and carry information relatively slowly. Using light beams, it is possible to connect all-optical components. There is no interference; lenses and mirrors can be used to communicate thousands of channels of information in parallel; the information-carrying capacity—the bandwidth—is enormous; and there is nothing faster than the speed of light in the known universe.

In 1969, Szöke et al. [24] proposed the principle of *optical bistability* and suggested that optical devices could be superior to their electronic counterparts.

The two essential ingredients for bistability are nonlinearity and feedback. For optical hysteresis, nonlinearity is provided by the medium as a refractive (or dispersive) nonlinearity or as an absorptive nonlinearity, or as both. Refractive nonlinearities alone will be considered in this chapter. The feedback is introduced through mirrors or fiber loops or by the use of an electronic circuit. The bistable optical effect was first observed in sodium vapor in 1976 at Bell Laboratories, and a theoretical explanation was provided by Felber and Marburger [7] in the same year. Nonlinearity was due to the Kerr effect (see Section 16.3), which modulated the refractive index of the medium.

Early experimental apparatus for producing optical bistability consisted of hybrid devices that contained both electronic and optical components. Materials used included indium antimonide (InSb), gallium arsenide (GaAs), and tellurium (Te). By 1979, micron-sized optical resonators had been constructed. A fundamental model of the nonlinear *Fabry-Perot interferometer* is shown in Figure 16.1.

Figure 16.1:   A Fabry-Perot resonator; I, R, and T stand for incident, reflected, and transmitted intensities, respectively.

An excellent introduction to nonlinearity in fiber optics is provided by the textbook of Agrawal [1]. Applications in nonlinear fiber optics are presented in [2] and [20]. In recent years, there has been the development of microfibers and resonators composed of these types of fiber [13]. Now the fiber diameter has been reduced down to the nanoscale, and resonator ring lengths are of the order of millimeters. Because of the narrowness of the fiber diameter, a significant proportion of the guided electric field can extend beyond the optical fiber core, known as the evanescent field, which makes them of interest in optical sensing applications.

A block diagram of the first electro-optic device is shown in Figure 16.2 and was constructed by Smith and Turner in 1977 [22]. Nonlinearity is induced by the Fabry–Perot interferometer, and a He-Ne (helium-neon) laser is used at 6328 Å. A bistable region is observed for a small range of parameter values. An isolated bistable region is shown in Figure 16.4(a). For input values between approximately 4 and 5 units, there are two possible output values. The output is dependent upon the history of the system, that is, whether the input power is increasing or decreasing.

In theoretical studies, Ikeda, Daido, and Akimoto [10] showed that optical circuits exhibiting bistable behavior can also contain temporal instabilities under certain conditions. The *cavity ring* (CR) resonator, first investigated by Ikeda, consists of a ring cavity comprising four mirrors that provide the feedback and

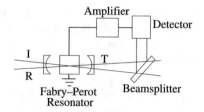

Figure 16.2:   The first electro-optic device to display bistability.

containing a nonlinear dielectric material (see Figure 16.3). Light circulates around the cavity in one direction, and the medium induces a nonlinear phase shift dependent on the intensity of the light. Mirrors M_1 and M_2 are partially reflective, while mirrors M_3 and M_4 are 100% reflective.

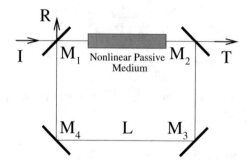

Figure 16.3:   The CR resonator containing a nonlinear dielectric medium.

Possible bifurcation diagrams for this device are shown in Figure 16.4. In Figure 16.4(a), the bistable region is isolated from any instabilities, but in Figure 16.4(b), instabilities have encroached upon the bistable cycle. These figures are similar to those that would be seen if the CR were connected to an oscilloscope. However, most of the dynamics are lost; mathematically, it is best to plot bifurcation diagrams using points alone. The length L is different in the two cases and hence so is the cavity round-trip time (the time it takes light to complete one loop in the cavity).

In recent years, there has been intense research activity in the field of fiber optics. Many school physics textbooks now provide an excellent introduction to the subject, and [4] provides an introduction to nonlinear optics. The interest in this chapter, however, lies solely in the application to all-optical bistability. A block diagram of the SFR resonator is shown in Figure 16.5. It has recently been shown that the dynamics of this device are the same as those for the CR resonator (over a limited range of initial time) apart from a scaling. The first all-optical experiment was carried out using a single-mode fiber in a simple loop arrangement, the fiber acting as the nonlinear medium [18]. In mathematical models, the input electric field is given as

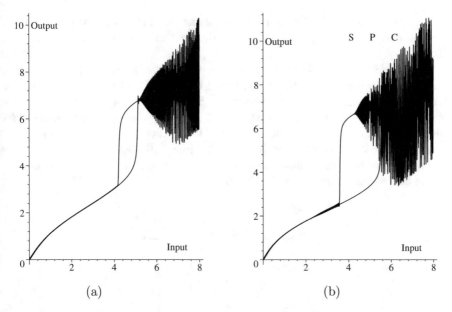

Figure 16.4: Possible bifurcation diagrams for the CR resonator: (a) an isolated bistable region and (b) instabilities within the bistable region. S represents stable behavior, P is period undoubling, and C stands for chaos.

$$E_{\text{in}}(t) = \xi_j(t)e^{i\omega t},$$

where ξ_j represents a complex amplitude (which may contain phase information) and ω is the circular frequency of the light.

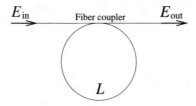

Figure 16.5: A schematic of the SFR resonator. The input electric field is E_{in} and the output electric field is E_{out}.

In experimental setups, for example, the light source could be a Q-switched YAG laser operating at 1.06 μm. The optical fiber is made of fused silica and is assumed to be lossless.

An analysis of the SFR resonator will be discussed in more detail in the next section, and the stability of the device will be investigated in Sections 16.5 and 16.6.

The *double-coupler fiber ring* resonator was investigated by Li and Ogusu [12] in 1998 (see Figure 16.6). It was found that there was a similarity between the dynamics displayed by this device and the Fabry–Perot resonator in terms of transmission and reflection bistability. It is possible to generate both clockwise and counterclockwise hysteresis loops using this device. An example of a counterclockwise bistable cycle is given in Figure 16.4(a). The reader will be asked to carry out some mathematical analysis for this device in the exercises at the end of the chapter (Section 16.8).

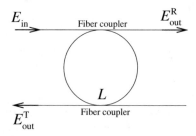

Figure 16.6: The double-coupler fiber ring resonator: E_{in} is the input field amplitude, $E_{\text{out}}^{\text{R}}$ is the reflected output, and $E_{\text{out}}^{\text{T}}$ is the transmitted output.

In 1994, Ja [11] presented a theoretical study of an *optical fiber double-ring* resonator, as shown in Figure 16.7. Ja predicted multiple bistability of the output intensity using the Kerr effect. However, instabilities were not discussed. It was proposed that this type of device could be used in new computer logic systems where more than two logic states are required. In principle, it is possible to link a number of loops of fiber, but instabilities are expected to cause some problems.

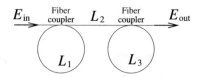

Figure 16.7: A fiber double-ring resonator with two couplers.

The *nonlinear optical loop mirror* (NOLM) with feedback, [6], [13], and [23], has been one of the most novel devices for demonstrating a wide range of all-optical processing functions including optical logic. The device is shown in

Figure 16.8. Note that the beams of light are counterpropagating in the large loop but not in the feedback section and that there are three couplers.

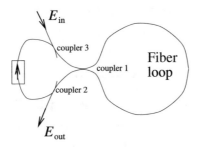

Figure 16.8: A schematic of a NOLM with feedback.

All of the devices discussed thus far can display bistability and instability leading to chaos. In order to understand some of these dynamics, the SFR resonator will now be discussed in some detail.

16.3 The Nonlinear SFR Resonator

Consider the all-optical fiber resonator as depicted in Figure 16.9 and define the slowly varying complex electric fields as indicated.

Note that the power P and intensity I are related to the electric field in the following way:

$$P \propto I \propto |E|^2.$$

If the electric field crosses the coupler, then a phase shift is induced, which is represented by a multiplication by i in the equations. Assume that there is

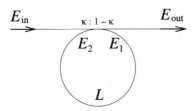

Figure 16.9: The SFR resonator. The electric field entering the fiber ring is labeled E_1 and the electric field leaving the fiber ring is labeled E_2. The coupler splits the power intensity in the ratio $\kappa : 1 - \kappa$.

no loss at the coupler. Then, across the coupler, the complex field amplitudes satisfy the following equations:

$$E_1 = \sqrt{\kappa}E_2 + i\sqrt{1 - \kappa}E_{\text{in}} \tag{16.5}$$

and

$$E_{\text{out}} = \sqrt{\kappa}E_{\text{in}} + i\sqrt{1 - \kappa}E_2, \tag{16.6}$$

where κ is the power-splitting ratio at the coupler. Consider the propagation from E_1 to E_2. Then

$$E_2 = E_1 e^{i\phi}, \tag{16.7}$$

where the total loss in the fiber is negligible (typically about 0.2 dB/km) and

$$\phi = \phi_L + \phi_{NL}.$$

The linear phase shift is ϕ_L, and the nonlinear phase shift due to propagation is given by

$$\phi_{NL} = \frac{2\pi r_2 L}{\lambda_0 A_{\text{eff}}}|E_1|^2,$$

where λ_0 is the wavelength of propagating light in a vacuum, A_{eff} is the effective core area of the fiber, L is the length of the fiber loop, and r_2 is the *nonlinear refractive index coefficient* of the fiber. It is well known that when the optical intensity is large enough, the constant r_2 satisfies the equation

$$r = r_0 + r_2 I = r_0 + \frac{r_2 r_0}{2\eta_0}|E_1|^2 = r_0 + r_2\frac{P}{A_{\text{eff}}},$$

where r is the refractive index of the fiber, r_0 is the linear value, I is the instantaneous optical intensity, and P is the power. If the nonlinearity of the fiber is represented by this equation, then the fiber is said to be of *Kerr type*. In most applications, it is assumed that the response time of the *Kerr effect* is much less than the time taken for light to circulate once in the loop.

Substitute (16.7) into equations (16.5) and (16.6). Simplify to obtain

$$E_1(t) = i\sqrt{1 - \kappa}E_{\text{in}}(t) + \sqrt{\kappa}E_1(t - t_R)e^{i\phi(t - t_R)},$$

where $t_R = \frac{rL}{c}$ is the time taken for the light to complete one loop, r is the refractive index, and c is the velocity of light in a vacuum. Note that this is an iterative formula for the electric field amplitude inside the ring. Take time steps of length equal to t_R. This expression can be written more conveniently as an iterative equation of the form

$$E_{n+1} = A + BE_n \exp\left(i\left(\frac{2\pi r_2 L}{\lambda_0 A_{\text{eff}}}|E_n|^2 + \phi_L\right)\right), \qquad (16.8)$$

where $A = i\sqrt{1-\kappa}E_{\text{in}}$, $B = \sqrt{\kappa}$, and E_j is the electric field amplitude at the jth circulation around the fiber loop. Typical fiber parameters chosen for this system are $\lambda_0 = 1.55 \times 10^{-6}$ m, $r_2 = 3.2 \times 10^{-20}$ m$^2 W^{-1}$, $A_{\text{eff}} = 30$ μm^2, and $L = 80$ m.

Equation (16.8) may be scaled without loss of generality to the simplified equation

$$E_{n+1} = A + BE_n \exp\left[i(|E_n|^2 + \phi_L)\right]. \qquad (16.9)$$

Some of the dynamics of equation (16.9) will be discussed in the next section.

16.4 Chaotic Attractors and Bistability

Split equation (16.9) into its real and imaginary parts by setting $E_n = x_n + iy_n$, and set $\phi_L = 0$. The equivalent real two-dimensional system is given by

$$\begin{aligned} x_{n+1} &= A + B\left(x_n \cos|E_n|^2 - y_n \sin|E_n|^2\right) \\ y_{n+1} &= B\left(x_n \sin|E_n|^2 + y_n \cos|E_n|^2\right), \end{aligned} \qquad (16.10)$$

where $|B| < 1$. This system is one version of the so-called *Ikeda map*. As with the Hénon map, introduced in Chapter 14, the Ikeda map can have fixed points of all periods. In this particular case, system (16.10) can have many fixed points of period one depending on the parameter values A and B.

Example 1. Determine and classify the fixed points of period one for system (16.10) when $B = 0.15$ and

(i) $A = 1$; .

(ii) $A = 2.2$.

Solution. The fixed points of period one satisfy the simultaneous equations

$$x = A + Bx\cos(x^2 + y^2) - By\sin(x^2 + y^2)$$

and

$$y = Bx\sin(x^2 + y^2) + By\cos(x^2 + y^2).$$

(i) When $A = 1$ and $B = 0.15$, there is one solution at $x_{1,1} \approx 1.048$, $y_{1,1} \approx$ 0.151. The solution is given graphically in Figure 16.10(a). To classify the critical point $P^* = (x_{1,1}, y_{1,1})$, consider the Jacobian matrix

$$J(P^*) = \begin{pmatrix} \frac{\partial P}{\partial x} & \frac{\partial P}{\partial y} \\ \frac{\partial Q}{\partial x} & \frac{\partial Q}{\partial y} \end{pmatrix}\Bigg|_{P^*}.$$

The eigenvalues of the Jacobian matrix at P^* are $\lambda_1 \approx -0.086 + 0.123i$ and $\lambda_2 \approx -0.086 - 0.123i$. Therefore, P^* is a stable fixed point of period one.

(ii) When $A = 2.2$ and $B = 0.15$, there are three points of period one, as the graphs in Figure 16.10(b) indicate. The fixed points occur approximately at the points $U = (2.562, 0.131)$, $M = (2.134, -0.317)$, and $L = (1.968, -0.185)$. Using the Jacobian matrix, the eigenvalues for U are $\lambda_{1,2} = -0.145 \pm 0.039i$; the eigenvalues for M are $\lambda_1 = 1.360$, $\lambda_2 = 0.017$; and the eigenvalues for L are $\lambda_1 = 0.555$, $\lambda_2 = 0.041$.

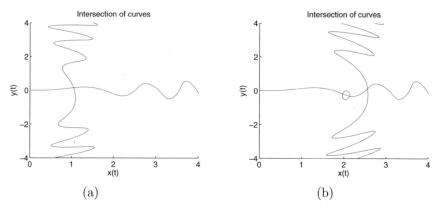

(a)   (b)

Figure 16.10: [Mathematica] The fixed points of period one are determined by the intersections of the two curves, $x = A + 0.15x\cos(x^2+y^2) - 0.15y\sin(x^2+y^2)$ and $y = 0.15x\sin(x^2+y^2) + 0.15y\cos(x^2+y^2)$; (a) $A = 1$ and (b) $A = 2.2$. Note in case (b) that the small closed curve and the vertical curve form one solution set.

Therefore, U and L are stable fixed points of period one, while M is an unstable fixed point of period one. These three points are located within a bistable region of the bifurcation diagram given later in this chapter. The point U lies on the upper branch of the hysteresis loop, and the point L lies on the lower branch. Since M is unstable, it does not appear in the bifurcation diagram but is located between U and L.

As the parameter A changes, the number of fixed points and the dynamics of the system change. For example, when $A = 1$, there is one fixed point of period one; when $A = 2.2$, there are two stable fixed points of period one and one unstable fixed point of period one; when $A = 2.4$, there are two stable fixed points of period two. As A increases, the system displays chaotic behavior (see Example 2). All of the information can be summarized on a bifurcation diagram that will be shown later in this chapter.

Example 2. Plot iterative maps for system (16.10) when $B = 0.15$ and

(a) $A = 5$;

(b) $A = 10$.

Solution. Two chaotic attractors for system (16.10) are shown in Figure 16.11.

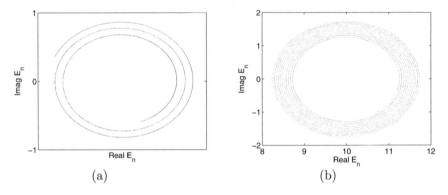

(a)   (b)

Figure 16.11: [Mathematica] The chaotic attractors when (a) $A = 5$ (5000 iterates) and (b) $A = 10$ (5000 iterates).

Theorem 1. *The circle of radius $\frac{|AB|}{1-B}$ centered at A is invariant for system (16.10).*

Proof. Suppose that a general initial point in the Argand diagram is taken to be E_n; then the first iterate is given by

$$E_{n+1} = A + BE_n e^{i|E_n|^2}.$$

The second iterate can be written as

$$E_{n+2} = A + BE_{n+1}e^{i|E_{n+1}|^2} = A + B\left(A + BE_n e^{i|E_n|^2}\right)e^{i|E_{n+1}|^2}.$$

Thus

$$E_{n+2} = A + ABe^{i|E_{n+1}|^2} + B^2 E_n e^{i(|E_n|^2+|E_{n+1}|^2)}.$$

Using a similar argument, the third iterate is

$$E_{n+3} = A + B\left(A + ABe^{i|E_{n+1}|^2} + B^2 E_n e^{i(|E_n|^2+|E_{n+1}|^2)}\right)e^{i|E_{n+2}|^2}.$$

Therefore,

$$E_{n+3} = A + ABe^{i|E_{n+2}|^2} + AB^2 e^{i(|E_{n+1}|^2+|E_{n+2}|^2)} + B^3 E_n e^{i(|E_n|^2+|E_{n+1}|^2+|E_{n+2}|^2)}.$$

A general expression for the Nth iterate E_{n+N} is not difficult to formulate. Hence

$$E_{n+N} = A + ABe^{i|E_{n+N-1}|^2} + AB^2 e^{i(|E_{n+N-2}|^2+|E_{n+N-1}|^2)} + \ldots$$

$$+ AB^{N-1}\exp\left(i\sum_{j=1}^{N-1}|E_{n+j}|^2\right) + B^N E_n \exp\left(i\sum_{j=0}^{N-1}|E_{n+j}|^2\right).$$

As $N \to \infty$, $B^N \to 0$, since $0 < B < 1$. Set $R_j = |E_{n+N-j}|^2$. Then

$$|E_{n+N} - A| = |ABe^{iR_1} + AB^2 e^{i(R_1+R_2)} + \ldots + AB^{N-1}e^{i(R_1+R_2+\ldots+R_{N-1})}|.$$

Since $|z_1 + z_2 + \ldots + z_m| \le |z_1| + |z_2| + \ldots + |z_m|$ and $|e^{i\theta}| = 1$,

$$|E_{n+N} - A| \le |AB| + |AB^2| + \ldots + |AB^{N-1}|.$$

This forms an infinite geometric series as $N \to \infty$. Therefore

$$|E_{n+N} - A| \le \frac{|AB|}{1 - B}.$$

The disc given by $|E - A| = AB/(1 - B)$ is positively invariant for system (16.10). The invariant disks in two cases are easily identified in Figures 16.11(a) and (b).

16.5 Linear Stability Analysis

To investigate the stability of the nonlinear SFR resonator, a linear stability analysis (see Chapter 2) will be applied. A first-order perturbative scheme is used to predict the values of a parameter where the stationary solutions become unstable. Briefly, a small perturbation is added to a stable solution, and a Taylor series expansion is carried out, the nonlinear terms are ignored, and a linear stability analysis is applied.

It was shown in Section 16.3 that the following simplified complex iterative equation can be used to model the electric field in the fiber ring:

$$E_{n+1} = A + BE_n \exp\left[i\left(|E_n|^2 - \phi_L\right)\right], \qquad (16.11)$$

where E_n is the slowly varying field amplitude; $A = i\sqrt{1 - \kappa}E_{\text{in}}$ is related to the input; $B = \sqrt{\kappa}$, where κ is the power coupling ratio; and ϕ_L is the linear phase shift suffered by the electric field as it propagates through the fiber loop. To simplify the linear stability analysis, there is assumed to be no loss at the coupler and the phase shift ϕ_L is set to zero. The effect of introducing a linear phase shift will be discussed later in this chapter.

Suppose that E_S is a stable solution of the iterative equation (16.11). Then

$$E_S = A + BE_S e^{i|E_S|^2}.$$

Therefore,

$$A = E_S\left[1 - B\left(\cos(|E_S|^2) + i\sin|E_S|^2\right)\right].$$

Using the relation $|z|^2 = zz^*$, where z^* is the conjugate of z,

$$\begin{aligned}
|A|^2 = &\left(E_S\left[1 - B\left(\cos(|E_S|^2) + i\sin|E_S|^2\right)\right]\right) \times \\
&\left(E_S^*\left[1 - B\left(\cos(|E_S|^2) - i\sin|E_S|^2\right)\right]\right).
\end{aligned}$$

Hence,

$$|A|^2 = |E_S|^2\left(1 + B^2 - 2B\cos(|E_S|^2)\right). \qquad (16.12)$$

The stationary solutions of system (16.11) are given as a multivalued function of A satisfying equation (16.12). This gives a bistable relationship equivalent to the *graphical method*, which is well documented in the literature; see, for example, [3], and [16].

Differentiate equation (16.12) to obtain

$$\frac{d|A|^2}{d|E_S|^2} = 1 + B^2 + 2B\left(|E_S|^2\sin(|E_S|^2) - \cos(|E_S|^2)\right). \qquad (16.13)$$

To establish where the stable solutions become unstable, consider a slight perturbation from the stable situation in the fiber ring, and let

$$E_n(t) = E_S + \xi_n(t) \quad \text{and} \quad E_{n+1}(t) = E_S + \xi_{n+1}(t), \qquad (16.14)$$

where $\xi_n(t)$ is a small time-dependent perturbation to E_S. Substitute (16.14) into (16.11) to get

$$E_S + \xi_{n+1} = A + B(E_S + \xi_n)\exp\left[i(E_S + \xi_n)(E_S^* + \xi_n^*)\right],$$

so

$$E_S + \xi_{n+1} = A + B(E_S + \xi_n)\exp[i|E_S|^2]\exp[i(E_S\xi_n^* + \xi_n E_S^* + |\xi_n|^2)]. \quad (16.15)$$

Take a Taylor series expansion of the exponential function to obtain

$$\exp\left[i(E_S\xi_n^* + \xi_n E_S^* + |\xi_n|^2)\right] = 1 + i(E_S\xi_n^* + \xi_n E_S^* + |\xi_n|^2) +$$
$$\frac{i^2(E_S\xi_n^* + \xi_n E_S^* + |\xi_n|^2)^2}{2} + \dots .$$

Ignore the nonlinear terms in ξ_n. Equation (16.15) then becomes

$$E_S + \xi_{n+1} = A + B(E_S + \xi_n)\exp[i|E_S|^2]\left(1 + iE_S\xi_n^* + \xi_n E_S^*\right).$$

Since $A = E_S - BE_S\exp[i|E_S|^2]$, the equation simplifies to

$$\xi_{n+1} = B\left(\xi_n + i|E_S|^2\xi_n + i(E_S)^2\xi_n^*\right)\exp\left(i|E_S|^2\right). \quad (16.16)$$

Since ξ is real, it may be split into its positive and negative frequency parts as follows:

$$\xi_n = E_+e^{\lambda t} + E_-e^{\lambda^* t} \quad \text{and} \quad \xi_{n+1} = E_+e^{\lambda(t+t_R)} + E_-e^{\lambda^*(t+t_R)}, \quad (16.17)$$

where $|E_+|, |E_-|$ are much smaller than $|E_S|$, t_R is the fiber ring round-trip time, and λ is the amplification rate of a small fluctuation added to a stable solution. Substitute equation (16.17) into (16.16). Then, the validity of (16.16) at all times t requires that

$$E_+e^{\lambda t_R} = B\left(E_+ + i|E_S|^2E_+ + iE_S^2E_-^*\right)\exp\left(i|E_S|^2\right),$$
$$E_-^*e^{\lambda t_R} = B\left(E_-^* - i|E_S|^2E_-^* - i(E_S^*)^2E_+\right)\exp\left(-i|E_S|^2\right)$$

or, equivalently,

$$\begin{pmatrix} \beta\left(1 + i|E_S|^2\right) - e^{\lambda t_R} & i\beta E_S^2 \\ -i\beta^*(E_S^*)^2 & \beta^*\left(1 - i|E_S|^2\right) - e^{\lambda t_R} \end{pmatrix}\begin{pmatrix} E_+ \\ E_-^* \end{pmatrix} = \begin{pmatrix} 0 \\ 0 \end{pmatrix},$$

where $\beta = B \exp \left(i|E_S|^2 \right)$. To obtain a valid solution, the characteristic equation must be solved:

$$e^{2\lambda t_R} - 2e^{\lambda t_R} B \left(\cos |E_S|^2 - |E_S|^2 \sin |E_S|^2 \right) + B^2 = 0.$$

Substituting from equation (16.13), the characteristic equation becomes

$$e^{2\lambda t_R} - e^{\lambda t_R} \left(1 + B^2 - \frac{d|A|^2}{d|E_S|^2} \right) + B^2 = 0. \tag{16.18}$$

Let $D = \frac{d|A|^2}{d|E_S|^2}$. The stability edges for E_S occur where $e^{\lambda t_R} = +1$ and $e^{\lambda t_R} = -1$, since this is a discrete mapping. Using equation (16.18), this yields the conditions

$$D_{+1} = 0 \quad \text{and} \quad D_{-1} = 2 \left(1 + B^2 \right).$$

Thus, the system is stable as long as

$$0 < D < 2 \left(1 + B^2 \right). \tag{16.19}$$

The condition $D = 0$ marks the boundary between the branches of positive and negative slope on the graph of $|E_S|^2$ versus $|A|^2$ and hence defines the regions where the system is bistable. Thus, the results from the graphical method match with the results from the linear stability analysis. The system becomes unstable at the boundary where $D = D_{-1}$.

It is now possible to apply four different methods of analysis to determine the stability of the electric field amplitude in the SFR resonator. Linear stability analysis may be used to determine both the unstable and bistable regions, and bifurcation diagrams can be plotted. The graphical method [1] is redundant in this case.

There are two methods commonly used to plot bifurcation diagrams—the first and second iterative methods.

The First Iterative Method. A parameter is fixed, and one or more initial points are iterated forward. Transients are ignored and, a number of the final iterates are plotted. The parameter is then increased by a suitable step length, and the process is repeated. There are many points plotted for each value of the parameter. For example, the bifurcation diagrams plotted in Chapter 14 were all generated using the first iterative method.

The Second Iterative Method. A parameter is varied, and the solution to the previous iterate is used as the initial condition for the next iterate. In this

way, a feedback mechanism is introduced. In this case, there is a history associated with the process, and only one point is plotted for each value of the
parameter. For example, most of the bifurcation diagrams plotted in Section
16.6 were plotted using the second iterative method.

The first and second iterative methods are used in other chapters of the book.

16.6 Instabilities and Bistability

In the previous section, the results from the linear stability analysis established
that system (16.11) is stable as long as equation (16.19) is satisfied. A possible
stability diagram for system (16.11) is given in Figure 16.12, which shows the
graph of $D = \frac{d|A|^2}{d|E_S|^2}$ and the bounding lines $D_{+1} = 0$ and $D_{-1} = 2\left(1 + B^2\right)$
when $B = 0.15$.

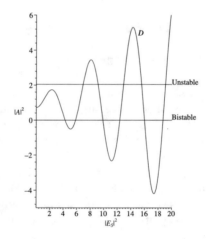

Figure 16.12: Stability diagram for the SFR resonator when $B = 0.15$ ($\kappa = 0.0225$). The system is stable as long as $0 < D < 2\left(1 + B^2\right)$.

Table 16.1 lists the first two bistable and unstable intensity regions for the
SFR resonator (in Watts per meter squared in physical applications) for a range
of fixed values of the parameter B.

The dynamic behavior of system (16.11) may also be investigated by plotting
bifurcation diagrams using either the first or second iterative methods. In order
to observe any hysteresis, one must, of course, use the second iterative method,
which involves a feedback. The method developed by Bischofberger and Shen
[3] in 1979 for a nonlinear Fabry–Perot interferometer is modified and used
here for the SFR resonator. The input intensity is increased to a maximum

Table 16.1: The first two regions of bistability and instability computed for the SFR resonator to three decimal places using a linear stability analysis

B	First bistable region A^2/Wm^{-2}	First unstable region A^2/Wm^{-2}	Second bistable region A^2/Wm^{-2}	Second unstable region A^2/Wm^{-2}
0.05	$10.970 - 11.038$	$12.683 - 16.272$	$16.785 - 17.704$	$17.878 - 23.561$
0.15	$4.389 - 4.915$	$5.436 - 12.007$	$9.009 - 12.765$	$9.554 - 20.510$
0.3	$3.046 - 5.951$	$1.987 - 4.704$	$6.142 - 16.175$	$3.633 - 15.758$
0.6	$1.004 - 8.798$	$1.523 - 7.930$	$2.010 - 24.412$	$1.461 - 24.090$
0.9	$0.063 - 12.348$	$1.759 - 11.335$	$0.126 - 34.401$	$0.603 - 34.021$

and then decreased back to zero, as depicted in Figure 16.13. In this case, the simulation consists of a triangular pulse entering the ring configuration, but it is not difficult to modify the Mathematica program to investigate *Gaussian input* pulses. The input intensity is increased linearly up to $16\,\text{Wm}^{-2}$ and then decreased back down to zero. Figure 16.13 shows the output intensity and input intensity against the number of passes around the ring, which in this particular case was 4000. To observe the bistable region, it is necessary to display the ramp-up and ramp-down parts of the diagram on the same graph, as in Figure 16.14(b).

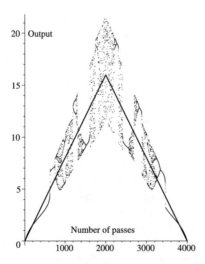

Figure 16.13: Bifurcation diagram when $B = 0.15$ using the second iterative method showing a plot of triangular input and output intensities against number of ring passes for the SFR resonator.

Figure 16.14 shows a gallery of bifurcation diagrams, corresponding to some of the parameter values used in Table 16.1 produced using the second iterative method. The diagrams make interesting comparisons with the results displayed in Table 16.1.

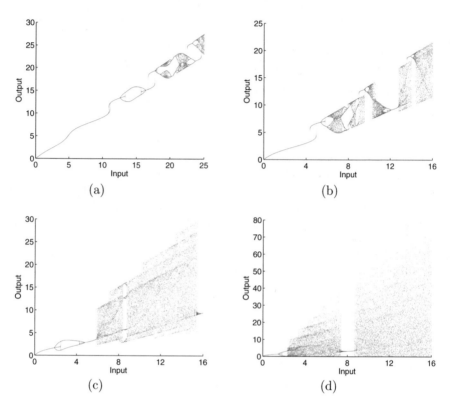

Figure 16.14:   A gallery of bifurcation diagrams for the SFR resonator (equation 16.11) when (a) $B = 0.05$, (b) $B = 0.15$, (c) $B = 0.3$, and (d) $B = 0.6$. In each case, 10000 iterations were carried out.

A numerical investigation has revealed that for a small range of values close to $B = 0.15$, (see Figure 16.14(b)), the SFR resonator could be used as a bistable device. Unfortunately, for most values of B, instabilities overlap with the first bistable region. For example, when $B = 0.3$ (Figure 16.14(c)), the first unstable region between 1.987 Wm^{-2} and 4.704 Wm^{-2} intersects with the first bistable region between 3.046 Wm^{-2} and 5.951 Wm^{-2}. Clearly, the instabilities have affected the bistable operation. In fact, the hysteresis cycle has failed to materialize. Recall that $B = \sqrt{\kappa}$, where κ is the power coupling ratio. As the parameter B gets larger, more of the input power is circulated in the ring, and this causes the system to become chaotic for low input intensities.

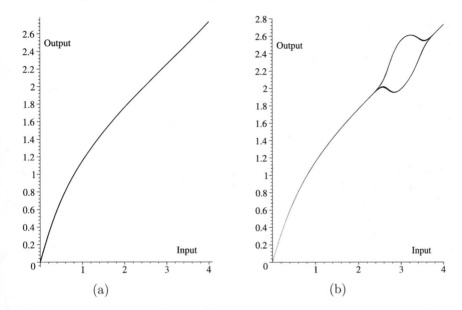

Figure 16.15: Bifurcation diagrams when $B = 0.225$ (a) using the second iterative method with feedback and (b) using the first iterative method without feedback.

The first iterative method can be employed to show regions of instability. Note, however, that bistable regions will not be displayed since there is no feedback in this method. It is sometimes possible for a small unstable region to be missed using the second iterative method. The steady state remains on the unstable branch until it becomes stable again. Thus, in a few cases, the first iterative method gives results which may be missed using the second iterative method. As a particular example, consider system (16.10) where $B = 0.225$. Results from a linear stability analysis indicate that there should be an unstable region in the range $2.741 - 3.416$ Wm^{-2}. Figure 16.15(a) shows that this region is missed using the second iterative method, whereas the first iterative method (Figure 16.15(b)) clearly displays period-two behavior. In physical applications, one would expect relatively small unstable regions to be skipped, as in the former case.

Consider the complex iterative equation

$$E_{n+1} = i\sqrt{1 - \kappa}E_{\text{in}} + \sqrt{\kappa}E_n \exp\left[i\left(\frac{2\pi n_2 L}{\lambda_0 A_{\text{eff}}}|E_n|^2 - \phi_L\right)\right],\qquad (16.20)$$

which was derived earlier. Equation (16.20) is the iterative equation that models the electric field in the SFR resonator. Typical *fiber parameters* chosen for this system are $\lambda_0 = 1.55 \times 10^{-6}$ m; $n_2 = 3.2 \times 10^{-20}$ m^{2}W^{-1}; A$_{\text{eff}} = 30\mu$m^2 and $L = 80$ m. Suppose that equation (16.11) was iterated 10000 times. This would equate to hundredths of a second of elapsed time in physical applications using these values for the fiber parameters.

In the work considered so far, the linear phase shift due to propagation ϕ_L has been set to zero. Figure 16.16 shows how the bistable region is affected when ϕ_L is nonzero and $B = 0.15$. As the linear phase shift increases from zero to $\frac{\pi}{4}$, the first bistable region gets larger and shifts to the right slightly, as depicted in Figure 16.16(b). When $\phi_L = \frac{\pi}{2}$, an instability has appeared between 20 Wm^{-2} and 40 Wm^{-2} and a second unstable region has encroached on the first bistable region, as shown in Figure 16.16(c). When $\phi_L = \pi$, instabilities appear at both ends of the bistable region, as shown in Figure 16.16(d). Therefore, the linear phase shift can affect the bistable operation of the SFR resonator. Should such systems be used for bistable operation, then the results indicate the need to control the feedback phase to prevent any instabilities from entering the power range in the hysteresis loop.

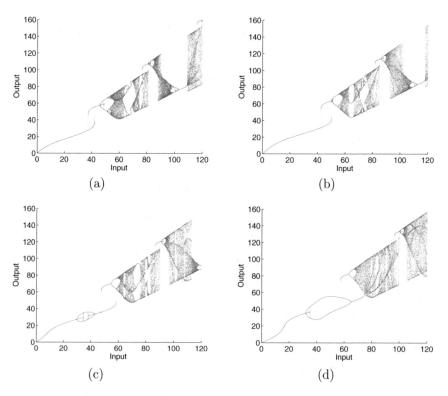

Figure 16.16: [Mathematica] Bifurcation diagrams for the SFR resonator using equation (16.8) when $\kappa = 0.0225$ and (a) $\phi_L = 0$, (b) $\phi_L = \frac{\pi}{4}$, (c) $\phi_L = \frac{\pi}{2}$, and (d) $\phi_L = \pi$.

In conclusion, the dynamic properties of a nonlinear optical resonator have been analyzed using a graphical method, a linear stability analysis, and bifurcation diagrams. The bifurcation diagrams give a clearer insight into the dynamics than the results from the linear stability analysis and graphical method, but all four used in conjunction provide useful results.

16.7 Mathematica Commands

```
(* Finding fixed points. See Figure 16.10(b). *)
A=2.2;B=0.15;
ContourPlot[{A+B*x*Cos[x^2+y^2]-B*y*Sin[x^2+y^2]-x,
B*x*Sin[x^2+y^2]+B*y*Cos[x^2+y^2]-y},{x,0,4},{y,-4,4},
Axes->True,FrameLabel->{"Real(Subscript[E, n])",
"Imag(Subscript[E, n])"},BaseStyle->{FontSize->18}]
```

```
(* Iteration of the Ikeda map. See Figure 16.11(b). *)
A=10;B=0.15;Itermax=10000;
Ikeda[{x_,y_}]={A+B (x Cos[x^2+y^2]-y Sin[x^2+y^2]),
B (x Sin[x^2+y^2]+y Cos[x^2+y^2])};
IkedaMap=Drop[NestList[Ikeda,{A,0},Itermax],100];
ListPlot[IkedaMap,AxesLabel->{" "," "},Axes->False,
PlotStyle->{PointSize[0.001],RGBColor[0,0,1]},Frame->True,
AspectRatio->Automatic,FrameLabel->{"Real(Subscript[E, n])",
"Imag(Subscript[E, n])"},BaseStyle->{FontSize->14}]
```

```
(* Bifurcation diagram for Ikeda Map. See Figure 16.16(a). *)
(* Ramp up output power is blue and ramp down output power is red. *)
MaxPower=120;kappa=0.0225;Const=0.345913;Initial=0+0 I;
HalfN=19999;phi=0;N1=1+HalfN;E1[1]=0;Itermax=2 HalfN+1;Esqr[1]=0;
For[n=1,n<=HalfN,n++,
{E2[n+1]=E1[n] Exp[I (Abs[Const E1[n]]^2-phi)],
E1[n+1]=I Sqrt[1-kappa] Sqrt[n MaxPower/N1]+
Sqrt[kappa] E2[n+1],Esqr[n+1]=Abs[E1[n+1]]^2}];
p1=ListPlot[Table[Flatten[{n*MaxPower/N1,Esqr[n]}],{n,1,HalfN}],
PlotStyle->PointSize[0.001]];
For[n=N1,n<=Itermax,n++,
{E2[n+1]=E1[n] Exp[I (Abs[Const E1[n]]^2-phi)],
E1[n+1]=I Sqrt[1-kappa] Sqrt[2 MaxPower-n MaxPower/N1]+
Sqrt[kappa] E2[n+1],Esqr[n+1]=Abs[E1[n+1]]^2}];
p2=ListPlot[Table[Flatten[{n*MaxPower/N1,Esqr[Itermax+1-n]}],{n,1,HalfN}],P
lotStyle->{PointSize[0.001],RGBColor[1,0,0]}];
Show[{p1,p2},PlotRange->All,AxesLabel->{"Input","Output"},
TextStyle->{FontSize->18},AspectRatio->Automatic]
```

16.8 Exercises

1. Determine the number of fixed points of period one for system (16.10) when $B = 0.4$ and $A = 3.9$ by plotting the graphs of the simultaneous equations.

2. Plot iterative maps for equation (16.8), using the parameter values given in the text, when $\kappa = 0.0225$ and (i) $E_{\text{in}} = 4.5$, (ii) $E_{\text{in}} = 6.3$, and (iii) $E_{\text{in}} = 11$.

3. Given that
$$E_{n+1} = A + BE_n e^{i|E_n|^2},$$
prove that the inverse map is given by
$$E_{n+1} = \left(\frac{E_n - A}{B} \right) \exp \left(\frac{-i|E_n - A|^2}{B^2} \right).$$

4. Given the complex Ikeda mapping
$$E_{n+1} = A + BE_n \exp \left[i \left(\phi - \frac{C}{1 + |E_n|^2} \right) \right],$$
where A, B, and C are constants, show that the steady-state solution, say, $E_{n+1} = E_n = E_S$, satisfies the equation
$$\cos \left(\frac{C}{1 + |E_S|^2} - \phi \right) = \frac{1}{2B} \left(1 + B^2 - \frac{A^2}{|E_S|^2} \right).$$

5. Consider the double-coupler nonlinear fiber ring resonator as shown in Figure 16.17.

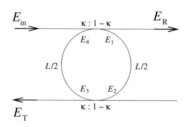

Figure 16.17:   Schematic of a double-coupler fiber ring resonator.

Suppose that

$$E_R(t) = \sqrt{\kappa}E_{\text{in}}(t) + i\sqrt{1-\kappa}E_4(t);$$
$$E_1(t) = i\sqrt{1-\kappa}E_{\text{in}}(t) + \sqrt{\kappa}E_4(t);$$
$$E_2(t) = E_1(t-t_R)e^{i\phi_1(t-t_R)};$$
$$\phi_1(t-t_R) = \frac{\pi r_2 L}{\lambda A_{\text{eff}}}|E_1(t-t_R)|^2;$$
$$E_3(t) = \sqrt{\kappa}E_2(t);$$
$$E_T(t) = i\sqrt{1-\kappa}E_2(t);$$
$$E_4(t) = E_3(t-t_R)e^{i\phi_2(t-t_R)};$$
$$\phi_2(t-t_R) = \frac{\pi r_2 L}{\lambda A_{\text{eff}}}|E_3(t-t_R)|^2;$$

where the fiber loop is of length L; both halves are of length $L/2$; t_R is the time taken for the electric field to complete half a fiber loop; and both couplers split the power in the ratio $\kappa : 1-\kappa$. Assuming that there are no losses in the fiber, show that

$$E_T(t) = -(1-\kappa)E_{\text{in}}(t-t_R)e^{i\phi_1(t-t_R)} + \kappa E_T(t-2t_R)e^{i(\phi_1(t-t_R)+\phi_2(t-2t_R))}.$$

6. Consider the complex iterative equation

$$E_{n+1} = A + BE_n \exp\left[i\left(|E_n|^2\right)\right],$$

used to model the SFR resonator. Use a linear stability analysis to determine the first bistable and unstable regions when (a) $B = 0.1$, (b) $B = 0.2$, and (c) $B = 0.25$ to three decimal places, respectively.

7. Plot bifurcation diagrams for Exercise 6, parts (a)–(c), when the maximum input intensity is 25 Wm^{-2} and the input pulse is triangular.

8. Plot the bifurcation diagram for the iterative equation in Exercise 6 for $B = 0.15$ when the input pulse is Gaussian with a maximum of 25 Wm^{-2}. How is the bistable region affected by the width of the pulse?

9. Consider the complex iterative equation

$$E_{n+1} = A + BE_n \exp\left[i\left(|E_n|^2 - \phi_L\right)\right],$$

where $B = 0.15$ and ϕ_L represents a linear phase shift. Plot bifurcation diagrams for a maximum input intensity of $A = 3$ units when

(a) $\phi_L = \frac{\pi}{4}$,

(b) $\phi_L = \frac{\pi}{2}$,

(c) $\phi_L = \frac{3\pi}{4}$,

(d) $\phi_L = \pi$,

(e) $\phi_L = \frac{5\pi}{4}$,

(f) $\phi_L = \frac{3\pi}{2}$,

(g) $\phi_L = \frac{7\pi}{4}$.

10. Apply the linear stability analysis to the iterative equation

$$E_{n+1} = i\sqrt{1 - \kappa}\,E_{\text{in}} + \sqrt{\kappa}\,E_n \exp\left[i\left(\frac{2\pi n_2 L}{\lambda_0 A_{\text{eff}}}|E_n|^2\right)\right],$$

for the parameter values given in this chapter. Compare the results with the bifurcation diagrams.

Bibliography

[1] G.P. Agrawal, *Nonlinear Fiber Optics*, 5th ed., Academic Press, New York, London, 2012.

[2] G.P. Agrawal, *Applications in Nonlinear Fiber Optics*, 2nd ed., Academic Press, New York, London, 2008.

[3] T. Bischofberger and Y.R. Shen, Theoretical and experimental study of the dynamic behavior of a nonlinear Fabry-Perot interferometer, *Phys. Rev. A* **19**, (1979), 1169–1176.

[4] R.W. Boyd, *Nonlinear Optics*, 3rd ed., Academic Press, New York, London, 2008.

[5] Chao-Xiang Shi, Nonlinear fiber loop mirror with optical feedback, *Optics Comm.* **107**, (1994), 276–280.

[6] N.J. Doran and D. Wood, Nonlinear-optical loop mirror, *Optics Lett.* **13**, (1988), 56–58.

[7] F.S. Felber and J.H. Marburger, Theory of nonresonant multistable optical devices, *Appl. Phys. Lett.* **28**, (1976), 731.

[8] W.J. Firth, Stability of nonlinear Fabry-Perot resonators, *Optics Comm.* **39**-5, (1981), 343–346.

[9] H.M. Gibbs, *Optical bistability: Controlling light with light*, Academic Press, New York, London, 1985.

[10] K. Ikeda, H. Daido, and O. Akimoto, Optical turbulence: chaotic behavior of transmitted light from a ring cavity, *Phys. Rev. Lett.* **45**-9, (1980), 709–712.

[11] Y.H. Ja, Multiple bistability in an optical-fiber double-ring resonator utilizing the Kerr effect, *IEEE J. Quantum Electron.* **30**-2, (1994), 329–333.

[12] H. Li and K. Ogusu, Analysis of optical instability in a double-coupler nonlinear fiber ring resonator, *Optics Comm.* **157**, (1998), 27–32.

[13] S. Lynch and A.L. Steele, Nonlinear Optical Fibre Resonators with Applications in Electrical Engineering and Computing, in *Applications of Nonlinear Dynamics and Chaos in Engineering*, Santo Banerjee, Mala Mitra, Lamberto Rondoni (Eds.), Springer, 1, (2011) 65–84.

[14] S. Lynch, A.L. Steele, and J.E. Hoad, Stability analysis of nonlinear optical resonators, *Chaos, Solitons and Fractals*, **9**-6, (1998) 935–946.

[15] P. Mandel, *Theoretical Problems in Cavity Nonlinear Optics*, Cambridge University Press, Cambridge, UK, 2005.

[16] J.H. Marburger and F.S. Felber, Theory of a lossless nonlinear Fabry-Perot interferometer, *Phys. Rev. A* **17**, (1978), 335–342.

[17] R. Matthews, Catch the wave, *New Scientist*, **162**-2189, (1999), 27–32.

[18] H. Natsuka, S. Asaka, H. Itoh, K. Ikeda, and M. Matouka, Observation of bifurcation to chaos in an all-optical bistable system, Phys. Rev. Lett. **50**, (1983), 109–112.

[19] K. Ogusu, A.L. Steele, J.E. Hoad, and S. Lynch, Corrections to and comments on "Dynamic behavior of reflection optical bistability in a nonlinear fiber ring resonator", *IEEE J. Quantum Electron.*, **33**, (1997), 2128–2129.

[20] T. Schneider, *Nonlinear Optics in Telecommunications*, Springer-Verlag, New York, 2004.

[21] S.D. Smith, Towards the optical computer, *Nature* **307**, 26 January, (1984), 315–316.

[22] P.W. Smith and E.H. Turner, *Appl. Phys. Lett.* **30**, (1977), 280–281.

[23] A.L. Steele, S. Lynch, and J.E. Hoad, Analysis of optical instabilities and bistability in a nonlinear optical fiber loop mirror with feedback, *Optics Comm.* **137**, (1997), 136–142.

[24] A. Szöke, V. Daneu, J. Goldhar, and N.A. Kirnit, Bistable optical element and its applications, *Appl. Phys. Lett.* **15**, (1969), 376.

Chapter 17
Fractals and Multifractals

Aims and Objectives

- To provide a brief introduction to fractals.

- To introduce the notion of fractal dimension.

- To provide a brief introduction to multifractals and define a multifractal formalism.

- To consider some very simple examples.

On completion of this chapter, the reader should be able to

- plot early-stage generations of certain fractals using either graph paper, pencil, and rule, or Mathematica;

- determine the fractal dimension of some mathematical fractals;

- estimate the fractal dimension using simple box-counting techniques;

- distinguish between homogeneous and heterogeneous fractals;

- appreciate how multifractal theory is being applied in the real world;

- construct multifractal Cantor sets and Koch curves and plot graphs of their respective multifractal spectra.

Fractals are introduced by means of some simple examples, and the fractal dimension is defined. Box-counting techniques are used to approximate the fractal dimension of certain early-stage generation fractals, which can be generated using pencil, paper, and rule.

© Springer International Publishing AG 2017
S. Lynch, *Dynamical Systems with Applications using Mathematica®*,
DOI 10.1007/978-3-319-61485-4_17

A multifractal formalism is introduced that avoids some of the more abstract pure mathematical concepts. The theory is explained in terms of box-counting dimensions, which are introduced in this chapter. This is potentially a very complicated topic, and readers new to this field are advised to look at Example 4 before attempting to understand the formalism.

Some applications of multifractal analysis to physical systems in the real world are also discussed. A few simple self-similar multifractals are constructed, and the analysis is applied to these objects.

17.1 Construction of Simple Examples

Definition 1. A *fractal* is an object that displays self-similarity under magnification and can be constructed using a simple motif (an image repeated on ever-reduced scales).

Fractals have generated a great deal of interest since the advent of the computer. Many shops now sell colorful posters and T-shirts displaying fractals, and some color fractals have been plotted in Chapter 15. Although the Julia sets and the Mandelbrot set are not true fractals, they do have fractal structure. Many objects in nature display this self-similarity at different scales; for example, cauliflower, ferns, trees, mountains, clouds, and even blood vessel networks in our own bodies have some fractal structure. These objects cannot be described using the geometry of lines, planes, and spheres. Instead, *fractal geometry* is required. Fractal analysis is being applied in many branches of science—for example, to computer graphics and image compression (e.g., take a closer look at the images on the Web) and to oil extraction from rocks using viscous fingering—and multifractal analysis has expanded rapidly over recent years (see later in this chapter). The reader is directed to Falconer's text [8] for a simple introduction to fractals and their many applications and reference [9] gives a more mathematical perspective.

It is important to note that all of the fractals appearing in this textbook are early-generation fractals. However, there is nothing to stop scientists from imagining an ideal mathematical fractal that is constructed to infinity. Some of these fractals will now be investigated.

The Cantor Set. The Cantor fractal was first considered by Georg Cantor in 1870. It is constructed by removing the middle third of a line segment at each stage of construction. Thus, at stage 0, there is one line segment of unit length. At stage 1, the middle third is removed to leave two segments each of length $\frac{1}{3}$. At stage 2, there will be four segments each of length $\frac{1}{9}$. Continuing in this way, it is not difficult to see that at the kth stage, there will be $N = 2^k$ segments each of length $l = 3^{-k}$. An early-stage construction (up to stage 3) is shown in Figure 17.1.

Figure 17.1: An early generation of the Cantor set.

If this process is continued to infinity, then

$$\lim_{k\to\infty} 2^k = \infty \quad \text{and} \quad \lim_{k\to\infty} 3^{-k} = 0.$$

The Cantor set will therefore consist of an infinite number of discrete points that, unfortunately, is impossible to generate on a computer screen. However, all is not lost. By using the ternary number system, it is possible to classify which points in the unit interval belong to the Cantor set and which do not. Recall that ternary proper fractions can be expanded by applying a simple algorithm: Treble the numerator of the proper fraction concerned; when this number is larger than or equal to the denominator, subtract the denominator, noting down the ternary factor above the line, and continue with the remainder. For example, $\frac{4}{7} = 0.\underline{120102}$, since

$$
\begin{array}{c|cccccccc}
 & 1 & 2 & 0 & 1 & 0 & 2 & 1 & \cdots \\
\hline
4 & 12 & & & & & & & \\
 & 5 & 15 & & & & & & \\
 & & 1 & 3 & 9 & & & & \\
 & & & & 2 & 6 & 18 & & \\
 & & & & & 4 & 12 & & \\
 & & & & & & 5 & \cdots, &
\end{array}
$$

where the underlining after the decimal point represents a recurring decimal. It is not too difficult to show that the Cantor set can be identified by points whose ternary fractions consist of zeroes and twos only. Thus, $p_1 = 0.20202$ will belong to the Cantor set, whereas $p_2 = 0.\underline{120102}$ will not.

The Koch Curve. Helge von Koch first imagined the Koch curve in 1904. It is constructed by replacing a unit line segment with a motif consisting of four line segments each of length $\frac{1}{3}$, as depicted in Figure 17.2.

A simple Mathematica program is given in Section 17.5 to plot early generations of the Koch curve. Note that at the kth stage there are $N = 4^k$

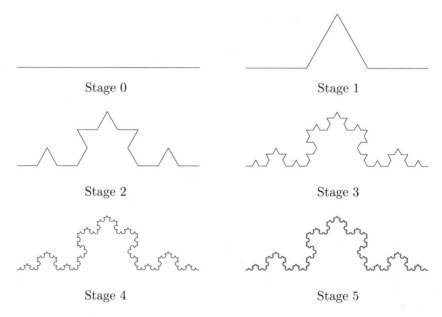

Figure 17.2: [Mathematica] Construction of the Koch curve up to stage 5.

line segments each of length $l = 3^{-k}$. Thus for the mathematical fractal constructed to infinity,

$$\lim_{k \to \infty} 4^k = \infty \quad \text{and} \quad \lim_{k \to \infty} 3^{-k} = 0,$$

so the mathematical Koch curve consists of a curve that is infinitely long.

The Koch Square. Consider a variation of the Koch curve that is constructed by replacing one line segment with five line segments each of length $\frac{1}{3}$. Furthermore, suppose that these curves are attached to the outer edge of a unit square. The first five stages of construction are shown in Figure 17.3.

It is possible to determine the area and perimeter bounded by the Koch square in the following way. Suppose that at stage 0 that the square has area $A_0 = 1$ unit^2 and that the area at stage k is A_k. Then,

$$A_1 = 1 + 4(3^{-2}) \text{ unit}^2.$$

At stage 2, the area is given by

$$A_2 = 1 + 4(3^{-2}) + 4 \times 5 \times (3^{-4}) \text{ unit}^2.$$

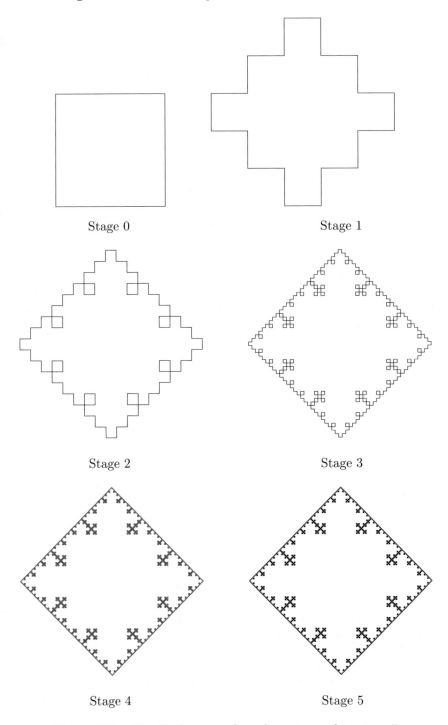

Stage 0   Stage 1

Stage 2   Stage 3

Stage 4   Stage 5

Figure 17.3: The Koch square fractal constructed to stage 5.

Figure 17.4:   The inverted Koch square at stage 5.

Continuing in this way, the area at the kth stage is given by

$$A_k = 1 + 4(3^{-2}) + 4 \times 5 \times (3^{-4}) + 4 \times 5^2 \times (3^{-6}) + \ldots + 4 \times 5^{k-1} \times (3^{-2k}) \text{ unit}^2.$$

Take the limit $k \to \infty$. Then,

$$A_\infty = 1 + \frac{4}{9} + \sum_{i=1}^{\infty} 4 \times 5^i \times (9^{-(i+1)}) \text{ unit}^2.$$

This is the sum of an infinite geometric series, and hence

$$A_\infty = 1 + \frac{4}{9} + \frac{\frac{4 \times 5}{9^2}}{1 - \frac{5}{3^2}} = 2 \text{ unit}^2.$$

It is not difficult to show that the perimeter P_k at the kth stage is given by

$$P_k = 4 \times \left(\frac{5}{3}\right)^k,$$

and $P_\infty = \infty$. Therefore, the Koch square has infinite perimeter and finite area.

It is possible to construct an inverted Koch square fractal by attaching the Koch curves to the inner edge of the unit square. The result up to stage 5 is shown in Figure 17.4.

The Sierpiński Triangle. This fractal may be constructed in a number of ways; see the exercises at the end of the chapter (Section 17.6). One way is

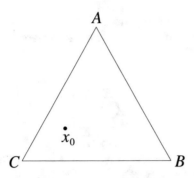

Figure 17.5: A triangle used in the chaos game with an initial point x_0.

to play a so-called chaos game with a die. Consider an equilateral triangle with vertices A, B, and C, as depicted in Figure 17.5.

The rules of the chaos game are very simple. Start with an initial point x_0 somewhere inside the triangle.

Step 1. Cast an ordinary cubic die with six faces.

Step 2. If the number is either 1 or 2, move half way to the point A and plot a point.

Step 2. Else, if the number is either 3 or 4, move half way to the point B and plot a point.

Step 2. Else, if the number is either 5 or 6, move half way to the point C and plot a point.

Step 3. Starting with the new point generated in Step 2, return to Step 1.

The die is cast again and again to generate a sequence of points $\{x_0, x_1, x_2, x_3, \ldots\}$. As with the other fractals considered here, the mathematical fractal would consist of an infinite number of points. In this way, a chaotic attractor is formed, as depicted in Figure 17.6. A Mathematica program is given in Section 17.5.

The first few initial points are omitted to reveal the chaotic attractor. This object is known as the Sierpiński triangle.

Stochastic processes can be introduced to obtain fractals that look more like objects in nature. We restrict ourselves to two-dimensional figures only in this chapter.

Definition 2. An *iterated function system* (IFS) is a finite set $T_1, T_2, T_3,$ $\ldots, T_n$ of affine linear transformations of $\Re^2$, where

Figure 17.6: [Mathematica] An early-stage-generation Sierpiński triangle plotted using the chaos game. There are 50,000 points plotted.

$$T_j(x, y) = (a_j x + b_j y + c_j, d_j x + e_j y + f_j).$$

Furthermore, a *hyperbolic iterated function system* is a collection of affine linear transformations that are also contractions.

The IFSs follow basic rules, as in the case of the chaos game used to generate the Sierpiński triangle. The rules of the chaos game can be generalized to allow greater freedom as follows:

Step 1. Create two or more affine linear transformations.

Step 2. Assign probabilities to each of the transformations.

Step 3. Start with an initial point.

Step 4. Select a random transformation to get a second point.

Step 5. Repeat the process.

An IFS consisting of four transformations was used to generate Figure 17.7. This figure resembles a fern in nature and is known as *Barnsley's fern*. A Mathematica program is listed in Section 17.5.

The affine linear transformations may be found by taking reflections, rotations, scalings, and translations of triangles that represent the fronds of the fern.

Figure 17.7: [Mathematica] A fractal attractor of an IFS. Barnsley's fern, generated using 60,000 points.

17.2 Calculating Fractal Dimensions

Definition 3. A self-similar fractal has fractal dimension (or *Hausdorff index*) D_f given by

$$D_f = \frac{\ln N(l)}{-\ln l},$$

where l represents a scaling and $N(l)$ denotes the number of segments of length l. Thus, the relationship

$$N(l) \propto (l)^{-D_f} \tag{17.1}$$

is also valid. The number D_f, which need not be an integer, gives a measure of how the density of the fractal object varies with respect to length scale.

Definition 4. A fractal is an object that has noninteger fractal dimension. (This is an alternative to Definition 1)

Example 1. Determine the fractal dimension of
    (i) the Cantor set,

   (ii) the Koch curve,

   (iii) the Koch square, and

   (iv) the Sierpiński triangle.

Solution. (i) A construction of the Cantor set up to stage 3 is depicted in Figure 17.1. At each stage, one segment is replaced with two segments that are $\frac{1}{3}$ the length of the previous segment. Thus, in this case, $N(l) = 2$ and $l = \frac{1}{3}$. The mathematical self-similar Cantor set fractal constructed to infinity will therefore have dimension given by

$$D_f = \frac{\ln 2}{\ln 3} \approx 0.6309.$$

Note that a point is defined to have dimension zero and a line dimension one. Hence, the Cantor set is more dense than a point, but less dense than a line.

(ii) The Koch curve is constructed up to stage 5 in Figure 17.2. In this case, one segment is replaced with four segments which are scaled by $\frac{1}{3}$; therefore, $N(l) = 4$ and $l = \frac{1}{3}$. The mathematical self-similar Koch fractal generated to infinity will have dimension

$$D_f = \frac{\ln 4}{\ln 3} \approx 1.2619.$$

Thus, the Koch curve is more dense than a line, but less dense than a plane, which is defined to have dimension two.

(iii) The Koch square generated to stage 5 is shown in Figure 17.3. Note that this object is not strictly self-similar; magnification will not reveal smaller Koch squares. However, it is possible to define a fractal dimension, since there is a scaling behavior. For the Koch square,

$$D_f = \frac{\ln 5}{\ln 3} \approx 1.4650.$$

Hence, the Koch square is more dense than the Koch curve but is still less dense than the plane. Note that the inverted Koch square will have exactly the same fractal dimension.

(iii) The mathematical Sierpiński triangle fractal (see Figure 17.6) may be constructed by removing the central triangle from equilateral triangles to infinity. A motif is shown in Figure 17.8.

It is important to note that the scaling l referred to in Definition 2 is linear. Thus, the linear scale is $\frac{1}{2}$ since the sides of the smaller triangles are half as long as the sides of the original triangle in the motif. At each stage, one triangle is replaced with three triangles, so $l = \frac{1}{2}$ and $N(l) = 3$. The fractal dimension of the mathematical Sierpiński triangle generated to infinity is

$$D_f = \frac{\ln 3}{\ln 2} \approx 1.5850.$$

The Sierpiński triangle has the highest dimension in examples (i)–(iv) and is therefore the most dense.

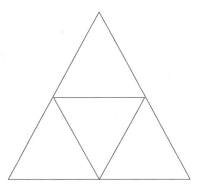

Figure 17.8: The motif used to generate the Sierpiński triangle.

Box-Counting Dimensions. The fractal dimensions calculated so far have been for hypothetical fractal objects that cannot exist in the real world. Mandelbrot [17] shows how fractals appear throughout science and nature. Trees, clouds, rocks, and the fractals generated in earlier chapters can display a certain type of scaling and self-similarity. Mandelbrot showed that these objects obey a power law as described in equation (17.1) over a certain range of scales. By covering the object with boxes of varying sizes and counting the number of boxes that contain the object, it is possible to estimate a so-called box-counting dimension, which is equivalent to the fractal dimension. Mandelbrot defined the fractal dimension to be

$$D_f = \lim_{l \to 0} \frac{\ln N(l)}{-\ln l},$$

where $N(l)$ boxes of length l cover the fractal object. These boxes need not be square.

Consider the following two examples.

Example 2. The Koch curve is covered with boxes of varying scales, as shown in Figure 5.9. Use a box-counting technique to show that the object obeys the power law given in equation (17.1) and hence estimate the box-counting dimension (Figure 17.9).

Solution. Table 17.1 gives the box count $N(l)$ for the different scalings l, and the natural logs are calculated.

Using the least-squares method of regression, the line of best fit on a log-log plot is given by $y \approx 1.2246x - 0.2817$, and the correlation coefficient is approximately 0.9857. The line of best fit is shown in Figure 17.10.

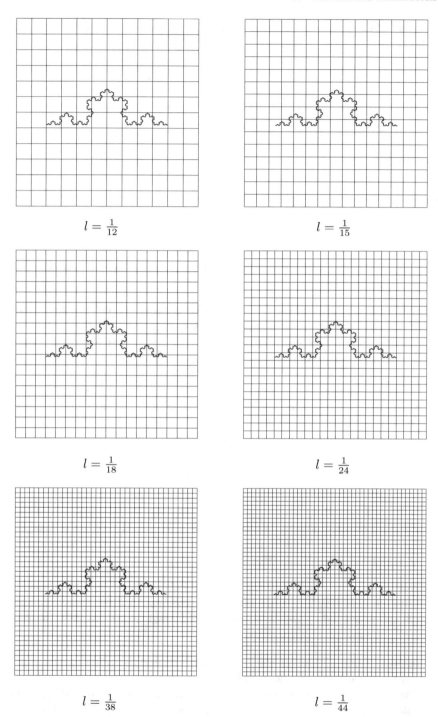

$l = \frac{1}{12}$

$l = \frac{1}{15}$

$l = \frac{1}{18}$

$l = \frac{1}{24}$

$l = \frac{1}{38}$

$l = \frac{1}{44}$

Figure 17.9: Different coarse coverings of the Koch curve generated to stage 6.

Table 17.1: Box-count data for the Koch curve generated to stage 6

l	12^{-1}	15^{-1}	18^{-1}	24^{-1}	38^{-1}	44^{-1}
$N(l)$	14	24	28	34	60	83
$-\ln l$	2.4849	2.7081	2.8904	3.1781	3.6376	3.7842
$\ln N(l)$	2.6391	3.1781	3.3322	3.5264	4.0943	4.4188

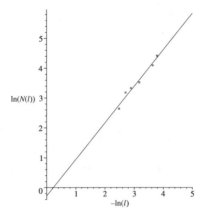

Figure 17.10: The line of best fit on a log-log plot for the early-generation Koch curve. The correlation coefficient is 0.9857.

Therefore, the box-counting dimension of the Koch curve generated to stage 6 is approximately 1.2246. There is obviously a scaling restriction with this object since the smallest segment is of length $3^{-6} \approx 0.0014$ units and the box-counting algorithm will break down as boxes approach this dimension. There is always some kind of scaling restriction with physical images as there are a limited number of pixels on a computer screen. It is interesting to note that the mathematical Koch curve has a higher dimension of approximately 1.2619. This is to be expected as true mathematical fractal is a lot more dense.

Example 3. A chaotic attractor comprising 5000 points for the Hénon map

$$x_{n+1} = 1.2 + 0.4y_n - x_n^2, \quad y_{n+1} = x_n$$

is covered with boxes of varying scales, as shown in Figure 17.11. Use a box-counting technique to show that the object obeys the power law given in equation (17.1) and hence estimate the box-counting dimension.

Solution. Table 17.2 gives the box count $N(l)$ for the different scalings l, and the natural logs are calculated.

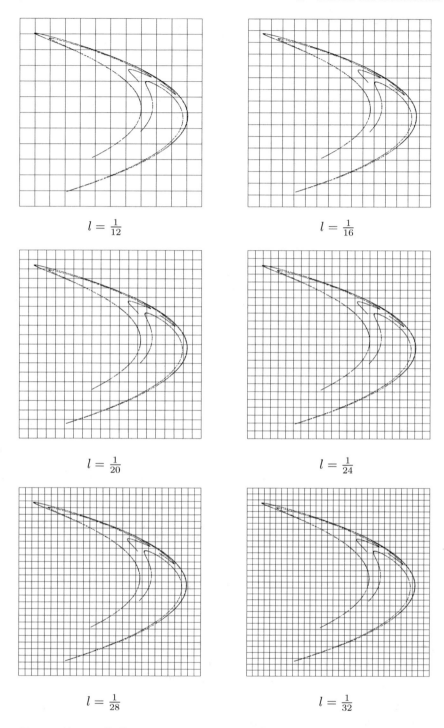

Figure 17.11: Different coarse coverings of the Hénon chaotic attractor.

Table 17.2: Box-count data for the Hénon map with 5000 points

l	12^{-1}	16^{-1}	20^{-1}	24^{-1}	28^{-1}	32^{-1}
$N(l)$	47	58	76	93	109	131
$-\ln l$	2.4849	2.7726	2.9957	3.1781	3.3322	3.4657
$\ln N(l)$	3.8501	4.0604	4.3307	4.5326	4.6914	4.8752

Using the least-squares method of regression, the line of best fit on a log-log plot is given by $y \approx 1.0562x + 1.1810$, and the correlation coefficient is approximately 0.9961. The line of best fit is shown in Figure 17.12.

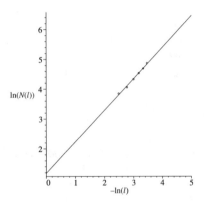

Figure 17.12: The line of best fit on a log-log plot for the early generation Hénon attractor. The correlation coefficient is 0.9961.

Therefore, the box-counting dimension of the Hénon attractor with 5000 points is approximately 1.0562. There is a scaling restriction in this case as there are only 5000 data points. Once more, the dimension of the mathematical fractal with an infinite number of data points will be larger.

The Hénon map is not self-similar and is, in fact, a multifractal. See the next section. Recent applications of fractals are presented in [3].

17.3 A Multifractal Formalism

In the previous section, it was shown that a fractal object can be characterized by its fractal dimension D_f, which gives a measure of how the density varies with respect to length scale. Most of the fractals appearing earlier in this chapter can be constructed to the infinite stage in the minds of mathematicians. They are *homogeneous* since the fractals consist of a geometrical figure repeated on an ever-reduced scale. For these objects, the fractal dimension is the same on all scales. Unfortunately, in the real world, fractals are not homogeneous; there is rarely an identical motif repeated on all scales. Two objects might have the same fractal dimension and yet look completely different. It has been found that real-world fractals are *heterogeneous*; that is, there is a nonuniformity possessing rich scaling and self-similarity properties that can change from point to point. Put plainly, the object can have different dimensions at different scales. It should also be pointed out that there is always some kind of scaling restriction with physical fractals. These more complicated objects are known as *multifractals*, and it is necessary to define continuous spectra of dimensions to classify them.

There are many different ways in which a mathematician can define *dimension*, and the subject can become very complicated and abstract. For example, there is Hausdorff dimension, topological dimension, Euclidean dimension, and box-counting dimension to name but a few. More details on the pure mathematical approach to multifractals are presented in [9] and [23]. The most widely used method of determining multifractal spectra is that of Falconer [9], which is described briefly below.

Let μ be a self-similar probability measure defined on an object $S \subset \Re^d$, where $\mu(B)$ is a probability measure determined from the probability of hitting the object in the box $B_i(l)$ and $N \propto \frac{1}{l^2}$ is the number of boxes in the grid. The *generalized fractal dimensions* D_q or, alternatively, the $f(\alpha)$ *spectrum of singularities* may be computed using box-counting techniques. First, consider the generalized fractal dimensions. Cover the object S with a grid of boxes $(B_i(l))_{i=1}^N$ of size l as in Section 17.1. The qth moment (or *partition function*) Z_q is defined by

$$Z_q(l) = \sum_{\mu(B)\neq 0} [\mu(B)]^q = \sum_{i=1}^N p_i^q(l), \qquad (17.2)$$

For self-similar multifractals, given a real number q, $\tau(q)$ may be defined as the positive number satisfying

$$\sum_{i=1}^N p_i^q r_i^{\tau(q)} = 1, \qquad (17.3)$$

where p_i represents probabilities ($\sum_{i=1}^N p_i = 1$) with r_i fragmentation ratios. The function $\tau : \Re \to \Re$ is a decreasing real analytic function with

$$\lim_{q \to -\infty} \tau(q) = \infty \quad \text{and} \quad \lim_{q \to \infty} \tau(q) = -\infty.$$

The generalized dimensions D_q and the scaling function $\tau(q)$ are defined by

$$\tau(q) = D_q(1 - q) = \lim_{l \to 0} \frac{\ln Z_q(l)}{-\ln l}. \tag{17.4}$$

The generalized dimensions are obtained from an assumed power-law behavior of the partition function in the limit as $l \to 0$ and $N \to \infty$,

$$Z_q \propto l^{D_q(q-1)}.$$

Definition 5. The generalized (box-counting) fractal dimensions D_q, where $q \in \Re$, are defined by

$$D_q = \lim_{l \to 0} \frac{1}{1 - q} \frac{\ln \sum_{i=1}^{N} p_i^q(l)}{-\ln l}, \tag{17.5}$$

where the index i labels the individual boxes of size l and $p_i(l)$ denotes the relative weight of the ith box or the probability of the object lying in the box. Hence,

$$p_i(l) = \frac{N_i(l)}{N},$$

where $N_i(l)$ is the weight of the ith box and N is the total weight of the object. When $q = 0$,

$$D_0 = D_f = \lim_{l \to 0} \frac{\ln N(l)}{-\ln(l)},$$

where $N(l)$ is the number of boxes contained in the minimal cover. When $q = 1$, L'Hopital's Rule can be applied (see the exercises in Section 17.6) to give

$$D_1 = \lim_{l \to 0} \frac{\sum_{i=1}^{N} p_i \ln(p_i)}{-\ln(l)},$$

which is known as the *information dimension*. This gives an indication of how the morphology increases as $l \to 0$. The quantity D_2 is known as the *correlation dimension* and indicates the correlation between pairs of points in each box. The generalized dimensions $D_3, D_4, \ldots$ are associated with correlations between triples, quadruples, etc., of points in each box.

Now consider the so-called $f(\alpha)$ spectrum of dimensions. The weight p_s of segments of type s scales with the size l of a box as follows:

$$p_s(l) \propto (l)^{\alpha_s},$$

where α_s is the so-called *coarse Hölder exponent* defined by

$$\alpha_s = \frac{\ln p_s(l)}{\ln l}.$$

The number of segments N_s of type s scales with the size l of a box according to

$$N_s(l) \propto (l)^{-f_s}.$$

The exponents α_s and f_s can then be used to determine $f(\alpha)$, as demonstrated in the examples in the next section.

In many cases, $f(\alpha) = \dim_H S_\alpha$ is related to the Hausdorff-Besicovich dimension of the set $\mathbf{x} \in S$; see [9] for more information. In most cases, a multifractal spectrum $f(\alpha)$ may be obtained from $\tau(q)$ by a so-called *Legendre transformation*, which is described here briefly for completeness. Hence,

$$f(\alpha) = \inf_{-\infty < q < \infty} (\tau(q) + \alpha q).$$

The $f(\alpha)$ can be derived from $\tau(q)$, and vice versa, by the identities

$$f(\alpha(q)) = q\alpha(q) + \tau(q) \quad \text{and} \quad \alpha = -\frac{\partial \tau}{\partial q}. \tag{17.6}$$

It is known that the function $f(\alpha)$ is strictly cap convex (see Figure 17.13(c)) and that $\alpha(q)$ is a decreasing function of q.

In practice, to compute $\tau(q)$ using the partition function, the following three steps are required:

- Cover the object with boxes $(B_i(l))_{i=1}^N$ of size l and compute the corresponding box-measures $\mu_i = \mu(B_i(l)) = p_i(l)$.

- Compute the partition function Z_q for various values of l.

- Check that the log-log plots for Z_q against l are straight lines. If so, then $\tau(q)$ is the slope of the line corresponding to the exponent q.

In summary, $\tau(q)$ and D_q can be obtained from equations (17.2) and (17.4), and the $f(\alpha)$ values can be determined as above or computed (see [6]) using the expressions

$$f(q) = \lim_{l \to 0} \frac{\sum_{i=1}^N \mu_i(q, l) \ln \mu_i(q, l)}{\ln l} \tag{17.7}$$

and

$$\alpha(q) = \lim_{l \to 0} \frac{\sum_{i=1}^{N} \mu_i(q,l) \ln p_i(l)}{\ln l}, \tag{17.8}$$

where $\mu_i(q,l)$ are the normalized probabilities

$$\mu_i(q,l) = \frac{p_i^q(l)}{\sum_{j=1}^{N} p_j^q(l)}.$$

In physical applications, an image on a computer screen of 512×512 pixels is typically used. A problem arises with negative values of q; boxes with very low measure may contribute disproportionately. Several papers have been published addressing this *clipping problem*; see [2], for example. This is not a problem with some of the physical applications discussed here since most of the useful results are obtained for $0 \le q \le 5$.

The multifractal functions $\tau(q)$, D_q, and $f(\alpha)$ have typical forms for self-similar measures. For example, consider $f : [\alpha_{\min}, \alpha_{\max}] \to \Re$, then $-\alpha_{\min}$ and $-\alpha_{\max}$ are the slopes of the asymptotes of the strictly convex function τ. The geometry of the Legendre transform determines that f is continuous on $[\alpha_{\min}, \alpha_{\max}]$ and $f(\alpha_{\min}) = f(\alpha_{\max}) = 0$. It is not difficult to show that $\tau(0) = D_0$ and $q = 0$ correspond to the maximum of $f(\alpha)$. When $q = 1$, $\tau(q) = 0$, and so $f(\alpha) = \alpha$. Moreover, $\frac{d}{d\alpha}(f(\alpha) - \alpha) = q - 1 = 0$. Thus $f(\alpha)$ is tangent to $f(\alpha) = \alpha$ at $q = 1$.

Typical $\tau(q)$, D_q, and $f(\alpha)$ curves and some of their properties are shown in Figure 5.13. Note that in Figure 17.13(a), the line asymptotic to the curve as $q \to \infty$ has slope $-\alpha_{\min}$, and line asymptotic to the curve as $q \to -\infty$ has slope $-\alpha_{\max}$.

There are major limitations associated with this so-called *fixed-size box-counting algorithm*, and in many applications, results are only reliable for a narrow range of q values, typically $0 \le q \le 5$. In [16], Mach, Mas, and Sagués also consider a *fixed-weight box-counting algorithm*, where the measure quantities p_i are fixed and the size factors r_i vary; see equation (17.3). They show that the fixed-size box-counting algorithm gives good results for small positive q, and the fixed-weight box-counting algorithm can be used to give good results for small negative q. Recently, Alber and Peinke [2] developed an improved multifractal box-counting algorithm using so-called fuzzy discs and a symmetric scaling-error compensation. They apply their method to the Hénon map with great success.

Some simple multifractals are constructed in the next section, and a multifractal analysis is applied to determine multifractal spectra.

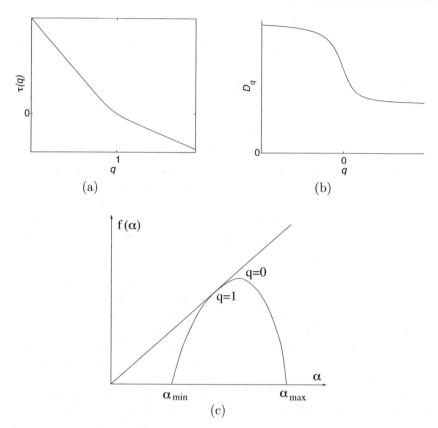

Figure 17.13:   Typical curves of (a) the $\tau(q)$ function, (b) the D_q spectrum, and (c) the $f(\alpha)$ spectrum. In case (c), points on the curve near α_{min} correspond to values of $q \to \infty$, and points on the curve near α_{max} correspond to values of $q \to -\infty$.

17.4 Multifractals in the Real World and Some Simple Examples

Since the publication of the seminal paper by Halsey et al. [12] on multifractals, there has been intense research activity, and numerous papers have been published in many diverse fields of science. A small selection of this research material will be discussed in order to demonstrate how the analysis is being applied in the physical world.

In 1989, Chhabra et al. [6] used equations (17.7) and (17.8) to determine the $f(\alpha)$ spectrum for fully developed turbulence in laboratory and atmospheric flows directly from experimental data. The same methods were employed by Blacher et al. [4] in 1993 and Mills et al. [18] and [19], when characterizing the morphology of multicomponent polymer systems. They found that there was a correlation between the mechanical properties of the samples and their respective $f(\alpha)$ curves. There have been many other studies on the mechanical properties of plastics and rubbers using image analysis techniques. A very useful tool is the multifractal analysis of density distributions. The analysis is usually applied to elemental dot maps produced by scanning electron microscopy coupled with energy dispersive X-ray spectroscopy. The analysis is used to produce generalized dimensions, and it has been found that $w = D_0 - D_5$ is related to factors such as tensile strength, elongation at break, and energy to break. The quantity w is a measure of the nonuniformity of the structure. The smaller the value of w, the more homogeneous the structure and the stronger the material. Multifractals are being applied in image compression techniques and signal processing. In [24], Sarkar and Chaudhuri estimate fractal and multifractal dimensions of gray-tone digital images, and in [11] a multifractal approach is used to extract relevant information on textural areas in satellite meteorological images. Generalized dimensions are being applied extensively in the geosciences to classify sedimentary rocks. In [21], Muller, Huseby, and Saucier relate porosity and permeability to the multifractal spectra of the relevant samples. The analysis is also often applied to *diffusion-limited aggregates* (DLA) clusters. For example, Mach, Mas, and Sagués [16] consider the electrodeposition of zinc sulfate on an electrode and apply the fixed-size and fixed-weight box algorithms to obtain the generalized dimensions. Multifractal characteristics are displayed by propagating cracks in brittle materials, as reported by Silberschmidt in [26]. In physics, the box-counting method was applied to show the multifractality of secondary-electron emission sites in silicon in [14]. In economics, Calvet and Fisher [5] provide a unified treatment on the use of multifractal techniques in finance. An accessible text on fractals and multifractals applied to ecology and aquatic science is given in [25].

Other examples of multifractal phenomena can be found in, for example, stock market analysis, rainfall, and even the distribution of stars and galaxies in the universe. Multifractal phenomena in chemistry and physics are presented in [27]. The examples listed above are by no means exhaustive, but the author hopes that the reader will be encouraged to look for more examples in his/her own particular field of specialization.

In the following examples, simple multifractals are constructed using nonuniform generalizations of the Cantor set and the Koch curve. Multifractal spectra curves are plotted in both cases.

Example 4. A Cantor multifractal set is constructed by removing the middle third segment at each stage and distributing a weight so that each of the remaining two segments receive a fraction p_1 and p_2 units, respectively, and such that $p_1 + p_2 = 1$. Illustrate how the weight is distributed after the first two stages of construction. Plot $\tau(q)$ curves, D_q spectra, and $f(\alpha)$ spectra when

(i) $p_1 = \frac{1}{3}$ and $p_2 = \frac{2}{3}$,

(ii) $p_1 = \frac{1}{9}$ and $p_2 = \frac{8}{9}$.

Which of the multifractals is more heterogeneous?

Solution. Figure 17.14 illustrates how the weight is distributed up to the second stage of construction.

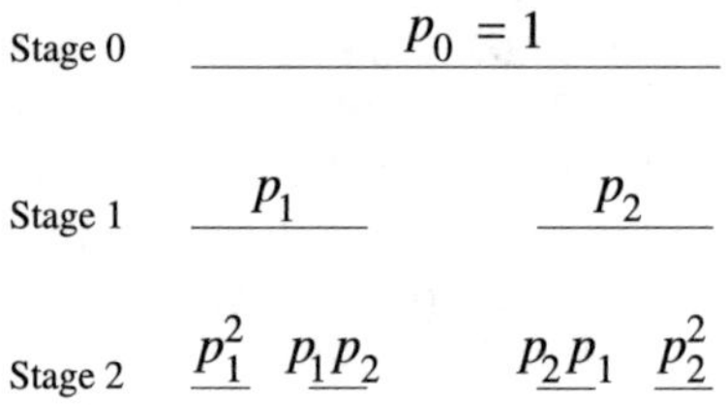

Figure 17.14:   The weight distribution on a Cantor multifractal set up to stage 2.

At stage k, each segment is of length $(\frac{1}{3})^k$ and there are $N = 2^k$ segments. Assign a unit weight to the original line. Then, for $k = 1$, one line segment has weight p_1, and the other has weight p_2. For $k = 2$, there are four segments: one with weight p_1^2, two with weight $p_1 p_2$, and one with weight p_2^2. At stage 3, there are eight segments: one with weight p_1^3, three with weight $p_1^2 p_2$, three with weight $p_1 p_2^2$, and one with weight p_2^3. It is not difficult to see that at stage k, there will be

$$N_s(l) = \binom{k}{s}$$

segments of weight $p_1^s p_2^{k-s}$. From equation (17.2), the partition function $Z_q(l)$ is given by

$$Z_q(3^{-k}) = \sum_{s=0}^{k} \binom{k}{s} p_1^{qs} p_2^{q(k-s)} = (p_1^q + p_2^q)^k,$$

from the binomial theorem. Therefore, from equation (5.4),

$$\tau(q) = D_q(1 - q) = \lim_{l \to 0} \frac{\ln(p_1^q + p_2^q)^k}{-\ln 3^{-k}},$$

so

$$\tau(q) = \frac{\ln(p_1^q + p_2^q)}{\ln 3}.$$

The D_q spectrum can be plotted using continuity at $q = 1$.

To construct an $f(\alpha)$ spectrum, consider how the weight p_s and the number of segments N_s each of type s scales with segment size l. Now

$$p_s(l) \propto (l)^{\alpha_s}, \quad \text{and} \quad N_s(l) \propto (l)^{-f_s},$$

where $s = 0, 1, \ldots, k$. Now $p_s = p_1^s p_2^{k-s}$ and $l = 3^{-k}$. Hence

$$\alpha_s = \frac{s \ln p_1 + (k - s) \ln p_2}{\ln 3^{-k}}.$$

The number of segments of weight p_s at the kth stage is

$$N_s = \binom{k}{s}.$$

Hence

$$-f_s = \frac{\ln \binom{k}{s}}{\ln 3^{-k}}.$$

These parametric curves may be plotted to produce $f(\alpha)$ using Mathematica. The programs are listed in the next section.

(i) Suppose that $p_1 = \frac{1}{3}$ and $p_2 = \frac{2}{3}$. The multifractal curves are given in Figure 17.15.
(ii) Suppose that $p_1 = \frac{1}{9}$ and $p_2 = \frac{8}{9}$. The multifractal curves are given in Figure 17.16.
Notice that in all cases, $D_0 = D_f = \frac{\ln 2}{\ln 3} \approx 0.63$. The multifractal in case (ii) is more heterogeneous. The $f(\alpha)$ curve is broader, and the generalized dimensions D_q cover a wider range of values.

The following images and plots have been supplied by one of the author's Ph.D. students, Dr. Steve Mills, working for BICC cables in Wrexham, for which the author is very grateful. This work is beyond the scope of the book, but the results are included as a matter of interest. Note that these are early-generation fractals and that the results have been obtained using powerful computers and image analysis techniques.

Example 5. Consider the image in Figure 17.17(a), produced by applying the weight distribution as indicated. Using the computer algorithms described in various papers, it is possible to compute the D_q and $f(\alpha)$ spectra. The theoretical multifractal spectra may be derived analytically using methods similar to those used in [9].

Solution. The computed multifractal spectra are plotted in Figure 17.18. For the motif displayed in Figure 17.17(b), it is not difficult to show that

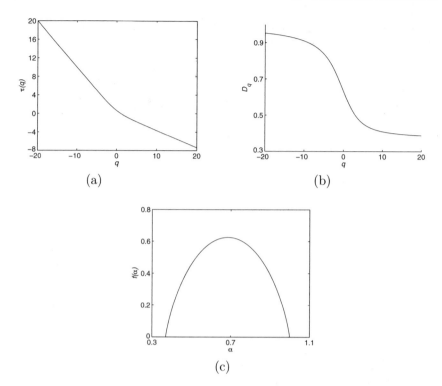

Figure 17.15: Multifractal spectra for part (i) of Example 1 when $p_1 = 1/3$ and $p_2 = 2/3$. (a) $\tau(q)$ curve; (b) D_q spectrum; (c) $f(\alpha)$ spectrum when $k = 500$.

$$\tau = \frac{\ln\left(\frac{1}{3}^q + \frac{1}{4}^q + \frac{1}{4}^q + \frac{1}{6}^q\right)}{\ln(2)},$$

and then, the theoretical $f(\alpha)$ spectrum can be plotted using the relations

$$\alpha = -\frac{d\tau}{dq}, \quad f = q\alpha + \tau.$$

This is left as an exercise for the reader.

Example 6. Consider the image in Figure 17.17(c), produced by applying the weight distribution as indicated. Using the computer algorithms described in various papers, it is possible to compute the D_q and $f(\alpha)$ spectra.

Solution. The computed multifractal spectra are plotted in Figure 17.19.

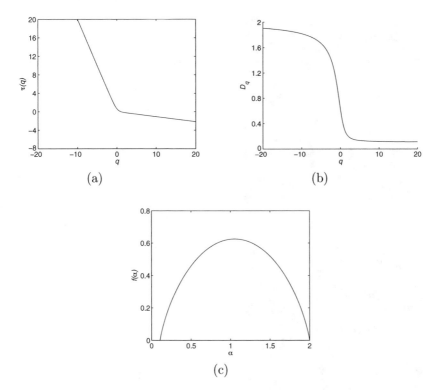

Figure 17.16: [Mathematica] Multifractal spectra for part (ii) of Example 1 when $p_1 = 1/9$ and $p_2 = 8/9$: (a) $\tau(q)$ curve; (b) D_q spectrum; (c) $f(\alpha)$ spectrum when $k = 500$.

The plots in Figures 17.18 and 17.19 are typical of those displayed in the research literature. Note that the latter image is more nonuniformly distributed and is a more heterogeneous fractal.

The quantity $w = D_0 - D_5$ may be used to measure dispersion. Thus, for Example 5, $w \approx 0.13$ and for Example 6, $w \approx 0.6$. Therefore, the dispersion is better in Example 5 since the fractal is more homogeneous.

As a final example, consider the image in Figure 17.20(a), the image was generated by Dr Kathryn Whitehead from the School of Biological Sciences at Manchester Metropolitan University. The image was generated using a scanning electron microscope (SEM), and it is the image of microbes distributed on a food surface. Figure 17.20(b) was obtained using simple image analysis techniques and is a black and white version of Figure 17.20(a). A multifractal analysis has been carried out to generate the f-α curve shown

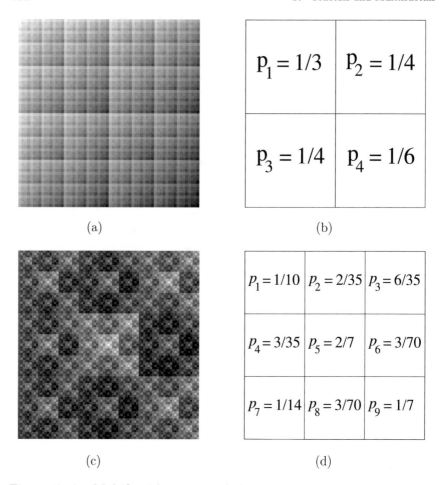

Figure 17.17: Multifractal images and the weight distribution motifs. The weights are related to the gray scale; for example, $p_1 = 1$ would be white and $p_1 = 0$ would be black on this scale.

in Figure 17.21. Measurements for microbial clustering, density, and distribution can be read from the f-α curve. Interested readers should consult our papers [20], [28], and [29].

Multifractal generalized Sierpiński triangles are considered in [10], and a multifractal spectrum of the Hénon map is discussed in [2].

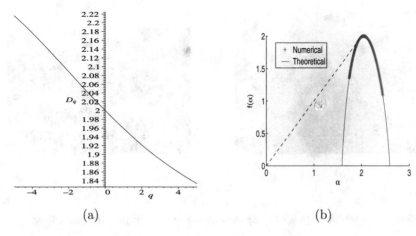

(a)   (b)

Figure 17.18:   Plots for Example 5: (a) the D_q spectrum when $-5 \le q \le 5$; (b) the $f(\alpha)$ spectrum showing both theoretical and numerical curves.

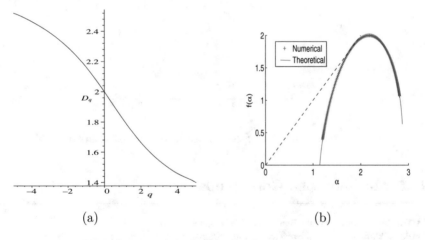

(a)   (b)

Figure 17.19:   Plots for Example 6: (a) the D_q spectrum when $-5 \le q \le 5$ and (b) the $f(\alpha)$ spectrum showing both theoretical and numerical curves.

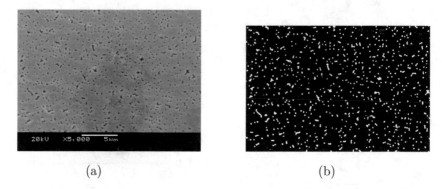

(a)   (b)

Figure 17.20:   Microbial distribution on a food surface: (a) SEM image; (b) a black and white version of (a) obtained from a simple image analysis.

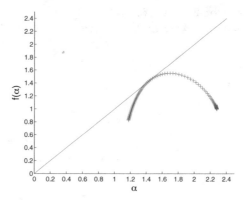

Figure 17.21:   The f-α spectrum for Figure 17.20(b).

17.5 Mathematica Commands

```
(* The Koch curve up to Stage 6. See Figure 17.2. *)
(* Note that from Mathematica 11, KochCurve[n] gives the line segments *)
(* representing the n'th-step Koch curve.*)
In[1]:=k=6;Itermax=4^k;h=3^(-k);
In[2]:=x[1]=0;y[1]=0;
In[3]:=angle[1]=0;angle[2]=Pi/3;angle[3]=-Pi/3;angle[4]=0;
In[4]:=For[a=1,a<=Itermax,a++,m=a-1;ang=0;
      For[b=1,b<=k,b++,segment[b]=Mod[m,4];m=Floor[m/4];
      r=segment[b]+1;ang=ang+angle[r]];x[a+1]=x[a]+h Cos[ang];
      y[a+1]=y[a]+h Sin[ang]];
      ListPlot[Table[{x[i],y[i]},{i,1,Itermax+1}],Joined->True,
      Axes->False,AspectRatio->Automatic]

In[5]:=Graphics[KochCurve[8]]
```

```
(* The Sierpinski triangle. See Figure 17.6. *)
(* There is a SierpinskiCurve[n] command in Mathematica. *)
In[6]:=V1={0,0};V2={4,0};V3={2,2 Sqrt[3]}; (*Vertices.*)
In[7]:=Itermax=100000;Half=1/2;P[1]={0,0};
In[8]:=P[n_]=For[n=1,n<=Itermax,n++,r=Random[ ];
        If[0<=r<=1/3,P[n+1]=N[P[n]+(V1-P[n]) Half],
        If[1/3<r<=2/3,P[n+1]=N[P[n]+(V2-P[n]) Half],
        P[n+1]=N[P[n]+(V3-P[n]) Half]]]];
In[9]:=Sierpinski=Table[P[i],{i,Itermax}];
In[10]:=ListPlot[Sierpinski,Axes->False,AspectRatio->Automatic,
        PlotStyle->{PointSize[0.001],RGBColor[1,0,0]}]
```

```
(* Barnsley's fern and IFSs. See Figure 17.7. *)
In[11]:=Itermax=100000;P[1]={0.5,0.5};
In[12]:=T[{x_,y_,a_,b_,c_,d_,e_,f_}]={a*x+b*y+c,d*x+e*y+f};
In[13]:=P[n_]=For[n=1,n<=Itermax,n++,r=Random[ ];
        If[0<=r<0.05,
        P[n+1]=T[{P[n][[1]],P[n][[2]],0,0,0,0,0.2,0}],
        If[r<0.86,
        P[n+1]=T[{P[n][[1]],P[n][[2]],0.85,0.05,0,-0.04,0.85,1.6}],
        If[r<0.93,
        P[n+1]=T[{P[n][[1]],P[n][[2]],0.2,-0.26,0,0.23,0.22,1.6}],
        P[n+1]=T[{P[n][[1]],P[n][[2]],-0.15,0.28,0,0.26,0.24,0.44}]]]]]
In[14]:=BarnsleyFern=Table[P[i],{i,Itermax}];
In[15]:=ListPlot[BarnsleyFern,Axes->False,AspectRatio->Automatic,
        PlotStyle->{PointSize[0.001],RGBColor[0,1,0]}]
```

```
(* The tau curve. See Figure 17.16(a). *)
In[16]:=Plot[Log[(1/9)^x+(8/9)^x]/Log[3],{x, -20, 20},
        AxesLabel->{"q","tau(q)"}]
```

```
(*The D_q curve. See Figure 17.16(b). *)
In[17]:=Plot[(Log[(1/9)^x+(8/9)^x]/Log[3]) 1/(1-x),{x,-20,20}]
```

```
(* The f(alpha) curve. See Figure 17.16(c). *)
In[18]:=k=500;
In[19]:=ParametricPlot[{(s Log[1/9]+(k-s) Log[8/9])/(Log[3^(-k)]),
        -Log[Binomial[k,s]]/(Log[3^(-k)])},{s,0,500},
        AxesLabel->{"alpha","f(alpha)"}];
```

17.6 Exercises

1. (a) Consider the unit interval. A variation of the Cantor set is con-
        structed by removing two line segments each of length $\frac{1}{5}$. Thus,
        at stage 1, remove the segments between $\{\frac{1}{5}..\frac{2}{5}\}$ and $\{\frac{3}{5}..\frac{4}{5}\}$ from
        the unit interval, leaving three line segments remaining. Contin-
        uing in this way, construct the fractal up to stage three either
        on graph paper or on a computer screen. Find the length of seg-
        ment remaining at stage k. Determine the fractal dimension of
        the mathematical fractal constructed to infinity.

    (b) A Lévy fractal is constructed by replacing a line segment with a
        try square. Thus, at each stage, one line segment of length, 1, say,
        is replaced by two of length $\frac{1}{\sqrt{2}}$. Construct the fractal up to stage
        7 either on graph paper or on a computer screen. When using
        graph paper, it is best to draw a skeleton (dotted line) of the
        previous stage. What is the true fractal dimension of the object
        generated to infinity?

    (c) A *Koch snowflake* is constructed by adjoining the Kock curve to
        the outer edges of a unit length equilateral triangle. Construct
        this fractal up to stage 4 either on graph paper or on a computer
        screen and show that the area bounded by the true fractal A_∞ is
        equal to

        $$A_\infty = \frac{2\sqrt{3}}{5} \text{ units}^2.$$

    (d) The inverted Koch snowflake is constructed in the same way as in
        Exercise 1(c), but the Koch curve is adjoined to the inner edges
        of an equilateral triangle. Construct the fractal up to stage 4 on
        graph paper or stage 6 on the computer.

2. Consider Pascal's triangle given below. Cover the odd numbers with
    small black discs (or shade the numbers). What do you notice about the
    pattern obtained?

```
                              1

                           1     1

                        1     2     1

                     1     3     3     1

                  1     4     6     4     1

               1     5    10    10     5     1

            1     6    15    20    15     6     1

         1     7    21    35    35    21     7     1

      1     8    28    56    70    56    28     8     1

   1     9    36    84   126   126    84    36     9     1

1    10    45   120   210   252   210   120    45    10    1

1   11    55   165   330   462   462   330   165   55    11   1

1   12    66   220   495   792   924   792   495   220   66   12   1

1   13    78   286   715    x0    x1    x1    x0   715   286   78   13   1

1   14    91   364    x2    x3    x4    x5    x4    x3    x2   364   91   14   1

1   15   105   455    x6    x7    x8    x9    x9    x8    x7    x6   455   105   15   1
```

where $x0 = 1287$, $x1 = 1716$, $x2 = 1001$, $x3 = 2002$, $x4 = 3003$, $x5 = 3432$, $x6 = 1365$, $x7 = 3003$, $x8 = 5005$, and $x9 = 6435$.

3. The Sierpiński triangle can be constructed by removing the central inverted equilateral triangle from an upright triangle; a motif is given in this chapter. Construct the Sierpiński triangle up to stage 4 on graph paper using this method.

4. A Sierpiński square is constructed by removing a central square at each stage. Construct this fractal up to stage 3 and determine the fractal dimension of the theoretical object generated to infinity.

5. Use the box-counting algorithm to approximate the fractal dimension of Barnsley's fern. The Mathematica program for plotting the fern is given in Section 17.5.

6. Consider the map defined by $x_{n+1} = f(x_n)$, where $f(x)$ is defined by

$$f(x) = \begin{cases} 1 - 4x & x \leq \frac{1}{2} \\ 4x - 3 & x > \frac{1}{2} \end{cases}.$$

Plot the function on graph paper. Consider the sets, S_n say, which remain in the interval $[0,1]$ after n iterations. List the intervals in S_1 and S_2. The set of points that never escape from the interval $[0,1]$ form a Cantor set. What is the fractal dimension of this Cantor set?

7. Plot $\tau(q)$ curves, D_q and $f(\alpha)$ spectra for the multifractal Cantor set described in Example 1 when (i) $p_1 = \frac{1}{2}$ and $p_2 = \frac{1}{2}$, (ii) $p_1 = \frac{1}{4}$ and $p_2 = \frac{3}{4}$, and (iii) $p_1 = \frac{2}{5}$ and $p_2 = \frac{3}{5}$.

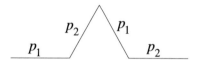

Figure 17.22: The motif used to construct the Koch curve multifractal, where $2p_1 + 2p_2 = 1$.

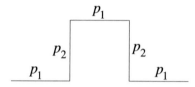

Figure 17.23: The motif used to construct the Koch curve multifractal, where $3p_1 + 2p_2 = 1$.

8. A multifractal Koch curve is constructed, and the weight is distributed as depicted in Figure 17.22. Plot the $f(\alpha)$ spectrum when $p_1 = \frac{1}{3}$ and $p_2 = \frac{1}{6}$.

9. A multifractal square Koch curve is constructed, and a weight is distributed as depicted in Figure 17.23. Plot the $\tau(q)$ curve, the D_q, and $f(\alpha)$ spectra when $p_1 = \frac{1}{9}$ and $p_2 = \frac{1}{3}$.

10. A multifractal Koch curve is constructed, and a weight is distributed as depicted in Figure 17.24, where $p_1 + p_2 + p_3 + p_4 = 1$. Determine α_s and f_s.

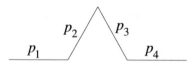

Figure 17.24: The motif used to construct the Koch curve multifractal, where $p_1 + p_2 + p_3 + p_4 = 1$.

Bibliography

[1] P.S. Addison, *Fractals and Chaos: An Illustrated Course*, Institute of Physics, London, UK, 1997.

[2] M. Alber and J. Peinke, Improved multifractal box-counting algorithm, virtual phase transitions, and negative dimensions, *Phys. Rev. E* **57**-5, (1998), 5489–5493.

[3] C. Bandt, M. Barnsley, R. Devaney, et al., *Fractals, Wavelets, and their Applications: Contributions from the International Conference and Workshop on Fractals and Wavelets*, Springer, New York, 2016.

[4] S. Blacher, F. Brouers, R. Fayt and P. Teyssié, Multifractal analysis. A new method for the characterization of the morphology of multicomponent polymer systems, *J. Polymer Sci. B: Polymer Physics* **31**, (1993), 655–662.

[5] L.E. Calvet and A.J. Fisher, *Multifractal Volatility: Theory, Forecasting, and Pricing*, Academic Press, New York, London, 2008.

[6] A.B. Chhabra, C. Meneveau, R.V. Jensen, and K.R. Sreenivasan, Direct determination of the $f(\alpha)$ singularity spectrum and its application to fully developed turbulence, *Phys. Rev. A* **40**-9, (1989), 5284–5294.

[7] R.M. Crownover, *Introduction to Fractals and Chaos*, Jones and Bartlett Publishers, 1995.

[8] K. Falconer, *Fractals: A Very Short Introduction*, Oxford University Press, Oxford, 2013.

[9] K. Falconer, *Fractal Geometry: Mathematical Foundations and Applications*, John Wiley and Sons, New York, 2003.

[10] K.J. Falconer and B. Lammering, Fractal properties of generalized Sierpiński triangles, *Fractals* **6**-1, (1998), 31–41.

[11] J. Grazzini, A. Turiel, H. Yahia, and I. Herlin, *A multifractal approach for extracting relevant textural areas in satellite meteorological images, (An article from: Environmental Modelling and Software), (HTML, Digital)*, Elsevier, 2007.

[12] T.C. Halsey, M.H. Jensen, L.P. Kadanoff, I. Procaccia, and B.I. Shraiman, Fractal measures and their singularities, *Phys. Rev. A* **33**, (1986), 1141.

[13] D. Harte, *Multifractals: Theory and Applications*, Chapman and Hall, London, UK, 2001.

[14] Li Hua, D. Ze-jun, and Wu Ziqin, Multifractal analysis of the spatial distribution of secondary-electron emission sites, *Phys. Rev. B* **53**-24, (1996), 16631–16636.

[15] N. Lesmoir-Gordon, *Introducing Fractal Geometry*, 3rd ed., Totem Books, 2006.

[16] J. Mach, F. Mas, and F. Sagués, Two representations in multifractal analysis, *J. Phys. A: Math. Gen.* **28**, (1995), 5607–5622.

[17] B.B. Mandelbrot, *The Fractal Geometry of Nature*, W.H. Freeman and Co., New York, 1983.

[18] S.L. Mills, G.C. Lees, C.M. Liauw, R.N. Rothon and S. Lynch, Prediction of physical properties following the dispersion assessment of flame retardant filler/polymer composites based on the multifractal analysis of SEM images, *J. Macromolecular Sci. B- Physics* **44**-6, (2005), 1137–1151.

[19] S.L. Mills, G.C. Lees, C.M. Liauw and S. Lynch, An improved method for the dispersion assessment of flame retardent filler/polymer systems based on the multifractal analysis of SEM images, *Macromolecular Materials and Engineering* **289**-10, (2004), 864–871.

[20] J.M.R. Moreira, L.C. Gomes, K.A. Whitehead, S. Lynch, L. Tetlow and F.J. Mergulhao Effect of surface conditioning with cellular extracts on Escherichia coli adhesion and initial biofilm formation, *Food and Bioproducts Processing* **104**, (2017), 1–12.

[21] J. Muller, O.K. Huseby and A. Saucier, Influence of multifractal scaling of pore geometry on permeabilities of sedimentary rocks, *Chaos, Solitons and Fractals* **5**-8, (1995), 1485–1492.

[22] H-O. Peitgen, H. Jürgens, and D Saupe, *Chaos and Fractals*, Springer-Verlag, 1992.

[23] H-O. Peitgen (ed.), E.M. Maletsky, H. Jürgens, T. Perciante, D. Saupe, and L. Yunker, *Fractals for the Classroom: Strategic Activities*, Volume 1, Springer-Verlag, New York, 1991.

[24] N. Sarkar and B.B. Chaudhuri, Multifractal and generalized dimensions of gray-tone digital images, *Signal Processing*, **42**, (1995), 181–190.

[25] L. Seuront, *Fractals and Multifractals in Ecology and Aquatic Science*, CRC Press, 2009.

[26] V. Silberschmidt, Fractal and multifractal characteristics of propagating cracks, *J. de Physique IV* **6**, (1996), 287–294.

[27] H.F. Stanley and P. Meakin, Multifractal phenomena in physics and chemistry, *Nature* **335**, (1988), 405–409.

[28] L. Tetlow, S. Lynch and K. Whitehead, The effect of surface properties on bacterial retention: a study utilising stainless steel and TiN/25.65at.%Ag substrata, *Food and Bioproducts Processing* **102** (2017), 332–339.

[29] D. Wickens, S. Lynch, P. Kelly, G. West, K. Whitehead and J. Verran, Quantifying the pattern of microbial cell dispersion, density and clustering on surfaces of differing chemistries and topographies using multifractal analysis, *Journal of Microbiological Methods*, 104, (2014), 101–108.

Chapter 18
Image Processing and Analysis with Mathematica

Aims and Objectives

- To provide a tutorial guide to Image Processing and Analysis.

- To show how to manipulate images.

- To provide tools to analyze images.

- To introduce fast Fourier transforms.

On completion of this chapter, the reader should be able to

- manipulate images interactively and in programs;

- load and save images;

- perform analysis on color, gray scale, and black and white images;

- plot power spectra of discrete and continuous dynamical systems;

- perform simple image processing.

As Wolfram states: "The Wolfram Language provides broad and deep built-in support for both programmatic and interactive modern industrial-strength image processing-fully integrated with the Wolfram Language's powerful mathematical and algorithmic capabilities. The Wolfram Language's unique symbolic architecture and notebook paradigm allow images in visual form to be included and manipulated directly, both interactively and in programs."

© Springer International Publishing AG 2017   417
S. Lynch, *Dynamical Systems with Applications using Mathematica*®,
DOI 10.1007/978-3-319-61485-4_18

18.1 Image Processing and Matrices

There is extensive online documentation accompanying Image Processing and Analysis. Probably the most popular textbooks specializing to date are [5], [8], and [9], whereas [1] and [10] specialize in biosignal and medical image processing. The reader will be shown how to read and write image files and perform image processing techniques on those images.

As a simple introduction, let us construct a multifractal image (see Chapter 17) using a simple motif.

```
(* Construction of a multifractal. *)
In[1]:=M={{0.3,0.4},{0.25,0.05}};
In[2]:=For[i = 1, i < 9, i++,
      M=Flatten[{{{0.3*M, 0.4*M},{0.25*M, 0.05*M}},{{1, 3},{2, 4}}]];
In[3]:=Dimensions[M]

Out[3]= {512, 512}

In[4]:=Multifractal=Image[M];ImAdj=ImageAdjust[Multifractal]
In[5]:= ImageAdjust[ImAdj, {0.045, 0.495, 0.335}]
```

You will note that the value of most pixels is close to zero, and this is why the screen initially looks black. One may alter the contrast, brightness, and gamma of the image by hitting the **Adjust** button as in Figure 18.1(b).

Suppose that one wanted to establish the number of red pixels in a color image. A simple program is listed below. The image of peppers.png is given in the Mathematica notebook and is shown in Figure 18.2(a).

```
(* A program to count the number of red pixels in an image. *)
In[9]:=Peppers = (* Paste the image here *)
In[10]:=ImageValue[Peppers, {100, 10}] (* The RGB value of a pixel *)
Out[10]= {0.411765, 0.25098, 0.458824}

In[11]:=Binarize[Peppers,#[[1]]>0.54 && #[[2]]<0.3 && #[[3]]<0.30 & ]
In[12]:=ImageLevels[
(* Gives a list of pixel values and counts for each channel in image. *)


Out[12]={{0,160940}, {1,25686}}
```

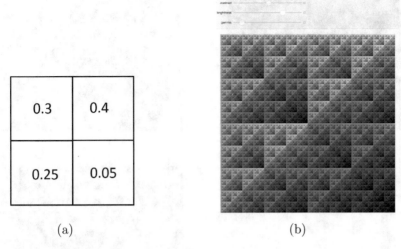

0.3	0.4
0.25	0.05

(a)   (b)

Figure 18.1: [Mathematica] (a) A weight distribution motif; (b) the multifractal image. The weights are related to the gray scale, for example, $p_1 = 1$ would be white and $p_1 = 0$ would be black on this scale. Adjust the contrast, brightness, and gamma in the Mathematica notebook.

(a)   (b)

Figure 18.2: [Mathematica] Using Mathematica to binarize a color image. (a) The image peppers.png. (b) The red pixels colored white. There are 25686 white pixels.

A truecolour image, or RGB image, is an image in which each pixel is defined by three numerical values for Red, Green, and Blue. Mathematica stores true-colour images as $m \times n \times 3$ matrices. See the Mathematica help pages for more information.

Note that the coordinate system used in the image processing is ordered from top to bottom and from left to right from the top left corner of the image.

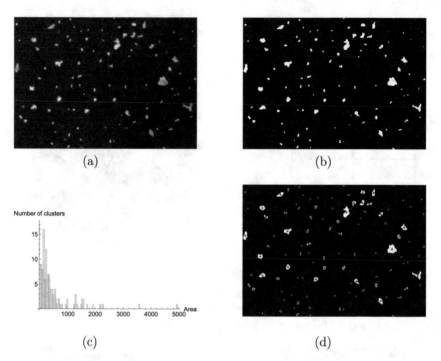

Figure 18.3: [Mathematica] (a) The image Microbes.png. (b) A binary image of Microbes.png. (c) Histogram of the areas of the clusters of microbes. (d) The centroids of the microbes clusters are shown as red dots.

So the pixel coordinate $(100, 10)$ refers to $x = 100$ (going left to right) and $y = 10$ (going top to bottom). Note that $ImageValue[Peppers, 100, 10]$ denotes the element in matrix $Peppers$ in row 100 and column 10. Check that the RGB value for the pixel coordinate $(100, 10)$ is $\{0.411765, 0.25098, 0.458824\}$.

The next example demonstrates how one may obtain a binary image from a grayscale image. Some statistical analysis is applied to the clusters of microbes, and the centroids of the clusters are also plotted, see Figure 18.3. The Microbes. png image is given in the Mathematica notebook and is shown in Figure 18.3(a).

```
In[13]:=Microbes= (* Paste image here. *)

In[14]:=ImageDimensions[Microbes]

Out[14]={1734,1212}

In[15]:=BWMicrobes=Binarize[Microbes, 0.4]

In[16]:=AreaData=ComponentMeasurements[BWMicrobes,{"Area"}][[All, 2, 1]];
```

```
In[17]:=Max[AreaData]
Min[AreaData]
Length[AreaData]
```

```
Out[17]=4921.88    3.    106
```

```
In[18]:=P1=Histogram[AreaData,Length[AreaData]];
        Show[P1,PlotRange->{{Min[AreaData],Max[AreaData]},{0, 17}}]
```

```
In[19]:=CentroidData=ComponentMeasurements[BWMicrobes,{"Centroid"}]
        [[All, 2, 1]];
        Centroids=ListPlot[CentroidData,PlotStyle->{Red,Large},
        PlotMarkers->"+",Axes -> None];
        Show[BWMicrobes, Centroids]
```

18.2 The Fast Fourier Transform

The Fourier transform is a mathematical transform with many applications in image processing, mathematics, engineering, and the physical sciences.

Definition 1. The continuous Fourier transform is defined by

$$F(\omega) = \int_{-\infty}^{\infty} f(t)e^{-2\pi i \omega t}dt,$$

which transforms a mathematical function of time, $f(t)$, into a function of frequency, $F(\omega)$. The new function is the Fourier transform or the Fourier spectrum of the function f.

Definition 2. The inverse Fourier transform is defined by

$$f(t) = \int_{-\infty}^{\infty} F(\omega)e^{2\pi i \omega t}d\omega.$$

The continuous Fourier transform converts an infinitely long time domain signal into a continuous spectrum of an infinite number of sinusoidal curves. In many physical applications, scientists deal with discretely sampled signals, usually at constant intervals. For such data, the discrete Fourier transform is appropriate.

Definition 3. The discrete Fourier transform and its inverse for vectors of length N are defined by

$$X_k = \sum_{n=1}^{N} t_n \omega_N^{(n-1)(k-1)},$$

and

$$x_n = \frac{1}{N} \sum_{k=1}^{N} X_k \omega_N^{-(n-1)(k-1)},$$

where

$$\omega_N = e^{(-2\pi i)/N},$$

and each X_k is a complex number that encodes both amplitude and phase of a sinusoidal component of function x_n.

A fast Fourier transform, or FFT, is an algorithm to compute the discrete Fourier transform. The FFT was first discovered by Gauss in 1805, but the modern incarnation is attributed to Cooley and Tukey [3] in 1965. Computing a set of N data points using the discrete Fourier transform requires $O(N^2)$ arithmetic operations, while a FFT can compute the same discrete Fourier transform in only $O(N \log N)$ operations.

FFT is a powerful signal analysis tool, applicable to a wide variety of fields including acoustics, applied mechanics, communications, digital filtering, instrumentation, medical imaging, modal analysis, numerical analysis, seismography, and spectral analysis.

Example 1. A common use of Fourier transforms is to find the frequency components of a signal buried in a noisy time domain signal. Consider data sampled at $10000Hz$. Form a signal containing a $4.5Hz$ sinusoid of amplitude 5 and $2Hz$ sinusoid of amplitude 2 and corrupt it with some zero-mean random noise. Use Mathematica to plot a graph of the signal and write a program that plots a single-sided amplitude spectrum for the signal.

Solution. Figure 18.4(a) shows the sum of a $5Hz$ sinusoid and a $2Hz$ sinusoid corrupted with zero-mean random noise, and 18.4(b) displays the single-sided amplitude spectrum of $y(t)$. The program for plotting the figures is listed below.

```
(* Adding noise to a signal. *)
In[1]:=time=0.5;Fs=10000; (* Fs is sample fequency *)
In[2]:=Amp1=5;Amp2=2;Freq1=4.5;Freq2=2;
In[3]:=SampleTime =1/Fs;
In[4]:=sampls=Table[Amp1*Sin[n*(2 Pi) Freq1] + Amp2*Sin[n*(2 Pi) Freq2],
       {n, 0, time, SampleTime}];
In[5]:=len=Length@sampls;
In[6]:=w=RandomReal[NormalDistribution[0,0.5],len];
In[7]:=ListLinePlot[sampls+w,DataRange->{0,time},PlotRange->{{0,time},All},
       AxesLabel->{"Time(ms)","y"},Axes->True,BaseStyle->{FontSize->15}]

(* Discrete Fourier transform of a noisy signal  *)
In[8]:=time=50;Fs=10000; (* Fs is sample fequency *)
```

```
Amp1=5;Amp2=2;Freq1=4.5;Freq2=2;SampleTime=1/Fs;
sampls=Table[Amp1*Sin[n*(2 Pi) Freq1]+Amp2*Sin[n*(2 Pi) Freq2],
{n, 0, time, SampleTime}];
len=Length@sampls;
w=RandomReal[NormalDistribution[0, 10],len];
ListLinePlot[Sqrt[4/len] Abs@Fourier[sampls+w],
PlotRange->{{0,10},All},DataRange->{0,(len-1)/time},
AxesLabel -> {"Frequency (Hz)",
"\[LeftBracketingBar]Y(f)\[RightBracketingBar]"},Axes->True,
BaseStyle->{FontSize->15}]
```

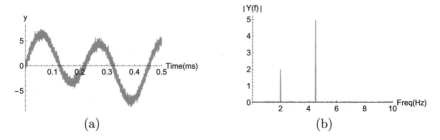

(a)   (b)

Figure 18.4: (a) Signal corrupted with zero-mean random noise. (b) Single-sided amplitude spectrum of y(t). You can read off the amplitude and frequencies.

The next example illustrates an application of FFT for finding the power spectra of time series data. Interested readers should consult Melbourne and Gottwald [7], who present results on the broadband nature of power spectra for diverse classes of discrete dynamical systems. For many years, the power spectrum has been used to distinguish periodic, quasiperiodic, and chaotic motions of certain dynamical systems from a broad range of fields. The power spectrum plotted for a periodic or quasiperiodic signal has discrete peaks at the harmonics and subharmonics, while the chaotic signal has a broadband component in its power spectrum. In order to illustrate these phenomena, consider the following simple example.

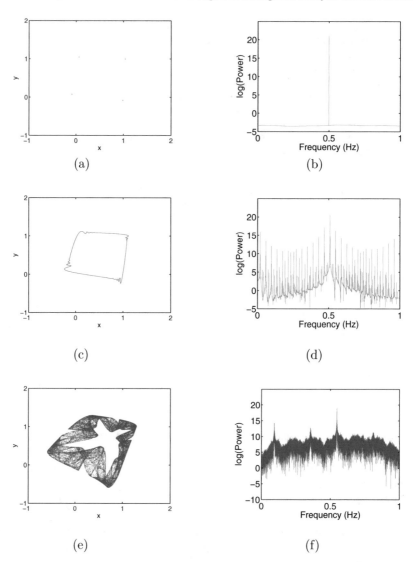

Figure 18.5:   Iterative plots and power spectra for system (18.1). (a) Periodic behavior when $\beta = 0.05$. (b) Power spectrum when $\beta = 0.05$. (c) Quasiperiodic behavior when $\beta = 0.12$. (d) Power spectrum when $\beta = 0.12$. (e) Chaotic behavior when $\beta = 0.3$. (f) Power spectrum when $\beta = 0.3$.

Example 2. Consider the two-dimensional discrete map defined by

$$x_{n+1} = 1 + \beta x_n - \alpha y_n^2$$
$$y_{n+1} = x_n, \tag{18.1}$$

where α and β are constants. Suppose that $\alpha = 1$, plot iterative plots and power spectra for system (18.1) when (i) $\beta = 0.05$; (ii) $\beta = 0.12$, and (iii) $\beta = 0.3$.

Solution. The Mathematica program for producing the plots in Figure 18.5 is listed below.

```
(* Iterative plot of chaotic system *)
In[1]:=a=1;b=0.3;Itermax=100100;
       TwoD[{x_,y_}]={1-a y^2+b x,x};
       TwoDMap=Drop[NestList[TwoD,{0.1,0.1},Itermax],101];
       ListPlot[TwoDMap,AxesLabel->{"x","y"},Axes->None,
       PlotStyle->{PointSize[0.001],RGBColor[0,0,1]},Frame->True,
       FrameLabel->{"x","y"},BaseStyle->{FontSize->18}]

(* Power spectrum of chaotic system *)
In[2]:=xvals=TwoDMap[[All,1]];
       ListLinePlot[Rest[Transpose[{Range[-Length[xvals]/2+1,
       Length[xvals]/2]*(2/Length[xvals]),
       Log[Abs[Fourier[xvals]^2]]}]],PlotRange->{{0,1},{-20,10}},
       AxesLabel->{"Freq(Hz)","log(Power)"},BaseStyle->{FontSize->14}]
```

18.3 The Fast Fourier Transform on Images

Among the many applications of the two-dimensional Fourier transform, there are some very interesting and useful image processing tools which include image compression, blurring and de-blurring, sharpening, noise removal, and edge detection, for example. Figure 18.6 depicts how to apply a low-pass filter to a JPEG image of Lena, and a Mathematica program to produce a fast Fourier transform is given below, see 18.6. Note that the ideal low-pass filter applies a Gaussian function (interested readers should consult some of the textbooks in the reference section of this chapter for more details). A low-pass filter is used to compress an image. By adopting the negated version of the low-pass filter, one may obtain a high-pass filter that can be used for edge detection. This is left as an exercise for the reader.

```
(* Using Fast Fourier transforms to compress an image. *)
img = ColorConvert[ExampleData[{"TestImage","Lena"}],"Grayscale"]
fft=RotateLeft[Fourier@ImageData[img],Floor[Reverse@ImageDimensions[img]/2]];
Image[Rescale[Log[Abs[fft]+10.^-10]]]
mask = DiskMatrix[64,Dimensions[fft]];
Image[Rescale[Log[Abs[mask]+10.^-10]]]
mfft = mask fft;
mimg=Image@Re@InverseFourier[RotateRight[mfft,Floor[Dimensions[fft]/2]]]
```

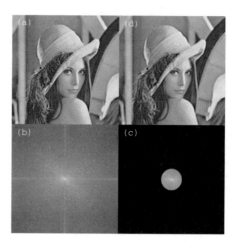

Figure 18.6: Low-pass filtering of Lena image. (a) Lena.jpg, (b) fast Fourier transform, (c) a circular low-pass filter, and (d) a compressed image of Lena. Use the Mathematica help pages for function definitions and syntax.

18.4 Exercises

1. Use the matrix motif $M = [0.1, 0.2; 0.2, 0.5]$ to produce a multifractal image up to stage 8. Use Mathematica to produce a figure representation of the multifractal.

2. Use the matrix $M = [0.1, 0.2, 0.05; 0.2, 0.05, 0.01; 0.3, 0.04, 0.05]$ to produce a multifractal image up to stage 5. Use Mathematica to produce a figure representation of the multifractal.

3. Use Mathematica to produce a binary image of the green pixels in peppers.png. Determine an approximate number of green pixels. The file peppers.png can be found in the Mathematica notebook.

4. Write a Mathematica program to detect the edges of the microbes for the file microbes.png. The file can be found in the Mathematica notebook.

5. Compute the first 10000 iterates of the logistic map

$$x_{n+1} = 4x_n(1 - x_n),$$

given that $x_0 = 0.1$. Use Mathematica to plot a power series spectrum.

6. Compute the first 10000 iterates of the Gaussian map

$$x_{n+1} = e^{-8x_n^2} - 0.6,$$

given that $x_0 = 0.1$. Use Mathematica to plot a power series spectrum.

7. Compute the first 10000 iterates of the Hénon map

$$x_{n+1} = 1 + y_n - 1.2x_n^2, \quad y_{n+1} = 0.4x_n$$

given that $x_0 = 0.1, y_0 = 0$. Use Mathematica to plot a power series spectrum.

8. Compute the first 10000 iterates of the minimal chaotic neuromodule

$$x_{n+1} = 2 + 3.5\phi_1(x_n) - 4\phi_2(y_n), \quad y_{n+1} = 3 + 5\phi_1(x_n),$$

where $\phi_1(x) = \phi_2(x) = 1/(1 + e^{-x})$, given that $x_0 = 1, y_0 = 0$. Use Mathematica to plot a power series spectrum.

9. Write Mathematica programs to produce (i) a circular high-pass filter of the Lena.jpg image as shown in Figure 18.6; (ii) an ideal low-pass filter of the Lena.jpg image using a suitable Gaussian function.

10. Carry out your own research to produce a high-pass filter used for edge detection on the Lena.jpg image.

Bibliography

[1] W. Birkfellner, *Applied Medical Image Processing, Second Edition: A Basic Course*, Taylor & Francis, New York, 2014.

[2] E. Brigham *Fast Fourier Transform and Its Applications*, Prentice-Hall, New Jersey, 1988.

[3] J.W. Cooley and J.W. Tukey, An algorithm for the machine calculation of complex Fourier series, *Math. Comput.* **19**, (1965), 297–301.

[4] Z. Elhadj and J.C. Sprott, A minimal 2-D quadratic map with quasiperiodic route to chaos, *Int. J. of Bifurcation and Chaos* **18**, (2008), 1567–1577.

[5] V. Hlavac, M. Sonka and R. Boyle, *Image Processing, Analysis and Machine Vision 4th ed.*, Nelson Engineering, Osprey, Florida, 2014.

[6] E.F. James *A Student's Guide to Fourier Transforms: With Applications in Physics and Engineering*, Cambridge University Press, Cambridge, 2011.

[7] I. Melbourne and G.A. Gottwald, Power spectra for deterministic chaotic dynamical systems, *Nonlinearity* **21** (2008), 279–292.

[8] M. Petrou and C. Petrou, *Image Processing: The Fundamentals 2nd ed.*, Wiley-Blackwell, Hoboken, New Jersey, 2010.

[9] J.C. Russ, *The Image Processing Cookbook 3rd ed.*, CreateSpace Independent Publishing Platform, 2016.

[10] J.L. Semmlow and B. Griffel, *Biosignal and Medical Image Processing 3rd ed.*, CRC Press, Boca Raton, Florida, 2014.

Chapter 19
Chaos Control and Synchronization

Aims and Objectives

- To provide a brief historical introduction to chaos control and synchronization.

- To introduce two methods of chaos control for one- and two-dimensional discrete maps.

- To introduce two methods of chaos synchronization.

On completion of this chapter, the reader should be able to

- control chaos in the logistic and Hénon maps;

- plot time series data to illustrate the control;

- synchronize chaotic systems;

- appreciate how chaos control and synchronization are being applied in the real world.

This chapter is intended to give the reader a brief introduction to the new and exciting field of chaos control and synchronization and to show how some of the theory is being applied to physical systems. There has been considerable research effort into chaos control in recent times, and practical methods have been applied in, for example, biochemistry, cardiology, communications, physics laboratories, and turbulence. Chaos control has been achieved using many different methods, but this chapter will concentrate on two procedures only. Chaos synchronization has applications in analog or digital communications and cryptography. For more background material on chaos control, the reader is directed to references [6], [8], [10], [12], [13], [15], [19], [21], [23], [24], and [25]. The other references are predominantly concerned with chaos synchronization.

© Springer International Publishing AG 2017
S. Lynch, *Dynamical Systems with Applications using Mathematica®*,
DOI 10.1007/978-3-319-61485-4_19

Control and synchronization of chaotic systems are possible for both discrete and continuous systems. Analysis of chaos control will be restricted to discrete systems in this chapter and synchronization will be restricted to continuous systems.

19.1 Historical Background

Even simple, well-defined discrete and continuous nonlinear dynamical systems without random terms can display highly complex, seemingly random behavior. Some of these systems have been investigated in this book, and mathematicians have labeled this phenomenon *deterministic chaos*. *Nondeterministic chaos*, where the underlying equations are not known, such as that observed in a lottery or on a roulette wheel, will not be discussed in this text. Throughout history, dynamical systems have been used to model both the natural and the technological sciences. In the early years of investigations, deterministic chaos was nearly always attributed to random external influences and was designed out if possible. The French mathematician and philosopher, Henri Poincaré laid down the foundations of the qualitative theory of dynamical systems at the turn of the century and is regarded by many as being the first *chaologist*. Poincaré devoted much of his life in attempting to determine whether or not the solar system is stable. Despite knowing the exact form of the equations defining the motions of just three celestial bodies, he could not always predict the long-term future of the system. In fact, it was Poincaré who first introduced the notion of sensitivity to initial conditions and long-term unpredictability.

In recent years, deterministic chaos has been observed when applying simple models to cardiology, chemical reactions, electronic circuits, laser technology, population dynamics, turbulence, and weather forecasting. In the past, scientists have attempted to remove the chaos when applying the theory to physical models, and it is only in the last thirty years that they have come to realize the potential uses for systems displaying chaotic phenomena. For some systems, scientists are replacing the maxim "stability good, chaos bad" with "stability good, chaos better." It has been found that the existence of chaotic behavior may even be desirable for certain systems.

Since the publication of the seminal paper of Ott, Grebogi, and Yorke [21] in 1990, there has been a great deal of progress in the development of techniques for the control of chaotic phenomena. Basic methods of controlling chaos along with several reprints of fundamental contributions to this topic may be found in the excellent textbooks of Kapitaniak [13], [14]. Some of these methods will now be discussed very briefly, and then, a selection of early applications of chaos control in the real world will be listed.

I. *Changing the systems parameters.* The simplest way to suppress chaos is to change the system parameters in such a way as to produce the desired result. In this respect, bifurcation diagrams can be used to determine the parameter values. For example, in Chapter 5, bifurcation diagrams were used to determine regions of bistability for nonlinear bistable optical resonators. It was found that isolated bistable regions existed for only a narrow range of parameter values. However, the major drawback with this procedure is that large parameter variations may be required, which could mean redesigning the apparatus and changing the dimensions of the physical system. In many practical situations, such changes are highly undesirable.

II. *Applying a damper.* A common method for suppressing chaotic oscillations is to apply some kind of damper to the system. In mechanical systems, this would be a shock absorber, and for electronic systems, one might use a shunt capacitor. Once more, this method would mean a large change to the physical system and might not be practical.

III. *Pyragas's method.* This method can be divided into two feedback-controlling mechanisms: linear feedback control and time-delay feedback control. In the first case, a periodic external force is applied whose period is equal to the period of one of the unstable periodic orbits contained in the chaotic attractor. In the second case, self-controlling delayed feedback is used in a similar manner. This method has been very successful in controlling chaos in electronic circuits such as the Duffing system and Chua's circuit. A simple linear feedback method has been applied to the logistic map in Section 19.2.

IV. *Stabilizing unstable periodic orbits (the Ott, Grebogi, and Yorke (OGY) method).* The method relies on the fact that the chaotic attractors contain an infinite number of unstable periodic orbits. By making small time-dependent perturbations to a control parameter of the system, it is possible to stabilize one or more of the unstable periodic orbits. The method has been very successful in applications, but there are some drawbacks. This method will be discussed in some detail at the end of this section.

V. *Occasional proportional feedback* (OPF). Developed by Hunt [12] in 1991, this is one of the most promising control techniques for real applications. It is a one-dimensional version of the OGY method and has been successful in suppressing chaos for many physical systems. The feedback consists of a series of kicks, whose amplitude is determined from the difference of the chaotic output signal from a relaxation oscillation embedded in the signal, applied to the input signal at periodic intervals.

VI. *Synchronization.* The possibility of synchronization of two chaotic systems was first proposed by Pecora and Carroll [22] in 1990 with applications in communications. By feeding the output from one chaotic oscillator (the transmitter) into another chaotic oscillator (the receiver), they were able to synchronize certain chaotic systems for certain parameter choices. The method opens up the possibilities for secure information transmission. More historical information and examples of chaos synchronization are presented in Section 19.4.

Before summarizing the OGY method, it is worthwhile to highlight some of the other major results not mentioned above. The first experimental suppression of chaos was performed by Ditto, Rausseo, and Spano [8] using the OGY algorithm. By making small adjustments to the amplitude of an external magnetic field, they were able to stabilize a gravitationally buckled magnetostrictive ribbon that oscillated chaotically in a magnetic field. They produced period-one and period-two behavior, and the procedure proved to be remarkably robust. Using both experimental and theoretical results, Singer, Wang, and Bau [25] applied a simple on-off strategy in order to laminarize (suppress) chaotic flow of a fluid in a thermal convection loop. The on-off controller was applied to the Lorenz equations, and the numerical results were in good agreement with the experimental results. Shortly afterward, Hunt [12] applied a modified version of the OGY algorithm called occasional proportional feedback (OPF) to the chaotic dynamics of a nonlinear diode resonator. Small perturbations were used to stabilize orbits of low period, but larger perturbations were required to stabilize orbits of high periods. By changing the level, width, and gain of the feedback signal, Hunt was able to stabilize orbits with periods as high as 23. Using the OPF algorithm developed by Hunt, Roy et al. [23] were able to stabilize a weakly chaotic green laser. In recent years, the implementation of the control algorithm has been carried out electronically using either digital signals or analog hardware. The hope for the future is that all-optical processors and feedback can be used in order to increase speed. The first experimental control in a biological system was performed by Garfinkel et al. [10] in 1992. They were able to stabilize arrhythmic behavior in eight out of eleven rabbit hearts using a feedback-control mechanism. It has been reported in [5] that a company has been set up to manufacture small defibrillators that can monitor the heart and deliver tiny electrical pulses to move the heart away from fibrillation and back to normality. It was also conjectured in the same article that the chaotic heart is more healthy than a regularly beating periodic heart. The OGY algorithm was implemented theoretically by the author and Steele [19] to control the chaos within a hysteresis cycle of a nonlinear bistable optical resonator using the real and imaginary parts of the electrical field amplitude. The same authors have recently managed to control the chaos using feedback of the electric field. This quantity is easy to continuously monitor and measure and could lead to physical applications in the future.

Methods I–VI and results given above are by no means exhaustive. This section is intended to provide a brief introduction to the subject and to encourage further reading.

The OGY Method

Following the paper of Ott, Grebogi, and Yorke [21], consider the n-dimensional map

$$\mathbf{Z}_{n+1} = \mathbf{f}(\mathbf{Z}_n, p), \tag{19.1}$$

where p is some accessible system parameter that can be changed in a small neighborhood of its nominal value, say, p_0. In the case of continuous-time systems, such a map can be constructed by introducing a transversal surface of section and setting up a Poincaré map.

It is well known that a chaotic attractor is densely filled with unstable periodic orbits, and that ergodicity guarantees that any small region on the chaotic attractor will be visited by a chaotic orbit. The OGY method hinges on the existence of stable manifolds around unstable periodic points. The basic idea is to make small time-dependent linear perturbations to the control parameter p in order to nudge the state toward the stable manifold of the desired fixed point. Note that this can only be achieved if the orbit is in a small neighborhood, or *control region*, of the fixed point.

Suppose that $\mathbf{Z}_S(p)$ is an unstable fixed point of equation (19.1). The position of this fixed point moves smoothly as the parameter p is varied. For values of p close to p_0 in a small neighborhood of $\mathbf{Z}_S(p_0)$, the map can be approximated by a linear map given by

$$\mathbf{Z}_{n+1} - \mathbf{Z}_S(p_0) = \mathbf{J}(\mathbf{Z}_n - \mathbf{Z}_S(p_0)) + \mathbf{C}(p - p_0), \tag{19.2}$$

where $\mathbf{J}$ is the Jacobian and $\mathbf{C} = \frac{\partial \mathbf{f}}{\partial p}$. All partial derivatives are evaluated at $\mathbf{Z}_S(p_0)$ and p_0.

Assume that in a small neighborhood around the fixed point

$$p - p_0 = -\mathbf{K}(\mathbf{Z}_n - \mathbf{Z}_S(p_0)), \tag{19.3}$$

where $\mathbf{K}$ is a constant vector of dimension n to be determined. Substitute (19.3) into (19.2) to obtain

$$\mathbf{Z}_{n+1} - \mathbf{Z}_S(p_0) = (\mathbf{J} - \mathbf{C}\mathbf{K})(\mathbf{Z}_n - \mathbf{Z}_S(p_0)). \tag{19.4}$$

The fixed point is then stable as long as the eigenvalues, or *regulator poles*, have modulus less than unity. The pole-placement technique from control theory can be applied to find the vector $\mathbf{K}$. A specific example is given in Section 19.3.

A simple schematic diagram is given in Figure 19.1 to demonstrate the action of the OGY algorithm. Physically, one can think of a marble placed on a saddle. If the marble is rolled toward the center (where the fixed point lies), then it will roll off as depicted in Figure 19.1(a). If, however, the saddle is moved slightly from side to side, by applying small perturbations, then the marble can be made to balance at the center of the saddle, as depicted in Figure 19.1(b).

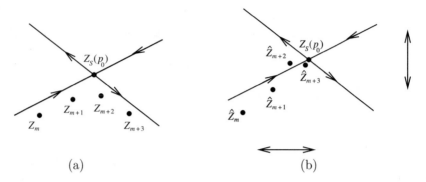

Figure 19.1: Possible iterations near the fixed point (a) without control and (b) with control. The double-ended arrows are supposed to represent small perturbations to the system dynamics. The iterates $\hat{Z}_j$ represent perturbed orbits.

Some useful points to note:

- The OGY technique is a feedback-control method.

- If the equations are unknown, sometimes delay-coordinate embedding techniques using a single-variable time series can be used (the map can be constructed from experimental data).

- There may be more than one control parameter available.

- Noise may affect the control algorithm. If the noise is relatively small, the control algorithm will still work in general.

It should also be pointed out that the OGY algorithm can only be applied once the orbit has entered a small control region around the fixed point. For certain nonlinear systems, the number of iterations required—and hence the time— for the orbit to enter this control region may be too many to be practical. Shinbrot et al. [24] solved this problem by targeting trajectories to the desired control regions in only a small number of iterations. The method has also been successfully applied in physical systems.

19.2 Controlling Chaos in the Logistic Map

Consider the logistic map given by

$$x_{n+1} = f_\mu(x_n) = \mu x_n(1 - x_n) \tag{19.5}$$

as introduced in Chapter 3. There are many methods available to control the chaos in this one-dimensional system, but the analysis is restricted to periodic proportional pulses in this section. For more details on the method and its application to the Hénon map, the reader is directed to [6]. To control the chaos in this system, instantaneous pulses will be applied to the system variables x_n once every p iterations such that

$$x_i \to kx_i,$$

where k is a constant to be determined and p denotes the period.

Recall that a fixed point of period one, say, x_S, of equation (19.5) satisfies the equation

$$x_S = f_\mu(x_S),$$

and this fixed point is stable if and only if

$$\left| \frac{df_\mu(x_S)}{dx} \right| < 1.$$

Define the composite function $F_\mu(x)$ by

$$F_\mu(x) = k f_\mu^p(x).$$

A fixed point of the function F_μ satisfies the equation

$$k f_\mu^p(x_S) = x_S, \qquad (19.6)$$

where the fixed point x_S is stable if

$$\left| k \frac{df_\mu^p(x_S)}{dx} \right| < 1. \qquad (19.7)$$

Define the function $C^p(x)$ by

$$C^p(x) = \frac{x}{f_\mu^p(x)} \frac{df_\mu^p(x_S)}{dx}.$$

Substituting from (19.6), equation (19.7) becomes

$$|C^p(x_S)| < 1. \qquad (19.8)$$

A fixed point of this composite map is a stable point of period p for the original logistic map when the control is switched on, providing condition (19.8) holds. In practice, chaos control always deals with periodic orbits of low periods, say, $p = 1$ to 4, and this method can be easily applied.

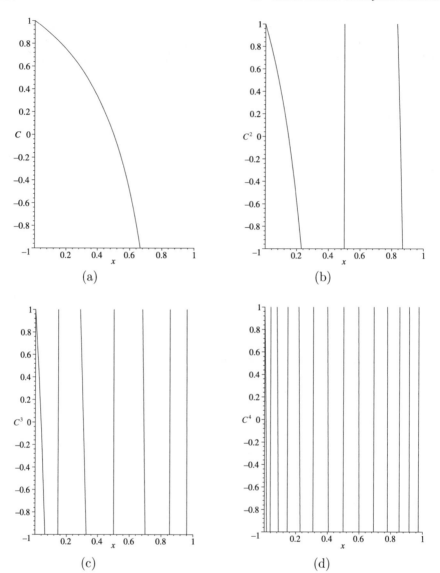

Figure 19.2:   Control curves $C^i, i = 1, 2, 3, 4$, for the logistic map when $\mu = 4$. The range is restricted to $-1 < C^p(x_S) < 1$ in each case.

To illustrate the method, consider the logistic map when $\mu = 4$ and the system is chaotic. The functions $C^1(x)$, $C^2(x)$, $C^3(x)$, and $C^4(x)$ are shown in Figure 19.2.

Figure 19.2(a) shows that fixed points of period one can be stabilized for every x_S in the range between zero and approximately 0.67. When $p = 2$,

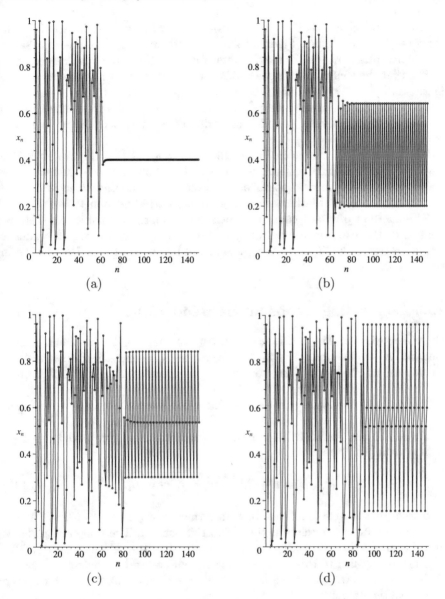

Figure 19.3: [Mathematica] Stabilization of points of periods one, two, three, and four for the logistic map when $\mu = 4$; (a) $x_S = 0.4$, $k = 0.417$; (b) $x_S = 0.2$, $k = 0.217$; (c) $x_S = 0.3$, $k = 0.302$; and (d) $x_S = 0.6$, $k = 0.601$. In each case k is computed to three decimal places.

Figure 19.2(b) shows that fixed points of period two can only be stabilized in three ranges of x_S values. Figures 19.2(c) and (d) indicate that there are seven and 14 acceptable ranges for fixed points of periods three and four, respectively. Notice that the control ranges are getting smaller and smaller as the periodicity increases.

Figure 19.3 shows time series data for specific examples when the chaos is controlled to period-one, period-two, period-three, and period-four behavior, respectively.

The values of x_S chosen in Figure 19.3 were derived from Figure 19.2. The values of k were calculated using equation (19.6). Note that the system can be stabilized to many different points on and even off the chaotic attractor (see the work of Chau [6]). A Mathematica program is listed in Section 19.5.

This method of chaos control by periodic proportional pulses can also be applied to the two-dimensional discrete Hénon map. The interested reader is again directed to [6]. The OGY algorithm will be applied to the Hénon map in the next section.

19.3 Controlling Chaos in the Hénon Map

Ott, Grebogi, and Yorke [21] used the Hénon map to illustrate the control method. A simple example will be given here. Consider the Hénon map as introduced in Chapter 3. The two-dimensional iterated map function is given by

$$X_{n+1} = 1 + Y_n - \alpha X_n^2, \quad Y_{n+1} = \beta X_n, \tag{19.9}$$

where $\alpha > 0$ and $|\beta| < 1$. Take a transformation $X_n = \frac{1}{\alpha} x_n$ and $Y_n = \frac{\beta}{\alpha} y_n$, then system (19.9) becomes

$$x_{n+1} = \alpha + \beta y_n - x_n^2, \quad y_{n+1} = x_n. \tag{19.10}$$

The proof that system (19.9) can be transformed into system (19.10) will be left to the reader in the exercises at the end of this chapter. The Hénon map is now in the form considered in [21], and the control algorithm given in Section 19.1 will now be applied to this map. Set $\beta = 0.4$, and allow the control parameter, in this case α, to vary around a nominal value, say, $\alpha_0 = 1.2$, for which the map has a chaotic attractor.

The fixed points of period one are determined by solving the simultaneous equations

$$\alpha_0 + \beta y - x^2 - x = 0 \quad \text{and} \quad x - y = 0.$$

In Chapter 14, it was shown that the Hénon map has two fixed points of period one if and only if $(1 - \beta)^2 + 4\alpha_0 > 0$. In this particular case, the fixed points of period one are located approximately at $A = (x_{1,1}, y_{1,1}) = (0.8358, 0.8358)$ and $B = (x_{1,2}, y_{1,2}) = (-1.4358, -1.4358)$. The chaotic attractor and points of period one are shown in Figure 19.4.

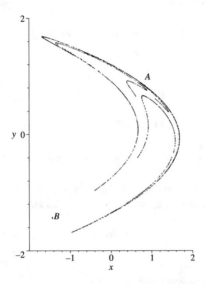

Figure 19.4: Iterative plot for the Hénon map (3000 iterations) when $\alpha_0 = 1.2$ and $\beta = 0.4$. The two fixed points of period one are labeled A and B.

The Jacobian matrix of partial derivatives of the map is given by

$$J = \begin{pmatrix} \frac{\partial P}{\partial x} & \frac{\partial P}{\partial y} \\ \frac{\partial Q}{\partial x} & \frac{\partial Q}{\partial y} \end{pmatrix},$$

where $P(x, y) = \alpha_0 + \beta y - x^2$ and $Q(x, y) = x$. Thus

$$J = \begin{pmatrix} -2x & \beta \\ 1 & 0 \end{pmatrix}.$$

Consider the fixed point at A; the fixed point is a saddle point. Using the notation introduced in Section 19.1, for values of α close to α_0 in a small neighborhood of A, the map can be approximated by a linear map

$$\mathbf{Z}_{n+1} - \mathbf{Z}_S(\alpha_0) = \mathbf{J}(\mathbf{Z}_n - \mathbf{Z}_S(\alpha_0)) + \mathbf{C}(\alpha - \alpha_0), \tag{19.11}$$

where $\mathbf{Z}_n = (x_n, y_n)^T$, $A = \mathbf{Z}_S(\alpha_0)$, $\mathbf{J}$ is the Jacobian, and

$$\mathbf{C} = \begin{pmatrix} \frac{\partial P}{\partial \alpha} \\ \frac{\partial Q}{\partial \alpha} \end{pmatrix},$$

and all partial derivatives are evaluated at α_0 and $\mathbf{Z}_S(\alpha_0)$. Assume in a small neighborhood of A,

$$\alpha - \alpha_0 = -\mathbf{K}(\mathbf{Z}_n - \mathbf{Z}_S(\alpha_0)), \qquad (19.12)$$

where

$$\mathbf{K} = \begin{pmatrix} k_1 \\ k_2 \end{pmatrix}.$$

Substitute (19.12) into (19.11) to obtain

$$\mathbf{Z}_{n+1} - \mathbf{Z}_S(\alpha_0) = (\mathbf{J} - \mathbf{C}\mathbf{K})(\mathbf{Z}_n - \mathbf{Z}_S(\alpha_0)).$$

Therefore, the fixed point at $A = \mathbf{Z}_S(\alpha_0)$ is stable if the matrix $\mathbf{J} - \mathbf{C}\mathbf{K}$ has eigenvalues (or regulator poles) with modulus less than unity. In this particular case,

$$\mathbf{J} - \mathbf{C}\mathbf{K} \approx \begin{pmatrix} -1.671563338 - k_1 & 0.4 - k_2 \\ 1 & 0 \end{pmatrix},$$

and the characteristic polynomial is given by

$$\lambda^2 + \lambda(1.671563338 + k_1) + (k_2 - 0.4) = 0.$$

Suppose that the eigenvalues (regulator poles) are given by λ_1 and λ_2; then

$$\lambda_1\lambda_2 = k_2 - 0.4 \quad \text{and} \quad -(\lambda_1 + \lambda_2) = 1.671563338 + k_1.$$

The lines of marginal stability are determined by solving the equations $\lambda_1 = \pm 1$ and $\lambda_1\lambda_2 = 1$. These conditions guarantee that the eigenvalues λ_1 and λ_2 have modulus less than unity. Suppose that $\lambda_1\lambda_2 = 1$. Then

$$k_2 = 1.4.$$

Suppose that $\lambda_1 = +1$. Then

$$\lambda_2 = k_2 - 0.4 \quad \text{and} \quad \lambda_2 = -2.671563338 - k_1.$$

Therefore,

$$k_2 = -k_1 - 2.271563338.$$

If $\lambda_1 = -1$, then

$$\lambda_2 = -(k_2 - 0.4) \quad \text{and} \quad \lambda_2 = -0.671563338 - k_1.$$

Therefore,

$$k_2 = k_1 + 1.071563338.$$

The stable eigenvalues (regulator poles) lie within a triangular region as depicted in Figure 19.5.

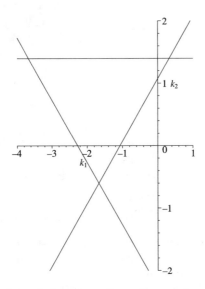

Figure 19.5:   The bounded region where the regulator poles are stable.

Select $k_1 = -1.5$ and $k_2 = 0.5$. This point lies well inside the triangular region as depicted in Figure 19.5. The perturbed Hénon map becomes

$$x_{n+1} = (-k_1(x_n - x_{1,1}) - k_2(y_n - y_{1,1}) + \alpha_0) + \beta y_n - x_n^2, \quad y_{n+1} = x_n. \tag{19.13}$$

Applying equations (19.10) and (19.13) without and with control, respectively, it is possible to plot time series data for these maps. Figure 19.6(a) shows a time series plot when the control is switched on after the 200th iterate; the control is left switched on until the 500th iterate. In Figure 19.6(b), the control is switched on after the 200th iterate and then switched off after the 300th iterate. Remember to check that the point is in the control region before switching on the control.

Once again, the Mathematica program is listed in Section 19.5.

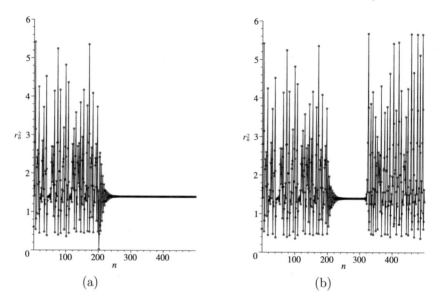

Figure 19.6: [Mathematica] Time series data for the Hénon map with and without control, $r^2 = x^2 + y^2$. In case (a), the control is activated after the 200th iterate, and in case (b), the control is switched off after the 300th iterate.

19.4 Chaos Synchronization

The first recorded experimental observation of synchronization is attributed to Huygens in 1665. Huygens was attempting to increase the accuracy of time measurement, and the experiment consisted of two huge pendula connected by a beam. He recorded that the imperceptible motion of the beam caused mutual anti-phase synchronization of the pendula. Synchronization phenomena were also observed by van der Pol (1927) and Rayleigh (1945) when investigating radio communication systems and acoustics in organ pipes, respectively. For other interesting examples of synchronization without chaos, the reader is directed to the excellent book of Strogatz [26].

This section is concerned with chaos synchronization, where two, or more, coupled chaotic systems (which may be equivalent or nonequivalent) exhibit a common, but still chaotic, behavior. Boccaletti et al. [4] present a review of the major methods of chaotic synchronization including complete synchronization, generalized synchronization, lag synchronization, phase, and imperfect phase synchronization. However, examples and theory of complete and generalized synchronization alone are presented here. The reader is directed to the textbooks [20] and [27] for more information.

Since the pioneering work of Pecora and Carroll [22], the most popular area of study is probably in secure communications. Electronic and optical circuits have been developed to synchronize chaos between a transmitter and a receiver. Cuomo and Oppenheim [7] built electronic circuits consisting of resistors, capacitors, operational amplifiers, and analog multiplier chips in order to mask and retrieve a message securely. Optically secure communications using synchronized chaos in lasers were discussed by Luo et al. in [18]. More recently, many papers have appeared on chaos synchronization with cryptographic applications, see [16], for example. Other examples of chaotic synchronization can be found in chemical kinetics [17], physiology [11], neural networks [30], and economics [28].

Complete Synchronization

Pecora and Carroll [22] consider chaotic systems of the form

$$\dot{\mathbf{u}} = \mathbf{f}(\mathbf{u}), \tag{19.14}$$

where $\mathbf{u} \in \Re^n$ and $\mathbf{f} : \Re^n \to \Re^n$. They split system (19.14) into two subsystems: one the driver system and the other the response.

$$\dot{\mathbf{x}} = \mathbf{d}(\mathbf{x}(t)) \quad \text{driver,}$$
$$\dot{\mathbf{y}} = \mathbf{r}(\mathbf{y}(t), \mathbf{x}(t)) \quad \text{response,}$$

where $\mathbf{x} \in \Re^k$, $\mathbf{y} \in \Re^m$ and $k + m = n$. The vector $\mathbf{x}(t)$ represents the driving signal. Some of the outputs from the driver system are used to drive the response system. Consider the following simple example involving a Lorenz system (see Section 13.4). The driver Lorenz system is

$$\dot{x}_1 = \sigma(x_2 - x_1), \quad \dot{x}_2 = rx_1 - x_2 - x_1 x_3, \quad \dot{x}_3 = x_1 x_2 - bx_3, \tag{19.15}$$

and the response is given by

$$\dot{y}_2 = -x_1 y_3 + rx_1 - y_2, \quad \dot{y}_3 = x_1 y_2 - by_3. \tag{19.16}$$

Note that the response Lorenz system is a subsystem of the driver, and in this case, $x_1(t)$ is the driving signal. Choose the parameter values $\sigma = 16$, $b = 4$, and $r = 45.92$, then the driver system (19.15) is chaotic. Pecora and Carroll [22] establish that synchronization can be achieved as long as the *conditional Lyapunov exponents* of the response system, when driven by the driver, are negative. However, the negativity of the conditional Lyapunov exponents gives only a necessary condition for stability of synchronization, see reference [4]. To prove stability of synchronization, it is sometimes possible to use a suitable Lyapunov function (see Chapter 6). Suppose, in this case, that

$$\mathbf{e} = (x_2, x_3) - (y_2, y_3) = \text{error signal}, \tag{19.17}$$

then we can prove that $\mathbf{e}(t) \to 0$ as $t \to \infty$, for any set of initial conditions for the coupled systems (19.15) and (19.16). Consider the following example:

Example 1. Find an appropriate Lyapunov function to show that $\mathbf{e}(t) \to 0$ as $t \to \infty$, for the driver-response system (19.15) and (19.16). Use Mathematica to show that the system synchronizes.

Solution. The equations governing the error dynamics (19.17) are given by

$$\dot{e}_2 = -x_1(t)e_3 - e_2$$
$$\dot{e}_3 = x_1(t)e_2 - be_3.$$

Multiply the first equation by e_2 and the second equation by e_3 and add to give

$$e_2\dot{e}_2 + e_3\dot{e}_3 = -e_2^2 - be_3^2,$$

and the chaos terms have canceled out. Note that

$$e_2\dot{e}_2 + e_3\dot{e}_3 = \frac{1}{2}\frac{d}{dt}\left(e_2^2 + e_3^2\right).$$

Define a Lyapunov function

$$V(e_2, e_3) = \frac{1}{2}\left(e_2^2 + e_3^2\right),$$

then

$$V(e_2, e_3) \geq 0 \quad \text{and} \quad \frac{dV}{dt} = -e_2^2 - be_3^2 < 0,$$

since $b > 0$. Therefore, $V(e_1, e_2)$ is a Lyapunov function and $(e_2, e_3) = (0, 0)$ is globally asymptotically stable. A Mathematica program for system (19.15) is listed in Section 19.5, and Figures 19.7(a) and (b) show synchronization of $x_2(t)$ with $y_2(t)$, and $x_3(t)$ with $y_3(t)$.

The choice of driving signal is crucial in complete synchronization, and some conditional Lyapunov exponents can be positive. A different choice of driving signal can lead to unstable synchronized states, see [4], for example. An alternative coupling configuration that addresses this problem is the *auxiliary system approach* which leads to generalized synchronization.

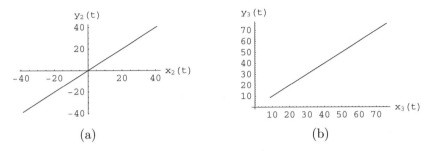

Figure 19.7: [Mathematica] Synchronization between (19.15) and (19.16): (a) $x_2(t)$ and $y_2(t)$, (b) $x_3(t)$ and $y_3(t)$.

Generalized Synchronization

Abarbanel et al. [1] introduce the auxiliary system approach which utilizes a second, identical response system to monitor the synchronized motions. They take system (19.14) and split it into three subsystems, first the driver system, second the response, and the third an auxiliary system, which is identical to the response system.

$$\dot{\mathbf{x}} = \mathbf{d}(\mathbf{x}(t)) \quad \text{driver,}$$
$$\dot{\mathbf{y}} = \mathbf{r}(\mathbf{y}(t), \mathbf{g}, \mathbf{x}(t)) \quad \text{response,}$$
$$\dot{\mathbf{z}} = \mathbf{a}(\mathbf{z}(t), \mathbf{g}, \mathbf{x}(t)) \quad \text{auxiliary,}$$

where $\mathbf{x} \in \Re^k$, $\mathbf{y} \in \Re^m$, $\mathbf{z} \in \Re^l$, $k + m + l = n$, and $\mathbf{g}$ represents the coupling strength. They state that two systems are generally synchronized if there is a transformation, say, $\mathbf{T}$, so that $\mathbf{y}(t) = \mathbf{T}(\mathbf{x}(t))$. When the response and auxiliary are driven by the same signal, then $\mathbf{y}(t) = \mathbf{T}(\mathbf{x}(t))$ and $\mathbf{z}(t) = \mathbf{T}(\mathbf{x}(t))$, and it is clear that a solution of the form $\mathbf{y}(t) = \mathbf{z}(t)$ exists as long as the initial conditions lie in the same basin of attraction. They further show that when the manifold $\mathbf{y} = \mathbf{z}$ is linearly stable, then the conditional Lyapunov exponents for the response system, driven by $\mathbf{x}(t)$, are all negative.

As a specific example, they consider generalized synchronization of chaotic oscillations in a three-dimensional Lorenz system that is driven by a chaotic signal from a Rössler system. The driver Rössler system is

$$\dot{x}_1 = -(x_2 + x_3), \quad \dot{x}_2 = x_1 + 0.2x_2, \quad \dot{x}_3 = 0.2 + x_3(x_1 - \mu), \qquad (19.18)$$

the response Lorenz system is

$$\dot{y}_1 = \sigma(y_2 - y_1) - g(y_1 - x_1), \quad \dot{y}_2 = ry_1 - y_2 - y_1 y_3, \quad \dot{y}_3 = y_1 y_2 - by_3, \qquad (19.19)$$

and the auxiliary Lorenz system is

$$\dot{z}_1 = \sigma(z_2 - z_1) - g(z_1 - x_1), \quad \dot{z}_2 = rz_1 - z_2 - z_1 z_3, \quad \dot{z}_3 = z_1 z_2 - b z_3. \quad (19.20)$$

Consider

$$\mathbf{e} = \mathbf{y}(t) - \mathbf{z}(t) = \text{error signal.} \quad (19.21)$$

The function

$$V(e_1, e_2, e_3) = \frac{1}{2} \left(4e_1^2 + e_2^2 + e_3^2 \right)$$

can be used as a Lyapunov function for the coupled system (19.19) and (19.20) as long as the coupling parameter g satisfies the inequality

$$g < \left(\frac{1}{4}\sigma + r - z_3 \right)^2 + \frac{z_2^2}{b} - \sigma.$$

The $z_i(t)$, $i = 1, 2, 3$, are bounded on a chaotic attractor, and so this condition can be satisfied when g is large enough. The numerical solutions to the nine-dimensional differential equations are easily computed with Mathematica. A program is listed in Section 19.5. Figure 19.8(a) shows synchronization between $y_2(t)$ and $z_2(t)$ when $g = 8$. Figure 19.8(b) shows that $y_2(t)$ and $z_2(t)$ are not synchronized when $g = 4$.

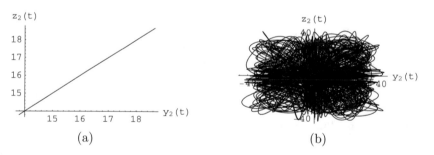

(a)   (b)

Figure 19.8: [Mathematica] (a) Synchronization between $y_2(t)$ and $z_2(t)$ when the coupling coefficient is $g = 8$ between systems (19.18), (19.19), and (19.20). (b) When $g = 4$, the system is not synchronized. The coupling is not strong enough.

More examples of chaos control and synchronization are presented in Chapter 21, and open problems are considered in [29].

19.5 Mathematica Commands

```
(* Chaos Control in the Logistic Map, see Figure 19.3(b) *)
mu=4;switchon=60;itermax=200;x[0]=0.6;k=0.217;
f[x_]=mu x (1-x);
Do[{x[i+1]=f[x[i]]},{i,0,switchon-1}]
Do[{x[i+1]=k f[x[i]],x[i+2]=f[x[i+1]]},{i,switchon,itermax,2}]
p1=ListPlot[Table[{i,x[i]},{i,0,itermax}],Joined->True,PlotRange->All];
p2=ListPlot[Table[{i,x[i]},{i,0,itermax}],PlotStyle->PointSize[0.015]];
Show[{p1,p2}]
```

```
(* Chaos Control in the Henon Map, see Figure 19.6(a)*)
alpha=1.2;beta=0.4;
k1=-1.8;k2=1.2;
xstar=0.8358;ystar=xstar;
itermax=400;x[0]=0.5;y[0]=0.6;r[0]=(x[0])^2+(y[0])^2;switch=196;
Henon[{x_,y_}]={alpha+beta y-x^2,x};
ControlHenon[{x_,y_}]={(-k1 (x-xstar)-k2 (y-ystar)+alpha)+beta y-x^2,x};
Do[{x[i+1],y[i+1]}=Henon[{x[i],y[i]}],{i,0,switch-1}]
Do[{x[i+1],y[i+1]}=ControlHenon[{x[i],y[i]}],{i,switch,itermax}]
Do[r[i+1]=(x[i+1])^2+(y[i+1])^2,{i,0,itermax}]
p1=ListPlot[Table[{i,r[i]},{i,0,itermax}],Joined->True,PlotRange->All];
p2=ListPlot[Table[{i,r[i]},{i,0,itermax}],PlotStyle->PointSize[0.015]];
Show[{p1,p2},PlotRange->All]
```

```
(* Chaos Synchronization Between Two Lorenz Systems, see Figure 19.7(b) *)
sigma=16;b=4;r=45.92;
Lorenz=NDSolve[{x1'[t]==sigma (x2[t]-x1[t]),x2'[t]==-x1[t] x3[t]+r x1[t]-x2[t],
x3'[t]==x1[t] x2[t]-b x3[t],y2'[t]==-x1[t] y3[t]+r x1[t]-y2[t],
y3'[t]==x1[t] y2[t]-b y3[t],x1[0]==15,x2[0]==20,x3[0]==30,y2[0]==10,y3[0]==20},
{x1,x2,x3,y2,y3},{t,0,100},MaxSteps->Infinity];
ParametricPlot3D[Evaluate[{x1[t],x2[t],x3[t]}/.Lorenz],{t,0,100},
PlotPoints->10000]
ParametricPlot[Evaluate[{x3[t],y3[t]}/.Lorenz],{t,50,100},PlotPoints->10000,
AxesLabel->{Subscript[x, 3],Subscript[y, 3]}]
```

```
(* Generalized Synchronization: The Auxiliary Approach, see Figure 19.8(a) *)
mu=5.7;sigma=16;b=4;r=45.92;g=4; (*No Synchronization*)
RosslerLorenz=NDSolve[{x1'[t]==-(x2[t]+x3[t]),x2'[t]==x1[t]+0.2 x2[t],
x3'[t]==0.2+x3[t] (x1[t]-mu),y1'[t]==sigma (y2[t]-y1[t])-g(y1[t]-x1[t]),
y2'[t]==-y1[t] y3[t]+r y1[t]-y2[t],
y3'[t]==y1[t] y2[t]-b y3[t],z1'[t]==sigma (z2[t]-z1[t])-g(z1[t]-x1[t]),
z2'[t]==-z1[t] z3[t]+r z1[t]-z2[t],
z3'[t]==z1[t] z2[t]-b z3[t],x1[0]==2,x2[0]==-10,x3[0]==44,y1[0]==30,y2[0]==10,
y3[0]==20,z1[0]==31,z2[0]==11,z3[0]==22},{x1,x2,x3,y1,y2,y3,z1,z2,z3},{t,0,200},
MaxSteps->Infinity];
ParametricPlot[Evaluate[{y2[t],z2[t]}/.RosslerLorenz],{t,50,200},
PlotPoints->10000,AxesLabel->{"Subscript[y, 2](t)","Subscript[z, 2](t)"}]
```

```
mu=5.7;sigma=16;b=4;r=45.92;g=8;  (*Synchronization*)
RosslerLorenz=NDSolve[{x1'[t]==-(x2[t]+x3[t]),x2'[t]==x1[t]+0.2 x2[t],
x3'[t]==0.2+x3[t] (x1[t]-mu),y1'[t]==sigma (y2[t]-y1[t])-g(y1[t]-x1[t]),
y2'[t]==-y1[t] y3[t]+r y1[t]-y2[t],
y3'[t]==y1[t] y2[t]-b y3[t],z1'[t]==sigma (z2[t]-z1[t])-g(z1[t]-x1[t]),
z2'[t]==-z1[t] z3[t]+r z1[t]-z2[t],
z3'[t]==z1[t] z2[t]-b z3[t],x1[0]==2,x2[0]==-10,x3[0]==44,
y1[0]==30,y2[0]==10,y3[0]==20,z1[0]==31,
z2[0]==11,z3[0]==22},
{x1,x2,x3,y1,y2,y3,z1,z2,z3},{t,0,500},MaxSteps->Infinity];
ParametricPlot[Evaluate[{y2[t],z2[t]}/.RosslerLorenz],{t,50,500},
PlotPoints->10000,AxesLabel->{"Subscript[y, 2](t)","Subscript[z, 2](t)"}]
```

19.6 Exercises

1. Show that the map defined by

$$x_{n+1} = 1 + y_n - ax_n^2, \quad y_{n+1} = bx_n$$

   can be written as

$$u_{n+1} = a + bv_n - u_n^2, \quad v_{n+1} = u_n$$

   using a suitable transformation.

2. Apply the method of chaos control by periodic proportional pulses (see Section 19.2) to the logistic map

$$x_{n+1} = \mu x_n(1 - x_n)$$

   when $\mu = 3.9$. Sketch the graphs $C^i(x), i = 1$ to 4. Plot time series data to illustrate control of fixed points of periods one, two, three, and four.

3. Find the points of periods one and two for the Hénon map given by

$$x_{n+1} = a + by_n - x_n^2, \quad y_{n+1} = x_n$$

   when $a = 1.4$ and $b = 0.4$, and determine their type.

4. Apply the method of chaos control by periodic proportional pulses (see Section 19.2) to the two-dimensional Hénon map

$$x_{n+1} = a + by_n - x_n^2, \quad y_{n+1} = x_n,$$

   where $a = 1.4$ and $b = 0.4$. (In this case, you must multiply x_m by k_1 and y_m by k_2, say, once every p iterations). Plot time series data to illustrate the control of points of periods one, two, and three.

5. Use the OGY algorithm given in Section 19.3 to stabilize a point of period one in the Hénon map

$$x_{n+1} = a + by_n - x_n^2, \quad y_{n+1} = x_n$$

when $a = 1.4$ and $b = 0.4$. Display the control using a time series graph.

6. Consider the Ikeda map, introduced in Chapter 3, given by

$$E_{n+1} = A + BE_n e^{i|E_n|^2}.$$

Suppose that $E_n = x_n + iy_n$, rewrite the Ikeda map as a two-dimensional map in x_n and y_n. Plot the chaotic attractor for the Ikeda map

$$E_{n+1} = A + BE_n e^{i|E_n|^2}$$

when $A = 2.7$ and $B = 0.15$. How many points are there of period one? Indicate where these points are with respect to the attractor.

7. Plot the chaotic attractor for the Ikeda map

$$E_{n+1} = A + BE_n e^{i|E_n|^2}$$

when

(i) $A = 4$ and $B = 0.15$;

(ii) $A = 7$ and $B = 0.15$.

How many points are there of period one in each case? Indicate where these points are for each of the attractors on the figures.

8. Use the OGY method (see Section 19.3 the parameter A to control the chaos to a point of period one in the Ikeda map

$$E_{n+1} = A + BE_n e^{i|E_n|^2}$$

when $A_0 = 2.7$ and $B = 0.15$. Display the control on a time series plot. (N.B.: Use a two-dimensional map).

9. Try the same procedure of control to period one for the Ikeda map as in Exercise 8 but with the parameters $A_0 = 7$ and $B = 0.15$. Investigate the size of the control region around one of the fixed points in this case and state how it compares to the control region in Exercise 8. What can you say about flexibility and controllability?

10. Use the methods described in Section 19.4 to demonstrate synchronization of chaos in Chua's circuit.

Bibliography

[1] H.D.I. Abarbanel, N.F. Rulkov and M.M. Sushchik, Generalized synchronization of chaos: the auxiliary system approach, *Phys. Rev. E* **53**(5), (1996), 4528–4535.

[2] A.T. Azar (Editor), *Fractional Order Control and Synchronization of Chaotic Systems (Studies in Computational Intelligence)*, Springer, New York, 2017.

[3] A. Balanov, N. Janson, D. Postnov, and O. Sosnovtseva, *Synchronization: From Simple to Complex*, Springer-Verlag, New York, 2008.

[4] S. Boccaletti, J. Kurths, G. Osipov, D. L. Valladares and C. S. Zhou, The synchronization of chaotic systems, *Physics Reports*, **366** (2002), 1–101.

[5] M. Buchanan, Fascinating rhythm, *New Scientist*, 3 Jan., (1998), 20–25.

[6] N.P. Chau, Controlling chaos by periodic proportional pulses, *Phys. Lett. A* **234**, (1997), 193–197.

[7] K.M. Cuomo and A.V. Oppenheim, Circuit implementation of synchronized chaos with applications to communications, *Phys. Rev. Lett.*, **71**, (1993), 65–68.

[8] W.L. Ditto, S.N. Rausseo, and M.L. Spano, Experimental control of chaos. *Phys. Rev. Lett.* **65**, (1990), 3211–3214.

[9] R. Femat and G. Solis-Perales, *Robust Synchronization of Chaotic Systems via Feedback*, Springer-Verlag, New York, 2008.

[10] A. Garfinkel, M.L. Spano, W.L. Ditto, and J.N. Weiss, Controlling cardiac chaos, *Science* **257**, (1992), 1230–1235.

[11] L. Glass, Synchronization and rhythmic processes in physiology, *Nature* **410**, (2001), 277–284.

[12] E.R. Hunt, Stabilizing high-period orbits in a chaotic system - the diode resonator, *Phys. Rev. Lett.* **67**, (1991), 1953–1955.

[13] T. Kapitaniak, *Controlling Chaos: Theoretical and Practical Methods in Non-linear Dynamics*, Academic Press, 1996.

[14] T. Kapitaniak, *Chaos for Engineers: Theory, Applications & Control*, Springer-Verlag, New York, Second ed., 2000.

[15] M.A. Khan, *Chaos Control, Chaos Synchronization and Its Application*, LAP Lambert Academic Publishing, Saabrücken, 2016.

[16] E. Klein, R. Mislovaty, I. Kanter, and W. Kinzel, Public-channel cryptography using chaos synchronization, *Phys. Rev. E* **72**(1): Art. No. 016214 Part 2, 2005.

[17] Y.N. Li, L. Chen, Z.S. Cai, and X.Z. Zhao, Experimental study of chaos synchronization in the Belousov-Zhabotinsky chemical system, *Chaos Solitons and Fractals* **22**(4), (2004), 767–771.

[18] L. Luo and P.L. Chu, Optical secure communications with chaotic erbium-doped fiber lasers, *J. Opt. Soc. Amer. B* **15**, (1998), 2524–2530.

[19] S. Lynch and A.L. Steele, Controlling chaos in nonlinear bistable optical resonators, *Chaos, Solitons and Fractals*, **11**-5, (2000), 721–728.

[20] E. Mosekilde, Y. Maistrenko, and D. Postnov, *Chaotic Synchronization*, World Scientific, Singapore, 2002.

[21] E. Ott, C. Grebogi, and J.A. Yorke, Controlling chaos, *Phys. Rev. Lett.* **64**, (1990), 1196–1199.

[22] L.M. Pecora and T.L. Carroll, Synchronization in chaotic systems, *Phys. Rev. Lett.* **64**, (1990), 821–824.

[23] R. Roy, T.W. Murphy, T.D. Maier, Z. Gills, and E.R. Hunt, Dynamical control of a chaotic laser: Experimental stabilization of a globally coupled system, *Phys. Rev. Lett.* **68**, (1992), 1259–1262.

[24] T. Shinbrot, C. Grebogi, E. Ott, and J.A. Yorke, Using chaos to direct trajectories to targets, *Phys. Rev. Lett.* **65**, (1990), 3215–3218.

[25] J. Singer, Y-Z. Wang, and H.H. Bau, Controlling a chaotic system, *Phys. Rev. Lett.* **66**, (1991), 1123–1125.

[26] S.H. Strogatz, *Sync: The Emerging Science of Spontaneous Order*, Theia, New York, 2003.

[27] C.W. Wu, *Synchronization in Coupled Chaotic Circuits and Systems*, World Scientific, Singapore, 2002.

[28] S. Yousefi, Y. Maistrenko, and S. Popovych, Complex dynamics in a simple model of interdependent open economies, *Discrete dynamics in nature and society*, **5**(3), (2000), 161–177.

[29] E. Zeraoulia *Chaos Control and Synchronization: Advancing Perspectives*, LAP LAMBERT Academic Publishing, Saarbrcken, 2012.

[30] X.H. Zhang and S.B. Zhou, Chaos synchronization for bi-directional coupled two-neuron systems with discrete delays, *Lecture notes in Computer Science* **3496**, (2005), 351–356.

Chapter 20
Neural Networks

Aims and Objectives

- To provide a brief historical background to neural networks.

- To investigate simple neural network architectures.

- To consider applications in the real world.

- To present working Mathematica program files for some neural networks.

- To introduce neurodynamics.

On completion of this chapter, the reader should be able to

- use the generalized delta learning rule with backpropagation of errors to train a network;

- determine the stability of Hopfield networks using a suitable Lyapunov function;

- use the Hopfield network as an associative memory;

- study the dynamics of a neuromodule in terms of bistability, chaos, periodicity, quasiperiodicity, and chaos control.

Neural networks are being used to solve all kinds of problems from a wide range of disciplines. Some neural networks work better than others on specific problems, and the models are run using continuous, discrete, and stochastic methods. For more information on stochastic methods, the reader is directed to the textbooks at the end of this chapter. The topic is highly interdisciplinary in nature, and so it is extremely difficult to develop an introductory and comprehensive treatise on the subject in one short chapter of a textbook. A brief historical introduction is given in Section 20.1 and the fundamentals are reviewed. Real-world applications are then discussed. The author has decided to concentrate on three types of neural network—the feedforward multilayer network and backpropagation of errors using the generalized delta learning rule,

© Springer International Publishing AG 2017
S. Lynch, *Dynamical Systems with Applications using Mathematica®*,
DOI 10.1007/978-3-319-61485-4_20

the recurrent Hopfield neural network, and the minimal chaotic neuromodule. The first network is probably the most widely used in applications in the real world; the second is a much studied network in terms of stability and Lyapunov functions; and the third provides a useful introduction to neurodynamics.

For a more detailed historical introduction and review of the theory of neural networks, the reader is once more directed to the textbooks in the reference section of this chapter, see [2]–[7], [8]–[18], [22], and [24]–[29], for example.

Some of the Mathematica programs listed in Section 20.5 are quite long. Remember that you can download the Mathematica notebook from Wolfram. Those readers already familiar with neural networks should read the Mathematica Help pages for more advanced features.

As Wolfram state: "The Wolfram Language has state-of-the-art capabilities for the construction, training and deployment of neural network machine learning systems. Many standard layer types are available and are assembled symbolically into a network, which can then immediately be trained and deployed on available CPUs and GPUs."

20.1 Introduction

This textbook has thus far been concerned with deterministic dynamical systems where the underlying equations are known. This chapter provides a means of tackling nondeterministic systems, where the equations used to model the system are not known. Unfortunately, many real-world problems do not come prepackaged with mathematical equations, and often, the equations derived might not be accurate or suitable. Throughout history, scientists have attempted to model physical systems using mathematical equations. This has been quite successful in some scientific fields, but not in all. For example, what equations would a doctor use to diagnose an illness and then prescribe a treatment? How does a bank manager determine whether to issue a mortgage? How can we tell whether somebody is telling the truth? These questions have been successfully dealt with by the adoption of *neural networks*, or *artificial neural networks*, as they are sometimes referred to, using machine learning or data mining. Applications of this theory will be dealt with in more detail at the end of this section.

Definition 1. A neural network is a parallel information-processing system that has certain characteristics in common with certain brain functions. It is composed of *neurons* and *synaptic weights* and performs complex computations through a *learning process.*

The brain is a highly complex nonlinear information-processing system. It is a parallel computer, infinitely more powerful than traditional, electronic, sequential, logic-based digital computers and powerful parallel and vector computers on the market today. The average human brain consists of some 10^{11} neurons, each about $100\,\mu m$ in size, and approximately 10^{14} synapses. The synapses, or dendrites, are mainly chemical, converting electrical signals into chemical signals and back to electrical again. The synapses connecting neurons store acquired knowledge and can be excitatory or inhibitory. It should be pointed out that the numbers of neurons and synaptic weights do not remain constant in the human brain. Scientists are attempting to incorporate some features of the way the brain works into modern computing.

Network Architecture

The neuronal model is made up of four basic components: an input vector, a set of synaptic weights, a summing junction with an *activation*, or (*transfer*), *function*, and an output. The *bias* increases or decreases the net input of the activation function. Synapses receive input signals that they send to the neural cell body; the soma (summing junction) sums these signals; and the axon transmits the signal to synapses of connecting neurons. A schematic illustrating a simple mathematical model of a neuron is shown in Figure 20.1.

The neuron has bias b_k, which is added to the summation of the products of weights with inputs to give

$$v_k = \mathbf{w}_k \mathbf{x} + b_k,$$

where v_k is the *activation potential*. The neuron output is written as

$$y_k = \phi(v_k).$$

Note in this case that $\mathbf{w}_k$ is a vector. The activation function $\phi(.)$ typically ranges from -1 to $+1$ (is *bipolar*) in applications and has an antisymmetric form with respect to the origin. This textbook will be concerned mainly with bipolar activation functions. There are *unipolar* activation functions, where the function ranges from 0 to $+1$, but bipolar functions are predominantly used in

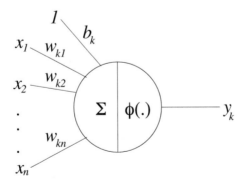

Figure 20.1: A simple nonlinear model of a single neuron k. The vector $\mathbf{x} = (x_1, x_2, \ldots, x_n)^T$ represents the input; the synaptic weights are denoted by $\mathbf{w}_k = w_{kj}$, $j = 1, 2, \ldots, n$; b_k is the bias; $\phi(.)$ is the activation function applied after a summation of the products of weights with inputs; and y_k is the output of neuron k.

applications. Some bipolar activation functions are shown in Figure 20.2. They
are defined by the following equations:

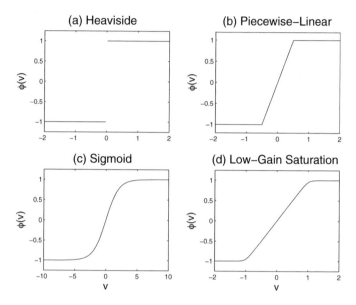

Figure 20.2: Some activation functions: (a) a Heaviside function; (b) a piecewise linear function; (c) a sigmoid function; (d) a low-gain saturation function.

$$\text{(a) } \phi(v) = \begin{cases} 1, & v \geq 0 \\ -1, & v < 0; \end{cases}$$

$$\text{(b) } \phi(v) = \begin{cases} 1, & v \geq 0.5 \\ v, & -0.5 < v < 0.5 \\ -1, & v \leq -0.5; \end{cases}$$

$$\text{(c) } \phi(v) = \tanh(av);$$

$$\text{(d) } \phi(v) = \frac{1}{2a} \log \frac{\cosh(a(v+1))}{\cosh(a(v-1))}.$$

The all-or-none law model of a neuron devised by McCulloch and Pitts [17]
in the early 1940s is widely acknowledged as the origin of the modern theory of
neural networks. They showed, in principle, that the neuron could compute any
arithmetic or logical function. Indeed, even today, the McCulloch–Pitts neuron
is the one most widely used as a logic circuit. In 1949, Hebb [10] proposed
the first learning law for neural networks used to modify synaptic weights. He
suggested that the strength of the synapse connecting two simultaneously active

neurons should be increased. There are many variations of Hebb's learning law, and they are being applied to a variety of neural network architectures; see Section 20.3, for example. In 1958, Rosenblatt [22] introduced a class of neural network called the *perceptron*. A typical architecture is shown in Figure 20.3. It was found that the perceptron learning rule was more powerful than the Hebb rule. Unfortunately, shortly afterward it was shown that the basic perceptron could only solve problems that were linearly separable. One simple example of a problem that is not linearly separable is the exclusive or (XOR) gate. An XOR gate is a circuit in a computer that fires only if one of its inputs fire.

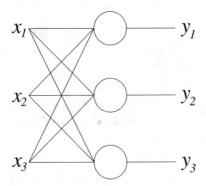

Figure 20.3: A feedforward single layer network.

Training

In 1960, Widrow and Hoff [29] introduced the ADALINE (ADAptive LInear NEuron) network, and a learning rule labeled as the *delta learning rule* or the *least mean squared* (LMS) algorithm. The perceptron learning rule adjusts synaptic weights whenever the response is incorrect, whereas the delta learning rule adjusts synaptic weights to reduce the error between the output vector and the target vector. This led to an improved ability of the network to generalize. Neither the ADALINE nor the perceptron was able to solve problems that were not linearly separable, as reported in the widely publicized book of Minsky and Papert [18]. Rumelhart and McClelland [24] edited a book that brought together the work of several researchers on backpropagation of errors using multilayer feedforward networks with hidden layers (see Figure 20.4). This algorithm partially addressed the problems raised by Minsky and Papert in the 1960s. Nowadays, over 90% of the applications to real-world problems use the backpropagation algorithm with *supervised learning*. Supervised learning is achieved by presenting a sequence of training vectors to the network, each with a corresponding known target vector. A complete set of input vectors with known targets is known as an *epoch*; it is usually loaded as a data file. A backpropagation algorithm using a supervised generalized delta learning rule is discussed in more detail in Section 20.2. Throughout the 1980s, Kohonen [15] developed self-organizing feature maps to form clusters for *unsupervised learning*. No target vectors are required for this algorithm—similar input vectors are assigned the same output cluster.

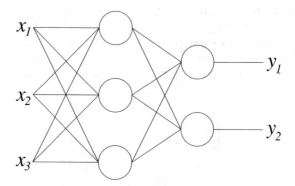

Figure 20.4:   A feedforward neural network with one hidden layer; there are three neurons in the hidden layer and two in the output layer.

The seminal paper of Hopfield [13] published in 1982 used statistical mechanics to explain the operation of a recurrent neural network used as an associative memory. The architecture of a recurrent Hopfield neural network comprising three neurons is shown in Figure 20.5. The main difference between a feedforward network and a recurrent network is that there is feedback in the latter case. Figure 20.5 illustrates the multiple-loop feedback for a three-neuron module. Note that the output of each neuron is fed back to each of the other neurons in the network.

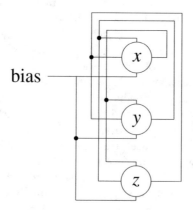

Figure 20.5:   A recurrent Hopfield neural network with feedback. Note that there is no self-feedback in this case.

The network operation can be analyzed using Lyapunov functions (see Chapter 6). Both continuous and discrete recurrent Hopfield networks are discussed in more detail in Section 20.3.

Applications. The field of neural networks has generated a phenomenal amount of interest from a broad range of scientific disciplines. One of the reasons for this is adaptability. Innovative architectures and new training rules have been tested on powerful computers, and it is difficult to predict where this research will take us in the future. As mentioned earlier, the vast majority of real-world applications have relied on the backpropagation algorithm for training multilayer networks, and recently, *kernel machines* have proved to be useful for a wide range of applications, including document classification and gene analysis, for example. In general, more than one network is required and each network is designed to perform a specific task. Some well-known applications are listed, and a more in-depth account is given for the research carried out on psychological profiling in the Department of Computing and Mathematics at Manchester Metropolitan University. The list is by no means exhaustive, and it will not be difficult for the reader to find examples applied in their own research area.

Neural networks are being used extensively in the fields of aeronautics, banking, defense, engineering, finance, insurance, marketing, manufacturing, medicine, robotics, psychology, security, and telecommunications. One of the early applications was in signal processing; the ADALINE was used to suppress noise on a telephone line. Many neural networks are being used as associative memories for pattern and speech production and recognition, for example. Simple networks can be set up as instant physicians. The expertise of many general practitioners can be used to train a network using symptoms to diagnose an illness and even suggest a possible treatment. In engineering, neural networks are being used extensively as controllers, and in banking, they are being used in mortgage assessment. Scientists find them very useful as function approximators. They can test whether the mathematical equations (which could have been used for many years) used to model a system are correct.

The Artificial Intelligence Group at Manchester Metropolitan University has developed a machine for automatic psychological profiling. The work has generated a huge amount of interest and recently was reported on national television in many countries around the world. Bandar et al. [1] have patented the machine, and the expectations are high for future applications. The machine could be used in police questioning, at airport customs, and by doctors diagnosing schizophrenia, depression, and stress. A short article on using the machine as a lie detector has recently appeared in *New Scientist* [23]. The group claims that the lie detector is accurate in 80% of test cases. Their machine uses about 20 independent neural networks, each one using the generalized delta learning rule and backpropagation of errors. Some of the channels used in the machine include eye gaze, blinking, head movement forward, hand movement, and blushing.

The same group has also carried out extensive work on conversational agents. It will not be long before we are all able to have conversations with our computers.

This introductory section has given a brief overview of neural networks. For more detailed information, the reader is directed to the many neural networks textbooks listed in the reference section of this chapter.

20.2 The Delta Learning Rule and Backpropagation

Widrow and Hoff [29] generalized the perceptron training algorithm to continuous inputs and outputs and presented the delta rule (or LMS rule). Consider a single neuron as in Figure 20.1. If the activation function is linear, then

$$y_k = \sum_j w_{kj} x_j + b_k.$$

Define an *error function* by the mean squared error, so

$$E = \frac{1}{2N} \sum_{\mathbf{x}} (E_k^{\mathbf{x}})^2 = \frac{1}{2N} \sum_{\mathbf{x}} (t_k - y_k)^2,$$

where the index $\mathbf{x}$ ranges over all input vectors, N is the number of neurons, $E^{\mathbf{x}}$ is the error on vector $\mathbf{x}$, and t_k is the target (or desired) output when vector $\mathbf{x}$ is presented. The aim is to minimize the error function E with respect to the weights w_{kj}. It is an unconstrained optimization problem; parameters w_{kj} are sought to minimize the error. The famous *method of steepest descent* is applied to the error function. Theorem 1 gives the delta rule when the activation function is linear. There are two ways to update the synaptic weights using the generalized delta rule. One is instantaneously (a weight is updated on each iteration) and the other is batch (where the weights are updated based on the average error for one epoch).

Theorem 1. *The iterative method of steepest descent for adjusting the weights in a neural network with a linear activation function is given by*

$$w_{kj}(n + 1) = w_{kj}(n) - \eta g_{kj},$$

where n is the number of iterates, $g_{kj} = -(t_k - y_k) x_j$ is the gradient vector, and η is a small positive constant called the learning rate.

Proof. Partially differentiating the error with respect to the weight vector gives

$$\frac{\partial E(w_{kj})}{\partial w_{kj}} = \frac{\partial E}{\partial E_k^{\mathbf{x}}} \frac{\partial E_k^{\mathbf{x}}}{\partial y_k} \frac{\partial y_k}{\partial w_{kj}}.$$

Now

$$\frac{\partial E}{\partial E_k^{\mathsf{x}}} = E_k^{\mathsf{x}} = (t_k - y_k),$$

and

$$\frac{\partial E_k^{\mathsf{x}}}{\partial y_k} = -1,$$

and

$$\frac{\partial y_k}{\partial w_{kj}} = x_j.$$

An estimate for the gradient vector is

$$g_{kj} = (y_k - t_k)\, x_j.$$

The delta rule for a linear activation function is thus formulated as

$$w_{kj}(n+1) = w_{kj}(n) - \eta g_{kj},$$

where η is the learning rate parameter. The choice of η is important in applications. If it is too large, the algorithm can become unstable. One normally experiments with η; it is not desirable for the algorithm to converge too slowly. □

Note that there are other optimization methods available, such as Newton's method and the Gauss–Newton method, which converge quicker and are less sensitive to the choice of η.

Theorem 2. *When the activation function is nonlinear, say, $y_k = \phi(v_k)$, the generalized delta rule can be formulated as*

$$w_{kj}(n+1) = w_{kj}(n) - \eta g_{kj}, \tag{20.1}$$

where

$$g_{kj} = (y_k - t_k)\frac{\partial \phi}{\partial v_k} x_j. \tag{20.2}$$

Proof. The proof will be left as an exercise for the reader in Section 20.6. □

The Backpropagation Algorithm

If neuron k is an output neuron, then Theorem 2 can be applied to adjust the weights of the synapses. However, if neuron j is a hidden neuron in a layer below neuron k, as depicted in Figure 20.6, then a new algorithm is required.

Theorem 3. *When neuron j is in a hidden layer, the error backpropagation rule is formulated as*

$$w_{ji}(n+1) = w_{ji}(n) - \eta g_{ji}, \qquad (20.3)$$

where

$$g_{ji} = \sum_k \left((y_k - t_k) \frac{\partial \phi}{\partial v_k} w_{kj} \right) \frac{\partial \phi}{\partial v_j} u_i. \qquad (20.4)$$

Proof. The proof is left as an exercise for the reader. The error is backpropagated through the network, layer by layer—back to the input layer, using gradient descent.   □

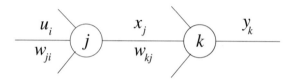

Figure 20.6:   An output neuron k connected to a hidden neuron j.

The generalized delta rule and backpropagation will now be applied to examples for estimating the value of owner-occupied homes in Boston, Massachusetts, in the 1970s.

The Boston housing data were downloaded from the UCI Machine Learning Repository on the Web at

http://www.ics.uci.edu/~mlearn/MLRepository.html.

The data can be found in the file housing.txt that can be downloaded from the Mathematica Wolfram Library Archive. Other databases at the site include arrhythmia data, automobile miles per gallon data, breast cancer data, and credit screening data.

The Boston housing data were created by D. Harrison and D.L. Rubinfeld (Hedonic prices and the demand for clean air, *J. Environmental Economics and Management*, **5** (1978), 81–102). They reported on housing values in the suburbs of Boston. There are 506 input vectors and 14 attributes including per capita crime rate by town, average number of rooms per dwelling, and pupil–teacher ratio by town.

Example 1. Write a Mathematica program to apply the generalized delta learning rule to the Boston housing data for three attributes: columns six (average number of rooms), nine (index of accessibility to radial highways), and 13 (percentage lower status of population), using the target data presented in column 14 (median value of owner-occupied homes in thousands of dollars). Use the activation function $\phi(v) = \tanh(v)$ and show how the weights are adjusted as the number of iterations increases. This is a simple three-neuron feedforward network; there are no hidden layers, and there is only one output (see Figure 20.1).

Solution. The Mathematica program file is listed in Section 20.5. A summary of the algorithm is listed below to aid in understanding the program:

1. Scale the data to zero mean, unit variance, and introduce a bias on the input.

2. Set small random weights.

3. Set the learning rate, say, η, and the number of epochs.

4. Calculate model outputs y_k, the error $t_k - y_k$, the gradients g, and perform the gradient descent to evaluate $w_{kj}(n+1) = w_{kj}(n) - \eta g_{kj}$ for each weight, see (20.1).

5. Plot a graph of weight values versus number of iterations.

Note that $\phi'(v) = 1 - (\phi(v))^2$, since $\phi(v) = \tanh(v)$. The reader will be asked to verify this in the exercises. The synaptic weights converge to the following approximate values: $b_1 \approx -0.27, w_{11} \approx 0.2, w_{12} \approx -0.04$, and $w_{13} \approx -0.24$, as shown in Figure 20.7.

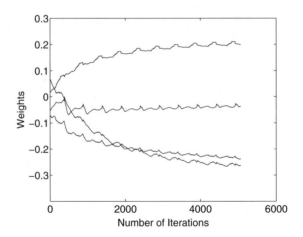

Figure 20.7: [Mathematica] Updates of the four weights (including the bias) against the number of iterations.

Example 2. Use the generalized delta rule with batch backpropagation of errors on the full data set listed in housing.txt for the Boston house data. Use the same activation function as in Example 1 and introduce one hidden layer in the neural network. Compare performance for one and two neurons in the hidden layer, when $\eta = 0.05$. One epoch consists of 506 input vectors, each with one target, and there are 13 input vectors.

Solution. A summary of the algorithm is listed below to aid in producing the program (which is left as an exercise for the reader):

1. Scale the data to zero mean, unit variance, and introduce a bias on the input.
2. Iterate over the number of neurons in the hidden layer.
3. Set random weights for the hidden and output layers.
4. Iterate over a number of epochs using batch error backpropagation.

   (a) Compute model outputs and the error.
   (b) Compute output and hidden gradients and perform gradient descent.
   (c) Determine the mean squared error for each epoch.

5. Plot a graph of mean squared error versus the number of epochs for each number of neurons in the hidden layer.

Note that it is possible to work with any number of hidden layers, but in general one hidden layer suffices. Indeed, it has been shown that one hidden layer is sufficient to approximate any continuous function. Often, the functionality that comes from extra hidden layers causes the network to overfit. The results on the full data set are shown in Figure 20.8.

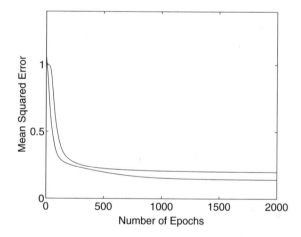

Figure 20.8: Number of epochs versus mean squared error for the Boston housing data. The upper curve is the error with one hidden neuron (settles to approximately 0.2); the lower curve is the error with two hidden neurons (stabilizes to approximately 0.14). The learning rate used in this case was $\eta = 0.05$.

20.3 The Hopfield Network and Lyapunov Stability

This section is concerned with recurrent neural networks that have fixed synaptic weights but where the activation values undergo relaxation processes through feedback. A primary application of the Hopfield network is as an associative memory, where the network is used to store patterns for future retrieval. The synaptic weights are set such that the stable points of the system correspond with the input patterns to be stored. One can think of these states as local minima in energy space. When a noisy or incomplete test pattern is input, the system should settle onto a stable state that corresponds to a stored pattern. A discrete Hopfield network is discussed in some detail later in this section, where it is used as an associative memory on some patterns. It should be noted that another famous problem addressed by Hopfield and Tank [11] was in optimization and is known as the traveling salesman problem. Simple continuous Hopfield networks are considered before the applications in order to highlight stability properties using Lyapunov functions.

The Continuous Hopfield Model

A Hopfield network does not require training data with targets. A network consisting of three neurons is shown in Figure 20.5, and a two-neuron module is shown in Figure 20.6. In 1984, Hopfield [12] showed how an analog electrical circuit could behave as a small network of neurons with graded response. He derived a Lyapunov function for the network to check for stability and used it as a content-addressable memory. The differential equations derived by Hopfield for the electrical circuit using Kirchhoff's laws could be reduced to the following system of differential equations

$$\frac{d}{dt}\mathbf{x}(t) = -\mathbf{x}(t) + \mathbf{W}\mathbf{a}(t) + \mathbf{b}, \tag{20.5}$$

where $\mathbf{x}(t)$ is a vector of neuron activation levels, $\mathbf{W}$ is the weight matrix representing synaptic connections, $\mathbf{b}$ are the biases, and $\mathbf{a}(t) = \phi(\mathbf{x}(t))$ are the nonlinear input/output activation levels. Hopfield derived the following theorem for stability properties.

Theorem 4. *A Lyapunov function for the n-neuron Hopfield network defined by equation (20.5) is given by*

$$\mathbf{V}(\mathbf{a}) = -\frac{1}{2}\mathbf{a}^T\mathbf{W}\mathbf{a} + \sum_{i=1}^{n}\left(\int_0^{a_i} \phi^{-1}(u)du\right) - \mathbf{b}^T\mathbf{a} \tag{20.6}$$

as long as

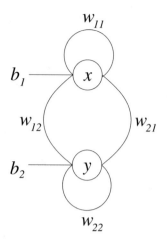

Figure 20.9:   A simple recurrent Hopfield neural network, a two-neuron module.

1. $\phi^{-1}(a_i)$ is an increasing function, that is,

$$\frac{d}{da_i}\phi^{-1}(a_i) > 0, \quad and$$

2. the weight matrix **W** is symmetric.

Proof. The proof is left as an exercise for the reader (see Section 20.6).   □

Consider the following two-neuron module taken from Hopfield's original paper [13].

Example 3. A schematic of the two-neuron module is shown in Figure 20.9. The differential equations used in Hopfield's model are given by

$$\dot{x} = -x + \frac{2}{\pi}\tan^{-1}\left(\frac{\gamma\pi y}{2}\right), \quad \dot{y} = -y + \frac{2}{\pi}\tan^{-1}\left(\frac{\gamma\pi x}{2}\right),$$

where the activation functions are arctan. Determine the stable critical points and derive a Lyapunov function.

Solution. In this case

$$\mathbf{W} = \begin{pmatrix} 0 & 1 \\ 1 & 0 \end{pmatrix}, \ \mathbf{b} = \begin{pmatrix} 0 \\ 0 \end{pmatrix}, \ a_1 = \frac{2}{\pi}\tan^{-1}\left(\frac{\gamma\pi x}{2}\right), \ a_2 = \frac{2}{\pi}\tan^{-1}\left(\frac{\gamma\pi y}{2}\right).$$

A Lyapunov function, derived using equation (20.6), is given by

$$V(\mathbf{a}) = -\frac{1}{2}(a_1\ a_2) \begin{pmatrix} 0 & 1 \\ 1 & 0 \end{pmatrix} \begin{pmatrix} a_1 \\ a_2 \end{pmatrix} + \int_0^{a_1} \phi^{-1}(u)du +$$

$$\int_0^{a_2} \phi^{-1}(u)du - (0\ 0)\begin{pmatrix} a_1 \\ a_2 \end{pmatrix}.$$

Therefore,

$$V(\mathbf{a}) = -a_1 a_2 - \frac{4}{\gamma\pi^2}\left(\log\left(\cos(\pi a_1/2)\right) + \log\left(\cos(\pi a_2/2)\right)\right).$$

Vector field plots for the differential equations are shown in Figure 20.10. The corresponding Lyapunov functions can be plotted using Mathematica when γ is given (see Section 6.2). Plot the surface for $|a_i| \le 1, i = 1, 2$.

When $0 < \gamma \le 1$, there is one stable critical point at the origin (see Figure 20.10(a)). As γ passes through one, two stable critical points bifurcate from the origin and the critical point at the origin becomes unstable (see Figure 20.10(b)). As $\gamma \to \infty$, the stable critical points approach corners of the unit square as depicted in Figure 20.10(c).

Example 4. Consider the recurrent Hopfield network modeled using the differential equations

$$\dot{x} = -x + 2\left(\frac{2}{\pi}\tan^{-1}\left(\frac{\gamma\pi x}{2}\right)\right), \quad \dot{y} = -y + 2\left(\frac{2}{\pi}\tan^{-1}\left(\frac{\gamma\pi y}{2}\right)\right).$$

Plot a vector field portrait and derive a suitable Lyapunov function.

Solution. In this case,

$$\mathbf{W} = \begin{pmatrix} 2 & 0 \\ 0 & 2 \end{pmatrix} \quad \text{and} \quad \mathbf{b} = \begin{pmatrix} 0 \\ 0 \end{pmatrix}.$$

A vector field plot is shown in Figure 20.11. There are four stable critical points and five unstable critical points.
A Lyapunov function is given by

$$V(\mathbf{a}) = -(a_1^2 + a_2^2) - \frac{4}{\gamma\pi^2}\left(\log\left(\cos(\pi a_1/2)\right) + \log\left(\cos(\pi a_2/2)\right)\right).$$

You can plot the Lyapunov function using Mathematica.

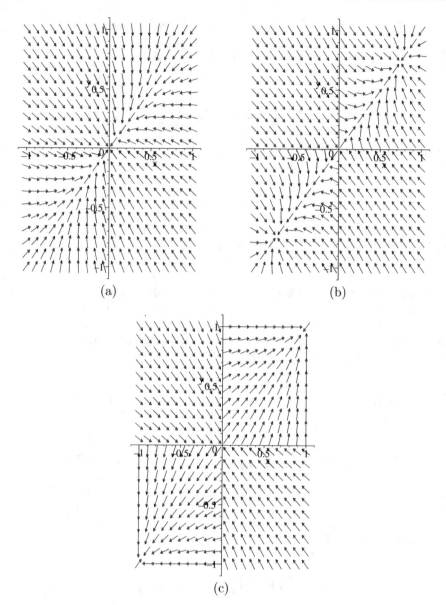

Figure 20.10: Vector field plots when (a) $0 < \gamma \le 1$, (b) $\gamma > 1$, and (c) $\gamma \to \infty$.

Continuous Hopfield networks with self-feedback loops can be Lyapunov stable. However, discrete systems must have no self-feedback to guarantee Lyapunov stability.

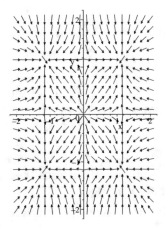

Figure 20.11: A vector field plot for Example 4 when $\gamma = 0.7$. There are nine critical points.

The Discrete Hopfield Model

Hopfield [11], [12], [13] used his network as a content-addressable memory using fixed points as attractors for certain fundamental memories. The Hopfield model can be summarized using the following four-step algorithm. There is no self-feedback in this case.

1. **Hebb's Postulate of Learning.** Let $x_1, x_2, \ldots, x_M$ denote a set of N-dimensional fundamental memories. The synaptic weights of the network are determined using the formula

$$\mathbf{W} = \frac{1}{N} \sum_{r=1}^{M} \mathbf{x}_r \mathbf{x}_r^T - \frac{M}{N} \mathbf{I}_n$$

   where $\mathbf{I}_n$ is the $N \times N$ identity matrix. Once computed, the synaptic weights remain fixed.

2. **Initialization.** Let $\mathbf{x}_p$ denote the unknown probe vector to be tested. The algorithm is initialized by setting

$$x_i(0) = x_{ip}, \quad i = 1, 2, \ldots, N,$$

   where $x_i(0)$ is the state of neuron i at time $n = 0$, x_{ip} is the ith element of vector $\mathbf{x}_p$, and N is the number of neurons.

3. **Iteration.** The elements are updated asynchronously (i.e., one at a time in a random order) according to the rule

$$x_i(n+1) = \text{hsgn} \left(\sum_{j=1}^{N} w_{ij}x_j(n) \right), i = 1, 2, \ldots, N,$$

where

$$\text{hsgn}(v_i(n+1)) = \begin{cases} 1, & v_i(n+1) > 0 \\ x_i(n), & v_i(n+1) = 0 \\ -1, & v_i(n+1) < 0 \end{cases}$$

and $v_i(n+1) = \sum_{j=1}^{N} w_{ij}x_j(n)$. The iterations are repeated until the vector converges to a stable value. Note that at least N iterations are carried out to guarantee convergence.

4. **Result.** The stable vector, say, $\mathbf{x}_{\text{fixed}}$, is the result.

The algorithm above uses asynchronous updating of synaptic weights. *Synchronous updating* is the procedure by which weights are updated simultaneously. The fundamental memories should first be presented to the Hopfield network. This tests the network's ability to recover the stored vectors using the computed synaptic weight matrix. The desired patterns should be recovered after one iteration; if not, then an error has been made. Distorted patterns or patterns missing information can then be tested using the above algorithm. There are two possible outcomes.

1. The network converges to one of the fundamental memories.

2. The network converges to a *spurious steady state*. Spurious steady states include the following:

   (a) *Reversed fundamental memories*—e.g., if $\mathbf{x}_f$ is a fundamental memory, then so is $-\mathbf{x}_f$.

   (b) *Mixed fundamental memories*—a linear combination of fundamental memories.

   (c) *Spin-glass states*—local minima not correlated with any fundamental memories.

Before looking at an application of a Hopfield network as a content-addressable memory, a simple example is shown below to illustrate the algorithm.

Example 5. A five-neuron discrete Hopfield network is required to store the following fundamental memories:

$$\mathbf{x}_1 = (1, 1, 1, 1, 1)^T, \quad \mathbf{x}_2 = (1, -1, -1, 1, -1)^T, \quad \mathbf{x}_3 = (-1, 1, -1, 1, 1)^T.$$

(a) Compute the synaptic weight matrix $\mathbf{W}$.

(b) Use asynchronous updating to show that the three fundamental memories are stable.

(c) Test the following vectors on the Hopfield network (the random orders affect the outcome):

$$\mathbf{x}_4 = (1, -1, 1, 1, 1)^T, \quad \mathbf{x}_5 = (0, 1, -1, 1, 1)^T, \quad \mathbf{x}_6 = (-1, 1, 1, 1, -1)^T.$$

Solution. (a) The synaptic weight matrix is given by

$$\mathbf{W} = \frac{1}{5} \left(\mathbf{x}_1 \mathbf{x}_1^T + \mathbf{x}_2 \mathbf{x}_2^T + \mathbf{x}_3 \mathbf{x}_3^T \right) - \frac{3}{5} \mathbf{I}_5,$$

so

$$\mathbf{W} = \frac{1}{5} \begin{pmatrix} 0 & -1 & 1 & 1 & -1 \\ -1 & 0 & 1 & 1 & 3 \\ 1 & 1 & 0 & -1 & 1 \\ 1 & 1 & -1 & 0 & 1 \\ -1 & 3 & 1 & 1 & 0 \end{pmatrix}.$$

(b) Step 1. First input vector, $\mathbf{x}_1 = \mathbf{x}(0) = (1, 1, 1, 1, 1)^T$.

Step 2. Initialize $x_1(0) = 1, x_2(0) = 1, x_3(0) = 1, x_4(0) = 1, x_5(0) = 1$.

Step 3. Update in random order $x_3(1), x_4(1), x_1(1), x_5(1), x_2(1)$, one at a time.

$$x_3(1) = \mathrm{hsgn}(0.4) = 1,$$
$$x_4(1) = \mathrm{hsgn}(0.4) = 1,$$
$$x_1(1) = \mathrm{hsgn}(0) = x_1(0) = 1,$$
$$x_5(1) = \mathrm{hsgn}(0.8) = 1,$$
$$x_2(1) = \mathrm{hsgn}(0.8) = 1.$$

Thus $\mathbf{x}(1) = \mathbf{x}(0)$ and the net has converged.

Step 4. The net has converged to the steady state $\mathbf{x}_1$.

Step 1. Second input vector, $\mathbf{x}_2 = \mathbf{x}(0) = (1, -1, -1, 1, -1)^T$.

Step 2. Initialize $x_1(0) = 1, x_2(0) = -1, x_3(0) = -1, x_4(0) = 1, x_5(0) = -1$.

Step 3. Update in random order $x_5(1), x_3(1), x_4(1), x_1(1), x_2(1)$, one at a time.

$$x_5(1) = \text{hsgn}(-0.8) = -1,$$
$$x_3(1) = \text{hsgn}(-0.4) = -1,$$
$$x_4(1) = \text{hsgn}(0) = x_4(0) = 1,$$
$$x_1(1) = \text{hsgn}(0.4) = 1,$$
$$x_2(1) = \text{hsgn}(-0.8) = -1.$$

Thus $\mathbf{x}(1) = \mathbf{x}(0)$ and the net has converged.

Step 4. The net has converged to the steady state $\mathbf{x}_2$.

Step 1. Third input vector, $\mathbf{x}_3 = \mathbf{x}(0) = (-1, 1, -1, 1, 1)^T$.

Step 2. Initialize $x_1(0) = -1, x_2(0) = 1, x_3(0) = -1, x_4(0) = 1, x_5(0) = 1$.

Step 3. Update in random order $x_5(1), x_1(1), x_4(1), x_2(1), x_3(1)$, one at a time.

$$x_5(1) = \text{hsgn}(0.8) = 1,$$
$$x_1(1) = \text{hsgn}(-0.4) = -1,$$
$$x_4(1) = \text{hsgn}(0.4) = 1,$$
$$x_2(1) = \text{hsgn}(0.8) = 1,$$
$$x_3(1) = \text{hsgn}(0) = x_3(0) = -1.$$

Thus $\mathbf{x}(1) = \mathbf{x}(0)$ and the net has converged.

Step 4. The net has converged to the steady state $\mathbf{x}_3$.

(c) Step 1. Fourth input vector, $\mathbf{x}_4 = \mathbf{x}(0) = (1, -1, 1, 1, 1)^T$.

Step 2. Initialize $x_1(0) = 1, x_2(0) = -1, x_3(0) = 1, x_4(0) = 1, x_5(0) = 1$.

Step 3. Update in random order $x_2(1), x_4(1), x_3(1), x_5(1), x_1(1)$, one at a time.

$$x_2(1) = \text{hsgn}(0.8) = 1,$$
$$x_4(1) = \text{hsgn}(0.4) = 1,$$
$$x_3(1) = \text{hsgn}(0.4) = 1,$$
$$x_5(1) = \text{hsgn}(0.8) = 1,$$
$$x_1(1) = \text{hsgn}(0) = x_1(0) = 1.$$

Thus $\mathbf{x}(1) = \mathbf{x}_1$ and the net has converged.

Step 4. The net has converged to the steady state $\mathbf{x}_1$.

Step 1. Fifth input vector, $\mathbf{x}_5 = \mathbf{x}(0) = (0, 1, -1, 1, 1)^T$, information is missing in the first row.

Step 2. Initialize $x_1(0) = 0, x_2(0) = 1, x_3(0) = -1, x_4(0) = 1, x_5(0) = 1$.

Step 3. Update in random order $x_4(1), x_5(1), x_1(1), x_2(1), x_3(1)$, one at a time.

$$x_4(1) = \text{hsgn}(0.6) = 1,$$
$$x_5(1) = \text{hsgn}(0.6) = 1,$$
$$x_1(1) = \text{hsgn}(-0.4) = -1,$$
$$x_2(1) = \text{hsgn}(0.8) = 1,$$
$$x_3(1) = \text{hsgn}(0) = x_3(0) = -1.$$

Thus $\mathbf{x}(1) = \mathbf{x}_3$ and the net has converged.

Step 4. The net has converged to the steady state $\mathbf{x}_3$.

Step 1. Sixth input vector, $\mathbf{x}_6 = \mathbf{x}(0) = (-1, 1, 1, 1, -1)^T$.

Step 2. Initialize $x_1(0) = -1, x_2(0) = 1, x_3(0) = 1, x_4(0) = 1, x_5(0) = -1$.

Step 3. Update in random order $x_3(1), x_2(1), x_5(1), x_4(1), x_1(1)$, one at a time.

$$x_3(1) = \text{hsgn}(-0.4) = -1,$$
$$x_2(1) = \text{hsgn}(-0.4) = -1,$$
$$x_5(1) = \text{hsgn}(-0.4) = -1,$$
$$x_4(1) = \text{hsgn}(-0.4) = -1,$$
$$x_1(1) = \text{hsgn}(0) = x_1(0) = -1.$$

Step 3 (again). Update in random order $x_2(1), x_1(1), x_5(1), x_4(1), x_3(1)$, one at a time.

$$x_2(2) = \text{hsgn}(-0.8) = -1,$$
$$x_1(2) = \text{hsgn}(0) = x_1(1) = -1,$$
$$x_5(2) = \text{hsgn}(-0.8) = -1,$$
$$x_4(2) = \text{hsgn}(-0.4) = -1,$$
$$x_3(2) = \text{hsgn}(-0.4) = -1.$$

Thus $\mathbf{x}(2) = \mathbf{x}(1)$ and the net has converged.

Step 4. The net has converged to the spurious steady state $-\mathbf{x}_1$.

Example 6. Write a Mathematica program that illustrates the behavior of the discrete Hopfield network as a content-addressable memory using $N = 81$ neurons and the set of handcrafted patterns displayed in Figure 20.12.

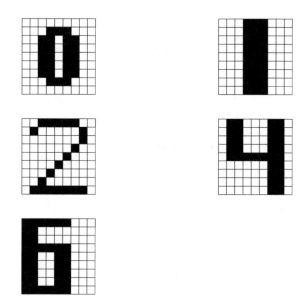

Figure 20.12: Patterns to be used as fundamental memories for the discrete Hopfield model.

Solution. See the program listed in Section 20.6 as a guide. Set a noise level to $\frac{1}{3}$. On average, the network will converge after $\frac{1}{3} \times 81 = 27$ iterations. In order for this algorithm to work, the vectors defining the patterns have to be as orthogonal as possible. If some patterns are similar, the network will not perform very well.

20.4 Neurodynamics

It is now understood that chaos, oscillations, synchronization effects, wave patterns, and feedback are present in higher-level brain functions and on different levels of signal processing. In recent years, the disciplines of neuroscience and nonlinear dynamics have increasingly coalesced, leading to a new branch of science called *neurodynamics*. This section will concentrate on a minimal chaotic *neouromodule*, studied in some detail by Pasemann and his group [19] and [20]. They have considered chaos control and synchronization effects for this simple model.

A Minimal Chaotic Neuromodule

The discrete two-dimensional system investigated by Pasemann is defined by the map

$$x_{n+1} = b_1 + w_{11}\phi_1(x_n) + w_{12}\phi_2(y_n), \quad y_{n+1} = b_2 + w_{21}\phi_1(x_n) + w_{22}\phi_2(y_n),$$
(20.7)

where its activity at time n is given by (x_n, y_n), b_1, b_2 are biases, w_{ij} are the synaptic weights connecting neurons, and ϕ represents the transfer function defined by

$$\phi_1(x) = \phi_2(x) = \frac{1}{1 + e^{-x}}.$$
(20.8)

The simple network architecture of this recurrent module with an excitatory neuron and an inhibitory neuron with self-connection is shown in Figure 20.9. Pasemann and Stollenwerk (see reference in Chapter 14) considered the model with the following parameter values

$$b_1 = -2, b_2 = 3, w_{11} = -20, w_{21} = -6, w_{12} = 6, \text{ and } w_{22} = 0.$$
(20.9)

Figure 20.13 shows the chaotic attractor for system (20.7) using the transfer function in equation (20.8) and the parameters listed in (20.9).

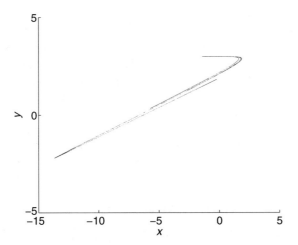

Figure 20.13: [Mathematica] Chaotic attractor for a minimal chaotic neuro-module.

The fixed points of periods one and two may be found in the usual way. Fixed points of period one satisfy the simultaneous equations $x_{n+1} = x_n = x$, and $y_{n+1} = y_n = y$. There is one fixed point of period one at $P_{11} = (-1.2804, 1.6951)$, working to four decimal places. The stability of this fixed point is determined by considering the eigenvalues of the Jacobian matrix given by

$$J = \begin{pmatrix} w_{11}\frac{\partial}{\partial x}\phi_1(x) & w_{12}\frac{\partial}{\partial y}\phi_2(y) \\ w_{21}\frac{\partial}{\partial x}\phi_1(x) & 0 \end{pmatrix}.$$

The eigenvalues for the fixed point of period one are given by $\lambda_1 = -3.1487, \lambda_2 = -0.2550$, and the fixed point is a saddle point. Hence P_{11} is unstable.

The fixed points of period two are found by solving the equations $x_{n+2} = x_n = x$ and $y_{n+2} = y_n = y$, which has two solutions at $P_{21} = (-7.8262, -0.4623)$ and $P_{22} = (0.3107, 2.9976)$. These fixed points are also unstable.

A Bistable Neuromodule

As with many nonlinear dynamical systems, higher-level brain functions can be subject to feedback. The author and Bandar have investigated system (20.7) with the following choice of parameters

$$b_1 = 2, b_2 = 3, w_{11} = 7, w_{21} = 5, w_{12} = -4, \text{ and } w_{22} = 0, \tag{20.10}$$

and using the transfer functions

$$\phi_1(x) = \tanh(ax) \text{ and } \phi_2(y) = \tanh(\alpha y), \tag{20.11}$$

with $a = 1$ and $\alpha = 0.3$. Using numerical techniques, there are three fixed points of period one at $P_{11} = (-2.8331, -1.9655)$, $P_{12} = (0.2371, 4.1638)$, and $P_{13} = (5.0648, 7.9996)$. Using the Jacobian matrix, point P_{11} has eigenvalues $\lambda_1 = 0.0481 + 0.2388i$, $\lambda_2 = 0.0481 - 0.2020i$. The fixed point is stable since $|\lambda_1| < 1$ and $|\lambda_2| < 1$. Points P_{12} and P_{13} have eigenvalues $\lambda_1 = 6.3706$, $\lambda_2 = 0.2502$ and $\lambda_1 = 0.0006 + 0.0055i$, $\lambda_2 = 0.0006 - 0.0055i$, respectively. Therefore, point P_{12} is an unstable saddle point, and point P_{13} is stable, since both eigenvalues have modulus less than one. We conclude that system (20.7) with the parameter values given above (20.10) and the transfer functions defined by equations (20.11) is multistable. That is, there are two stable fixed points for one set of parameter values, and the fixed point attained is solely dependent upon the initial conditions chosen.

Now introduce a feedback mechanism. In the first case, we vary the parameter α, which determines the gradient of the transfer function $\phi_2(y)$. The other parameters are fixed as above (20.10). The parameter α is increased linearly from $\alpha = -5$ to $\alpha = 5$ and then decreased back down to $\alpha = -5$. Figure 20.14 shows the bifurcation diagrams for the activity of neuron x. Similar bifurcation diagrams may be plotted for the neuron y. The upper figure shows the activity against the number of iterations. The lower figure shows the activity level of neuron x as the parameter α is increased then decreased. As α is increased from -5, the steady state is on the lower branch until $\alpha \approx 1$, where there is a sudden jump to the other steady state. As α increases further, the steady state remains at $x_n \approx 5$. As α is decreased, the steady state remains at $x_n \approx 5$ until $\alpha \approx 0$ where it jumps to $x_n \approx 15$. There is a large bistable region for $-5 < \alpha < 1$, approximately.

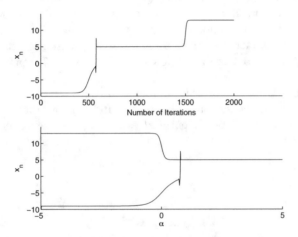

Figure 20.14: Bifurcation diagrams for system (20.7) under conditions (20.10) and (20.11) as α varies. The initial conditions chosen at $\alpha = -5$ were $x_0 = -10$ and $y_0 = -3$.

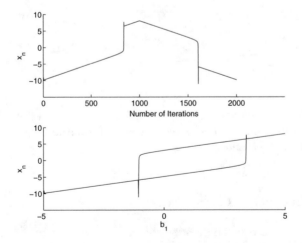

Figure 20.15: Bifurcation diagrams for system (20.7) under conditions (20.10) and (20.11) as b_1 varies. The initial conditions chosen at $b_1 = -5$ were $x_0 = -10$ and $y_0 = -3$.

In the second case, fix the parameters and vary b_1, which is the bias for neuron x. The parameter b_1 is ramped up from $b_1 = -5$ to $b_1 = 5$ and then ramped back down to $b_1 = -5$. There is an isolated counterclockwise bistable region for $-1 < b_1 < 3.5$, approximately. Note the ringing at both ends of the bistable region; see Figure 20.15.

In the final case, fix the parameters and vary w_{11}, which is the synaptic weight connecting neuron x to itself. The parameter is decreased from $w_{11} = 7$ down to zero and then increased back up to $w_{11} = 7$. The activity of neuron x is on the lower branch until $w_{11} \approx 5.5$, where it jumps to the upper branch. As w_{11} decreases, the system descends into regions of quasiperiodicity and periodicity. As the parameter is increased from zero, the steady state remains on the upper branch, and there is a bistable region for $5.5 < w_{11} < 7$, approximately; see Figure 20.16.

Clearly, the dynamics of this simple two-neuron module are dependent upon the history of the system. The author and his coworkers at Manchester Metropolitan University are currently investigating areas of application for this research.

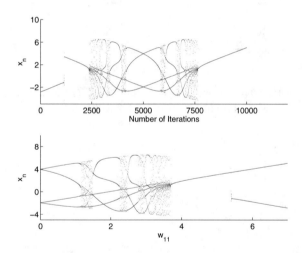

Figure 20.16: [Mathematica] Bifurcation diagrams for system (20.7) under conditions (20.10) and (20.11) as w_{11} varies. The initial conditions chosen at $w_{11} = 7$ were $x_0 = -3$ and $y_0 = -2$.

20.5 Mathematica Commands

Using the Help Browser, the reader should type Import to see how data are loaded in Mathematica.

```
(* Generalized Delta Learning Rule, see Figure 20.7 *)
(* N.B. ENSURE THAT housing.txt IS IN THE CORRECT DIRECTORY. *)
(* EXPORT A FILE TO SEE WHERE IT IS EXPORTED TO! *)

X=Import["housing.txt","Data"];
t=X[[All,14]];
tmean=(Max[t]+Min[t])/2;
tstd=(Max[t]-Min[t])/2;
t=N[(t-tmean)/tstd];
Y={X[[All,6]],X[[All,9]],X[[All,13]]};
Y=Transpose[Y];
MnX=List[Mean[Y]];
Std=List[StandardDeviation[Y]];
Dim=Part[Dimensions[Y],1];
ones=Table[{1},{i,Dim}];
X=N[(Y-ones.MnX)/ones.Std];
X=Transpose[X];
X=Prepend[X,Flatten[Table[1,{i,Dim}]]];
SeedRandom[123456];
ww[1]=0.1 Table[{Random[]},{4}];
Epochs=10;
eta=0.001;k=1;
For[n=1,n<=Epochs,n++,For[j=1,j<=Dim,j++,
yk=Flatten[Tanh[List[X[[All,j]]].ww[k]]];
err=yk-t[[j]];g=Transpose[List[X[[All,j]]]].((1-yk.yk) err);
ww[k+1]=ww[k]-eta g;k=k+1]];
For[i=1,i<=4,i++,pts[i]=Table[Flatten[ww[j]][[i]],{j,1,k}]]
For[i=1,i<=4,i++,p[i]=ListPlot[pts[i],PlotStyle->PointSize[0.001]]]
Show[{p[1],p[2],p[3],p[4]},AxesLabel->{"Iterations","Weights"},
PlotRange->All]
```

```
(* Example 5 The Discrete Hopfield Network *)
X={{1,1,1,1,1},{1,-1,-1,1,-1},{-1,1,-1,1,1}};
W=Transpose[X].X/5-3 IdentityMatrix[5]/5
{{0,-(1/5),1/5,1/5,-(1/5)},{-(1/5),0,1/5,1/5,3/5},{1/5,1/5,0,-(1/5),1/5},
{1/5,1/5,-(1/5),0,1/5},{-(1/5),3/5,1/5,1/5,0}}
RandomRelist[x_List]:=Block[{n=x,p},Do[p=Random[Integer,{1,i}];
n[[{p,i}]]=n[[{i,p}]],{i,Length[x]}];n];
n=RandomRelist[{1,2,3,4,5}]

xinput={-1,-1,1,1,1};
xtest=xinput;
hsgn[{v_,x_}]=Which[v>0,1,v==0,x,v<0,-1];
For[j=1,j<=5,j++,xtest[[n[[j]]]]=hsgn[{W[[n[[j]]]].xtest,xtest[[n[[j]]]]}]];
If[xtest==X[[1]],Print["Net has converged to X1"],If[xtest==X[[2]],
Print["Net has converged to X2"],If[xtest==X[[3]],
Print["Net has converged to X3"],
Print["Iterate again: May have converged to a spurious steady state"]]]]
```

"Net has converged to X3"

```
(* The Minimal Chaotic Neuromodule, see Figure 20.8. *)

b1=-2;b2=3;w11=-20;w21=-6;w12=6;
Itermax=10000;
Neuromodule[{x_,y_}]={b1+w11/(1+Exp[-x])+w12/(1+Exp[-y]),b2+w21/(1+Exp[-x])};
NeuroMap=N[Drop[NestList[Neuromodule,{1,0.2},Itermax],50]];
ListPlot[NeuroMap,AxesLabel->{"x","y"},PlotStyle->{PointSize[0.001]},
Axes->None,Frame->True]
```

```
(* Bifurcation Diagram for a Bistable Neuromodule, see Figure 20.15 *)
Start=-5;
Maximum=10;b2=3;w11=7;w12=-4;w21=5;
HalfN=4999;phi=0;N1=1+HalfN;Itermax=2 HalfN+1;x[1]=-10;y[1]=-3;

For[n=1,n<=HalfN,n++,{b1=(Start+n Maximum/HalfN),
x[n+1]=N[b1+w11 Tanh[x[n]]+w12 Tanh[0.3 y[n]]],
y[n+1]=N[b2+w21 Tanh[x[n]]]}];
p1=ListPlot[Table[Flatten[{Start+n*Maximum/N1,x[n]}],{n,1,HalfN}],
PlotStyle->{PointSize[0.001],RGBColor[0,1,0]}];
For[n=N1,n<=Itermax,n++,{b1=(Start+2 Maximum-n Maximum/HalfN),
x[n+1]=N[b1+w11 Tanh[x[n]]+w12 Tanh[0.3 y[n]]],
y[n+1]=N[b2+w21 Tanh[x[n]]]}];
p2=ListPlot[Table[Flatten[{Start+n*Maximum/N1,x[Itermax+1-n]}],
{n,1,HalfN}],
PlotStyle->{PointSize[0.001],RGBColor[0,1,0]}];
Show[{p1,p2},PlotRange->All,AxesLabel->{"Subscript[b,1]","Subscript[x,n]"},
Frame->True,TextStyle->{FontSize->15},AspectRatio->Automatic]
```

20.6 Exercises

1. For the following activation functions, show that

   (a) if $\phi(v) = 1/(1 + e^{-av})$, then $\phi'(v) = a\phi(v)(1 - \phi(v))$;

   (b) if $\phi(v) = a\tanh(bv)$, then $\phi'(v) = \frac{b}{a}(a^2 - \phi^2(v))$;

   (c) if $\phi(v) = \frac{1}{2a} \log \frac{\cosh(a(v+1))}{\cosh(a(v-1))}$, then

   $$\phi'(v) = (\tanh(a(v+1)) - \tanh(a(v-1)))/2.$$

2. Prove Theorem 2, showing that when the activation function is nonlinear, say, $y_k = \phi(v_k)$, the generalized delta rule can be formulated as

   $$w_{kj}(n+1) = w_{kj}(n) - \eta g_{kj},$$

where

$$g_{kj} = (y_k - t_k) \frac{\partial \phi}{\partial v_k} x_j.$$

3. By editing the programs listed in Section 20.5:

   (a) Investigate what happens to the mean squared error for varying eta values of your choice.

   (b) Investigate what happens to the mean squared error as the number of hidden neurons increases to five.

4. Use another data set of your choice from the URL

   http://www.ics.uci.edu/~mlearn/MLRepository.html

   using an edited version of the programs listed in Section 20.5 to carry out your analysis.

5. (a) Prove Theorem 3 regarding Lyapunov functions of continuous Hopfield models.

   (b) Consider the recurrent Hopfield network modeled using the differential equations

   $$\dot{x} = -x + 7 \left(\frac{2}{\pi} \tan^{-1} \left(\frac{\gamma \pi x}{2} \right) \right) + 6 \left(\frac{2}{\pi} \tan^{-1} \left(\frac{\gamma \pi y}{2} \right) \right),$$

   $$\dot{y} = -y + 6 \left(\frac{2}{\pi} \tan^{-1} \left(\frac{\gamma \pi x}{2} \right) \right) - 2 \left(\frac{2}{\pi} \tan^{-1} \left(\frac{\gamma \pi y}{2} \right) \right).$$

   Plot a vector field portrait and derive a suitable Lyapunov function.

   (c) Plot surface plots for the Lyapunov functions for Examples 3 and 4 and Exercise 5(b). Plot the surfaces for $|a_i| \leq 1, i = 1, 2$.

6. Consider the discrete Hopfield model investigated in Example 5. Test the vector $\mathbf{x}_7 = (-1, -1, 1, 1, 1)^T$, update in the following orders, and determine to which vector the algorithm converges:

   (a) $x_3(1), x_4(1), x_5(1), x_2(1), x_1(1)$;

   (b) $x_1(1), x_4(1), x_3(1), x_2(1), x_5(1)$;

   (c) $x_5(1), x_3(1), x_2(1), x_1(1), x_4(1)$;

   (d) $x_3(1), x_5(1), x_2(1), x_4(1), x_1(1)$.

7. Add suitable characters '3' and '5' to the fundamental memories shown in Figure 20.12. You may need to increase the grids to 10×10 and work with 100 neurons.

8. A simple model of a neuron with self-interaction is described by Pasemann [19]. The difference equation is given by

$$a_{n+1} = \gamma a_n + \theta + w\sigma(a_n), \quad 0 \le \gamma < 1,$$

where a_n is the activation level of the neuron, θ is a bias, w is a self-weight, γ represents dissipation in a neuron, and the output is given by the sigmoidal transfer function

$$\sigma(x) = \frac{1}{1 + e^{-x}}.$$

(a) Determine an equation for the fixed points of period one and show that the stability condition is given by $|\gamma + w\sigma'(a)| < 1$, where a is a fixed point of period one.

(b) Show that the system is bistable in the region bounded by the parametric equations:

$$\theta(a) = (1 - \gamma)a - \frac{(1 - \gamma)}{(1 - \sigma(a))}, \quad w(a) = \frac{(1 - \gamma)}{\sigma'(a)}.$$

(c) Show that the system is unstable in the region bounded by the parametric equations:

$$\theta(a) = (1 - \gamma)a + \frac{(1 + \gamma)}{(1 - \sigma(a))}, \quad w(a) = -\frac{(1 + \gamma)}{\sigma'(a)}.$$

(d) Use the first iterative method to plot a bifurcation diagram when $\theta = 4$ and $w = -16$ for $0 < \gamma < 1$.

(e) Use the second iterative method to plot a bifurcation diagram when $\theta = -2.4$ and $\gamma = 0$ for $3 < w < 7$. Ramp w up and down.

9. Consider the neuromodule defined by the equations

$$x_{n+1} = 2 + 3.5 \tanh(x) - 4 \tanh(0.3y), \quad y_{n+1} = 3 + 5 \tanh(x).$$

Iterate the system and show that it is quasiperiodic.

10. Use the OGY method to control chaos in the minimal chaotic neuromodule.

Bibliography

[1] Z. G. Bandar, D. A. McLean, J. D. O'Shea, and J. A. Rothwell, Analysis of the behaviour of a subject, *International Publication Number* WO 02/087443 A1, (2002).

[2] P. Dayan and L. F. Abbott, *Theoretical Neuroscience: Computational and Mathematical Modeling of Neural Systems*, MIT Press, Cambridge, MA, 2001.

[3] C. Dawson (Editor), *Applied Artificial Neural Network*, MDPI AG, Basel, 2016.

[4] M. Di Marco, M. Forti, and A. Tesi, Existence and characterization of limit cycles in nearly symmetric neural networks. *IEEE Trans. circuits & systems-1: Fundamental theory and applications*, **49** (2002), 979–992.

[5] W. J. Freeman, *Neurodynamics: An Exploration in Mesoscopic Brain Dynamics (Perspectives in Neural Computing)*, Springer-Verlag, New York, 2000.

[6] M. T. Hagan, H. B. Demuth, and M.H. Beale, *Neural Network Design*, Brooks-Cole, Pacific Grove, CA, 1995.

[7] H. Haken, *Brain Dynamics: An Introduction to Models and Simulations*, Springer-Verlag, New York, 2008.

[8] S. O. Haykin, *Neural Networks and Learning Machines*, 3rd ed. Prentice-Hall, Upper Saddle River, NJ, 2008.

[9] J. Heaton, *Introduction to the Math of Neural Networks (Kindle Edition)*, Heaton Research Inc., 2012.

[10] D. O. Hebb, *The Organization of Behaviour*, John Wiley, New York, 1949.

[11] J. J. Hopfield and D. W. Tank, Neural computation of decisions in optimization problems, *Biological Cybernetics*, **52** (1985), 141–154.

[12] J. J. Hopfield, Neurons with graded response have collective computational properties like those of two-state neurons, *Proc. National Academy of Sciences*, **81** (1984), 3088–3092.

[13] J. J. Hopfield, Neural networks and physical systems with emergent collective computational abilities. *Proc. National Academy of Sciences*, **79** (1982), 2554–2558.

[14] E.M. Izhikevich, *Dynamical Systems in Neuroscience: The Geometry of Excitability and Bursting (Computational Neuroscience)*, MIT Press, 2006.

[15] T. Kohonen, Self-organized formation of topologically correct feature maps, *Biological Cybernetics*, **43** (1982), 59–69.

[16] B. Kosko, *Neural Networks and Fuzzy Systems: A Dynamical Systems Approach to Machine Intelligence*, Prentice-Hall, Upper Saddle River, NJ, 1999.

[17] W. McCulloch and W. Pitts, A logical calculus of the ideas immanent in nervous activity, *Bulletin of Mathematical Biophysics*, **5** (1943), 115–133.

[18] M. Minsky and S. Papert, *Perceptrons*, MIT Press, Cambridge, MA, 1969.

[19] F. Pasemann, Driving neuromodules into synchronous chaos, *Lecture Notes in Computer Science*, **1606** (1999), 377–384.

[20] F. Pasemann, A simple chaotic neuron, *Physica D*, **104** (1997), 205–211.

[21] T. Rashid, *Make Your Own Neural Network*, CreateSpace Independent Publishing Platform, 2016.

[22] F. Rosenblatt, The perceptron: A probabalistic model for information storage and organization in the brain, *Psychological Review*, **65** (1958), 386–408.

[23] J. A. Rothwell, The word liar, *New Scientist*, March (2003), 51.

[24] D. E. Rumelhart and J. L. McClelland, eds., *Parallel Distributed Processing: Explorations in the Microstructure of Cognition*, Cambridge, MA: MIT Press, **1**, 1986.

[25] I. W. Sandberg (ed.), J. T. Lo, C. L. Fancourt, J. Principe, S. Haykin, and S. Katargi, *Nonlinear Dynamical Systems: Feedforward Neural Network Perspectives (Adaptive Learning Systems to Signal Processing, Communications and Control)*, Wiley-Interscience, New York, 2001.

[26] S. Samarasinghe, *Neural Networks for Applied Sciences and Engineering*, Auerbach, 2006.

[27] S. J. Schiff, K. Jerger, D. H. Doung, T. Chang, M. L. Spano, and W. L. Ditto, Controlling chaos in the brain, *Nature*, **370** (1994), 615.

[28] A.M.F. Souza and F.M. Soares, *Neural Network Programming with Java*, Packt Publishing, 2016.

[29] B. Widrow and M. E. Hoff, Adaptive switching circuits, 1960 IRE WESCON Convention Record, New York, IRE Part 4 (1960), 96–104.

Chapter 21
Binary Oscillator Computing

Aims and Objectives

- To provide a brief historical introduction to binary oscillator computing.

- To review basic operations of neurons.

- To introduce threshold oscillatory logic and memory.

On completion of this chapter, the reader should be able to

- perform simple binary logic operations using threshold oscillators;

- plot time series data to illustrate the functionality;

- appreciate the potential applications of the devices in the real world.

The work presented in this chapter is inspired by brain dynamics and has led to the submission of International, UK and Taiwanese patents [15, 16, 18]. The author and co-inventor Jon Borresen are currently working with collaborators toward building super fast binary oscillator computers as well as assays for electrochemical cell degradation.

21.1 Brain Inspired Computing

As with neural networks, the subject of Chapter 20, the main ideas in this chapter are inspired by biological brain dynamics which will now be briefly discussed for completeness. Figure 21.1 shows a schematic of a neuron which is

© Springer International Publishing AG 2017   485
S. Lynch, *Dynamical Systems with Applications using Mathematica®*,
DOI 10.1007/978-3-319-61485-4_21

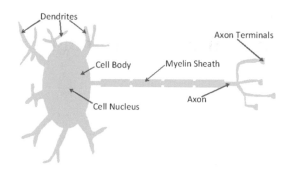

Figure 21.1: Schematic of a neuron. Notice how this figure is similar to that shown in Figure 20.1 for the neuronal mathematical model.

comprised of typical parts of cells with a few specialized structures that make it unique. The main part of the cell is the cell body (or soma) which contains the cell nucleus comprising genetic material in the form of chromosomes. Dendrites branch out from the cell body, and it is primarily these spikes that receive chemical signals from other neurons. If the neuron fires, an electrochemical signal is transmitted along the axon to the axon terminals. Note that longer axons are usually covered with a myelin sheath that act in a similar manner to insulation around an electrical wire. In order for signals (or action potentials) to be transmitted from neuron to neuron, between the axon terminal and the dendrite of a connecting neuron there exists a very tiny membrane junction or gap called the synaptic gap (or cleft). As the signal reaches the axon terminal, tiny bubbles of chemicals called synaptic vesicles release their contents which diffuse across the gap to bind with specific receptors in the membrane of the adjoining neuron. The endogeneuos chemicals transmitted are called neurotransmitters which may be excitatory or inhibitory. Examples of excitatory neurotransmitters include glutamate (the most prevalent neurotransmitter in the brain), acetylcholine, aspartate, histamine, and noradrenaline, while GABA (γ-aminobutyric acid, the second most prevalent neurotransmitter in the brain), glycine, and seratonin are inhibitory. Among the many neurotransmitters, note that certain neurotransmitters such as acetylcholine and dopamine have both excitatory and inhibitory receptors, so it is an oversimplification to label them in this way. It has been estimated that the typical human brain has approximately 80% excitatory and 20% inhibitory neurotransmitters.

A neuron's membrane forms a barrier between the extracellular space around the cell and its intracellular fluid and is selectively permeable to ions such as sodium (Na^+), potassium (K^+), and chlorine (Cl^-). It is mostly permeable to K^+ ions, less so to Cl^- ions, and a lot less to Na^+ ions. The voltage difference between the extracellular and intracellular spaces is typically between $-60mV$ and $-80mV$ for a neuron in a resting state. If a stimulus causes the membrane potential to reach $-50mV$ or above, then an action potential develops. A depolarization occurs whereby the Na^+ channels open and Na^+ begins to enter the cell, further depolarizing the cell. At the end of the depolarization phase, the Na^+ channels become refractory and no more Na^+ ions enter the cell. The K^+ channels are then activated and K^+ ions start to leave the cell, a process called repolarization, and the membrane potential falls below the level of the resting potential where the membrane is actually hyperpolarized. The K^+ channels close and the Na^+ channels reset, while the extra K^+ ions in the extracellular space diffuse away and the resting membrane potential is finally re-established. If the stimulus remains, then a series of action potentials (known as a spike train) is generated as shown in Figure 21.2.

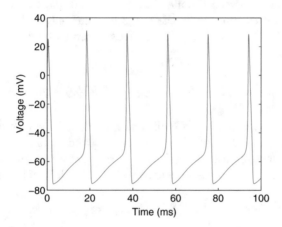

Figure 21.2: [Mathematica] Spike train of action potentials that travel down the axon. At the beginning of the action potential, the Na^+ channels open and Na^+ ions move into the axon causing depolarization. Repolarization occurs when the K^+ channels open and K^+ ions move out of the axon. The signal travels down the axon to the axon terminal (see Figure 21.1) where it can trigger other neurons.

In the simplest sense, neurons are either firing or not firing. Once the neuron has been sufficiently excited above some threshold (typically $-55\,\mathrm{mV}$), the cell fires, if the neuron does not reach this threshold, it will not depolarize or create an action potential. If the stimulus does not reach threshold, then the neuron does not fire. As the stimulus passes the threshold value and continues to rise, the neuron starts to fire and the amplitude of oscillation remains constant, hence the All or None principle of neuron firing. Note, however, as the stimulus increases (up to a limit), the frequency of oscillation increases. The reader can verify this by attempting one of the exercises in Section 21.5.

The Hodgkin–Huxley Equations. In 1952, Alan Lloyd Hodgkin and Andrew Huxley were modeling the ionic mechanisms underlying the initiation and propagation of action potentials in the giant squid axon [13]. By treating each component of the excitable cell as an electrical element, and applying the conservation of electric charge on a piece of membrane, they were able to derive the following equation for membrane current density:

$$I = C\frac{dV}{dt} + I_{Na} + I_K + I_L, \tag{21.1}$$

where I is the total membrane current density, C is the membrane capacitance, V is the difference between the membrane potential and the resting potential, I_{Na} is the sodium current, I_K is the potassium current, and I_L is the leakage current. Using Ohm's law, Hodgkin and Huxley were able to expand equation (21.1) to give:

$$C\frac{dV}{dt} = I - g_{Na}m^3h\left(V - V_{Na}\right) - g_Kn^4\left(V - V_K\right) - g_L\left(V - V_L\right), \tag{21.2}$$

where V_{Na}, V_K, V_L, C, and g_L are all constants determined from experimental data, and g_{Na} and g_K are both functions of time and membrane potential. The three dimensionless quantities m, h, and n represent sodium, potassium, and leakage gating variables and evolve according to the differential equations:

$$\begin{aligned} \frac{dm}{dt} &= \alpha_m(1-m) - \beta_m m \\ \frac{dh}{dt} &= \alpha_h(1-h) - \beta_h h \\ \frac{dn}{dt} &= \alpha_n(1-n) - \beta_n n, \end{aligned} \tag{21.3}$$

where α_i and β_i are the transition rate constants for the i-th ion channel. The individual gates act in a similar manner to first-order chemical reactions with two states. The rate constant α_i represents the number of times per second that a shut gate opens, and similarly, β_i represents the number of times per second that an open gate shuts. Based on experimental data, the following parameter values have been chosen to generate Figures 21.2 and 21.3:

$$\alpha_m = \frac{0.1(V + 40)}{1 - \exp(-0.1(V + 40))}, \quad \beta_m = 4\exp(-0.0556(V + 65)),$$

$$\alpha_h = 0.07\exp(-0.05(V + 65)), \quad \beta_h = \frac{1}{1 + \exp(-0.1(V + 35))},$$

$$\alpha_n = \frac{0.01(V + 55)}{1 - exp(-0.1(V + 55))}, \quad \beta_n = 0.125\exp(-0.0125(V + 65)), \tag{21.4}$$

and additionally,

$$C = 1 \, \mu \text{Fcm}^{-2},$$

$$g_L = 0.3 \, \text{mmhocm}^{-2}, g_K = 36 \, \text{mmhocm}^{-2}, g_{Na} = 120 \, \text{mmhocm}^{-2},$$

$$V_L = -54.402 \, \text{mV}, V_K = -77 \, \text{mV}, V_{Na} = 50 \, \text{mV}. \tag{21.5}$$

The Mathematica program for producing Figures 21.2 and 21.3 is listed in Section 21.4.

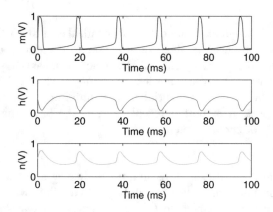

Figure 21.3: [Mathematica] The gating variables m, h, and n for equations and parameters listed in (21.2), (21.3), (21.4), and (21.5).

In 1994, Destexhe et al. [8] derived an efficient method for computing synaptic conductances based on chemical kinetics. Figure 21.4 shows the results of modeling chemical excitation and inhibition. In Figure 21.4(a), the upper green trace is the action potential of an excitatory neuron, the middle magenta trace depicts the ratio of excitatory conductance, and the lower blue curve shows that the post synaptic neuron is firing. In Figure 21.4(b), the upper red trace is the action potential of an inhibitory neuron, the middle magenta trace depicts the ratio of excitatory conductance, and the lower blue curve shows that the post synaptic neuron is switched off. Readers can reproduce these results in Mathematica.

In order to simplify the work to follow in this chapter the Fitzhugh–Nagumo system [11, 23], which is essentially a reduction of the Hodgkin–Huxley equations [13], will be used to model the action potential of a spiking neuron. The describing equations are:

$$\dot{v} = C + v(v - \alpha)(1 - v) - w, \quad \dot{w} = \epsilon(v - \gamma w), \tag{21.6}$$

where v is a fast variable (in biological terms - the action potential) and w represents a slow variable (biologically - the sodium gating variable). The parameters

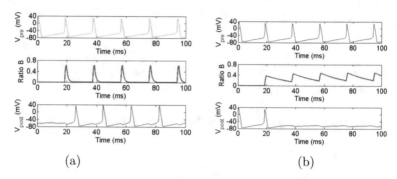

(a)   (b)

Figure 21.4:   (a) Mathematical modeling of chemical excitation. The green trace depicts the action potential in the excitatory neuron, the magenta trace is the proportion of bound receptors, and the blue trace is the action potential of the post synaptic neuron. (b) Mathematical modeling of chemical inhibition. The red trace depicts the action potential in the inhibitory neuron, the magenta trace is the proportion of bound receptors, and the blue trace is the action potential of the post synaptic neuron.

α, γ, and ϵ dictate the threshold, oscillatory frequency and the location of the critical points for v and w. A neuron will begin to oscillate when the input current C is above a critical threshold C_T, say. Figure 21.5 shows a typical phase portrait and time series solution for the integrate and fire neuron when the critical point is at the origin. When the input current is below the threshold C_T, the solution quickly settles to the stable critical point at the origin and there is no oscillation. When the input current exceeds the threshold then the neuron displays oscillatory behavior as in Figure 21.5(b).

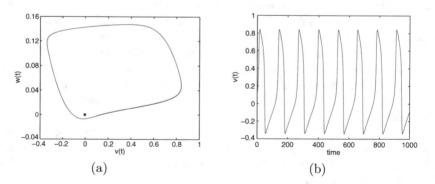

(a)   (b)

Figure 21.5:   (a) Typical phase portrait of a stable limit cycle for the Fitzhugh–Nagumo equation (21.6), there is also a critical point at the origin in this case. (b) Typical time series of an integrate and fire neuron for the variable $v(t)$.

21.2 Oscillatory Threshold Logic

Computing using oscillators is not a new concept, indeed the first modern computers were made using vacuum tube oscillators, and oscillators in a variety of forms are integral components in many devices. The use of neural oscillators has also been widely studied; however, in all cases, the method of computation is derived from concepts of biological neural encoding. Current research into encoding using neural oscillators is therefore spatio-temporal, rate, or more usually synchronization based [2], [30], and [35]. Borresen and Lynch [3], [4], have proposed using oscillators as the fundamental components of computing devices (with all the inherent dynamical richness that this provides) and designing them in such a way as to perform binary logic in an equivalent manner to standard transistor logic. In implementation, the oscillator will provide a binary output (1 equivalent to an oscillator firing or 0 where the oscillator does not fire) and the output from a single oscillator can be interpreted in exactly the same way as that of a transistor.

Threshold logic has been studied as an alternative to Boolean logic for some time. For many implementations this is advantageous, allowing for reduced component counts and/or number of logic levels, as the implementation of complex logical operations may be achieved using a single gate [32]. Threshold logic gates [22] have a set of inputs $\{I_1, I_2, ...I_n\}$, weights $\{w_1, w_2, ...w_n\}$ and a binary output y. The output y is typically described by:

$$y = \phi \left(\sum_{i=1}^{n} I_i w_i \right),$$

where the function ϕ is an activation function (eg Heaviside, tanh, sigmoid, piecewise linear, low gain saturation, see Chapter 20) and the binary output 1 is defined at some threshold, $y > T$, say.

Threshold logic implementation has not supplanted standard logic implementation in CMOS due to sensitivity to parameter changes and variable connection weights requiring very low tolerance engineering. Recent advances in nanotechnology, in particular, Resonant Tunneling Devices (RTD) [33] and memristor devices [28] have the potential to overcome such concerns.

A threshold oscillator is an oscillatory device that will begin oscillating when the input to the device is above a certain threshold. Below this level, the oscillator remains in a resting state and gives no output. It is possible to use the output of one threshold oscillator as the input of another oscillator to cause the second oscillator to operate (excitation) and under certain circumstances, it is also possible to cause the input of one oscillator to suppress the output of another oscillator (inhibition), see Figure 21.4.

There are numerous viable methods for implementing binary computation using threshold oscillators. In order to perform the logical operations, it is necessary that either oscillators with differing thresholds be used or the connections to the oscillators be of differing weights. The latter method is used here as this mimics more closely biological neural systems, from where the idea originated.

Logical operations can be performed in a similar manner to standard logic circuits; however, due to the threshold nature, it is possible to formulate logical operations as solutions of sets of linear inequalities. For instance, the AND function can be replicated by a threshold oscillator with two inputs, where the input strengths are scaled such that the total input is only above threshold if both the inputs are on. For a single input or for no input, the total input would be below threshold. Defining the inputs to the logical circuits in vector form and scaling the input strength to a binary 1 or 0, we write $\sum I = I_{1,1} + I_{2,1}$ as the total input to the circuit. The threshold equations may be thus written as:

$$\text{for } I = \begin{pmatrix} 0 \\ 0 \end{pmatrix}, \begin{pmatrix} 1 \\ 0 \end{pmatrix}, \begin{pmatrix} 0 \\ 1 \end{pmatrix} \qquad \sum Iw < T$$

$$\text{for } I = \begin{pmatrix} 1 \\ 1 \end{pmatrix} \qquad \sum Iw > T, \qquad (21.7)$$

where T is the oscillator threshold and w the coupling weight between the inputs and the oscillator performing the AND operation. Clearly, the solution to the above system (21.7) is $\frac{T}{2} < w < T$. For the logical OR operation, the solution $w > T$ would suffice.

Using threshold oscillators in this manner, it is straightforward to implement the logical NOT operation using a negative coupling strength; however, as the logical NOT is effectively redundant in more complex logically complete circuit design where NAND and XOR operations are used, all models using the latter formulations will be implemented.

One of the simplest computing circuits is the binary half adder. The binary half adder gives the sum of two binary inputs as a two-bit binary output. Standard transistor implementation of a binary half adder uses one XOR gate (to give the sum) and one AND gate (to give the carry). Implementation of this circuit using threshold oscillators can be achieved via a similar design, with two oscillators replicating the logical functions. The AND operation is implemented as described above, and the XOR operation can be achieved using an OR operation (as above) with an additional connection from the AND oscillator, which in some way inhibits the operation of the OR oscillator if the AND oscillator is active. The method by which inhibition occurs would be dependent upon the oscillators being used to form the circuitry.

Figure 21.6(a) demonstrates a viable circuit schematic for half adder implementation using two oscillators O_1 and O_2 and two inputs I_1 and I_2, which may themselves be the output from other oscillators in a more complex circuit. Schematically, the circuit design is not dissimilar to standard threshold logic half adders [21]; however, due to the nature of the connections between oscillators, implementation may be markedly different. If we consider oscillators with identical thresholds, we will require that the coupling strength, w_1, say, from

I_1 and I_2 to O_1 be sufficient to cause O_1 to oscillate for only one input and for the coupling strength, w_2, say, from I_1 and I_2 to O_2 to be sufficient for it to oscillate for two inputs. The additional connection x_1, say, from O_2 to O_1 is inhibitory such that if O_2 is oscillating it suppresses O_1. Denoting the output from O_2 as $\hat{O}_2$, the total input to O_1 and O_2 are thus given by:

$$O_1 = \sum I w_1 - \hat{O}_2 x_1$$
$$O_2 = \sum I w_2. \tag{21.8}$$

We can consider such a system as a set of linear inequalities with normalized input vectors I and threshold T requiring solutions of the form:

$$
\text{for } I = \begin{pmatrix} 0 \\ 0 \end{pmatrix} \qquad
\begin{cases} \sum I w_1 - \hat{O}_2 x_1 < T \\ \sum I w_2 < T \end{cases}
$$

$$
\text{for } I = \begin{pmatrix} 1 \\ 0 \end{pmatrix}, \begin{pmatrix} 0 \\ 1 \end{pmatrix} \qquad
\begin{cases} \sum I w_1 - \hat{O}_2 x_1 > T \\ \sum I w_2 < T \end{cases}
$$

$$
\text{for } I = \begin{pmatrix} 1 \\ 1 \end{pmatrix} \qquad
\begin{cases} \sum I w_1 - \hat{O}_2 x_1 < T \\ \sum I w_2 > T. \end{cases} \tag{21.9}
$$

Thus, for instance, for a total input of $\sum I = 1$, only O_1 will be above threshold causing oscillation giving a binary equivalent output of 1. If both I_1 and I_2 are active, O_2 will oscillate, but O_1 is suppressed if $\hat{O}_2 x_1 > T/2 + w_1$, giving a binary output $1 + 1 = 10$, as required.

It is possible to couple the oscillators together via various methods. For biological neural systems, where there is synaptic coupling between neurons the coupling function is complex, relying on diffusion of neurotransmitters across a synaptic gap. The connections between neurons may either depolarize (excite) or hyperpolarize (inhibit) the post synaptic neuron.

Crucially, the hyperpolarizing inhibitory effect has a temporal component such that if inhibition occurs, the post synaptic neuron remains inhibited for some period of time after the pre-synaptic neuron fires. It is not straightforward to simulate such a system using the Fitzhugh–Nagumo model without either integration of the signal pulse or introducing arbitrary conditions on oscillators receiving an inhibitory pulse—which would not be viable from an implementation perspective. As such a method which is phenomenologically similar to neural hyperpolarization is employed but is not necessarily consistent with any biological process.

Implementation by coupling through either the fast v variable or the slow w variable are equally viable. Any coupling function to be used must take into account the specific dynamics of whichever variable is used. As is common in such biologically inspired models, a sigmoidal transfer function is applied between oscillators of the form:

$$S(x) = \frac{1}{1 + e^{m(-x+c)}},\tag{21.10}$$

where c is the threshold at which the output begins to rise and m denotes the steepness of the curve of the function $S(x)$. In biological systems, neural connections can exhibit plastic responses and become "tuned" (via some Hebbian learning rule [12]) allowing for more reliable excitation and inhibition. Choosing suitable values of m and c would in many respects replicate such a process.

Numerical simulations for systems of Fitzhugh–Nagumo oscillators coupled as in Figure 21.6(a) will now be discussed. The inputs to the logical circuits are oscillatory, being provided by Fitzhugh–Nagumo oscillators with similar coupling and parameter values to the computational oscillators. Oscillatory inputs of this form have been chosen over continuous inputs, as this demonstrates the necessary robustness of signal integrity which would be required for larger computational circuits. Continuous inputs to the computational oscillators would be equally viable and present no difficulties in implementation. As such the matrix form for the input weights for each oscillator is 4×4 rather than 2×2 as two additional oscillators are used as inputs. One solution, in matrix form, to the inequalities (21.9) for the binary half adder would be:

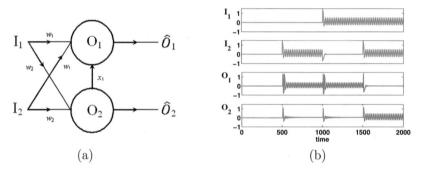

(a)   (b)

Figure 21.6:   [Mathematica] (a) Schematic of a binary oscillator half adder comprising two inputs I_1 and I_2, two oscillators O_1 and O_2, and a set of excitatory synaptic connections with weights w_1, w_2, and an inhibitory connection with weight x_1. The sum oscillator O_1 will oscillate if either I_1 or I_2 are active. The carry oscillator O_2 will oscillate if both I_1 and I_2 are active. The inhibitory connection x_1, from O_2 to O_1 suppresses oscillator O_1 if O_2 is active. (b) Time series showing that the half adder is functioning correctly when the oscillations are simulated using Fitzhugh–Nagumo systems. Oscillations are equivalent to a binary one in these simulations and no oscillation is zero.

$$W = \begin{pmatrix} 0 & 0 & 0.8 & 0.45 \\ 0 & 0 & 0.8 & 0.45 \\ 0 & 0 & 0 & 0 \\ 0 & 0 & -1.5 & 0 \end{pmatrix}, \tag{21.11}$$

where $C = 0.5$ in equation (21.6) for the inputs I_1 and I_2. This would give the parameter values shown in Figure 21.6(a) as $w_1 = 0.8$, $w_2 = 0.45$ and $x_1 = 1.5$. The time series for such is shown in Fig. 21.6(b) and the Mathematica program is listed in Section 21.4.

A two-oscillator binary full adder can be constructed by simply introducing another input, I_3, say, as in Figure 21.7(a), and Figure 21.7(b) shows the time series for the Fitzhugh–Nagumo two-oscillator full adder. Deriving the threshold inequalities for the full adder is left as an exercise for the reader (see Section 21.5).

In order to more fully demonstrate the applicability of binary oscillator computing more complex circuits, such as the three oscillator seven input full adder and the 2×2, bit binary multiplier may be constructed [4]. This is again left as an exercise for the reader (see Section 21.5).

Figure 21.8(a) shows a schematic of a Set-Reset (SR) flip-flop circuit, the input I_1 is commonly referred to as the Set and input I_2 is referred to as the Reset. Output $\hat{O}_2$ is the complement of output $\hat{O}_1$. Note that both oscillators require a constant input I_C, say, for the circuit to function properly. This circuit

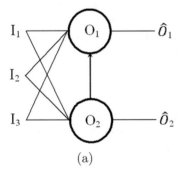

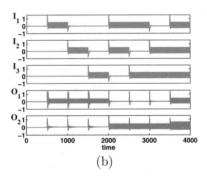

(a)   (b)

Figure 21.7: (a) Oscillator circuit diagram for a binary full adder comprising three inputs I_1, I_2, and I_3 and two oscillators O_1 and O_2. Oscillator O_1 will oscillate if either I_1, I_2, or I_3 are active. Oscillator O_2 will oscillate if any two of I_1, I_2, and I_3 are active. An inhibitory connection from O_2 to O_1 suppresses oscillator O_1 if O_2 is active; however, the inhibition is only sufficient to suppress O_1 for $\sum I = 2$. For inputs of $\sum I = 3$ the total input to O_1 is still sufficient to induce oscillation. (b) Time series for a Fitzhugh–Nagumo two-oscillator full adder. All binary combinations of oscillatory inputs I_1, I_2, and I_3 give the required binary outputs for O_1 and O_2.

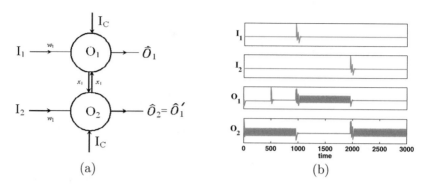

Figure 21.8: (a) Schematic of an SR flip-flop for memory. (b) Time series of an SR flip-flop using single input pulses (ballistic propagation) to switch based on Fitzhugh–Nagumo oscillations.

acts as a memory, storing a bit and presenting it on its output $\hat{O}_1$, as can be seen in Figure 21.8(b).

The SR flip-flop described here is an application of the "winnerless competition" principle. In the absence of coupling between the oscillators, both will remain active. However, a symmetric inhibitory coupling between them ensures that from an initial state, where only one oscillator is active, the other will remain suppressed in the absence of any external perturbation. When an input is given to the inactive oscillator, this is switched on, simultaneously suppressing the previously active oscillator. When the external input is turned off, the system remains in the switched state. Note that for a switch to occur, an input pulse of only one period is required (see Figure 21.8(b)). Switching using a single pulse in this way can open an opportunity to use ballistic propagation of signals between gates and memory cells, which could significantly reduce the energy required to operate memory circuits, where currently power intensive line charging is required to initiate memory switches. One important consideration, particularly with respect to flip-flop circuits is the ability to switch accurately in the presence of noise, the authors have demonstrated that oscillator-based SR flip-flops are highly noise resistant [4].

21.3 Applications and Future Work

In 1948, the world's first successful program was run on Manchester University's small-scale experimental machine the "Baby". To mark the 50th anniversary, the museum of Science and Industry based in Manchester, constructed a working replica which is still on display today. In 1951, Manchester University in partnership with Ferranti Limited built the world's first commercially available general-purpose computer, the Ferranti Mark 1. Given that one of the principal components in those machines was the vacuum tube oscillator (see Chapter

5) and the fact that the most powerful computer, the brain, also works with threshold oscillators, then the proposition of building modern computers using naturally oscillating devices should come as no surprise.

There are potentially five major avenues of research for binary oscillator computing which are listed below:

- **Biological Neuron Oscillators.** The inventors are currently working with cell biologists, stem cell researchers, and engineers in order to build the world's first assay for neuronal degradation. It is likely that further patents will be applied for as these biological logic circuits are built and further details will be published once the work is protected.

- **CMOS Oscillators.** In 2011, Manchester Metropolitan University employed a SPICE (Simulation Program with Integrated Circuit Emphasis) modeler to simulate the binary half adder circuits using CMOS-based oscillators and the simulations produced the required output [15]. Once more, the inventors are seeking industrial partners, and results will be published at a later date.

- **Josephson Junction (JJ) Oscillators.** JJs are superconducting natural threshold oscillators that cycle one hundred million times faster than biological neurons. The inventors are currently working with collaborators based at Colgate University and HYPRES Inc., both based in New York in the USA.

- **Memristors.** Memristor circuits can be built to make neuristors [27] (they mimic neurons), and axons and synapses are natural memristors. It is believed that highly connected neuronal circuits can be fabricated using memristors.

- **Optical Oscillators.** This avenue of research has yet to be pursued.

Currently, the inventors are pursuing two of the avenues of research highlighted above, namely biological and JJ neuron computing. The biological neuron oscillators will be used to make an assay for neuronal degradation and results and patents will follow. The oscillators depicted in Figures 21.6 to 21.8 could be fabricated using biological neurons, memristors, transistor circuits, all-optical circuits, or from JJ circuits [15]. JJs are natural threshold oscillators and unsurprisingly they can be designed to act like biological neurons with excitatory or inhibitory connections between them [19]–[25]. Crotty et al. [6] have even suggested that JJ neuronal circuits could be built to model parts of the brain. Superconductive computing based on Rapid Single Flux Quantum (RSFQ) technology is at an advanced stage and has already produced practical digital and mixed-signal circuits with world record processing speeds at exceptionally low power [20]–[10]. An 8-bit high frequency RSFQ-based Arithmetic Logic Unit (ALU) was fabricated with HYPRES' standard $4.5\,\text{kAcm}^{-2}$ process and consisted of 7950 JJs, including input and output interfaces. A high performance

ALU is a fundamental building block for any computer processor, and we now demonstrate how threshold oscillatory logic could help to further improve on this JJ performance. HYPRES foundry is producing complex digital RSFQ circuits operating at tens of gigahertz clock speed. Small RSFQ circuits were demonstrated to operate up to 770 GHz. As far as energy consumption is concerned, the current industry best for CMOS is approximately 1.7 GFLOPS/Watt compared with a potential 500 GFLOPS/Watt for JJ circuits. As well as a linear increase in components using binary oscillator logic described here [4], there are no migration issues and a proven radiation hardness with JJs. In 2005, the US National Security Agency published a report entitled "Superconducting Technology Assessment" [24] written by experts in the field. The authors concluded that transistors were rapidly approaching the limits of functionality and that the most likely successor to that technology would be based on JJs. They surmised that, given the investment required, a true petascale computer could be built by 2012. This paper demonstrates how coupled threshold oscillators may be used to perform both binary logic and memory in a manner entirely consistent with modern architectures. The benefits of using JJ-based components in terms of speed and energy usage are well documented. The additional benefits of JJ oscillator based computing include a doubling of processing power with a linear increase in components as well as ballistic propagation-based data interconnect between processor and memory. By utilizing some of the dynamics of the brain and using JJ circuitry, it should be possible to build a true exascale supercomputer based on this technology. In a recent development, Ken Segall et al. [29] have demonstrated synchronization dynamics on the pico-second timescale for physical JJ circuits acting like neurons. We are currently working with Ken in an attempt to build prototypes of our patented circuitry. We expect that the problem of fan-in and fan-out (connecting to and from JJ neurons) will be addressed using either low power memristor cross-bar lattices [31] or low power graphene nano-ribbon electronics [5].

To conclude this chapter, simple mathematical models of a JJ and a memristor (see Chapter 2) will be presented.

Mathematical Model of a JJ. A JJ with two superconducting layers sandwiching an insulating layer will be investigated. The differential equation used to model the resistively shunted JJ is written as

$$\frac{d^2\phi}{d\tau^2} + \beta_J \frac{d\phi}{d\tau} + \sin\phi = \kappa, \tag{21.12}$$

where ϕ is a phase difference, β_J is a parameter inversely related to the Josephson plasma frequency ω_J, κ is related to the total current across the junction and

$$\frac{d\phi}{dt} = \omega_J \frac{d\phi}{d\tau}.$$

Let $\frac{d\phi}{d\tau} = \Omega$, then the second-order ODE (21.12) can be written in the form

$$\frac{d\phi}{d\tau} = \Omega, \quad \frac{d\Omega}{d\tau} = \kappa - \beta_J\Omega - \sin\phi, \tag{21.13}$$

where $\eta = \beta_J\Omega$, is proportional to voltage. When $\kappa = \beta_J = 0$, then system (21.13) represents a Hamiltonian system given by

$$H(\phi, \Omega) = \frac{\Omega^2}{2} - \cos\phi.$$

Note that the Hamiltonian is very similar to that for the simple nonlinear pendulum depicted in Figure 6.1 and the trajectories are similar to those displayed in Figure 6.2. Note that the resistively shunted JJ acts as a threshold oscillator. Figure 21.9 shows a limit cycle when $\kappa = 2$, when $\beta_J = 1.2$. A Mathematica program for plotting Figure 21.9 is listed in Section 21.4. The tunneling JJ also displays hysteresis as shown in Figure 21.10, which shows a typical I-V (κ-$\langle\eta\rangle$) characteristic curve for a tunneling JJ as the voltage is increased and decreased. When $\langle\eta\rangle = 0$, Josephson current (up to a maximum threshold value, I_c) flows. A direct current (DC) Josephson supercurrent flows under $\langle\eta\rangle = 0$. When the current exceeds I_c, there is a bifurcation to an oscillating tunneling current. As $\langle\eta\rangle$ is increased further, the relation $\kappa = \langle\eta\rangle$ holds valid. As the voltage $\langle\eta\rangle$ is decreased, the relation $\kappa = \langle\eta\rangle$ still holds until a point where $\kappa_c \approx 0.6965$, and $\langle\eta\rangle = 2\Delta/e$, where Δ is an energy gap of the superconductor, where there is a bifurcation from oscillatory behavior back to the zero-voltage state. Note that the normalized DC voltage $\langle\eta\rangle = \beta_J\langle\Omega\rangle$, where $\langle\Omega\rangle$, is the average of the maximum and minimum values of Ω in the long τ region.

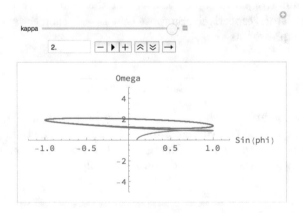

Figure 21.9: [Mathematica animation] Bifurcation of a limit cycle in a resistively shunted JJ. Readers can move the slider in the interactive notebook.

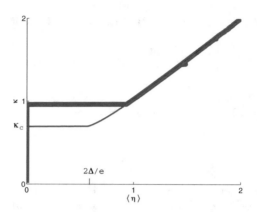

Figure 21.10:   A typical I-V characteristic curve usually observed in a tunneling JJ. The blue curve shows the current for increasing voltage, and the red curve depicts current for decreasing voltage. There is a clockwise hysteresis cycle. Note that $\kappa = \frac{I}{I_c}$ and the average voltage, $\langle \eta \rangle = \beta_J \langle \Omega \rangle$.

Mathematical Model of a Memristor. The memristor was briefly discussed in Chapter 2. A simple mathematical model will be presented in this section and a Mathematica program for plotting a pinched hysteresis loop will be listed in Section 21.4. Figure 21.11 depicts a titanium dioxide memristor which was first presented by Hewlett-Packard Laboratories in 2008 (see reference [13] in Chapter 2). The instantaneous resistance $M(w)$ of the memristor is given by

$$M(w) = \frac{w}{D} \mathrm{R_{ON}} + \left(1 - \frac{w}{D}\right) \mathrm{R_{OFF}}, \tag{21.14}$$

where $\mathrm{R_{ON}}$ and $\mathrm{R_{OFF}}$ are the resistances of the completely doped and the undoped memristor, respectively. Suppose that $D = 1$, then the speed of dopant movement is expressed as

$$\frac{dw}{dt} = \frac{\eta f(w(t), p) v_0 \sin\left(\frac{2\pi t}{T}\right)}{w \mathrm{R_{ON}} + (1-w)\mathrm{R_{OFF}}}, \tag{21.15}$$

where η is the polarity of the memristor (if $\eta = +1$, then w increases with positive voltage), v_0 is the voltage amplitude, and the function, $f(w(t), p) = 1 - (2w(t) - 1)^{2p}$, is the window function for the nonlinear dopant drift. The differential equation has initial condition $w_0 = w(0)$.

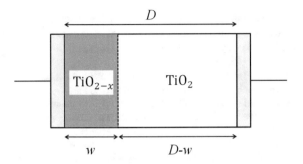

Figure 21.11: Abstract structure of a two terminal titanium dioxide (TiO₂) HP Labs memristor.

The differential equation (21.15) can be solved with Mathematica (see Section 21.4), and the voltage against current pinched hysteresis loop can be plotted. Two pinched hysteresis loops are shown in Figure 21.12.

The author and Borresen believe that JJ neurons can be connected together using memristors as axons and synapses. This is another avenue for research.

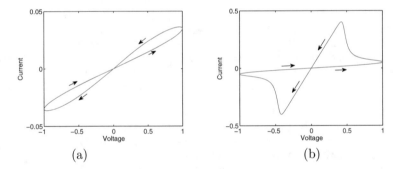

Figure 21.12: [Mathematica] Pinched hysteresis (voltage against current) of a memristor for a sinusoidal wave input voltage $v(t) = v_0 \sin\left(\frac{\pi t}{T}\right)$. The memristors parameters are $D = 1$, $R_{OFF} = 70$, $R_{ON} = 1$, $\eta = 1$, $p = 10$, $v_0 = 1$, $T = 20$, and $f(w,p) = 1 - (2w - 1)^{2p}$. (a) When $w_0 = 0.5$; (b) when $w_0 = 0.6$.

21.4 Mathematica Commands

```
(* Hodgkin--Huxley equations. See Figures 21.2 and 21.3. *)
CurrentI=6.3;GNa=120;GK=36;Gl=0.3;VNa=50;VK=-77;Vl=-54.402;Cap=1;TMax=100;
HH=NDSolve[{V'[t]==(CurrentI-GNa*m[t]^3*h[t]*(V[t]-VNa)-
GK*n[t]^4*(V[t]-VK)-Gl*(V[t]-Vl))/Cap,
m'[t]==0.1*(V[t]+40)/(1-Exp[-0.1*(V[t]+40)])*(1-m[t])-
```

```
4*Exp[-0.0556*(V[t]+65)]*m[t],
h'[t]==0.07*Exp[-0.05*(V[t]+65)]*(1-h[t])-1/(1+Exp[-0.1*(V[t]+35)])*h[t],
n'[t]==0.01*(V[t]+55)/(1-Exp[-0.1*(V[t]+55)])*(1-n[t])-
0.125*Exp[-0.0125*(V[t]+65)]*n[t],
V[0]==15,m[0]==0.01,h[0]==0.5,n[0]==0.4},{V,m,h,n},{t,0,TMax},
Method->{"StiffnessSwitching","NonstiffTest"->False}];
ParametricPlot[Evaluate[{t,V[t]}/.HH],{t,0,TMax},PlotPoints->1000,
PlotRange->{{0,TMax},{-80,40}},AspectRatio->1/2,AxesLabel->{t,V}]
ParametricPlot[Evaluate[{t,m[t]}/.HH],{t,0,TMax},PlotPoints->1000,
PlotRange->{{0,TMax},{0,1}},AspectRatio->1/2,AxesLabel->{t,"m(V)"}]
ParametricPlot[Evaluate[{t,h[t]}/.HH],{t,0,TMax},PlotPoints->1000,
PlotRange->{{0,TMax},{0,1}},AspectRatio->1/2,AxesLabel->{t,"h(V)"}]
ParametricPlot[Evaluate[{t,n[t]}/.HH],{t,0,TMax},PlotPoints->1000,
PlotRange->{{0,TMax},{0,1}},AspectRatio->1/2,AxesLabel->{t,"n(V)"}]
```

```
(* Fitzhugh--Nagumo half adder. See Figure 21.6. *)
Inp1 = Piecewise[{{0, 0 <= t < 500}, {1, 500 <= t < 1000}, {0,
    1000 <= t < 1500}, {1, 1500 <= t < 2000}}];
Inp2 = Piecewise[{{0, 0 <= t < 500}, {0, 500 <= t < 1000}, {1,
    1000 <= t < 1500}, {1, 1500 <= t < 2000}}];
theta = .1; gamma = 0.1; epsilon = .1; Tmax = 2000;
m = -100; c = 60;
sol = NDSolve[{u1'[t] == -u1[t] (u1[t] - theta) (u1[t] - 1) - v1[t] +
    Inp1, v1'[t] == epsilon (u1[t] - gamma v1[t]),
   u2'[t] == -u2[t] (u2[t] - theta) (u2[t] - 1) - v2[t] + Inp2,
   v2'[t] == epsilon (u2[t] - gamma v2[t]),
   u3'[t] == -u3[t] (u3[t] - theta) (u3[t] - 1) - v3[t] +
    0.8/(1 + Exp[m*v1[t] + c]) + 0.8/(1 + Exp[m*v2[t] + c]) -
    1.5/(1 + Exp[m*v4[t] + c]),
   v3'[t] == epsilon (u3[t] - gamma v3[t]),
   u4'[t] == -u4[t] (u4[t] - theta) (u4[t] - 1) - v4[t] +
    0.45/(1 + Exp[m*v1[t] + c]) + 0.45/(1 + Exp[m*v2[t] + c]),
   v4'[t] == epsilon (u4[t] - gamma v4[t]),
   u1[0] == .5, v1[0] == .1, u2[0] == 0.1, v2[0] == 0.1, u3[0] == 0,
   v3[0] == 0, u4[0] == 0, v4[0] == 0}, {u1[t], v1[t], u2[t], v2[t],
   u3[t], v3[t], u4[t], v4[t]}, {t, 0, Tmax}];
GraphicsColumn[
 {Plot[Evaluate[{u1[t]} /. sol], {t, 0, Tmax},
   PlotStyle -> {Blue, Thick}, PlotRange -> {{0, Tmax}, {-1, 1.5}},
   PlotPoints -> 100, AxesLabel -> {"t", "In 1"},
   AxesOrigin -> {0, -1}],
  Plot[Evaluate[{u2[t]} /. sol], {t, 0, Tmax},
   PlotStyle -> {Blue, Thick}, PlotRange -> {{0, Tmax}, {-1, 1.5}},
   PlotPoints -> 100, AxesLabel -> {"t", "In 2"},
   AxesOrigin -> {0, -1}],
  Plot[Evaluate[{u3[t]} /. sol], {t, 0, Tmax},
   PlotStyle -> {Green, Thick}, PlotRange -> {{0, Tmax}, {-1, 1.5}},
   PlotPoints -> 100, AxesLabel -> {"t", "Out 1"},
   AxesOrigin -> {0, -1}],
  Plot[Evaluate[{u4[t]} /. sol], {t, 0, Tmax},
```

```
    PlotStyle -> {Green, Thick}, PlotRange -> {{0, Tmax}, {-1, 1.5}},
     PlotPoints -> 100, AxesLabel -> {"t", "Out 2"},
     AxesOrigin -> {0, -1}]}, ImageSize -> 500, Spacings -> 10]
```

```
(* Bifurcation of a limit cycle for JJ circuit. See Figure 21.9. *)
Manipulate[BJ=1.2;T=100;
JJ=NDSolve[{phi'[t]==Omega[t],Omega'[t]==kappa-BJ*Omega[t]-Sin[phi[t]],
phi[0]==0.1,Omega[0]==0.1},{phi,Omega},{t,0,T} ];
ParametricPlot[Evaluate[{Sin[phi[t]],Omega[t]}/.JJ],{t,0,T},
PlotPoints->1000,PlotRange->{{-1.2,1.2},{-5,5}},AspectRatio->1/2,
AxesLabel->{"Sin(phi)","Omega"}],{kappa,0.1,2}]
```

```
(* Pinched hysteresis in the memristor. See Figure 21.12. *)
eta=1;L=1;Roff=70;Ron=1;p=10;T=20;w0=0.6;TMax=100;
Memristor=NDSolve[{w'[t]==(eta*(1-(2*w[t]-1)^(2*p))*
Sin[2*Pi*t/T])/(Roff-(Roff-Ron)*w[t]),
w[0]==w0},w,{t,0,T} ,Method -> {"StiffnessSwitching",
"NonstiffTest"->False}];
ParametricPlot[Evaluate[{Sin[2*Pi*t/T],
Sin[2*Pi*t/T]/(Roff-(Roff-Ron)*w[t])}/.Memristor],
{t,0,T},PlotPoints->1000,PlotRange->{{-1,1},{-0.5,0.5}},AspectRatio->1/2,
AxesLabel->{"Voltage","Current"}]
```

21.5 Exercises

1. Approximate the threshold input, I, for the Hodgkin–Huxley equations described by equations (21.2) to (21.5). Determine the frequency of spiking when (a) $I = 8\,\mathrm{mV}$ and (b) $I = 20\,\mathrm{mV}$.

2. Determine a Fitzhugh–Nagumo system (see equation (21.6)) that has a critical point at the origin.

3. Using a similar notation used in equation (21.9) for the half adder, determine the set of corresponding linear inequalities for the binary oscillator full adder depicted in Figure 21.7(a).

4. Write a Mathematica program to produce time series of the Fitzhugh–Nagumo two-oscillator full adder as depicted in Figure 21.7.

5. Write a Mathematica program to produce a time series of the Fitzhugh–Nagumo seven input three oscillator full adder as depicted in Figure 21.13.

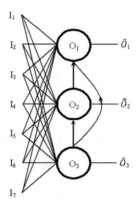

Figure 21.13:   Schematic of a seven input, three oscillator full adder.

6. Write down the truth table for a 2 × 2 bit binary multiplier and use the schematic shown in Figure 21.14 to produce a time series for a 2 × 2 bit Fitzhugh–Nagumo binary multiplier.

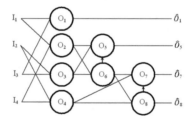

Figure 21.14:   Schematic of a 2 × 2 bit multiplier based on standard circuitry.

7. Write a Mathematica program to produce a time series of the Fitzhugh–Nagumo SR flip-flop as depicted in Figure 21.8.

8. Show that the Fitzhugh–Nagumo SR flip-flop modeled in the previous exercise is resistant to noise.

9. Plot the trajectories for the Hamiltonian modeling the resistively shunted JJ system (21.14) when $\kappa = \beta_J = 0$, given by

$$H(\phi, \Omega) = \frac{\Omega^2}{2} - \cos \phi.$$

10. Use Mathematica to plot the hysteresis curve displayed in Figure 21.10.

Bibliography

[1] P. Ashwin and J. Borresen, Encoding via conjugate symmetries of slow oscillations for globally coupled oscillators. *Phys Rev E* (2004) 70:026203.

[2] A.K. Barreiro, E. Shea-Brown, and E.L. Thilo, Time scales of spike-train correlation for neural oscillators with common drive. *Phys Rev E* (2010) 81:011916.

[3] J. Borresen and S. Lynch, Neuronal computers, *Nonlinear Anal. Theory, Meth. and Appl.*, 71 (2009), 2372–2376

[4] J. Borresen and S. Lynch, Oscillatory threshold logic, *PLoS ONE* **7** (11): e48498. doi:10.1371/journal.pone.0048498 (2012).

[5] Z. Chen, Y.M. Lin, M.J. Rooks, et al., Graphene nano-ribbon electronics, Physica E Low-Dimensional Systems and Nanostructures, **40** (2007), 228–232.

[6] P. Crotty, D. Schult and K. Segall, Josephson junction simulation of neurons, *Phys. Rev.*, **82**, 011914, 2010.

[7] S.K. Dana, D.C. Sengupta and Hu Chin Kun, Spiking and bursting in Josephson junction, *IEEE Transactions on Circuits and Systems 11-Express Briefs*, **10** (2006), 1031–1034.

[8] A. Destexhe, Z.F. Mainen and T.J. Sejnowski, An efficient method for computing synaptic conductances based on a kinetic model of receptor binding, *Neural Computation* **6**, (1994), 14–18.

[9] T. Filippov, M. Dorojevets, A. Sahu, A.F. Kirichenko, C. Ayala and O. Mukhanov, 8-bit asynchronous wave-pipelined RSFQ arithmetic-logic unit, *IEEE Trans. on Applied Superconductivity*, **21** (2011) pp 847–851.

[10] T. Filippov, A. Sahu, A.F. Kirichenko, I.V. Vernik, M. Dorojevets, C.L. Ayala and O. Mukhanov, 20 GHz operation of an asynchronous wave-pipelined RSFQ arithmetic-logic unit, *Physics Procedia*, **36** (2012), 59–65.

[11] R. Fitzhugh, Impulses and physiological states in theoretical models of nerve membranes, *Biophys.*, **1182**, (1961) 445–466.

[12] D.O. Hebb, *The organization of behaviour: A Neurophysiological Theory*, Wiley, New York, 1949.

[13] A.L. Hodgkin and A.F. Huxley, A qualitative description of membrane current and its application to conduction and excitation in nerve, *J. Physiol.* **117** (1952), 500–544, 1952. Reproduced in Bull. Math. Biol., vol. 52, (1990) pp 25–71.

[14] A.F. Kirichenko, S. Sarwana, D. Gupta, I. Rochwarger and O. Mukhanov, Multi-channel time digitizing system, *IEEE Trans. Appl. Supercond.*, **13** (2003), 454–458.

[15] S. Lynch and J. Borresen, Binary Half Adder using Oscillators, International Publication Number, WO 2012/001372 A1, (2012) 1–57.

[16] S. Lynch and J. Borresen, Binary Half Adder and Other Logic Circuits, UK Patent Number, GB 2481717 A, (2012) 1–57.

[17] S. Lynch and J. Borresen, Josephson junction binary oscillator computing, *Proceedings of the IEEE International Superconductive Electronics Conference, Cambridge, Massachusetts*, (2013), 1–3.

[18] S. Lynch, J. Borresen and M.A. Slevin (2016) US Patent: Assay utilising cellular binary half-adder system, Patent Application Number 14/230, 511, Publication Number US9274096 B2.

[19] O.A. Mukhanov and V.K. Semenov, Reproduction of the Single Flux Quantum pulses in Josephson junction systems. II. Neuristor lines and logic elements, *Mikroelektronika [Sov. Microelectronics]*, **17** (1988), 155–161.

[20] O.A. Mukhanov and S.V. Rylov, Time-to-Digital converters based on RSFQ digital counters, *IEEE Trans. Appl. Supercond.*, **7** (1997), 2669–2672.

[21] S. Muroga, *Threshold Logic and Its Applications*, Wiley, New York, 1971.

[22] W. McCulloch and W. Pitts, A logical calculus of the ideas imminent in nervous activity, *Bull Math Biophys* (1943) 5:115–133.

[23] J. Nagumo, S. Arimoto, and S. Yoshizawa, An active pulse transmission line simulating 1214-nerve axons, *Proc. IRL* **50**, (1970) 2061–2070.

[24] National Security Agency, Superconducting Technology Assessment report, available: www.nitrd.gov/PUBS/nsa/sta.pdf, 2005 [April 30, 2014].

[25] T. Onomi, Y. Maenami and K. Nakajima, Superconducting neural network for solving a combinatorial optimization problem, *IEEE Trans. Appl. Supercond.*, **21** (2011), 701–704.

[26] T. Ortlepp, O. Wetzstein, S. Engert, J. Kunert and H. Toepfer, Reduced power consumption in superconducting electronics, *IEEE Trans. on Applied Superconductivity*, **21** (2011), 770–775.

[27] M.D. Picket, G. Medeiros-Ribeiro and R.S. Williams, A scalable neuristor built with Mott memristors *Nature Materials*, **12**, 114, 2013.

[28] J. Rajendran, H. Manem, R. Karri, and G.S. Rose, An energy efficient memristive threshold logic circuit, *IEEE Trans on Computers*, (2012) 61:474–487.

[29] K. Segall, M. LeGro, S. Kaplan et al., Synchronization dynamics on the picosecond timescale in coupled Josephson junction neurons, preprint, arXiv:1608.04680, 2017.

[30] R.M. Smeal, G.B. Ermentrout, and J.A. White, Phase response curves and synchronized neural networks, *Phil Trans R Soc B.* 365 (1551) (2010) 2402–2407.

[31] S. Vaidyanathan (Editor) and C. Volos (Editor), *Advances in Memristors, Memristive Devices and Systems (Studies in Computational Intelligence)*, Springer, New York, 2017

[32] R. Waser R, *Nanoelectronics and Information Technology*, Wiley, New York, 2012.

[33] Y. Wei and J. Shen J, Novel universal threshold logic gate based on RTD and its application, *Microelectronics Journal*, (2011) 42:851–854.

[34] M. Zanin, F. Del Pozo, and S. Boccaletti, Computation emerges from adaptive synchronization of networking neurons, *PloS ONE* (2011) 11: e26467.

[35] X. Zhang, R. Wang, Z. Zhang, J. Qu, J. Cao, et al., Dynamic phase synchronization characteristics of variable high-order coupled neuronal oscillator population, *Neurocomputing*, **73** (13–15) (2010), 2665–2670.

Chapter 22
An Introduction to Wolfram SystemModeler

Aims and Objectives

- To introduce Wolfram SystemModeler™ to new users.

- To introduce both a-causal and causal models.

- To model, simulate, and analyze dynamical systems using SystemModeler.

On completion of this chapter, the reader should be able to

- build a Modelica (the language of SystemModeler) differential equations model;

- drag-and-drop components from the Modelica Standard Library in to a model and connect the icons using the Connection Line Tool;

- produce causal models from equations;

- run simulations and see how they are affected by input and parameter changes;

- analyze simulation results.

As Wolfram state: "Wolfram SystemModeler is an easy-to-use, next-generation modeling and simulation environment for cyber-physical systems. Using drag-and-drop from the large selection of built-in and expandable modeling libraries, you can build industrial strength, multidomain models of your complete system. Adding the power of Mathematica gives you a fully integrated environment for analyzing, understanding, and quickly iterating system designs. Driving insight, innovation, and results."

The package is used extensively by engineers and researchers around the world. It is easy to use and is the closest one can get to experimentation without the need for expensive physical apparatus and laboratories.

© Springer International Publishing AG 2017   509
S. Lynch, *Dynamical Systems with Applications using Mathematica*®,
DOI 10.1007/978-3-319-61485-4_22

This chapter is intended to introduce the reader to SystemModeler using models referred to in earlier sections of the book. Note that SystemModeler does not come as part of the Mathematica package. New users should read the online Wolfram document [3] and run the many examples that come with the package.

22.1 Introduction

To start SystemModeler simply click on the icon, the Wolfram SystemModeler Model Center window appears on the computer screen with a smaller window entitled "Welcome to Wolfram SystemModeler" in front, please see Figure 22.1. After watching the introductory videos in the "Welcome to Wolfram System-Modeler" window, users can then close that window. The Wolfram SystemModeler Model Center window is displayed in Figure 22.2.

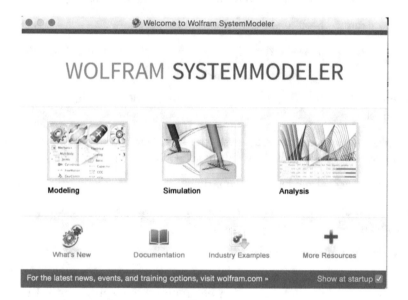

Figure 22.1: Welcome to Wolfram SystemModeler window. New users should watch the videos Modeling, Simulation, and Analysis before attempting to create their own models.

New users should now go to the Class Browser window on the left-hand side of the Wolfram SystemModeler Model Center window and click on Introductory Examples as shown in Figure 22.3. The first introductory example (HelloWorld) shows users how to create a simple model based on a differential equation. Readers will now be shown how to create a simple model based on the logistic

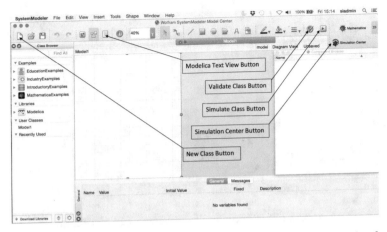

Figure 22.2: Wolfram SystemModeler Model Center window. The buttons Modelica Text View, Validate Class, Simulate Class, Simulation Center, and New Class are clearly labeled.

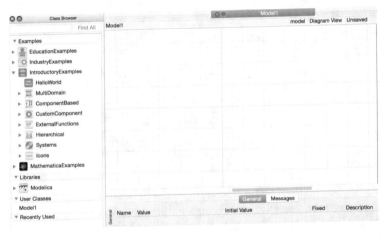

Figure 22.3: Class Browser window and Model1 window. New users should go through all of the introductory examples before attempting to create their own models.

differential equation introduced in Chapter 2. Click on the New Class button in the top left-hand corner of the Wolfram SystemModeler Model Center window and then click on the Modelica Text View button, see Figure 22.2. Users can type in the command lines as indicated in Figure 22.4 and save the file as Logistic.mo. To run the model, simply click on the Simulate Class button and click Simulate. A SimulationCenter window opens as shown in Figure 22.5. Initially, there is no plot, users must hit the Plot button to obtain a plot, and parameters, variables

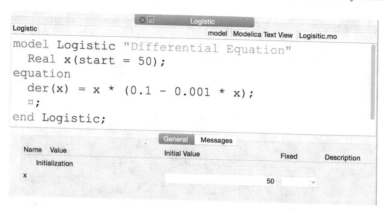

Figure 22.4:   Modelica Text View of Logistic.mo.

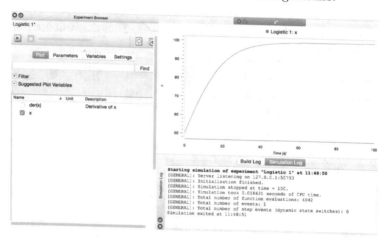

Figure 22.5:   SimulationCenter window. On the left-hand side is the Experiment Browser where users can Plot and change Parameters, Variables and Settings. On the right is a plot of the solution $x(t)$ for $0 \leq t \leq 100$, and beneath the plot is a Build Log.

and settings can be altered in the Experiment Browser window. Users check the boxes under Suggested Plot Variables to obtain a plot for that variable. In Figure 22.5, the x box has been ticked.

Figure 22.6(a) shows the Modelica text for the Fitzhugh–Nagumo system, introduced in Chapter 5, and Figure 22.6(b) shows time series plots obtained by running the model. There is a limit cycle plotted for x against y.

Figure 22.7(a) shows the Modelica text for the Lorenz system, introduced in Chapter 8, and Figure 22.7(b) shows time series plots obtained by running

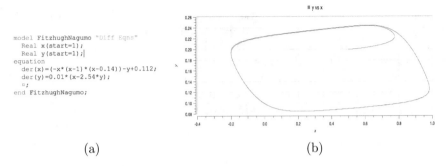

```
model FitzhughNagumo "Diff Eqns"
  Real x(start=1);
  Real y(start=1);|
equation
  der(x)=(-x*(x-1)*(x-0.14))-y+0.112;
  der(y)=0.01*(x-2.54*y);
  □;
end FitzhughNagumo;
```

(a)   (b)

Figure 22.6: (a) Modelica text for the Fitzhugh–Nagumo system, see Chapter 5. (b) Y(X) Plot Window for x versus y. Under Variables, set $x = 0.5$ and $y = 0.2$ as the initial conditions.

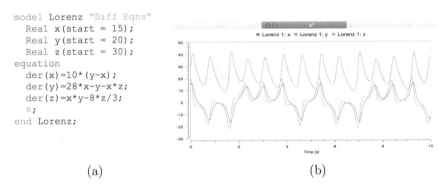

```
model Lorenz "Diff Eqns"
  Real x(start = 15);
  Real y(start = 20);
  Real z(start = 30);
equation
  der(x)=10*(y-x);
  der(y)=28*x-y-x*z;
  der(z)=x*y-8*z/3;
  □;
end Lorenz;
```

(a)   (b)

Figure 22.7: (a) Modelica text for the Lorenz system, see Chapter 8. (b) Time series plots for x, y, and z.

the model. One can clearly see chaotic behavior for this system. Readers will be shown how to save these Modelica models as components in the following sections.

22.2 Electric Circuits

Simple electric circuit theory was introduced in Chapter 2. The first example shows a simple series resistor-inductor (RL) circuit as displayed in Figure 22.8(a). To draw the RL circuit, one simply drags-and-drops components from the library of electrical models which is found by clicking on Libraries, then Modelica, Electrical, Analog, and finally Basic buttons in the Class Browser window. To obtain the applied electromotive force, or voltage source, one clicks on Libraries, Electrical, and then Sources in the Class Browser window. In this example, we have used the SineVoltage source. To set parameters for the model, simply click on each component and set the parameter values. In this case, we take $R = 10\Omega$ for resistor1 and $L = 2$H for inductor1. For the voltage source, we take an amplitude $V = 50$V and a frequency $freqHz = 0.1592$Hz. To simulate the model, click on the Simulate Class button and Simulate. Initially, the plot is empty. Under the Experiment Browser window, click on Plot, then the resistor1 button and then tick the i box. Under Settings, set the stop time to 40 seconds. The graph plotted is displayed in Figure 22.8(b) and shows the current flowing through resistor1.

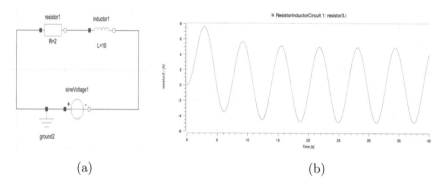

(a)   (b)

Figure 22.8: (a) Modelica RL circuit, see Chapter 2. (b) Time series plot for the current flowing through resistor1.

Chua's electric circuit was introduced in Chapter 8. This circuit is used as one of the SystemModeler examples for electrical analog circuits. Most of the components can be obtained as in Figure 22.8; however, one must click on Libraries, then Electrical, Analog, Examples, and finally Utilities to find the NonlinearResistor component. To set parameters for the model, simply click on each component and set the parameter values. In this case, we take $R = 12.5e - 3\Omega$ for resistor R_0, $L = 18$H for inductor L, $G = 10.565s$ for conductor G, $C = 10$F for conductor C1, $C = 100$F for conductor C2, $G_a = -0.757576s$ for the conductance in the inner voltage range, $G_b = -0.409091s$ for the conductance in the outer voltage range, and $V_e = 1$ for the inner voltage range limit. To simulate the model, click on the Simulate Class button and Simulate. Initially, the plot is empty. In the main toolbar, click on the New Y(X) Plot Window to plot one variable against another. Under the Experiment Browser window, click on Plot, then tick the $C1, v$, and $C2, v$ buttons, and under Variables set $C1, v = 2$, and $C2, v = 0$. Under Settings, set the tolerance to $1e - 10$; the graph plotted is displayed in Figure 22.10 and shows the voltage across $C1$ versus the voltage across $C2$. Click on the Simulate button to produce the graph.

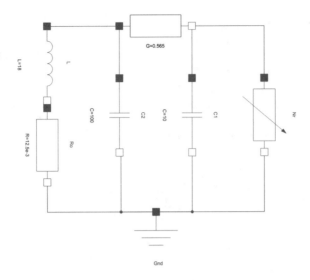

Figure 22.9:   Modelica Chua circuit.

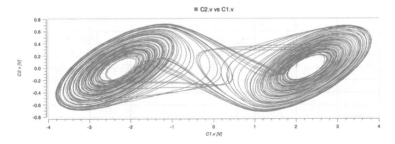

Figure 22.10:   Voltages across $C1$ and $C2$ for Chua's circuit. See Figure 22.9.

22.3 A Mechanical System

The double pendulum provides a wonderful example of chaos in a mechanical system and is a fantastic attraction at university open days and visit days. In this section, we shall simulate a double pendulum and plot a chaotic solution. A Modelica model of a double pendulum is shown in Figure 22.11. To draw the chaotic double pendulum model circuit, one simply drags-and-drops components from the library of mechanics models which is found by clicking on **Libraries**, then **Modelica**, and finally **Mechanics** buttons in the Class Browser window. The first component to drag-and-drop is the **World** component found under the **MultiBody** button. The World model provides an Earth coordinate system plus gravity field plus default animation definition. The next components to drag across are from the **Joints** which constrain the motion between two frames. Under **Joints**, drag-

and-drop two **Revolute** joints, labeled Revolute1 and Revolute2 in Figure 22.11. For Revolute1, set the **useAxisFlange** parameter to true, and this will enable the axis flange for joint Revolute1. Set the parameter $n[3] = \{0, 0, 1\}$, then the axis of rotation in $frame_a$ (= same as in $frame_b$). Under Initialization, set $phi = 0$, and the relative t = rotation angle from $frame_a$ to $frame_b$ to true. Set $w = 0$, and the first derivative of angle phi (relative angular velocity) to true. Finally, set $a = 0$. For Revolute2, set the **useAxisFlange** parameter to false, and the other parameters are the same as for joint Revolute1.

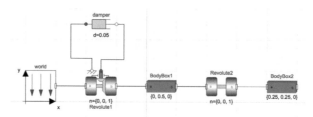

Figure 22.11: Modelica model of a double pendulum.

The next component to drag-and-drop can be found under **MultiBody** and **Forces** library and is the **Damper**, labeled as damper in Figure 22.11. Set the useHeatPort to false, and set $d = 0.05$ Nms rad^{-1} (Figure 22.12).

Name	Value				Name	Value		
Parameters					Parameters			
animation	true				animation	true		
r[3]	{0, 0.5, 0}	m			r[3]	{0.25, 0.25, 0}	m	
r_shape[3]	{0, 0, 0}	m			r_shape[3]	{0, 0, 0}	m	
lengthDirection[3]	to_unit1(r - r_shape)				lengthDirection[3]	to_unit1(r - r_shape)		
widthDirection[3]	{0, 1, 0}				widthDirection[3]	{0, 1, 0}		
length	Math.Vectors.length(r - r_shape)	m			length	Math.Vectors.length(r - r_shape)	m	
width	0.03	m			width	0.03	m	
height	width	m			height	0.03	m	
innerWidth	0	m			innerWidth	0	m	
innerHeight	innerWidth	m			innerHeight	innerWidth	m	
density	4	g/cm³			density	8	g/cm³	
color[3]	{0, 0, 200}				color[3]	{155, 0, 0}		
specularCoefficient	l.defaultSpecularCoefficient				specularCoefficient	l.defaultSpecularCoefficient		

(a)   (b)

Figure 22.12: (a) Parameter values for BodyBox1. (b) Parameter values for BodyBox2.

The final components to drag in to the model are found in the **Parts** library and are the **BodyBox** components, which depict rigid bodies with box shapes.

The parameter values chosen for the two **BodyBox** components are shown in Figure 22.13. Run the simulation by clicking on the **Simulate Class** button and the **Simulate**. To run an animation, simulate the model in the usual way, and then in the Wolfram SystemModeler Simulation Center window, users click the **Animation** button (it looks like a bouncing ball) in the toolbar. Simply click the **Play** button to watch the animation.

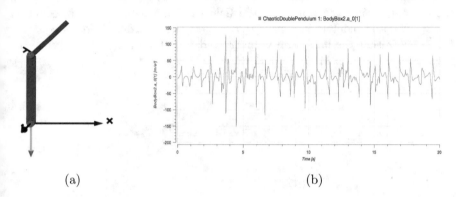

(a)   (b)

Figure 22.13:   Animation of a dampened double pendulum. (a) Initial position of the double pendulum. The red bar is half the length of the blue bar but is twice as dense. Click on the **Play** button to watch the animation in SystemModeler. (b) Chaotic time series plot for the acceleration of BodyBox2.

22.4 Causal (Block Based) Modeling

The examples presented in Sections 22.2 and 22.3 are a-causal (component based) models and are drawn using components that closely match the physical model. This section presents examples of causal (block based) models that are easily constructed if the user knows the equations of the system to be studied.

As a simple example, consider the RL electrical circuit modeled in Section 22.2. In Chapter 2, it was shown that this RL circuit can be modeled using the differential equation:

$$L\frac{dI(t)}{dt} + RI(t) = E(t), \tag{22.1}$$

where L is the inductance, R is resistance, E is the electromotive force, and I is the current flowing through the circuit. Using the parameters given in Section 22.2, equation (22.1) can be written as:

$$\frac{dI(t)}{dt} = 5\sin(t) - 0.2 * I(t). \tag{22.2}$$

Knowing the equation of the model makes it is easy to construct a causal model. Under Libraries and Modelica, go to the Blocks sublibrary. The Integrator block is found under the Continuous sublibrary, the Gain and Add blocks are found under the Math sublibrary, and finally, the Sine block is found under the Sources sublibrary. Readers familiar with Simulink, see [1] and [2], will understand the block diagram very well. For new users, consider Figure 22.14 and equation (22.2) together. The Integrator block integrates $\frac{dI}{dt}$ to give the current $I(t)$, and the current is then multiplied by the Gain to give the term $0.2I$ in equation (22.2). The Add block is then used to complete the right-hand side of equation (22.2). Note that the $0.2I$ term is subtracted from the $5\sin(t)$ term, hence the need for the -1 term in the Add block.

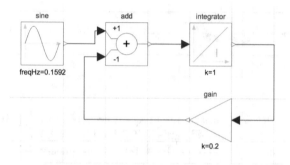

Figure 22.14: A causal model of the RL circuit using equation (22.2) to construct the model. See Figure 22.8(a) for the a-causal version. Running the model will produce the same graph as that shown in Figure 22.8(b).

Wolfram SystemModeler Link. It is possible to integrate Mathematica and Wolfram SystemModeler using a single command from within Mathematica as the two examples below illustrate.

Example 1. Use the Wolfram SystemModeler link to create a new function block for the function $f_\mu(x) = \mu x(1-x)$.

Solution. The Mathematica code for producing a new function block that can be used within Wolfram SystemModeler is listed below. Note that one must always start by typing in: Needs["WSMLink'"];

```
(* Mathematica commands to create a new block in SystemModeler. *)
Needs["WSMLink'"];
mmodel = WSMCreateModel["NewFunctionBlock", Function[x, 4*x*(1 - x)]];
```

Now, within Wolfram SystemModeler, a new block is created called New-FunctionBlock, and it can be used like any other block within the package.

Figure 22.15 shows the new block connected to a constant input with value 0.5. Readers can run the simulation and check that the resulting value from the NewFunctionBlock is 1.

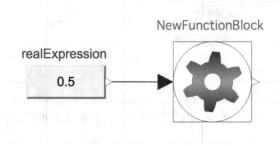

Figure 22.15:   realExpression block connected to NewFunctionBlock.

Example 2. Consider the water tank problem as depicted in Figure 22.16(a), which is solved using the ODEs:

$$p_1(t) = \rho g H1(t), p_2(t) = \rho g H2(t),$$

$$A1 \frac{dH1}{dt} = -\frac{(p_1(t) - p_2(t))\pi d^4}{128\mu L},$$

$$A2 \frac{dH2}{dt} = \frac{(p_1(t) - p_2(t))\pi d^4}{128\mu L} - A2 H2(t)/10, \qquad (22.3)$$

where p_1, p_2 are pressures, ρ, μ are constants, d, L are the diameters and lengths of the connecting pipe, respectively, g is the acceleration due to gravity, $A1, A2$ are cross-sectional areas, and $H1, H2$ are the heights of water in the tanks. The first tank is of cross-sectional area $A1 = 2m$, and there is height $H1$ of water contained in the tank. There is a pipe of diameter $d = 0.2m$ and length $L = 0.1m$, connecting the two pipes. The second tank is of cross-sectional area $A2 = 1m$, and there is a small pipe representing a leak in the second tank. Use Mathematica and Wolfram SystemModeler to solve the problem to give the heights of the water levels in both tanks for $0 \le t \le 100$ seconds.

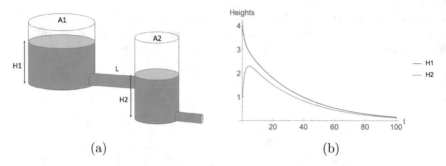

(a)   (b)

Figure 22.16: (a) The water tank problem modeled using equation 22.3. (b) Solutions for heights $H1$ and $H2$ for the water tank problem.

Solution. The Mathematica program for producing Figure 22.16(b) is listed below:

```
(* The water tank problem. See Figure 22.16(a). *)
Needs["WSMLink`"];
WaterEqns = {p1[t] == \[Rho]*g*H1[t], p2[t] == \[Rho]*g*H2[t]};
FlowRate = (p1[t] - p2[t])*(Pi*diam^4)/(128*\[Mu]*len);
massCons = {A1*H1'[t] == -FlowRate,
A2*H2'[t] == FlowRate - A2*H2[t]/10};
eqns = Join[WaterEqns, massCons, {H1[0] == 4, H2[0] == 1}];
pars = {len -> 0.1, diam -> 0.2, \[Rho] -> 0.2, \[Mu] -> 2*10^-3,
g -> 9.81, A1 -> 2, A2 -> 1};
WSMCreateModel["WaterFlow", eqns, t, WSMParameterValues -> pars];
sim = WSMSimulate["WaterFlow", 100];
WSMPlot[sim, {"H1", "H2"}, PlotRange -> All,
AxesLabel -> {"t", "Heights"}, Axes -> True,
BaseStyle -> {FontSize -> 15}]
```

22.5 Exercises

1. Create a Modelica text model for the Lotka–Volterra model. See Chapter 4 for the equations.

2. Create a Modelica text model for the van der Pol oscillator. See Chapter 5 for the equations.

3. Create a Modelica text model for Chua's circuit. See Chapter 8 for the equations.

4. Create an a-causal model of a periodically driven pendulum of your choice.

5. Change the electromotive force in the RL model, see equation (22.2), from $E(t) = 5\sin(t)$ to $E(t) = 10\sin\left(t + \frac{\pi}{3}\right)$, and run the causal model given that $I(0) = 0$.

6. Create a causal model to simulate the RLC circuit modeled by the differential equation

$$\frac{d^2 I}{dt^2} + 3\frac{dI}{dt} + 2I = 5\sin(t + \pi/4),$$

where $I(0) = \dot{I}(0) = 0$.

7. Use SystemModeler to reproduce the results of hysteresis in muscle, modeled using Hill's ODEs as demonstrated in our paper published in 2017 and listed in the Bibliography in Chapter 2.

8. Using the Wolfram SystemModeler link, extend the two-tank problem as depicted in Figure 22.16(a) into a three-tank problem with parameters of your choice.

9. Use SystemModeler to demonstrate synchronization of chaos in Chua's circuit.

10. Use SystemModeler to demonstrate the Fitzhugh–Nagumo-based binary oscillator half adder from Chapter 21.

Bibliography

[1] C.A. Kluever, *Dynamic Systems: Modeling, Simulation, and Control*, Wiley, New Jersey, 2015.

[2] S. Lynch, *Dynamical Systems with Applications using MATLAB, 2nd ed.*, Springer, New York, 2014.

[3] Web pages: An online pdf document for getting started with Wolfram SystemModeler: http://reference.wolfram.com/legacy/system-modeler/v3/getting-started/system-modeler-getting-started.pdf, last accessed, March 2017.

Chapter 23
Coursework and Examination-Type Questions

Aims and Objectives

- To model real-world problems.

- To investigate data generated by Mathematica.

- To use Mathematica in an examination environment.

On completion of this chapter, the reader should be able to

- write Mathematica programs to solve real-world problems;

- give a physical interpretation of the results;

- use Mathematica to solve examination questions.

This chapter provides examples of both coursework questions and examination questions that I have used with my students for nearly two decades. Most of the coursework questions require programming in Mathematica and are unsuitable for an examination environment. The examination questions require the use of Mathematica as a graphing calculator with some short programming. Short answers to the examination questions have been listed in Chapter 24. If any instructors require solutions to the coursework questions then they can email me directly. I still use these questions with my current students.

Note that the numbers given in square brackets denote the marks available for each part question.

© Springer International Publishing AG 2017   523
S. Lynch, *Dynamical Systems with Applications using Mathematica®*,
DOI 10.1007/978-3-319-61485-4_23

23.1 Examples of Coursework Questions

1. A very simple model of the spread of a mobile phone virus (such as Commwarrior-A) via Multimedia Messaging Services (MMS) and Bluetooth is represented by the smart phone state conversion schematic shown in Figure 23.1. We divide the phone modes into $SEIRD$ states and ten kinds of state conversions. Among them: $S \rightarrow I$, $I \rightarrow D$, $D \rightarrow I$ are related to Bluetooth spread mode; $S \rightarrow E$, $E \rightarrow S$ and $E \rightarrow R$ are related to MMS spread mode; and $S \rightarrow R$, $I \rightarrow R$, $E \rightarrow I$, and $I \rightarrow S$, are owned in common by Bluetooth and MMS two way.

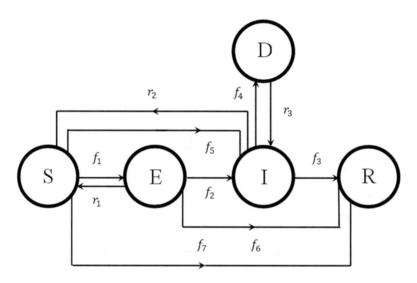

Figure 23.1: Smart phone state conversions. A Susceptible-Exposed-Infected-Recovered-Dormant (SEIRD) model for MMS and Bluetooth mixed virus spread.

(i) Assuming a simple linear model of the state vector $\underline{x} = [S, E, I, R, D]$, where f_i represents forward state conversions and r_i represents reverse state conversions, write down the differential equations that model this system.

(ii) Given that $f_1 = 0.09, f_2 = 0.02, f_3 = 0.01, f_4 = 0.04, f_5 = 0.01, f_6 = 0.01, f_7 = 0.01, r_1 = 0.01, r_2 = 0.006, r_3 = 0.01, S(0) = 1000$ and $E(0) = I(0) = D(0) = R(0) = 0$, use Mathematica to solve the differential equations and plot the time series on one graph for $0 \le t \le 500$.

(iii) Determine the maximum values of $E(t), I(t)$ and $D(t)$ for $0 \le t \le 500$.

[25]

2. The differential equations used to model the motion of the double pendulum are given by

$$\ddot{\theta}_1 = \frac{-g\left(2m_1 + m_2\right)\sin\theta_1 - m_2 g \sin\left(\theta_1 - 2\theta_2\right)}{L_1\left(2m_1 + m_2 - m_2\cos\left(2\theta_1 - 2\theta_2\right)\right)}$$
$$\frac{-2\sin\left(\theta_1 - \theta_2\right)m_2\left(\dot{\theta}_2{}^2 L_2 + \dot{\theta}_1{}^2 L_1 \cos\left(\theta_1 - \theta_2\right)\right)}{L_1\left(2m_1 + m_2 - m_2\cos\left(2\theta_1 - 2\theta_2\right)\right)},$$

$$\ddot{\theta}_2 = \frac{2\sin\left(\theta_1 - \theta_2\right)\left(\dot{\theta}_1{}^2 L_1\left(m_1 + m_2\right) + g\left(m_1 + m_2\right)\cos\theta_1 + \dot{\theta}_2{}^2 L_2 m_2 \cos\left(\theta_1 - \theta_2\right)\right)}{L_2\left(2m_1 + m_2 - m_2\cos\left(2\theta_1 - 2\theta_2\right)\right)}.$$

Use Mathematica to plot phase solutions, θ_1 against θ_2, for the following parameter values:

(a) $g = 9.8, m_1 = m_2 = 1, L_1 = 5, L_2 = 1, \theta_1(0) = 0.5, \dot{\theta}_1 = 1.5, \theta_2 = 0, \dot{\theta}_2 = 0$;

(b) $g = 9.8, m_1 = m_2 = 1, L_1 = 6, L_2 = 1, \theta_1(0) = 0.5, \dot{\theta}_1 = 1.5, \theta_2 = 0, \dot{\theta}_2 = 0$.

Vary the initial conditions for $\dot{\theta}_1$ slightly and run the simulations for parts (a) and (b) again. Give a physical interpretation of the results.

[25]

3. The differential equation used to calculate the precession of the perihelion of the planet Mercury using General Relativity is given by:

$$\frac{d^2}{d\theta^2}\left(\frac{1}{r(\theta)}\right) + \frac{1}{r(\theta)} = \frac{GM}{h^2} + \frac{3GM}{c^2 r(\theta)^2}, \tag{23.1}$$

where $r(\theta)$ is the path of Mercury around the sun (taken to be at the origin), M is the mass of the sun, G is the gravitational constant, h is related to the angular velocity of Mercury, and c is the speed of light in a vacuum.

(i) Consider a scaled model in which $M = 1, G = 1, h = 1$, and $c = 8$. Use Mathematica to solve the differential equation (23.1) given the initial conditions $r(0) = 2/3$ and $\frac{dr}{d\theta}(0) = 0$. Finally, plot a polar solution of $r(\theta)$ versus θ, for $0 \le \theta \le 10\pi$. HINT: Solve for $\frac{1}{r(\theta)}$.

(ii) Using the same set of parameters as in part (i), plot Cartesian plots of $r(\theta)$ against θ and $\frac{dr}{d\theta}$ against θ. HINT: Solve for $\frac{1}{r(\theta)}$.

(iii) The *perihelion* of an orbit is defined to be the point on the orbit where Mercury is closest to the sun. In this case, the perihelion rotates about the sun and describes a *precessing perihelion*. The perihelia occur when $r(\theta)$ is a local minimum, and the first perihelion occurs at $\theta_1 = 0$. Use the graph from part (ii) to determine the next three perihelia, $\theta_{2,3,4}$. Hence, deduce that the amount that the perihelion precesses per revolution is approximately $\delta\theta = \theta_{i+1} - \theta_i - 2\pi \approx 0.336$ radians per revolution.

[25]

4. The Hamiltonian function for a particle of mass m that bounces on a springy surface is approximated by

$$H(x,p) = \frac{1}{2m}p^2 + V(x),$$

where

$$V(x) = \begin{cases} \frac{1}{2}Cx^2 & x \leq 0 \\ mgx & x \geq 0, \end{cases}$$

and x is the position of the particle, p is the momentum of the particle, $V(x)$ is the potential energy, and C and g are positive constants.

(a) Write down the equations of motion for x and p in the cases where $x \geq 0$ and $x \leq 0$.

(b) Given that $m = 1$, $g = 10$, $C = 2$, and $E = 10$:

(i) sketch the contour of the Hamiltonian $H(x,p) = E$;

(ii) solve the equations of motion for $x(t)$ and $p(t)$ for this trajectory, giving expressions for both $x \leq 0$ and $x \geq 0$, separately;

Hint: for $x \geq 0$, assume $x(0) = 0, p(0) = 2\sqrt{5}$.

(iii) prove that the solution spends a time $T_1 = \frac{2}{\sqrt{5}}$ in the region $x \geq 0$, and a time $T_2 = \pi\sqrt{\frac{1}{2}}$ in the region $x \leq 0$.

[25]

5. The following five-dimensional system models a dispersive driven Jaynes-Cummings model from quantum optics:

$$\begin{aligned}
\dot{x}_1 &= \alpha x_2 + x_3 + \epsilon \\
\dot{x}_2 &= x_4 - \alpha x_1 \\
\dot{x}_3 &= \beta x_4 + x_1 x_5 \\
\dot{x}_4 &= x_2 x_5 - \beta x_3 \\
\dot{x}_5 &= -4\left(x_1 x_3 + x_2 x_4\right).
\end{aligned}$$

Use Mathematica to plot 3-D phase portraits (using x_1, x_2, and x_5 axes) and the corresponding power spectra when:

(i) $\epsilon = 2, \alpha = 2, \beta = 2$;

(ii) $\epsilon = 2, \alpha = 0.1, \beta = 30$.

Take initial conditions $x_1(0) = x_2(0) = x_3(0) = x_4(0) = 0$ and $x_5(0) = -1$. Describe the solutions in both cases.

[25]

6. Consider the following two-dimensional system:

$$\begin{aligned}
\dot{x} &= y \\
\dot{y} &= by - nx - m - rx^2 - x^2 y - x^3,
\end{aligned}$$

when $r = 0.87, m = -1, n = -1.127921667$, and $b = 0.897258546$.

(i) Determine the critical point(s) and their stability.

(ii) Given that there are four limit cycles in the region $-2 \le x \le 2$ and $-2 \le y \le 3$, use Mathematica to plot all four limit cycles.

Hint: Three of the limit cycles are very close to one another. Zoom in near the cusp point.

[25]

7. Use the Lindstedt-Poincaré technique to determine an $O(\epsilon^3)$ solution to the van der Pol equation:

$$\frac{d^2 x}{dt^2} + \epsilon \left(x^2 - 1\right) \frac{dx}{dt} + x = 0,$$

given that $x(0) = a$ and $\dot{x}(0) = 0$.

[25]

8. A preloaded two-bar linkage mechanism with joints P, Q, and R, preloaded by a stiffness k, is shown in Figure 23.2. A periodic force $F = \sin(\omega t)$ is applied at Q, where the two bars are joined by a frictionless pin. The angle θ denotes

the counterclockwise angle the left bar makes with the horizontal, q denotes the distance between P and R, and x is the distance between the joint Q and the horizontal dashed line. Then

$$x = l \sin \theta, \quad q = 2l \cos \theta. \tag{23.2}$$

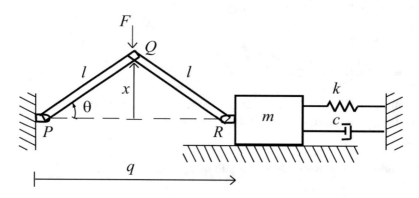

Figure 23.2: The preloaded two-bar linkage with a periodic force F acting at the joint Q. As the point Q moves vertically up and down, the mass m moves horizontally left and right.

The equations of dynamics for the preloaded two-bar linkage are given by

$$\left((2ml^2 + \frac{9}{8} m_{bar} l^2) \sin^2 \theta + \frac{5}{24} m_{bar} l^2 \right) \ddot{\theta} + (2ml^2 + \frac{9}{8} m_{bar} l^2) \dot{\theta}^2 \sin \theta \cos \theta$$
$$+ 2cl^2 \dot{\theta} \sin^2 \theta + 2kl^2 (\cos \theta_0 - \cos \theta) \sin \theta = -\frac{l \cos \theta}{2} F, \tag{23.3}$$

where $\theta_0 = \frac{\pi}{4}$.

(i) Rewrite equation (23.3) as a 2-D system of ODEs $\left(\text{let } \dot{\theta} = \phi \right)$ and use Mathematica to plot the solutions to these differential equations given that the parameters used are $\omega = 0.05$ (per second), $k = 1$ (N/m), $m = 1$ (kg), $c = 1$ (Ns/m), $m_{bar} = 0.5$ (kg), $l = 1$ (m), and $F = \sin(\omega t)$. Taking $\theta(0) = 0.8, \phi(0) = 0$, plot F, on the x-axis, against the vertical displacement x, on the y-axis, as F varies sinusoidally from $F = -1$ to $F = +1$. How would you describe this solution?

(ii) Given the same parameters as those used in part (i), use Mathematica to plot F, on the x-axis, against the horizontal displacement q, on the y-axis, as F varies sinusoidally from $F = -1$ to $F = +1$. How would you describe this solution?

(iii) Use equations (23.2) and (23.3) to prove that

$$((m + \frac{9}{16}m_{bar})(4l^2 - q^2) + \frac{5}{12}m_{bar}l^2)(4l^2 - q^2)\ddot{q} + \frac{5}{12}m_{bar}l^2 q\dot{q}^2$$
$$+ c\dot{q}(4l^2 - q^2)^2 + k(q - q_0)(4l^2 - q^2)^2 = \frac{1}{2}q(4l^2 - q^2)^{\frac{3}{2}} F, \qquad (23.4)$$

where $q_0 = 2l\cos(\theta_0)$.

[50]

9. In 1991, Tso et al. [1] published a paper on the energy exchange model for climate change and urban climatological studies. Since that publication, it has been shown that by adding 10% green cover to areas with little green, such as town centres and high-density residential areas, maximum surface temperatures in these areas can be kept below 1961–1990 baseline temperatures.

By linearizing the heat storage surface energy balance model, Tso et al. arrive at the following set of simultaneous differential equations for T_S, the surface temperature and T_L, the soil temperature:

The pre-dawn model. $0 \le t \le 4$.

$$\frac{dT_S}{dt} = B\left(b_1 + b_2 T_S + b_3 T_L + b_4\right), \quad \frac{dT_L}{dt} = B\left(b_5 T_S + b_6 T_L + b_7\right).$$

The daytime model. $4 \le t \le 20$.

$$\frac{dT_S}{dt} = B\left(c_1 \sin\left(wB(t-4)\right) + b_2 T_S + b_3 T_L + b_4\right), \quad \frac{dT_L}{dt} = B\left(b_5 T_S + b_6 T_L + b_7\right).$$

The nightime model. $20 \le t \le 24$.

$$\frac{dT_S}{dt} = B\left(b_1 + b_2 T_S + b_3 T_L + b_4\right), \quad \frac{dT_L}{dt} = B\left(b_5 T_S + b_6 T_L + b_7\right). \qquad (23.5)$$

[1] Tso C.P., Chan B.K., and Hashim M.A., Analytical solutions to the near-neutral atmospheric surface energy balance with and without heat storage for urban climatological studies, *American Meteorological Society*, **30** 4, 413–424, 1991.

(a) Solve the pre-dawn model equations (23.5) analytically using Laplace transforms.

(b) Given that, for Greater Manchester,

$$b_1 = -4.1706e - 004; b_2 = -0.0003453037152548344;$$
$$b_3 = 0.000048566764726985724; b_4 = 0.003747576508219105;$$
$$b_5 = 0.000050988700564971743; b_6 = -0.00010197740112994349;$$
$$b_7 = 0.001019774011299435; c_1 = 0.0035987838128722104;$$
$$\omega = 5.4542e - 005 \quad \text{and} \quad B = 3600;$$

and the initial conditions $T_S(0) = 12.1174, T_L(0) = 17.0565$, use Mathematica to plot the solution curves for T_S and T_L for $0 \le t \le 24$. Determine the maximum of T_S, and the maximum of T_L over this time interval.

[50]

10. Consider the following map:

$$x_{n+1} = (x_n + y_n + \mu \cos(2\pi y_n)) \bmod 1, \quad y_{n+1} = (x_n + 2y_n) \bmod 1.$$

(a) Given that $\mu = 0.1$ with initial conditions $x_0 = 0.1, y_0 = 0.1$, plot the first 10,000 iterates, ignoring the first 100. What can you deduce about the orbit?

(b) Determine the number of fixed points of period one when $\mu = 0.5$, and determine the stability of the fixed point at $(0.5, 0.5)$.

(c) Suppose that $\mu = 0.5$ and $d_0 = 10^{-10}$. Let $\mathbf{x}(n) = (x_n, y_n)$ be the n'th iterate of the initial point $\mathbf{x}(0) = (0.5 + d_0, 0.5)$. Furthermore, let

$$d(n) = \sqrt{(x_n - 0.5)^2 + (y_n - 0.5)^2}$$

be the distance of $\mathbf{x}(n)$ from the initial point. Given that

$$F(n) = \frac{1}{n} \ln\left(\frac{d(n)}{d_0}\right),$$

compute $F(4), F(20)$, and $F(100)$. What is the relationship between these values of $F(n)$ and the magnitude of the largest value of the Jacobian matrix for the fixed point $(0.5, 0.5)$?

[25]

11. Consider the following 2-dimensional mapping:

$$x_{n+1} = x_n^2 - y_n^2 + a_1 x_n + a_2 y_n,$$
$$y_{n+1} = 2 x_n y_n + a_3 x_n + a_4 y_n. \qquad (23.6)$$

(a) Given that $a_1 = 0.9, a_2 = -0.6, a_3 = 2$, and $a_4 = 0.5$:

(i) determine the fixed points of period one for system (23.7) and determine their stability;

(ii) obtain an iterative plot given that $x_0 = y_0 = 0.1$, explain the results;

(iii) suppose that $d_0 = 10^{-10}$, and let $\mathbf{x}(n) = (x_n, y_n)$ be the n'th iterate of the point $\mathbf{x}(0) = (0, d_0)$. Let

$$d(n) = \sqrt{x_n^2 + y_n^2}$$

be the distance of $\mathbf{x}(n)$ from the fixed point $(0, 0)$ and

$$F(n, m) = \frac{1}{n - m} \ln \left(\frac{d(n)}{d(m)} \right).$$

Use Mathematica to compute $F(4, 1)$, $F(20, 4)$, and $F(100, 20)$. Comment on the relationship between these values of $F(n, m)$ and the magnitude of the largest eigenvalue of the Jacobian matrix for the fixed point $(0, 0)$.

(b) Try to obtain iterative plots in the cases:

(i) $a_1 = 0.5, a_2 = -0.5, a_3 = 2$, and $a_4 = 0.3$, given $x(0) = y(0) = 0.1$, explain the results;

(ii) $a_1 = 0.9, a_2 = -0.6, a_3 = 2$ and $a_4 = 0.8$, given $x(0) = y(0) = 0.1$, explain the results.

[25]

12. A simple model of a two-neuron module with one self-interaction is described by the difference equations

$$x_{n+1} = b_1 + w_{11} \tanh (a x_n) + w_{12} \tanh (b y_n), \quad y_{n+1} = b_2 + w_{21} \tanh (a x_n)$$

where x_n, y_n are the activation levels of neurons x and y, b_1, b_2 are biases, w_{11} is a self-weight, and w_{12}, w_{21} are weights of synaptic connections (Fig. 23.3).

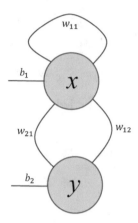

Figure 23.3:   A two-neuron module.

(a) Given that $a = 1, b = 0.3, b_2 = -3, w_{11} = -2, w_{12} = -1$, and $w_{21} = 5$, use Mathematica to determine the number, location, and stability of the fixed points of period one when: (i) $b_1 = -4$ and (ii) $b_1 = -2$.

(b) Using the same parameter values listed in part (a), edit the relevant Mathematica program from the notes to plot bifurcation diagrams for $-5 \leq b_1 \leq 5$, and give a physical interpretation of the results.

[25]

13. The Tinkerbell map is given by:

$$x_{n+1} = x_n^2 - y_n^2 + ax_n + by_n, \quad y_{n+1} = 2x_ny_n + cx_n + dy_n, \quad (23.7)$$

where $a, b, c,$ and d are all constants.

(a) Suppose that $b = -0.5$, $c = 2.3$, and $d = 0.5$ in system (23.7). Use Mathematica to obtain iterative plots when: (i) $a = 0.2$; (ii) $a = 0.4$; (iii) $a = 0.6$; (iv) $a = 0.8$; (v) $a = 1$. Describe the behavior of the system (23.7) for each value of a listed above.

(b) Using the same parameter values listed in part (a), plot a bifurcation diagram for system (23.7) for $0 \leq a \leq 1$.

[25]

14. Use Mathematica to plot a Newton fractal for the function $f(z) = (z^2 + 1)(z^2 - 5.29)$. If you have plotted the correct figure, you will notice that there

are regions in which the Newton method failed to converge to one of the roots. How would you explain this phenomenon?

[25]

15. Consider the weight distribution motif displayed in Figure 23.4. How will the weight be distributed at stage 2 of the construction of the multifractal?

| 0.02 | 0.05 |
| 0.03 | 0.9 |

Figure 23.4: Motif of a multifractal.

Given that:

$$\tau(q) = \frac{\ln\left(\sum_{i=1}^{N} p_i^q(\epsilon)\right)}{-\ln(\epsilon)},$$

$$\alpha = -\frac{\partial \tau}{\partial q},$$

and

$$f(\alpha(q)) = q\alpha(q) + \tau(q),$$

use Mathematica to plot an $f(\alpha)$ curve for the multifractal generated by the weight distribution motif given in Figure 23.4

[25]

23.2 Examination 1

Typically, students would be required to answer five out of 8 questions in three hours. The examination would take place in a computer laboratory with access to Mathematica.

1.   (a) Sketch a phase portrait for the following system showing all isoclines:

$$\frac{dx}{dt} = 3x + 2y, \quad \frac{dy}{dt} = x - 2y.$$

[8]

(b) Show that the system

$$\frac{dx}{dt} = xy - x^2 y + y^3, \quad \frac{dy}{dt} = y^2 + x^3 - xy^2$$

can be transformed into

$$\frac{dr}{dt} = r^2 \sin(\theta), \quad \frac{d\theta}{dt} = r^2 \left(\cos(\theta) - \sin(\theta)\right)\left(\cos(\theta) + \sin(\theta)\right)$$

using the relations $r\dot{r} = x\dot{x} + y\dot{y}$ and $r^2\dot{\theta} = x\dot{y} - y\dot{x}$. Sketch a phase portrait for this system given that there is one nonhyperbolic critical point at the origin.

[12]

2.   (a) Prove that the origin of the system

$$\frac{dx}{dt} = -\frac{x}{2} + 2x^2 y, \quad \frac{dy}{dt} = x - y - x^3$$

is asymptotically stable using the Lyapunov function $V = x^2 + 2y^2$.

[6]

(b) Solve the differential equations

$$\frac{dr}{dt} = -r^2, \quad \frac{d\theta}{dt} = 1,$$

given that $r(0) = 1$ and $\theta(0) = 0$. Hence show that the return map, say, $\mathbf{P}$, mapping points, say, r_n, on the positive x-axis to itself is given by

$$r_{n+1} = \mathbf{P}\left(r_n\right) = \frac{r_n}{1 + 2\pi r_n}.$$

[14]

3. (a) Find the eigenvalues of the following system and sketch a phase portrait in three-dimensional space

$$\frac{dx}{dt} = -2x - z, \quad \frac{dy}{dt} = -y, \quad \frac{dz}{dt} = x - 2z.$$

[12]

(b) Show that the origin of the following nonlinear system is not hyperbolic:

$$\frac{dx}{dt} = -2y + yz, \quad \frac{dy}{dt} = x - xz - y^3, \quad \frac{dz}{dt} = xy - z^3.$$

Prove that the origin is asymptotically stable using the Lyapunov function $V = x^2 + 2y^2 + z^2$. What does asymptotic stability imply for a trajectory $\gamma(t)$ close to the origin?

[8]

4. (a) Consider the 2-dimensional system

$$\frac{dr}{dt} = r\left(\mu - r\right)\left(\mu - r^2\right), \quad \frac{d\theta}{dt} = -1.$$

Show how the phase portrait changes as the parameter μ varies and draw a bifurcation diagram.

[10]

(b) Prove that none of the following systems has a limit cycle:

(i) $\frac{dx}{dt} = y - x^3, \quad \frac{dy}{dt} = x - y - x^4 y$;

(ii) $\frac{dx}{dt} = y^2 - 2xy + y^4, \quad \frac{dy}{dt} = x^2 + y^2 + x^3 y^3$;

(iii) $\frac{dx}{dt} = x + xy^2, \quad \frac{dy}{dt} = x^2 + y^2$.

[10]

5. (a) Let T be the function $T : [0, 1] \to [0, 1]$ defined by

$$T(x) = \begin{cases} \frac{7}{4}x & 0 \le x < \frac{1}{2} \\ \frac{7}{4}(1 - x) & \frac{1}{2} \le x \le 1. \end{cases}$$

Determine the fixed points of periods one, two, and three.

[12]

(b) Determine the fixed points of periods one and two for the complex mapping

$$z_{n+1} = z_n^2 - 3.$$

Determine the stability of the fixed points of period one.

[8]

6. (a) Starting with an equilateral triangle (each side of length 1 unit) construct the inverted Koch snowflake up to stage two on graph paper. At each stage, each segment is $\frac{1}{3}$ the length of the previous segment, and each segment is replaced by four segments. Determine the area bounded by the true fractal and the fractal dimension.

[14]

(b) Prove that

$$D_1 = \lim_{l \to 0} \frac{\sum_{i=1}^N p_i \ln(p_i)}{-\ln(l)},$$

by applying L'Hospital's rule to the equation

$$D_q = \lim_{l \to 0} \frac{1}{1-q} \frac{\ln \sum_{i=1}^N p_i^q(l)}{-\ln l}.$$

[6]

7. (a) Find and classify the fixed points of period one of the Hénon map defined by

$$x_{n+1} = 1 - \frac{9}{5}x_n^2 + y_n \quad y_{n+1} = \frac{1}{5}x_n.$$

[8]

(b) Consider the complex iterative equation

$$E_{n+1} = A + BE_n \exp\left(i\,|E_n|^2\right).$$

Derive the inverse map and show that

$$\frac{d|A|^2}{d|E_S|^2} = 1 + B^2 + 2B \left(|E_S|^2 \sin |E_S|^2 - \cos |E_S|^2 \right),$$

where E_S is a steady-state solution.

[12]

8. (a) A four-neuron discrete Hopfield network is required to store the following fundamental memories:

$$\mathbf{x}_1 = (1, 1, 1, 1)^T, \quad \mathbf{x}_2 = (1, -1, 1, -1)^T \quad \mathbf{x}_3 = (1, -1, -1, 1)^T.$$

(i) Compute the synaptic weight matrix $\mathbf{W}$.

(ii) Use asynchronous updating to show that the three fundamental memories are stable.

(iii) Test the vector $(-1, -1, -1, 1)^T$ on the Hopfield network.

Use your own set of random orders in (ii) and (iii).

[10]

(b) Derive a suitable Lyapunov function for the recurrent Hopfield network modeled using the differential equations

$$\dot{x} = -x + \left(\frac{2}{\pi} \tan^{-1} \left(\frac{\gamma \pi x}{2} \right) \right) + \left(\frac{2}{\pi} \tan^{-1} \left(\frac{\gamma \pi y}{2} \right) \right) + 6,$$

$$\dot{y} = -y + \left(\frac{2}{\pi} \tan^{-1} \left(\frac{\gamma \pi x}{2} \right) \right) + 4 \left(\frac{2}{\pi} \tan^{-1} \left(\frac{\gamma \pi y}{2} \right) \right) + 10.$$

[10]

23.3 Examination 2

Typically, students would be required to answer five out of 8 questions in three hours. The examination would take place in a computer laboratory with access to Mathematica.

1. (a) The radioactive decay of Polonium-218 to Bismuth-214 is given by

$$^{218}\text{Po} \rightarrow \,^{214}\text{Pb} \rightarrow \,^{214}\text{Bi},$$

where the first reaction rate is $k_1 = 0.5s^{-1}$, and the second reaction rate is $k_2 = 0.06s^{-1}$.

(i) Write down the differential equations representing this system. Solve the ODEs.

(ii) Determine the amount of each substance after 20 seconds given that the initial amount of ^{218}Po was one unit. Assume that the initial amounts of the other two substances was zero.

(iii) Plot solution curves against time for each substance.

(iv) Plot a trajectory in three-dimensional space.

[14]

(b) Plot the limit cycle of the system

$$\frac{dx}{dt} = y + 0.5x(1 - 0.5 - x^2 - y^2), \quad \frac{dy}{dt} = -x + 0.5y(1 - x^2 - y^2).$$

Find the approximate period of this limit cycle.

[6]

2. (a) Two solutes X and Y are mixed in a beaker. Their respective concentrations $x(t)$ and $y(t)$ satisfy the following differential equations:

$$\frac{dx}{dt} = x - xy - \mu x^2, \quad \frac{dy}{dt} = -y + xy - \mu y^2.$$

Find and classify the critical points for $\mu > 0$, and plot possible phase portraits showing the different types of qualitative behavior. Interpret the results in terms of the concentrations of solutes X and Y.

[14]

(b) Determine the Hamiltonian of the system

$$\frac{dx}{dt} = y, \quad \frac{dy}{dt} = x - x^2.$$

Plot a phase portrait.

[6]

3. (a) For the system

$$\frac{dx}{dt} = \mu x + x^3, \quad \frac{dy}{dt} = -y$$

sketch phase portraits for $\mu < 0$, $\mu = 0$, and $\mu > 0$. Plot a bifurcation diagram.

[10]

(b) Plot a phase portrait and Poincaré section for the forced Duffing system

$$\frac{dx}{dt} = y, \quad \frac{dy}{dt} = x - 0.3y - x^3 + 0.39\cos(1.25t).$$

Describe the behavior of the system.

[10]

4. (a) Given that $f(x) = 3.5x(1-x)$,

  (i) plot the graphs of $f(x), f^2(x), f^3(x)$, and $f^4(x)$;
  (ii) approximate the fixed points of periods one, two, three, and four, if they exist;
  (iii) determine the stability of each point computed in part (ii).

[14]

(b) Use Mathematica to approximate the fixed points of periods one and two for the complex mapping $z_{n+1} = z_n^2 + 2 + 3i$.

[6]

5. (a) Find and classify the fixed points of period one for the Hénon map

$$x_{n+1} = 1.5 + 0.2y_n - x_n^2, \quad y_{n+1} = x_n.$$

Find the approximate location of fixed points of period two if they exist. Plot a chaotic attractor using suitable initial conditions.

[14]

(b) Using the derivative method, compute the Lyapunov exponent of the logistic map $x_{n+1} = \mu x_n (1 - x_n)$, when $\mu = 3.9$.

[6]

6. (a) Edit the given program for plotting a bifurcation diagram for the logistic map to plot a bifurcation diagram for the tent map. (Students would have access to the Mathematica program given in Chapter 14).

[10]

(b) Write a program to plot a Julia set $J(0, 1.3)$, for the mapping $z_{n+1} = z_n^2 + 1.3i$.

[10]

7. (a) Given the complex mapping $E_{n+1} = A + BE_n e^{i|E_n|^2}$, determine the number and approximate location of fixed points of period one when $A = 3.2$ and $B = 0.3$.

[10]

(b) Edit the given program for producing a triangular Koch curve to produce a square Koch curve. At each stage, one segment is replaced by five segments and the scaling factor is $\frac{1}{3}$. (Students would have access to the Mathematica program given in Chapter 17).

[10]

8. (a) A six-neuron discrete Hopfield network is required to store the following fundamental memories:

$$\mathbf{x_1} = (1, 1, 1, 1, 1, 1)^T,$$

$$\mathbf{x_2} = (1, -1, 1, -1, -1, 1)^T,$$

$$\mathbf{x_3} = (1, -1, -1, 1, -1, 1)^T.$$

(i) Compute the synaptic weight matrix **W**.

(ii) Use asynchronous updating to show that the three fundamental memories are stable.

(iii) Test the vector $(-1, -1, -1, 1, 1, 1)^T$ on the Hopfield network. Use your own set of random orders in (ii) and (iii).

[10]

(b) Given that

$$\alpha_s = \frac{s \ln p_1 + (k - s) \ln p_2}{-k \ln 3}, \quad -f_s = \frac{\ln \left({}^k C_s \right)}{-k \ln 3},$$

write a short Mathematica program to plot the $f(\alpha)$ spectrum for the multifractal Cantor set constructed by removing the middle third segment at each stage and distributing the weight in the proportions $p_1 = \frac{1}{7}$ and $p_2 = \frac{6}{7}$. Sketch the $f(\alpha)$ curve and write down the Mathematica code in your answer booklet. What information does the width of the curve give?

[10]

23.4 Examination 3

Typically, students would be required to answer five out of 8 questions in three hours. The examination would take place in a computer laboratory with access to Mathematica.

1. (a) Sketch a phase portrait for the following system showing all isoclines:

$$\frac{dx}{dt} = x + 3y, \quad \frac{dy}{dt} = 2x - 4y. \qquad [8]$$

(b) Solve the differential equations

$$\frac{dr}{dt} = r - r^2, \quad \frac{d\theta}{dt} = 1,$$

given that $r(0) = 2$ and $\theta(0) = 0$. Hence determine the Poincaré return map mapping points, say, r_n, on the positive x-axis to itself.

[12]

2. (a) Plot phase portraits and a bifurcation diagram for the system

$$\frac{dx}{dt} = -x, \quad \frac{dy}{dt} = y(y - \mu + 1).$$

[14]

(b) Plot a bifurcation diagram for the system

$$\frac{dr}{dt} = r(\mu + r), \quad \frac{d\theta}{dt} = -1.$$

[6]

3. (a) An interacting species model of the Balsan fir tree, moose, and wolf at the Isle Royale National Park USA is given by

$$\frac{db}{dt} = b(1-b) - bm, \quad \frac{dm}{dt} = m(1-m) + bm - mw, \quad \frac{dw}{dt} = w(1-w) + mw,$$

where $b(t)$ represents the population of Balsan fir trees, $m(t)$ is the population of moose and $w(t)$ gives the population of wolves at time t. Determine the number and location of all critical points and show that there is a stable critical point for $b(t), m(t), w(t) > 0$.

[12]

(b) Find the fixed points of periods one and two for the complex mapping:

$$z_{n+1} = z_n^2 - 1 + i.$$

[8]

4. (a) Consider the mathematical model of glycolysis:

$$\dot{x} = -x + 0.1y + x^2y, \quad \dot{y} = 0.5 - 0.1y - x^2y,$$

where x and y represent the concentrations of ADP (adenosine diphosphate) and F6P (fructose 6-phosphate), respectively. Plot the isoclines given by $(\dot{x} = \dot{y} = 0)$ and show where the flow is vertical and horizontal. Given that, there is a critical point in the first quadrant at $(0.5, 1.4286)$ show that it is unstable.

[10]

(b) Show that there is an annular region which is positively invariant by considering the critical point from part (a) and the flow along the lines:

$$L_1 : y = 5, 0 \le x \le 0.5;$$
$$L_2 : x = 5.4, 0 \le y \le 0.1;$$
$$L_3 : x = 0, 0 \le y \le 5;$$
$$L_4 : y = 0, 0 \le x \le 5.4;$$
$$L_5 : y = -x + 5.5, 0.5 \le x \le 5.4.$$

What can you deduce from these results and the results in part (a)?

[10]

5. (a) Derive the inverse of the complex Ikeda mapping

$$E_{n+1} = A + BE_n \exp\left(i\left(\phi - \frac{C}{1 + |E_n|^2}\right)\right).$$ [8]

   (b) A highly simplified model for the gross national product (GNP) of a country is given by the iterative equation

$$k_{t+1} = f(k_t) = 0.5\frac{Bk_t^{0.3}(1 - k_t)^{0.2}}{1.2}.$$

   Plot the curves for $f(k)$ when $B = 1$ and $B = 4$. Use Mathematica to plot the corresponding curves for $\frac{df}{dk}$. Show that there is a stable fixed point of period one when $B = 1$ and an unstable fixed point of period one when $B = 4$. What happens when B is approximately 3.26?

[12]

6. (a) Find the fixed points of periods one and two for the Lozi map

$$x_{n+1} = 1 + y_n - 2|x_n|, \quad y_{n+1} = \frac{1}{2}x_n$$

   and classify the fixed points of period one.

[10]

   (b) Consider the map defined by $x_{n+1} = f(x_n)$, where $f(x)$ is defined by

$$f(x) = \begin{cases} -6x + 2 & x \le \frac{1}{2}, \\ 6x - 4 & x > \frac{1}{2}. \end{cases}$$

   Plot the function on graph paper. Consider the sets, S_n say, which remain in the interval $[0, 1]$ after n iterations. List the intervals in S_1

and S_2. The set of points that never escape from the interval $[0, 1]$ form a Cantor set. What is the fractal dimension of this Cantor set?

[10]

7. (a) A certain species of insect can be divided into three age classes: 0–6 months, 6–12 months, and 12–18 months. A Leslie matrix for the female population is given by

$$L = \begin{pmatrix} 0 & 10 & 20 \\ \frac{1}{2} & 0 & 0 \\ 0 & \frac{1}{3} & 0 \end{pmatrix}.$$

Determine the long-term distribution of the insect population. An insecticide is applied, which kills off 50% of the youngest age class. Determine the long-term distribution if the insecticide is applied every six months.

[10]

(b) Consider the Ikeda map given by:

$$E_{n+1} = 10 + 0.15E_n \exp\left(i|E_n|^2\right),$$

where E_n is the electric field strength of propagating light in a SFR resonator (see Chapter 16). Using Mathematica and taking $E(1) = 10$, iterate 10000 times and plot a power spectrum for $|E_n|^2$. Write down the Mathematica program in your answer booklet.

[10]

8. (a) The Lyapunov exponent, say λ, for the map

$$x_{n+1} = f(x_n)$$

is defined by

$$\lambda = \lim_{n \to \infty} \left(\frac{1}{n} \sum_{i=0}^{n-1} \ln |f'(x_i)| \right).$$

Use Mathematica to compute the Lyapunov exponent of the sine map

$$x_{n+1} = r \sin(\pi x_n),$$

for $0 \le x_n \le 1$, when $r = 1.3$ and $x_0 = 0.1$. Write down the value of the Lyapunov exponent and the Mathematica code in your answer booklet. What type of solution is defined by (i) $\lambda < 0$, (ii) $\lambda = 0$, and (iii) $\lambda > 0$?

[10]

(b) Motifs for the Koch curve and Lévy curve are shown in Figure 23.5 and a Mathematica program for plotting the Koch curve up to stage 7 is listed below. Edit this program (copy your program into the answer booklet) to plot the Lévy curve up to stage 7.

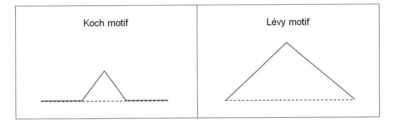

Figure 23.5:   Motifs for the Koch curve and Lévy curves.

```
(* The Koch curve up to Stage 7. *)
In[1]:=k=7;Itermax=4^k;h=3^(-k);
In[2]:=x[1]=0;y[1]=0;
In[3]:=angle[1]=0;angle[2]=Pi/3;angle[3]=-Pi/3;angle[4]=0;
In[4]:=For[a=1,a<=Itermax,a++,m=a-1;ang=0;
        For[b=1,b<=k,b++,segment[b]=Mod[m,4];m=Floor[m/4];
        r=segment[b]+1;ang=ang+angle[r]];x[a+1]=x[a]+h Cos[ang];
        y[a+1]=y[a]+h Sin[ang]];
In[5]:=ListPlot[Table[{x[i],y[i]},{i,1,Itermax+1}],PlotJoined->True,
        Axes->False,AspectRatio->Automatic]
```

[10]

END

Chapter 24
Solutions to Exercises

24.1 Chapter 1

1. (a) 3; (b) 531441; (c) 0.3090; (d) 151; (e) $-\frac{1}{10}$.

2. (a)

$$A + 4BC = \begin{pmatrix} 57 & 38 & 19 \\ 40 & 25 & 16 \\ 35 & 19 & 14 \end{pmatrix}.$$

   (b)

$$A^{-1} = \begin{pmatrix} 0.4 & -0.6 & 0.2 \\ 0 & 1 & 0 \\ -0.6 & 1.4 & 0.2 \end{pmatrix}, \quad B^{-1} = \begin{pmatrix} 0 & 1 & -1 \\ 2 & -2 & -1 \\ -1 & 1 & 1 \end{pmatrix}.$$

   The matrix C is singular.

   (c)

$$A^3 = \begin{pmatrix} -11 & 4 & -4 \\ 0 & 1 & 0 \\ 12 & 20 & -7 \end{pmatrix}.$$

   (d) Determinant of $C = 0$.

   (e) Eigenvalues and corresponding eigenvectors are

   $\lambda_1 = -0.3772$, $(0.4429, -0.8264, 0.3477)^T$;

   $\lambda_2 = 0.7261$, $(0.7139, 0.5508, -0.4324)^T$;

   $\lambda_3 = 3.6511$, $(0.7763, 0.5392, 0.3266)^T$.

3. (a) $-1 + 3i$; (b) $1 - 3i$; (c) $1.4687 + 2.2874i$; (d) $0.3466 + 0.7854i$; (e) $-1.1752i$.

4. (a) 1; (b) $\frac{1}{2}$; (c) 0; (d) ∞; (e) 0.

© Springer International Publishing AG 2017
S. Lynch, *Dynamical Systems with Applications using Mathematica®*,
DOI 10.1007/978-3-319-61485-4_24

5. (a) $9x^2 + 4x$; (b) $\frac{2x^3}{\sqrt{1+x^4}}$; (c) $e^x(\sin(x)\cos(x) + \cos^2(x) - \sin^2(x))$;
(d) $1 - \tanh^2 x$; (e) $\frac{2\ln x x^{\ln x}}{x}$.

6. (a) $-\frac{43}{12}$; (b) 1; (c) $\sqrt{\pi}$; (d) 2; (e) divergent.

7. See Section 1.3.

8. (a) $y(x) = \frac{1}{2}\sqrt{2x^2 + 2}$; (b) $y(x) = \frac{6}{x}$; (c) $y(x) = \frac{(108x^3 + 81)^{1/4}}{3}$;
(d) $x(t) = -2e^{-3t} + 3e^{-2t}$; (e) $\frac{16}{5}e^{-2t} - \frac{21}{10}e^{-3t} - \frac{1}{10}\cos t + \frac{1}{10}\sin t$.

9.

(a) When x(0):=0.2,

(b) when x(0):=0.2001,

x(91):=0.8779563852	x(91):=0.6932414820
x(92):=0.4285958836	x(92):=0.8506309185
x(93):=0.9796058084	x(93):=0.5082318360
x(94):=0.7991307420e-1	x(94):=0.9997289475
x(95):=0.2941078991	x(95):=0.1083916122e-2
x(96):=0.8304337709	x(96):=0.4330964991e-2
x(97):=0.5632540923	x(97):=0.1724883093e-1
x(98):=0.9839956791	x(98):=0.6780523505e-1
x(99):=0.6299273044e-1	x(99):=0.2528307406
x(100):=0.2360985855	x(100):=0.7556294285

10. (* Euclid's algorithm, greatest common divisor. *)

```
In[1]:=a=12348;b=14238;
In[2]:=While[b!=0,d=Mod[a,b];a=b;b=d]
In[3]:=a
Out[3]:=126
```

24.2 Chapter 2

1. (a) $y = \frac{C}{x}$; (b) $y = Cx^2$; (c) $y = C\sqrt{x}$; (d) $\frac{1}{y} = \ln\left(\frac{C}{x}\right)$; (e) $\frac{y^4}{4} + \frac{x^2 y^2}{2} = C$; (f) $y = Ce^{-\frac{1}{x}}$.

2. The fossil is 8.03×10^6 years old.

3. (a) $\dot{d} = k_f(a_0 - d)(b_0 - d)(c_0 - d) - k_r(d_0 + d)$;
(b) $\dot{x} = k_f(a_0 - 3x)^3 - k_r x$, where $a = [A], x = [A_3], b = [B], c = [C]$, and $d = [D]$.

4. (a) The current is $I = 0.733$ amps;
(b) the charge is $Q(t) = 50(1 - \exp(-10t - t^2))$ coulombs.

5. (a) Time 1.18 hours. (b) The concentration of glucose is

$$g(t) = \frac{G}{100kV} - Ce^{-kt}.$$

6. Set $x(t) = \sum_{n=0}^{\infty} a_n t^n$.

7. The differential equations are

$$\dot{A} = -\alpha A, \ \dot{B} = \alpha A - \beta B, \ \dot{C} = \beta B.$$

9. The differential equations are

$$\dot{H} = -aH + bI, \ \dot{I} = aH - (b+c)I, \ \dot{D} = cI.$$

The number of dead is given by

$$D(t) = acN \left(\frac{\alpha - \beta + \beta e^{\alpha t} - \alpha e^{\beta t}}{\alpha\beta(\alpha - \beta)} \right),$$

where α and β are the roots of $\lambda^2 + (a+b+c)\lambda + ac = 0$. This is not realistic as the whole population eventually dies. In reality people recover and some are immune.

10. (a) (i) Solution is $x^3 = 1/(1-3t)$, with maximal interval (MI) $-\infty < t < \frac{1}{3}$; (ii) $x(t) = (e^t + 3)/(3 - e^t)$, with MI $-\infty < t < \ln 3$; (iii) $x(t) = 6/(3 - e^{2t})$, with MI $-\infty < t < \ln\sqrt{3}$.

   (b) Solution is $x(t) = (t + x_0^{1/2} - t_0)^2$, with MI $t_0 - x_0^{1/2} < t < \infty$.

24.3 Chapter 3

1. (a) Eigenvalues and eigenvectors are $\lambda_1 = -10$, $(-2, 1)^T$; $\lambda_2 = -3$, $(\frac{3}{2}, 1)^T$. The origin is a stable node.

   (b) Eigenvalues and eigenvectors are $\lambda_1 = -4$, $(1, 0)^T$; $\lambda_2 = 2$, $(-\frac{4}{3}, 1)^T$. The origin is a saddle point.

2. (a) All trajectories are vertical and there are an infinite number of critical points on the line $y = -\frac{x}{2}$.

   (b) All trajectories are horizontal and there are an infinite number of critical points on the line $y = -\frac{x}{2}$.

   (c) Eigenvalues and eigenvectors are $\lambda_1 = 5$, $(2, 1)^T$; $\lambda_2 = -5$, $(1, -2)^T$. The origin is a saddle point.

   (d) Eigenvalues are $\lambda_1 = 3+i$, $\lambda_2 = 3-i$, and the origin is an unstable focus.

   (e) There are two repeated eigenvalues and one linearly independent eigenvector: $\lambda_1 = -1$, $(-1, 1)^T$. The origin is a stable degenerate node.

   (f) This is a nonsimple fixed point. There are an infinite number of critical points on the line $y = x$.

3. (a) $\dot{x} = y$, $\dot{y} = -25x - \mu y$;

    (b) (i) unstable focus, (ii) center, (iii) stable focus, (iv) stable node;

    (c) (i) oscillations grow (not physically possible), (ii) periodic oscil-
        lations, (iii) damping, (iv) *critical damping*.
    The constant μ is called the *damping coefficient*.

4. (a) There is one critical point at the origin which is a col. Plot the
        isoclines. The eigenvalues are $\lambda = \frac{-1 \pm \sqrt{5}}{2}$ with eigenvectors $\begin{pmatrix} 1 \\ \lambda_1 \end{pmatrix}$
        and $\begin{pmatrix} 1 \\ \lambda_2 \end{pmatrix}$.

    (b) There are two critical points at $A = (0,2)$ and $B = (1,0)$. A is
        a stable focus and B is a col with eigenvalues and corresponding
        eigenvectors given by $\lambda_1 = 1$, $\begin{pmatrix} 1 \\ -3 \end{pmatrix}$ and $\lambda_2 = -2$, $\begin{pmatrix} 1 \\ 0 \end{pmatrix}$.

    (c) There are two critical points at $A = (1,1)$ and $B = (1,-1)$. A is
        an unstable focus and B is a stable focus. Plot the isoclines where
        $\dot{x} = 0$ and $\dot{y} = 0$.

    (d) There are three critical points at $A = (2,0)$, $B = (1,1)$, and
        $C = (1,-1)$; A is a col and B and C are both stable foci.

    (e) There is one nonhyperbolic critical point at the origin. The solu-
        tion curves are given by $y^3 = x^3 + C$. The line $y = x$ is invariant,
        the flow is horizontal on $\dot{y} = x^2 = 0$, and the flow is vertical on
        the line $\dot{x} = y^2 = 0$. The slope of the trajectories is given by
        $\frac{dy}{dx} = \frac{x^2}{y^2}$.

    (f) There is one nonhyperbolic critical point at the origin. The solu-
        tion curves are given by $y = \frac{x}{1+Cx}$. The line $y = x$ is invariant.

    (g) There is one nonhyperbolic critical point at the origin. The solu-
        tion curves are given by $2y^2 = x^4 + C$. The slope of the orbits is
        given by $\frac{dy}{dx} = \frac{x^3}{y}$.

    (h) When $\mu < 0$ there are no critical points. When $\mu = 0$, the solution
        curves are given by $|x| = Ce^{\frac{1}{y}}$. When $\mu > 0$, there are two critical
        points at $A = (0, \sqrt{\mu})$ and $B = (0, -\sqrt{\mu})$; A is a col and B is an
        unstable node.

5. One possible system is

$$\dot{x} = y^2 - x^2, \ \dot{y} = x^2 + y^2 - 2,$$

    for example.

6. There are three critical points at $O = (0,0)$, $A = (1,0)$, and $B = (-1,0)$. If $a_0 > 0$, since det $J_O > 0$ and trace $J_O < 0$, the origin is stable and A and B are cols because det $J < 0$ for these points. If $a_0 < 0$, the origin is unstable and A and B are still cols. Therefore, if $a_0 > 0$, the current in the circuit eventually dies away to zero with increasing time. If $a_0 < 0$, the current increases indefinitely, which is physically impossible.

7. There are three critical points at $O = (0,0)$, $A = (\frac{a}{b},0)$, and $B = \left(\frac{c+a}{b}, \frac{c(c+a)}{b}\right)$. The origin is an unstable node and A is a col. The critical point at B is stable since det $J_B > 0$ and trace $J_B < 0$. Therefore, the population and birth rate stabilize to the values given by B in the long term.

8. When $\alpha\beta > 1$, there is one stable critical point at $A = (0, \frac{1}{\beta})$. When $\alpha\beta < 1$, A becomes a col and $B = (\sqrt{1-\alpha\beta}, \alpha)$ and $C = (-\sqrt{1-\alpha\beta}, \alpha)$ are both stable. When $\alpha\beta > 1$, the power goes to zero and the velocity of the wheel tends to $\frac{1}{\beta}$ and when $\alpha\beta < 1$, the power and velocity stabilize to the point B.

9. (a) There is one critical point at $\left(\frac{KG_0}{K-C}, \frac{G_0}{K-C}\right)$, which is in the first quadrant if $K > C$. When $C = 1$, the critical point is nonhyperbolic. The system can be solved and there are closed trajectories around the critical point. The economy oscillates (as long as $I(t), S(t) > 0$). If $C \neq 1$, then the critical point is unstable if $0 < C < 1$ and stable if $C > 1$.

   (b) The critical point is stable and the trajectory tends to this point. The choice of initial condition is important to avoid $I(t)$ or $S(t)$ from going negative, where the model is no longer valid.

10. Note that $\frac{d\eta}{d\tau} = e^t$ and $\frac{d^2\eta}{d\tau^2} = \frac{d\eta}{d\tau}\frac{dt}{d\tau}$. There are four critical points: $O = (0,0)$, an unstable node; $A = (-1,0)$, a col; $B = (0,2)$, a col; and $C = (-\frac{3}{2}, \frac{1}{2})$, a stable focus.

24.4 Chapter 4

1. This is a competing species model. There are four critical points in the first quadrant at $O = (0,0)$, $P = (0,3)$, $Q = (2,0)$, and $R = (1,1)$. The point O is an unstable node, P and Q are both stable nodes, and R is a saddle point. There is mutual exclusion and one of the species will become extinct depending on the initial populations.

2. This is a Lotka–Volterra model with critical points at $O = (0,0)$ and $A = (3,2)$. The system is structurally unstable. The populations oscillate but the cycles are dependent on the initial values of x and y.

3. This is a predator-prey model. There are three critical points in the first quadrant at $O = (0,0)$, $F = (2,0)$, and $G = (\frac{3}{2}, \frac{1}{2})$. The points O and F are saddle points and G is a stable focus. In terms of species behavior, the two species coexist and approach constant population values.

4. Consider the three cases separately.

   (i) If $0 < \mu < \frac{1}{2}$, then there are four critical points at $O = (0,0)$, $L = (2,0)$, $M = (0,\mu)$, and $N = \left(\frac{\mu-2}{\mu^2-1}, \frac{\mu(2\mu-1)}{\mu^2-1} \right)$. The point O is an unstable node, L and M are saddle points, and N is a stable point. To classify the critical points, consider det J and trace J. The two species coexist.

   (ii) If $\frac{1}{2} < \mu < 2$, then there are three critical points in the first quadrant, all of which lie on the axes. The point O is an unstable node, L is a stable node, and M is a saddle point. Species y becomes extinct.

   (iii) If $\mu > 2$, then there are four critical points in the first quadrant. The point O is an unstable node, L and M are stable nodes, and N is a saddle point. One species becomes extinct.

5. (a) A predator-prey model. There is coexistence; the populations stabilize to the point $(\frac{5}{4}, \frac{11}{4})$.

   (b) A competing species model. There is mutual exclusion; one species becomes extinct.

6. There are three critical points in the first quadrant if $0 \le \epsilon < 1$: at $O = (0,0)$, $A = (\frac{1}{\epsilon}, 0)$, and $B = (\frac{1+\epsilon}{1+\epsilon^2}, \frac{1-\epsilon}{1+\epsilon^2})$. There are two when $\epsilon \ge 1$. The origin is always a col. When $\epsilon = 0$, the system is Lotka–Volterra, and trajectories lie on closed curves away from the axes. If $0 < \epsilon < 1$, A is a col, and B is stable since the trace of the Jacobian is negative and the determinant is positive. When $\epsilon \ge 1$, A is stable.

7. There are three critical points at $O = (0,0)$, $P = (1,0)$, and $Q = (0.6, 0.24)$. Points O and P are cols and Q is stable. There is coexistence.

8. There is a limit cycle enclosing the critical point at $(0.48, 0.2496)$. The populations vary periodically and coexist.

9. One example would be the following. X and Y prey on each other; Y has cannibalistic tendencies and also preys on Z. A diagram depicting this behavior is plotted in Figure 24.1.

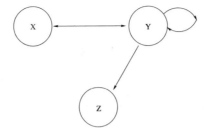

Figure 24.1: One possible interaction between three interacting insect species.

10. Let species X, Y, and Z have populations $x(t)$, $y(t)$, and $z(t)$, respectively. The interactions are as follows: X preys on Y; Z preys on X; Y and Z are in competition.

24.5 Chapter 5

1. Convert to polar coordinates to get

$$\dot{r} = r\left(1 - r^2 - \frac{1}{2}\cos^2\theta\right), \quad \dot{\theta} = -1 + \frac{1}{2}\cos\theta\sin\theta.$$

Since $\dot{\theta} < 0$, the origin is the only critical point. On $r = \frac{1}{2}$, $\dot{r} > 0$, and on $r = 2$, $\dot{r} < 0$. Therefore, there exists a limit cycle by the corollary to the Poincaré-Bendixson Theorem.

2. Plot the graph of $y = x - x^3\cos^3(\pi x)$ to prove that the origin is the only critical point inside the square. Linearize to show that the origin is an unstable focus. Consider the flow on the sides of the rectangle, for example, on $x = 1$, with $-1 \le y \le 1$, $\dot{x} = -y + \cos\pi \le 0$. Hence, the flow is from right to left on this line. Show that the rectangle is invariant and use the corollary to the Poincaré-Bendixson Theorem.

3. (a) Substitute for y from $\dot{x} = 0$ and plot the graph of $y = x^8 - 3x^6 + 3x^4 - 2x^2 + 2$ to prove that the origin is a unique critical point. Convert to polar coordinates to get

$$\dot{r} = r\left(1 - r^2(\cos^4\theta + \sin^4\theta)\right), \quad \dot{\theta} = 1 - r^2\cos\theta\sin\theta(\sin^2\theta - \cos^2\theta).$$

Now $\mathrm{div}(\mathbf{X}) = 2 - 3r^2$ and so $\mathrm{div}(\mathbf{X})$ is nonzero in the annulus $A = \{1 < r < 2\}$. On the circle $r = 1 - \epsilon$, $\dot{r} > 0$, and on the circle $r = 2 + \epsilon$, $\dot{r} < 0$. Therefore, there is a unique limit cycle contained in the annulus by Dulac's criteria. (b) Convert to polar coordinates and consider the annular region $\frac{1}{\sqrt{\mu+\rho}} < r < \frac{1}{\sqrt{\mu}}$.

4. Convert to polars and use the Poincaré-Bendixson theorem.

5. Consider the isocline curves. If the straight line intersects the parabola to the right of the maximum, then there is no limit cycle. If the straight line intersects the parabola to the left of the maximum, then there exists a limit cycle.

6. (a) The limit cycle is circular. (b) The limit cycle has fast and slow branches.

7. It will help if you draw rough diagrams.

(a) Now $\text{div}(\mathbf{X}) = -(1+x^2+x^4) < 0$. Hence, there are no limit cycles by Bendixson's criteria.

(b) Now $\text{div}(\mathbf{X}) = 2-x$. There are four critical points at $(0,0), (1,0)$, $(-1,1)$, and $(-1,-1)$. The x axis is invariant. On $x = 0$, $\dot{x} = 2y^2 \geq 0$. Hence, there are no limit cycles in the plane.

(c) Now $\text{div}(\mathbf{X}) = -6 - 2x^2 < 0$. Hence, there are no limit cycles by Bendixson's criteria.

(d) Now $\text{div}(\mathbf{X}) = -3 - x^2 < 0$. Hence, there are no limit cycles by Bendixson's criteria.

(e) Now $\text{div}(\mathbf{X}) = 3x - 2$, and $\text{div}(\mathbf{X}) = 0$ on the line $x = \frac{2}{3}$. There are three critical points at $(1,0), (-1,0)$, and $(2,3)$. The x-axis is invariant, and $\dot{x} < 0$ for $y > 0$ on the line $x = \frac{2}{3}$. Hence, there are no limit cycles by Bendixson's criteria.

(f) Now $\text{div}(\mathbf{X}) = -3x^2y^2$. Therefore, there are no limit cycles lying entirely in one of the quadrants. However, $\dot{x} = -y^2$ on the line $x = 0$ and $\dot{y} = x^5$ on the line $y = 0$. Hence, there are no limit cycles by Bendixson's criteria.

(g) Now $\text{div}(\mathbf{X}) = (x - 2)^2$. On the line $x = 2$, $\dot{x} = -y^2$, and so no limit cycle can cross this line. Hence, there are no limit cycles by Bendixson's criteria.

8. (a) The axes are invariant. Now $\text{div}(\psi\mathbf{X}) = \frac{1}{xy^2}(2 - 2x)$ and so $\text{div}(\psi\mathbf{X}) = 0$ when $x = 1$. There are four critical points and only one, $(-16, 38)$, lying wholly in one of the quadrants. Since the divergence is nonzero in this quadrant, there are no limit cycles.

(b) Now $\text{div}(\psi\mathbf{X}) = -\frac{\delta}{y} - \frac{d}{x}$ and so $\text{div}(\psi\mathbf{X}) = 0$ when $y = -\frac{\delta x}{d}$. Since $\delta > 0$ and $d > 0$, there are no limit cycles contained in the first quadrant.

9. (a) The one-term uniform expansion is $x(t, \epsilon) = a\cos(t)\left(1 - \epsilon\right.$
$\left(\frac{1}{2} + \frac{a^2}{8}\right) + \cdots\left.\right) + O(\epsilon)$, as $\epsilon \to 0$.

10. See Section 5.4.

24.6 Chapter 6

1. The Hamiltonian is $H(x,y) = \frac{y^2}{2} - \frac{x^2}{2} + \frac{x^4}{4}$. There are three critical points: $(0,0)$, which is a saddle point and $(1,0)$ and $(-1,0)$, which are both centers.

2. There are three critical points: $(0,0)$, which is a center, and $(1,0)$ and $(-1,0)$, which are both saddle points.

3. The critical points occur at $(n\pi, 0)$, where n is an integer. When n is odd, the critical points are saddle points, and when n is even the critical points are stable foci. The system is now damped and the pendulum swings less and less, eventually coming to rest at $\theta = 2n\pi$ degrees. The saddle points represent the unstable equilibria when $\theta = (2n+1)\pi$ degrees.

4. The Hamiltonian is $H(x,y) = \frac{y^4}{4} - \frac{y^2}{2} - \frac{x^2}{2} + \frac{x^4}{4}$. There are nine critical points.

5. (a) The origin is asymptotically stable.

   (b) The origin is asymptotically stable if $x < \alpha$ and $y < \beta$.

   (c) The origin is unstable.

6. The origin is asymptotically stable. The positive limit sets are either the origin or the ellipse $4x^2 + y^2 = 1$, depending on the value of p.

7. The function $V(x,y)$ is a Lyapunov function if $a > \frac{1}{4}$.

8. The basin of attraction of the origin is the circle $x^2 + y^2 < 4$.

9. Use Mathematica.

10. Now $\dot{V} = -2\epsilon y^4\left(1 + x^2\right) \leq 0$, and $\frac{dV}{dt} = 0$, when $y = 0$. On $y = 0$, $\dot{y} = -kx$ and is nonzero for $x \neq 0$, therefore, trajectories move off the x-axis and cross the level curves inwards. The origin is asymptotically stable.

24.7 Chapter 7

1. (a) There is one critical point when $\mu \leq 0$, and there are two critical points when $\mu > 0$. This is a saddle-node bifurcation.

   (b) When $\mu < 0$, there are two critical points and the origin is stable. When $\mu > 0$, there is one critical point at the origin which is unstable. The origin undergoes a transcritical bifurcation.

   (c) There is one critical point at the origin when $\mu \leq 0$, and there are three critical points—two are unstable—when $\mu > 0$. This is called a *subcritical pitchfork bifurcation*.

2. Possible examples include

   (a) $\dot{x} = \mu x (\mu^2 - x^2)$;

   (b) $\dot{x} = x^4 - \mu^2$; and

   (c) $\dot{x} = x(\mu^2 + x^2 - 1)$.

3. The critical points are given by $O = (0,0)$, $A = \frac{12+\sqrt{169-125h}}{5}$, and $B = \frac{12-\sqrt{169-125h}}{5}$. There are two critical points if $h \leq 0$, the origin is unstable, and A is stable (but negative harvesting is discounted). There are three critical points if $0 < h < 1.352$, the origin and A are stable, and B is unstable. There is one stable critical point at the origin if $h \geq 1.352$.

   The term $x(1 - \frac{x}{5})$ represents the usual logistic growth when there is no harvesting. The term $\frac{hx}{0.2+x}$ represents harvesting from h is zero up to a maximum of h, no matter how large x becomes (plot the graph). When $h = 0$, the population stabilizes to 5×10^5; when $0 < h < 1.352$, the population stabilizes to $A \times 10^5$; and when $h > 1.352$, the population decreases to zero. Use animation in Mathematica to plot $\dot{x}$ as h varies from zero to eight. The harvesting is *sustainable* if $0 < h < 1.352$, where the fish persist, and it is *unsustainable* if $h > 1.352$, when the fish become extinct from the lake.

4. (a) No critical points if $\mu < 0$. There is one nonhyperbolic critical point at $O = (0,0)$ if $\mu = 0$, and there are two critical points at $A = (0, \sqrt[4]{\mu})$ and $B = (0, -\sqrt[4]{\mu})$. Both A and B are unstable.

   (b) There are two critical points at $O = (0,0)$ and $A = (\mu^2, 0)$ if $\mu \neq 0$ (symmetry). O is stable and A is unstable. There is one nonhyperbolic critical point at $O = (0,0)$ if $\mu = 0$.

   (c) There are no critical points if $\mu < 0$. There is one nonhyperbolic critical point at $O = (0,0)$ if $\mu = 0$, and there are four critical points at $A = (2\sqrt{\mu}, 0)$, $B = (-2\sqrt{\mu}, 0)$, $C = (\sqrt{\mu}, 0)$, and $D = (-\sqrt{\mu}, 0)$ if $\mu > 0$. The points A and D are stable, while B and C are unstable.

5. (a) If $\mu < 0$, there is a stable critical point at the origin and an unstable limit cycle of radius $r = -\mu$. If $\mu = 0$, the origin is a center, and if $\mu > 0$, the origin becomes unstable. The flow is counterclockwise.

(b) If $\mu \leq 0$, the origin is an unstable focus. If $\mu > 0$, the origin is unstable, and there is a stable limit cycle of radius $r = \frac{\mu}{2}$ and an unstable limit cycle of radius $r = \mu$.

(c) If $\mu \neq 0$, the origin is unstable and there is a stable limit cycle of radius $|r| = \mu$. If $\mu = 0$, the origin is stable.

6. Take $\mathbf{x} = \mathbf{u} + \mathbf{f}_3(\mathbf{u})$. Then, if the eigenvalues of J are not resonant of order 3,

$$f_{30} = \frac{a_{30}}{2\lambda_1}, \quad f_{21} = \frac{a_{21}}{\lambda_1 + \lambda_2}, \quad f_{12} = \frac{a_{12}}{2\lambda_2}, \quad f_{03} = \frac{a_{03}}{3\lambda_2 - \lambda_1},$$

$$g_{30} = \frac{b_{30}}{3\lambda_1 - \lambda_2}, \quad g_{21} = \frac{b_{21}}{2\lambda_1}, \quad g_{12} = \frac{b_{12}}{\lambda_1 + \lambda_2}, \quad g_{03} = \frac{b_{03}}{2\lambda_2}$$

and all of the cubic terms can be eliminated from the system resulting in a linear normal form $\dot{\mathbf{u}} = J\mathbf{u}$.

7. See the book of Guckenheimer and Holmes referenced in Chapter 9.

8. (a) There is one critical point at the origin and there are at most two stable limit cycles. As μ increases through zero, there is a Hopf bifurcation at the origin. Next there is a saddle-node bifurcation to a large-amplitude limit cycle. If μ is then decreased back through zero, there is another saddle-node bifurcation back to the steady state at the origin.

(b) If $\mu < 0$, the origin is unstable, and if $\mu = 0$, $\dot{r} > 0$ if $r \neq 0$ the origin is unstable and there is a semistable limit cycle at $r = 1$. If $\mu > 0$, the origin is unstable, there is a stable limit cycle of radius $r = \frac{2+\mu-\sqrt{\mu^2+4\mu}}{2}$ and an unstable limit cycle of radius $r = \frac{2+\mu+\sqrt{\mu^2+4\mu}}{2}$. It is known as a fold bifurcation because a fold in the graph of $y = (r-1)^2 - \mu r$ crosses the r-axis at $\mu = 0$.

9. If $\mu < 0$, the origin is a stable focus and as μ passes through zero, the origin changes from a stable to an unstable spiral. If $\mu > 0$, convert to polars. The origin is unstable and a stable limit cycle bifurcates.

10. The critical points occur at $A = (0, -\frac{\alpha}{\beta})$ and $B = (\alpha + \beta, 1)$. Thus, there are two critical points everywhere in the (α, β) plane apart from along the line $\alpha = -\beta$ where there is only one. The eigenvalues for the matrix J_A are $\lambda_1 = \beta$ and $\lambda_2 = -\frac{(\alpha+\beta)}{\beta}$. The eigenvalues for the matrix J_B are $\lambda = \frac{-\alpha \pm \sqrt{\alpha^2 - 4(\alpha+\beta)}}{2}$. There is a codimension-2 bifurcation along the line $\alpha = -\beta$ and it is a transcritical bifurcation.

24.8 Chapter 8

1. Eigenvalues and eigenvectors given by $[3, (-2, -2, 1)^T], [-3, (-2, 1, -2)^T]$, and $[9, (1, -2, -2)^T]$. The origin is unstable; there is a col in two planes and an unstable node in the other.

2. Eigenvalues are $\lambda_{1,2} = 1 \pm i\sqrt{6}, \lambda_3 = 1$. The origin is unstable, and the flow is rotating. Plot solution curves using Mathematica.

3. There are two critical points at $O = (0, 0, 0)$ and $P = (-1, -1, -1)$. The critical points are both hyperbolic and unstable. The eigenvalues for O are $[1, 1, -1]$ and those for P are $[1, -1, -1]$.

4. Consider the flow on $x = 0$ with $y \geq 0$ and $z \geq 0$, etc. The first quadrant is positively invariant. The plane $x + y + 2z = k$ is invariant since $\dot{x} + \dot{y} + 2\dot{z} = 0$. Hence, if a trajectory starts on this plane, then it remains there forever. The critical points are given by $\left(\frac{\lambda y}{1+y}, y, y/2 \right)$. Now on the plane $x + y + 2z = k$, the critical point satisfies the equation $\frac{\lambda y}{1+y} + y + y = k$, which has solutions $y = \frac{(2-\lambda) \pm \sqrt{(2-\lambda)^2 + 32}}{4}$. Since the first quadrant is invariant, $\lambda^+(p)$ must tend to this critical point.

5. (a) Take $V = x^2 + y^2 + z^2$. Then $\dot{V} = -\left(x^2 + y^4 + (y - z^2)^2 + (z - x^2)^2 \right) \leq 0$. Now $\dot{V} = 0$ if and only if $x = y = z = 0$; hence, the origin is globally asymptotically stable.

   (b) Consider $V = ax^2 + by^2 + cz^2$. Now $\dot{V} = -2(a^2x^2 + b^2y^2 + c^2z^2) + 2xyz(ax + by + cz)$. Hence, $\dot{V} < \frac{V^2}{c} - 2cV$ and $\dot{V} < 0$ in the set $V < 2c^2$. Therefore, the origin is asymptotically stable in the ellipsoid $V < 2c^2$.

6. See the Mathematica program listed in Chapter 8.

7. There are eight critical points at $(0, 0, 0)$, $(0, 0, 1/2)$, $(0, 1/2, 0)$, $(0, 1, -1)$, $(1/2, 0, 0)$, $(-1/3, 0, 1/3)$, $(1/3, -1/3, 0)$, and $(1/14, 3/14, 3/14)$. The plane $x + y + z = 1/2$ is a solution plane since $\dot{x} + \dot{y} + \dot{z} = (x + y + z) - 2(x + y + z)^2 = 0$ on this plane. There are closed curves on the plane representing periodic behavior. The three species coexist, and the populations oscillate in phase. The system is structurally unstable.

8. (i) The populations settle on to a period-two cycle. (ii) The populations settle on to a period-four cycle.

9. Use Mathematica to plot a time series.

10. A Jordan curve lying wholly in the first quadrant exists, similar to the limit cycle for the Liénard system when a parameter is large. The choice of q and C is important.

24.9 Chapter 9

1. Starting with $r_0 = 4$, the returns are $r_1 = 1.13854$, $r_2 = 0.66373$, ...,
   $r_{10} = 0.15307$, to five decimal places.

2. The Poincaré map is given by $r_{n+1} = \mathbf{P}(r_n) = \frac{\mu r_n}{r_n + e^{-2\mu\pi}(\mu - r_n)}$.

3. Now $\left.\frac{d\mathbf{P}}{dr}\right|_{\mu} = e^{-2\mu\pi}$. Therefore, the limit cycle at $r = \mu$ is hyperbolic
   stable if $\mu > 0$ and hyperbolic unstable if $\mu < 0$. What happens when
   $\mu = 0$?

4. The Poincaré map is given by $r_{n+1} = \mathbf{P}(r_n) = \left(\frac{r_n^2}{r_n^2 + e^{-4\pi}(1 - r_n^2)}\right)^{\frac{1}{2}}$.

5. The limit cycle at $r = 1$ is stable since $\left.\frac{d\mathbf{P}}{dr}\right|_{r=1} = e^{-4\pi}$.

6. (a) The Poincaré section in the $p_1 q_1$ plane is crossed 14 times.
   (b) The trajectory is quasiperiodic.

7. Edit the Mathematica program listed in Section 9.4.

8. Edit the Mathematica program listed in Section 9.4.

9. A chaotic attractor is formed.

10. (a) See Figure 24.2(a).

    (b) See Figure 24.2(b). Take $\Gamma = 0.07$. For example, choose initial
    conditions (i) $x_0 = 1.16, y_0 = 0.112$ and (ii) $x_0 = 0.585, y_0 = 0.29$.

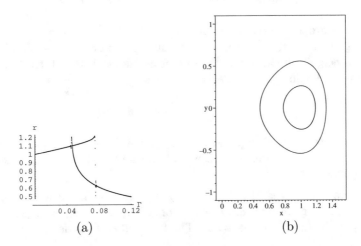

Figure 24.2:   (a) Bifurcation diagram. (b) Multistable behavior.

24.10 Chapter 10

1. Differentiate to obtain $2u\dot{u} = G'(x)\dot{x}$ and find $\frac{dy}{du}$.

2. Using Mathematica: $\{\{x^2 - 3xy + 9y^2, 0, -26y^2 - 36yz - 26z^2\}$, $-25z^3\}$ and $\{\{9, 9 + x^2 - 3xy, -27 + y^2 + yz + z^2\}, -27z + 2z^3\}$.

3. Lex $\{y^3 - y^4 - 2y^6 + y^9, x + y^2 + y^4 - y^7\}$; DegLex $\{-x^2 + y^3, -x + x^3 - y^2\}$; DegRevLex $\{-x^2 + y^3, -x + x^3 - y^2\}$. Solutions are $(0, 0)$, $(-0.471074, 0.605423)$, and $(1.46107, 1.28760)$.

4. The Lyapunov quantities are given by $L(i) = a_{2i+1}$, where $i = 0$ to 6.

5. See the Lloyd and Lynch paper in the Bibliography.

7. The Lyapunov quantities are given by $L(0) = -a_1, L(1) = -3b_{03} - b_{21}, L(2) = -3b_{30}b_{03} - b_{41}$, and $L(3) = b_{03}^3$.

8. The homoclinic loop lies on the curve $y^2 = x^2 + \frac{2}{3}x^3$.

10. There are three limit cycles when $\lambda = -0.9$.

24.11 Chapter 11

1. There is one critical point in the finite plane at the origin which is a stable node. The eigenvalues and eigenvectors are given by $\lambda_1 = -1$, $(1, -1)^T$ and $\lambda_2 = -4$, $(1, -4)^T$, respectively. The function $g_2(\theta)$ is defined as

$$g_2(\theta) = -4\cos^2\theta - 5\cos\theta\sin\theta - \sin^2\theta.$$

There are four critical points at infinity at $\theta_1 = \tan^{-1}(-1)$, $\theta_2 = \tan^{-1}(-1) + \pi$, $\theta_3 = \tan^{-1}(-4)$, and $\theta_4 = \tan^{-1}(-4) + \pi$. The flow in a neighborhood of a critical point at infinity is qualitatively equivalent to the flow on $X = 1$ given by

$$\dot{y} = -y^2 - 5y - 4, \quad \dot{z} = -yz.$$

There are two critical points at $(-1, 0)$, which is a col and $(-4, 0)$, which is an unstable node. Since n is odd, antinodal points are qualitatively equivalent.

2. There is one critical point in the finite plane at the origin which is a col. The eigenvalues and eigenvectors are given by $\lambda_1 = 1$, $(1, 1)^T$ and $\lambda_2 = -1$, $(2, 1)^T$, respectively. The function $g_2(\theta)$ is defined as

$$g_2(\theta) = -2\cos^2\theta + 6\cos\theta\sin\theta - 4\sin^2\theta.$$

There are four critical points at infinity at $\theta_1 = \tan^{-1}(1)$, $\theta_2 = \tan^{-1}(1) + \pi$, $\theta_3 = \tan^{-1}(1/2)$, and $\theta_4 = \tan^{-1}(1/2) + \pi$. The flow in a neighborhood of a critical point at infinity is qualitatively equivalent to the flow on $X = 1$ given by

$$\dot{y} = -4y^2 + 6y - 2, \quad \dot{z} = 3z - 4yz.$$

There are two critical points at $(1,0)$, which is a stable node and $(1/2, 0)$, which is an unstable node. Since n is odd, antinodal points are qualitatively equivalent.

3. There are no critical points in the finite plane. The function $g_3(\theta)$ is given by

$$g_3(\theta) = 4\cos^2\theta\sin\theta - \sin^3\theta.$$

The function has six roots in the interval $[0, 2\pi)$, at $\theta_1 = 0$, $\theta_2 = 1.10715$, $\theta_3 = 2.03444$, $\theta_4 = 3.14159$, $\theta_5 = 4.24874$, and $\theta_6 = 5.1764$. All of the angles are measured in radians. The behavior on the plane $X = 1$ is determined from the system

$$\dot{y} = 4y - 5z^2 - y^3 + yz^2, \quad \dot{z} = -z - zy^2 + z^3.$$

There are three critical points at $O = (0,0)$, $A = (2,0)$, and $B = (-2,0)$. Points A and B are stable nodes and O is a col. Since n is even, antinodal points are qualitatively equivalent, but the flow is reversed. All of the positive and negative limit sets for this system are made up of the critical points at infinity.

4. There is one critical point at the origin in the finite plane which is a stable focus. The critical points at infinity occur at $\theta_1 = 0$ radians, $\theta_2 = \frac{\pi}{2}$ radians, $\theta_3 = -\frac{\pi}{2}$ radians, and $\theta_4 = \pi$ radians. Two of the points at infinity are cols, and the other two are unstable nodes.

5. There is a unique critical point in the finite plane at the origin which is an unstable node. The critical points at infinity occur at $\theta_1 = 0$ radians, $\theta_2 = \frac{\pi}{2}$ radians, $\theta_3 = -\frac{\pi}{2}$ radians, and $\theta_4 = \pi$ radians. Two of the points at infinity are cols, and the other two are unstable nodes. There is at least one limit cycle surrounding the origin by the corollary to the Poincaré-Bendixson Theorem.

7. If $a_1a_3 > 0$, then the system has no limit cycles. If $a_1a_3 < 0$, there is a unique hyperbolic limit cycle. If $a_1 = 0$ and $a_3 \neq 0$, then there are no limit cycles. If $a_3 = 0$ and $a_1 \neq 0$, then there are no limit cycles. If $a_1 = a_3 = 0$, then the origin is a center by the classical symmetry argument.

8. When ϵ is small one may apply the Melnikov theory of Chapter 11 to establish where the limit cycles occur. The limit cycles are asymptotic

to circles centered at the origin. If the degree of F is $2m+1$ or $2m+2$, there can be no more than m limit cycles. When ϵ is large, if a limit cycle exists, it shoots across in the horizontal direction to meet a branch of the curve $y = F(x)$, where the trajectory slows down and remains near the branch until it shoots back across to another branch of $F(x)$ where it slows down again. The trajectory follows this pattern forever. Once more there can be no more than m limit cycles.

9. Use a similar argument to that used in the proof to Theorem 4. See Liénard's paper in Chapter 5.

10. The function F has to satisfy the conditions $a_1 > 0$, $a_3 < 0$, and $a_3^2 > 4a_1$, for example. This guarantees that there are five roots for $F(x)$. If there is a local maximum of $F(x)$ at say $(\alpha_1, 0)$, a root at $(\alpha_2, 0)$, and a local minimum at $(\alpha_3, 0)$, then it is possible to prove that there is a unique hyperbolic limit cycle crossing $F(x)$ in the interval (α_1, α_2) and a second hyperbolic limit cycle crossing $F(x)$ in the interval (α_3, ∞). Use similar arguments to those used in the proof of Theorem 4.

24.12 Chapter 12

1.,2. Work out the solution on $[-1, 2]$ by hand and then edit the Mathematica program listed in Section 12.1 to determine the analytical solution on $[-1, 4]$.

5. Edit the Mathematica program listed in Section 12.2. There is periodic, quasi periodic and possibly chaotic behavior.

7. When the global warming term W is small we see no discernible difference in the steady-state solutions, however, when the global warming gets too large, the oscillatory solution disappears.

8. It will help if you plot Poincaré sections. (i) Periodic; (ii) quasiperiodic; (iii) chaotic.

9. See the paper cited in the question.

10. See the paper cited in the question.

24.13 Chapter 13

1. The general solution is $x_n = \pi(4n + cn(n-1))$.

2. (a) $2 \times 3^n - 2^n$; (b) $2^{-n}(3n+1)$; (c) $2^{\frac{n}{2}}(\cos(n\pi/4) + \sin(n\pi/4))$;
   (d) $F_n = \frac{1}{2^n \sqrt{5}}[(1+\sqrt{5})^n - (1-\sqrt{5})^n]$;
   (e) (i) $x_n = 2^n + 1$;
      (ii) $x_n = \frac{1}{2}(-1)^n + 2^n + n + \frac{1}{2}$;

(iii) $x_n = \frac{1}{3}(-1)^n + \frac{5}{3}2^n - \frac{1}{6}e^n(-1)^n - \frac{1}{3}e^n 2^n + \frac{1}{2}e^n$.

3. The dominant eigenvalue is $\lambda_1 = 1.107$ and

(a)
$$X^{(15)} = \begin{pmatrix} 64932 \\ 52799 \\ 38156 \end{pmatrix};$$

(b)
$$X^{(50)} = \begin{pmatrix} 2.271 \times 10^6 \\ 1.847 \times 10^6 \\ 1.335 \times 10^6 \end{pmatrix};$$

(c)
$$X^{(100)} = \begin{pmatrix} 3.645 \times 10^8 \\ 2.964 \times 10^8 \\ 2.142 \times 10^8 \end{pmatrix}.$$

4. The eigenvalues are $\lambda_1 = 1$ and $\lambda_{2,3} = \frac{-1 \pm \sqrt{3}}{2}$. There is no dominant eigenvalue since $|\lambda_1| = |\lambda_2| = |\lambda_3|$. The population stabilizes.

5. The eigenvalues are $0, 0, -0.656 \pm 0.626i$, and $\lambda_1 = 1.313$. Therefore, the population increases by 31.3% every 15 years. The normalized eigenvector is given by
$$\hat{X} = \begin{pmatrix} 0.415 \\ 0.283 \\ 0.173 \\ 0.092 \\ 0.035 \end{pmatrix}.$$

7. Before insecticide is applied, $\lambda_1 = 1.465$, which means that the population increases by 46.5% every 6 months. The normalized eigenvector is
$$\hat{X} = \begin{pmatrix} 0.764 \\ 0.208 \\ 0.028 \end{pmatrix}.$$

After the insecticide is applied, $\lambda_1 = 1.082$, which means that the population increases by 8.2% every 6 months. The normalized eigenvector is given by
$$\hat{X} = \begin{pmatrix} 0.695 \\ 0.257 \\ 0.048 \end{pmatrix}.$$

8. For this policy, $d_1 = 0.1, d_2 = 0.4$, and $d_3 = 0.6$. The dominant eigenvalue is $\lambda_1 = 1.017$ and the normalized eigenvector is

$$\hat{X} = \begin{pmatrix} 0.797 \\ 0.188 \\ 0.015 \end{pmatrix}.$$

9. Without any harvesting the population would double each year since
$\lambda_1 = 2$.

(a) $\lambda_1 = 1$; $\hat{X} = \begin{pmatrix} 24/29 \\ 4/29 \\ 1/29 \end{pmatrix}.$

(b) $h_1 = 6/7$; $\hat{X} = \begin{pmatrix} 2/3 \\ 2/9 \\ 1/9 \end{pmatrix}.$

(c) $\lambda_1 = 1.558$; $\hat{X} = \begin{pmatrix} 0.780 \\ 0.167 \\ 0.053 \end{pmatrix}.$

(d) $h_1 = 0.604, \lambda_1 = 1.433$; $\hat{X} = \begin{pmatrix} 0.761 \\ 0.177 \\ 0.062 \end{pmatrix}.$

(e) $\lambda_1 = 1.672$; $\hat{X} = \begin{pmatrix} 0.668 \\ 0.132 \\ 0.199 \end{pmatrix}.$

10. Take $h_2 = h_3 = 1$, then $\lambda_1 = 1, \lambda_2 = -1$, and $\lambda_3 = 0$. The population
stabilizes.

24.14 Chapter 14

1. The iterates give orbits with periods (i) one, (ii) one, (iii) three, and
(iv) nine. There are two points of period one, two points of period two,
six points of period three, and twelve points of period four. In general,
there are 2^N-(sum of points of periods that divide N) points of period
N.

2. (a) The functions are given by

$$T^2(x) = \begin{cases} \frac{9}{4}x & 0 \le x < \frac{1}{3} \\ \frac{3}{2} - \frac{9}{4}x & \frac{1}{3} \le x < \frac{1}{2} \\ \frac{9}{4}x - \frac{3}{4} & \frac{1}{2} \le x < \frac{2}{3} \\ \frac{9}{4}(1 - x) & \frac{2}{3} \le x \le 1 \end{cases}$$

and

$$
T^3(x) = \begin{cases}
\frac{27}{8}x & 0 \le x < \frac{2}{9} \\
\frac{3}{2} - \frac{27}{8}x & \frac{2}{9} \le x < \frac{1}{3} \\
\frac{27}{8}x - \frac{3}{4} & \frac{1}{3} \le x < \frac{4}{9} \\
\frac{9}{4} - \frac{27}{8}x & \frac{4}{9} \le x < \frac{1}{2} \\
\frac{27}{8}x - \frac{9}{8} & \frac{1}{2} \le x < \frac{5}{9} \\
\frac{21}{8} - \frac{27}{8}x & \frac{5}{9} \le x < \frac{2}{3} \\
\frac{27}{8}x - \frac{15}{8} & \frac{2}{3} \le x < \frac{7}{9} \\
\frac{27}{8}(1 - x) & \frac{7}{9} \le x < 1.
\end{cases}
$$

There are two points of period one, two points of period two, and no points of period three.

(b) $x_{1,1} = 0$, $x_{1,2} = \frac{9}{14}$; $x_{2,1} = \frac{45}{106}$, $x_{2,2} = \frac{81}{106}$; $x_{3,1} = \frac{45}{151}$, $x_{3,2} = \frac{81}{151}$, $x_{3,3} = \frac{126}{151}$, $x_{3,4} = \frac{225}{854}$, $x_{3,5} = \frac{405}{854}$, $x_{3,6} = \frac{729}{854}$.

4. Use functions of functions to determine f_μ^N. There are two, two, six, and twelve points of periods one, two, three, and four, respectively.

5. A value consistent with period-two behavior is $\mu = 0.011$. Points of period two satisfy the equation

$$
\mu^2 x^2 - 100\mu^2 x - \mu x + 100\mu + 1 = 0.
$$

6. Edit a program from Section 14.6.

7. Points of period one are $(-3/10, -3/10)$ and $(1/5, 1/5)$. Two points of period two are given by $(x_1/2, (0.1 - x_1)/2)$, where x_1 is a root of $5x^2 - x - 1 = 0$. The inverse map is given by

$$
x_{n+1} = y_n, \quad y_{n+1} = \frac{10}{9}\left(x_n - \frac{3}{50} + y_n^2\right).
$$

8. (a) The eigenvalues are given by $\lambda_{1,2} = -\alpha x \pm \sqrt{\alpha^2 x^2 + \beta}$. A bifurcation occurs when one of the $|\lambda| = 1$. Take the case where $\lambda = -1$.

   (c) The program is listed in Section 14.6.

9. (a) (i) When $a = 0.2$, $c_{1,1} = 0$ is stable, $c_{1,2} = 0.155$ is unstable, and $c_{1,3} = 0.946$ is stable. (ii) When $a = 0.3$, $c_{1,1} = 0$ is stable, $c_{1,2} = 0.170$ is unstable, and $c_{1,3} = 0.897$ is unstable.

10. See the Ahmed paper in the Bibliography.

24.15 Chapter 15

1. (a) The orbit remains bounded forever, $z_{500} \approx -0.3829 + 0.1700i$;

   (b) the orbit is unbounded, $z_{10} \approx -0.6674 \times 10^{197} + 0.2396 \times 10^{197}$.

2. Fixed points of period one are given by

$$z_{1,1} = \frac{1}{2} + \frac{1}{4}\sqrt{10 + 2\sqrt{41}} - \frac{i}{4}\sqrt{2\sqrt{41} - 10},$$

$$z_{1,2} = \frac{1}{2} - \frac{1}{4}\sqrt{10 + 2\sqrt{41}} + \frac{i}{4}\sqrt{2\sqrt{41} - 10}.$$

Fixed points of period two are given by

$$z_{2,1} = -\frac{1}{2} + \frac{1}{4}\sqrt{2 + 2\sqrt{17}} - \frac{i}{4}\sqrt{2\sqrt{17} - 2},$$

$$z_{2,2} = -\frac{1}{2} - \frac{1}{4}\sqrt{2 + 2\sqrt{17}} + \frac{i}{4}\sqrt{2\sqrt{17} - 2}.$$

3. Use the Mathematica program given in Section 15.4; $J(0,0)$ is a circle and $J(-2,0)$ is a line segment.

4. There is one fixed point located approximately at $z_{1,1} = 1.8202 - 0.0284i$.

5. See the example in the text. The curves are again a cardioid and a circle but the locations are different in this case.

7. Fixed points of period one are given by

$$z_{1,1} = \frac{3 + \sqrt{9 - 4c}}{2}, \quad z_{1,2} = \frac{3 - \sqrt{9 - 4c}}{2}.$$

Fixed points of period two are given by

$$z_{2,1} = \frac{1 + \sqrt{5 - 4c}}{2}, \quad z_{2,2} = \frac{1 - \sqrt{5 - 4c}}{2}.$$

9. (i) Period four and (ii) period three.

10. There are regions where periodic points fail to converge. You should write your program so that these points are plotted in black.

24.16 Chapter 16

1. There are 11 points of period one.

2. See the programs in Section 16.7.

3. Find an expression for E_n in terms of E_{n+1}.

5. See the paper of Li and Ogusu in the Bibliography.

6. (a) Bistable: $4.765 - 4.766$ Wm^{-2}. Unstable: $6.377 - 10.612$ Wm^{-2}.
   (b) Bistable: $3.936 - 5.208$ Wm^{-2}. Unstable: $4.74 - 13.262$ Wm^{-2}.
   (c) Bistable: $3.482 - 5.561$ Wm^{-2}. Unstable: $1.903 - 3.995$ Wm^{-2}.

8. Use the function $G(x) = ae^{-bx^2}$ to generate the Gaussian pulse. The parameter b controls the width of the pulse.

24.17 Chapter 17

1. (a) The length remaining at stage k is given by

$$L = 1 - \frac{2}{5} - \frac{2 \times 3}{5^2} - \cdots - \frac{2 \times 3^{k-1}}{5^k}.$$

   The dimension is $D_f = \frac{\ln 3}{\ln 5} \approx 0.6826$.

   (b) $D_f = \frac{\ln 2}{\ln \sqrt{2}} = 2$. If the fractal were constructed to infinity there would be no holes and the object would have the same dimension as a plane. Thus, this mathematical object is not a fractal.

2. The figure is similar to the stage 3 construction of the Sierpiński triangle. In fact, this gives yet another method for constructing this fractal as Pascal's triangle is extended to infinity.

3. See Figure 17.8 as a guide.

4. The dimension is $D_f = \frac{\ln 8}{\ln 3} \approx 1.8928$.

6. $S_1 = \left[0, \frac{1}{4}\right] \cup \left[\frac{3}{4}, 1\right]$, $S_2 = \left[0, \frac{1}{16}\right] \cup \left[\frac{3}{16}, \frac{1}{4}\right] \cup \left[\frac{3}{4}, \frac{13}{16}\right] \cup \left[\frac{15}{16}, 1\right]$. $D_f = 0.5$.

7. (i) The fractal is homogeneous; (ii) $\alpha_{max} \approx 1.26$ and $\alpha_{min} \approx 0.26$; and (iii) $\alpha_{max} \approx 0.83$ and $\alpha_{min} \approx 0.46$. Take $k = 500$ in the plot commands.

8. Using the same methods as in Example 7:

$$D_0 = \frac{\ln 4}{\ln 3}, \quad \alpha_s = \frac{s \ln p_1 + (k - s) \ln p_2}{-k \ln 3}, \quad \text{and} \quad -f_s = \frac{\ln\left(2^k \binom{k}{s}\right)}{-k \ln 3}.$$

9. At the kth stage, there are 5^k segments of length 3^{-k}. A number

$$N_s = 3^{k-s} 2^s \binom{k}{s}$$

of these have weight $p_1^{k-s} p_2^s$. Use the same methods as in Example 7.

10. Using multinomials,

$$\alpha_s = \frac{n_1 \ln p_1 + n_2 \ln p_2 + n_3 \ln p_3 + n_4 \ln p_4}{\ln 3^{-k}} \quad \text{and} \quad -f_s = \frac{\ln \frac{4!}{n_1! n_2! n_3! n_4!}}{\ln 3^{-k}},$$

where $n_1 + n_2 + n_3 + n_4 = k$.

24.18 Chapter 18

For questions 1. 5. 6. 7. and 8., see the Mathematica programs in Chapter 18.

9. Gaussian function is

$$G(x, y) = \frac{1}{2\pi\sigma^2} \exp\left(-\frac{x^2 + y^2}{2\sigma^2}\right).$$

24.19 Chapter 19

1. Take the transformations $x_n = \frac{1}{a} u_n$ and $y_n = \frac{b}{a} v_n$.

2. There is one control range when $p = 1$, there are three control ranges when $p = 2$, seven control ranges when $p = 3$, and twelve control ranges when $p = 4$.

3. Points of period one are located at approximately $(-1.521, -1.521)$ and $(0.921, 0.921)$. Points of period two are located near $(-0.763, 1.363)$ and $(1.363, -0.763)$.

4. See the paper of Chau in the Bibliography.

5. See Section 19.3.

6. The two-dimensional mapping is given by

$$x_{n+1} = A + B(x_n \cos(x_n^2 + y_n^2) - y_n \sin(x_n^2 + y_n^2)),$$

$$y_{n+1} = B(x_n \sin(x_n^2 + y_n^2) + y_n \cos(x_n^2 + y_n^2)).$$

The one point of period one is located near $(2.731, 0.413)$.

7.   (i) There are three points of period one; (ii) there are nine points of period one.

8. See our research paper on chaos control in the Bibliography.

9. The control region is very small and targeting is needed in this case. The chaotic transients are very long. Targeting is not required in Exercise 9, where the control region is much larger. Although there is greater flexibility (nine points of period one) with this system, the controllability is reduced.

24.20 Chapter 20

2. Use the chain rule.

5. (a) Show that $\frac{d\mathbf{V}(\mathbf{a})}{dt} = -\sum_{i=1}^{n} \left(\frac{d}{da_i} \left(\phi^{-1}(a_i) \right) \right) \left(\frac{da_i}{dt} \right)^2$.

   (b)

$$\mathbf{V}(\mathbf{a}) = -\frac{1}{2} \left(7a_1^2 + 12a_1 a_2 - 2a_2^2 \right) - \frac{4}{\gamma \pi^2} \left(\log \left(\cos(\pi a_1/2) \right) \right.$$
$$\left. + \log \left(\cos(\pi a_2/2) \right) \right).$$

   There are two stable critical points, one at $(12.98, 3.99)$, and the other at $(-12.98, -3.99)$.

6. The algorithm converges to (a) $\mathbf{x}_2$; (b)$\mathbf{x}_1$; (c)$\mathbf{x}_3$; (d) $-\mathbf{x}_1$.

8. (a) Fixed points of period one satisfy the equation $a = \gamma a + \theta + w \sigma(a)$.

   (b–d) See Pasemann's paper referenced in Chapter 14.

   (e) There is a bistable region for $4.5 < w < 5.5$, approximately.

9. Iterate $10,000$ times. A closed loop starts to form, indicating that the system is quasiperiodic.

24.21 Chapter 21

1. The threshold voltage is approximately 6.3mV. (a) When $I = 8$mV, frequency is approximately 62.5Hz. (b) When $I = 20$mV, frequency is approximately 80Hz.

2. An example of a Fitzhugh–Nagumo system with a critical point at the origin is given by

$$\dot{x} = (x+0.1)*((x-0.039)(0.9-x))-0.0035-y, \quad \dot{y} = 0.008(x-2.54y).$$

3. The inequalities are given by:

$$\text{for } I = \begin{pmatrix} 0 \\ 0 \\ 0 \end{pmatrix} \qquad \begin{cases} \sum I w_1 - \hat{O}_2 x_1 < T \\ \sum I w_2 < T \end{cases}$$

$$\text{for } I = \begin{pmatrix} 1 \\ 0 \\ 0 \end{pmatrix}, \begin{pmatrix} 0 \\ 1 \\ 0 \end{pmatrix}, \begin{pmatrix} 0 \\ 0 \\ 1 \end{pmatrix} \qquad \begin{cases} \sum I w_1 - \hat{O}_2 x_1 > T \\ \sum I w_2 < T \end{cases}$$

$$\text{for } I = \begin{pmatrix} 1 \\ 1 \\ 0 \end{pmatrix}, \begin{pmatrix} 1 \\ 0 \\ 1 \end{pmatrix}, \begin{pmatrix} 0 \\ 1 \\ 1 \end{pmatrix} \qquad \begin{cases} \sum I w_1 - \hat{O}_2 x_1 < T \\ \sum I w_2 > T \end{cases}$$

$$\text{for } I = \begin{pmatrix} 1 \\ 1 \\ 1 \end{pmatrix} \qquad \begin{cases} \sum I w_1 - \hat{O}_2 x_1 > T \\ \sum I w_2 > T. \end{cases} \qquad (24.1)$$

6. The truth table and time series are shown in Figure 24.3.

Input				Output			
I_1	I_2	I_3	I_4	O_1	O_5	O_7	O_8
0	0	0	0	0	0	0	0
0	0	0	1	0	0	0	0
0	0	1	0	0	0	0	0
0	0	1	1	0	0	0	0
0	1	0	0	0	0	0	0
0	1	0	1	0	0	1	0
0	1	1	0	0	1	0	0
0	1	1	1	0	1	1	0
1	0	0	0	0	0	0	0
1	0	0	1	0	1	0	0
1	0	1	0	1	0	0	0
1	0	1	1	1	1	0	0
1	1	0	0	0	0	0	0
1	1	0	1	0	1	1	0
1	1	1	0	1	1	0	0
1	1	1	1	1	0	0	1

(a)

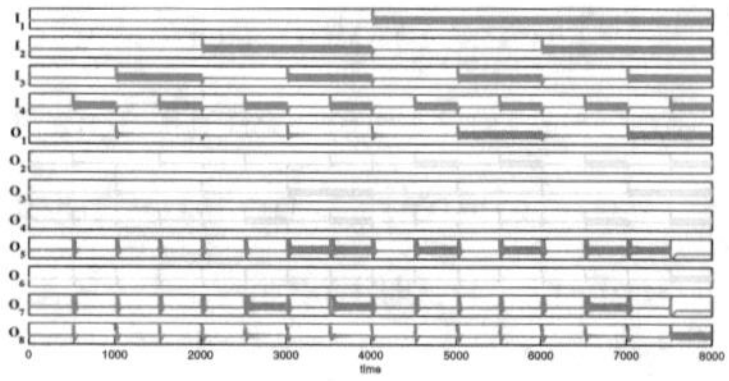

(b)

Figure 24.3:   (a) Truth table for a 2×2 bit binary multiplier. (b) Time series of a 2×2 bit binary multiplier based on Fitzhugh–Nagumo oscillations.

9. See Chapter 6.

10. Trajectories are either attracted to critical points on the $\Omega = 0$ axis, or they oscillate, depending on the initial conditions.

24.22 Chapter 22

1. See Section 4.2.

2. See Section 5.1.

3. See Section 8.4.

4. See Section 9.3.

7. See J. Ramos, S. Lynch, D.A. Jones and H. Degens, Hysteresis in muscle, International Journal of Bifurcation and Chaos 27, (2017), 1730003. The paper can be downloaded from ResearchGate.

8. There will be three differential equations. Extend Example 2.

10. See Section 21.1.

24.23 Chapter 23

Examination 1

1. (a) Eigenvalues and eigenvectors $\lambda_1 = 3.37$, $(1, 0.19)^T$; $\lambda_2 = -2.37$, $(1, -2.7)^T$. Saddle point, $\dot{x} = 0$ on $y = -\frac{3}{2}x$, $\dot{y} = 0$ on $y = \frac{1}{2}x$.

   (b) $\dot{r} > 0$ when $0 < \theta < \pi$, $\dot{r} < 0$ when $\pi < \theta < 2\pi$, $\dot{r} = 0$ when $\theta = 0, \pi$, $\dot{\theta} = 0$ when $\theta = \frac{(2n-1)}{4}\pi$, $n = 1, 2, 3, 4$.

2. (a) $\dot{V} = -(x-2y)^2$, $\dot{V} = 0$ when $y = \frac{x}{2}$. On $y = \frac{x}{2}$, $\dot{x}, \dot{y} \neq 0$, therefore, the origin is asymptotically stable.

   (b) $r = \frac{1}{t+1}$, $\theta = t + 2n\pi$.

3. (a) $\lambda_1 = -1$, $\lambda_2 = -2 + i$, $\lambda_3 = -2 - i$. Origin is globally asymptotically stable.

   (b) $\dot{V} = -4y^4 - 2z^4 < 0$, if $y, z \neq 0$. Therefore, the origin is asymptotically stable, trajectories approach the origin forever.

4. (a) One limit cycle when $\mu < 0$, three limit cycles when $\mu > 0$, $\mu \neq 1$, and two limit cycles when $\mu = 1$.

   (b) Use Bendixson's criteria:

   (i) $\text{div}\mathbf{X} = -(1 + 3x^2 + x^4) < 0$;

   (ii) $\text{div}\mathbf{X} = 3x^3y^2$, on $x = 0$, $\dot{x} \geq 0$, on $y = 0$, $\dot{y} \geq 0$, no limit cycles in the quadrants and axes invariant;

   (iii) $\text{div}\mathbf{X} = (1 + y)^2$. On $y = -1$, $\dot{y} > 0$.

5. (a) $x_{1,1} = 0$, $x_{1,2} = \frac{7}{11}$; $x_{2,1} = \frac{28}{65}$, $x_{2,2} = \frac{49}{65}$; $x_{3,1} = \frac{28}{93}$, $x_{3,2} = \frac{49}{93}$, $x_{3,3} = \frac{77}{93}$, $x_{3,4} = \frac{112}{407}$, $x_{3,5} = \frac{196}{407}$, $x_{3,6} = \frac{343}{407}$.

   (b) $z_{1,1} = \frac{1+\sqrt{13}}{2}$, $z_{1,2} = \frac{1-\sqrt{13}}{2}$; $z_{2,1} = 1$, $z_{2,2} = -2$. Fixed points of period one are unstable.

6. (a) Area of inverted Koch snowflake is $\frac{\sqrt{3}}{10}$ units^2, $D_f = 1.2619$.

   (b) Use L'Hospital.

7. (a) Period one $\left(\frac{5}{9}, \frac{1}{9}\right)$, $\left(-1, -\frac{1}{5}\right)$, both fixed points are unstable.

(b) See Chapter 6.

8. (a)

$$\mathbf{W} = \frac{1}{4} \begin{pmatrix} 0 & -1 & 1 & 1 \\ -1 & 0 & 1 & 1 \\ 1 & 1 & 0 & -1 \\ 1 & 1 & -1 & 0 \end{pmatrix}.$$

(b)

$$\mathbf{V(a)} = -\frac{1}{2} \left(a_1^2 + 2a_1a_2 + 4a_2^2 + 12a_1 + 20a_2 \right)$$

$$- \frac{4}{\gamma\pi^2} \left(\log\left(\cos(\pi a_1/2)\right) + \log\left(\cos(\pi a_2/2)\right) \right).$$

Examination 2

1. (a) $\dot{x} = -k_1 x$, $\dot{y} = k_1 x - k_2 y$, $\dot{z} = k_2 y$; $x(20) = 4.54 \times 10^{-5}$, $y(20) = 0.3422$, $z(20) = 0.6577$.

   (b) Period is approximately $T \approx -6.333$.

2. (a) See Section 10.5, Exercise 6.

   (b) $H(x,y) = \frac{y^2}{2} - \frac{x^2}{2} + \frac{x^3}{3}$, saddle point at origin, center at $(1,0)$.

3. (a) Three critical points when $\mu < 0$, one when $\mu \geq 0$.

   (b) Chaos.

4. (a) $x_{1,1} = 0, x_{1,2} = 0.716, x_{2,1} = 0.43, x_{2,2} = 0.858$, no points of period three, $x_{4,1} = 0.383, x_{4,2} = 0.5, x_{4,3} = 0.825, x_{4,4} = 0.877$.

   (b) $z_{1,1} = -0.428 + 1.616i, z_{1,2} = 1.428 - 1.616i$; $z_{2,1} = -1.312 + 1.847i, z_{2,2} = 0.312 - 1.847i$; $z_{3,1} = -1.452 + 1.668i, z_{3,2} = -1.269 + 1.800i, z_{3,3} = -0.327 + 1.834i, z_{3,4} = 0.352 - 1.891i, z_{3,5} = 0.370 - 1.570i, z_{3,6} = 1.326 - 1.845i$.

5. (a) Fixed points of period one $(0.888, 0.888), (-1.688, -1.688)$; fixed points of period two $(1.410, -0.610), (-0.610, 1.410)$.

   (b) Lyapunov exponent is approximately 0.4978.

6. (b) $J(0, 1.3)$: Scattered dust, totally disconnected.

7. (a) Period-one points $(2.76, 0.73), (3.21, -1.01), (3.53, 1.05), (4.33, 0.67)$.

8. (a)

$$\mathbf{W} = \frac{1}{6} \begin{pmatrix} 0 & -1 & 1 & 1 & -1 & 3 \\ -1 & 0 & 1 & 1 & 3 & -1 \\ 1 & 1 & 0 & -1 & 1 & 1 \\ 1 & 1 & -1 & 0 & 1 & 1 \\ -1 & 3 & 1 & 1 & 0 & -1 \\ 3 & -1 & 1 & 1 & -1 & 0 \end{pmatrix}.$$

(b) See Chapter 17.

Examination 3

1. (a) Eigenvalues and eigenvectors $\lambda_1 = 2, (3,1)^T; \lambda_2 = -5, (1,-2)^T$. Saddle point.

   (b) $r_{n+1} = \frac{r_n}{r_n - (r_n-1)e^{-2\pi}}$.

2. (a) Critical points at $(0,0)$ and $(0, \mu - 1)$. Critical point at origin is unstable for $\mu < 1$ and stable for $\mu > 1$. Other critical point has opposite stability.

   (b) One unstable limit cycle when $\mu < 0$. Critical point at origin is stable when $\mu < 0$ and unstable when $\mu > 0$.

3. (a) Critical points at $(0,0,0), (0,0,1), (0,1,0), (1,0,0), (1,0,1)$, and $(\frac{2}{3}, \frac{1}{3}, \frac{4}{3})$. Critical point away from axes is stable.

   (b) $z_{1,1} = 1.6939 - 0.4188i, z_{1,2} = -0.6939 + 0.4188i, z_{1,2} = -1.3002 + 0.6248i, z_{2,2} = 0.3002 - 0.6248i$.

4. (a) Critical point at $(0.5, 1.428)$ is an unstable focus.

   (b) There exists a limit cycle by the Poincaré-Bendixson theorem.

5. (a) Determine an expression for E_n in terms of E_{n+1}.

   (b) Fixed point at $k \approx 0.26$ is stable when $B = 1$, and fixed point $k \approx 0.92$ is unstable when $B = 4$. There is a bifurcation when $B \approx 0.36$.

6. (a) Period-one fixed points at $(\frac{2}{5}, \frac{1}{5})$ and $(-\frac{2}{3}, -\frac{1}{3})$. Period-two fixed points at $(\frac{10}{17}, -\frac{3}{17})$ and $(-\frac{6}{17}, \frac{5}{17})$.

   (b) $S_1 = [\frac{1}{6}, \frac{2}{6}] \cup [\frac{4}{6}, \frac{5}{6}]$. $D_f \approx 0.3869$.

7.   (a) Eigenvalues are $\lambda_1 = 2.5150, \lambda_2 = -1.7635, \lambda_3 = -0.7516$. Long-term population is $[0.8163; 0.1623; 0.0215]$.

     (b) Edit Mathematica program from Section 17.5.

8.   (a) Lyapunov exponent $= 0.940166$.

     (b) `Itermax=2^k;h=sqrt(2)^(-k);angle[1]=pi/4;`

         `angle[2]=-pi/4;segment[b]=Mod(m,2);m=Floor(m/2).`

Index

© Springer International Publishing AG 2017
S. Lynch, *Dynamical Systems with Applications using Mathematica®*,
DOI 10.1007/978-3-319-61485-4

Printed in the United States
By Bookmasters